Progress in Drug Research
Fortschritte der Arzneimittelforschung
Progrès des recherches pharmaceutiques
Vol. 29

Progress in Drug Research
Fortschritte der Arzneimittelforschung
Progrès des recherches pharmaceutiques
Vol. 29

Edited by / Herausgegeben von / Rédigé par
Ernst Jucker, Basel

Authors / Autoren / Auteurs
Joseph G. Cannon · D. Craig Brater and Michael R. Vasko · Eric J. Lien
R. Ludowyke and D. Lagunoff · E. Marshall Johnson · Neelima, B. K. Bhat
and A. P. Bhaduri · H. Rommelspacher and R. Susilo · Robert Saunders
George deStevens · G. M. Williams and J. H. Weisburger

1985

Birkhäuser Verlag
Basel · Boston · Stuttgart

ISBN 3-7643-1672-1
Printed in Western Germany

Foreword

Volume 29 of 'Progress in Drug Research' contains 10 articles, a subject index for this volume, an alphabetic index of articles for volumes 1–29, and an author und subject index for all the volumes which have so far been published. The contributions of volume 29 are particularly concerned with drugs in general, hypertension and cardiovascular drugs, atherosclerosis, teratogenic hazards and carcinogenecity, histamine, dopamine agonists, tetrahydroquinolines and β-carbolines, and meddicinal research.
The authors have tried, and I think they have succeded, not only to summarize the current status of particular fields of drug research, but also to provide leads for future research activity. The articles of this volume will be of special value to those actively engaged in drug research, and to those who wish to keep abrest of the latest developments influencing modern therapy. In addition, it is believed that the 29 volumes of 'Progress in Drug Research' now available represent a useful reference book of encyclopedic character. The editor would like to take the occasion of the publication of this volume to express his gratitude both to the authors and to the readers. The authors have willingly undertaken the great labor of writing significant topical contributions, and many readers have helped the editor with criticism and advise. With these thanks, the editor would like to express his gratitude to the publishers, Birkhäuser Verlag Basel, particularly to Messrs. H. J. Bender, C. Einsele and A. Gomm, and their associates for the excellent cooperation.

Bâle, October 1985 Dr. E. Jucker

Vorwort

Der 29. Band der «Fortschritte der Arzneimittelforschung» umfasst 10 Beiträge sowie ein Stichwortverzeichnis dieses Bandes, ein Verzeichnis der Artikel der Bände 1–29 nach Gebieten geordnet und ein alphabetisches Register aller Autoren und Artikel der Bände 1–29. Die Beiträge des vorliegenden Bandes befassen sich mit verschiedenen aktuellen Problemen der Arzneimittelforschung sowie mit Arzneimitteln im allgemeinen. Es wurde wiederum Wert gelegt auf Beiträge mit spezifischer und gezielter Richtung sowie auf solche mit einer das gesamte Gebiet der Arzneimittelforschung tangierender Thematik.

Die Autoren auch dieses Bandes haben wiederum versucht, ihr Fachgebiet prägnant und übersichtlich darzustellen, die neuesten Entwicklungen aufzuzeigen und darüber hinaus auch in die Zukunft weisende Betrachtungen anzustellen. So dürfte auch der 29. Band der Reihe «Fortschritte der Arzneimittelforschung» dem aktiven Forscher, sei es in der Industrie oder an der Hochschule, von Nutzen sein und demjenigen, der sich über die neuesten Entwicklungen auf dem laufenden halten will, eine gute Hilfe bieten. Die vorliegenden 29 Bände stellen sicherlich ein wertvolles Nachschlagewerk mit enzyklopädischem Charakter dar.

Der Herausgeber möchte hiermit den Autoren und den Lesern seinen Dank abstatten; den Autoren für die große Arbeit, die sie bei der Abfassung der Beiträge geleistet haben, den Lesern für ihre Kritik und Anregungen. Die vielen Zuschriften und die Rezensionen helfen entscheidend mit, die Reihe auf einem hohen Niveau zu halten und den heutigen, sich stets verändernden Bedingungen der Forschung anzupassen. Dank sei auch dem Birkhäuser Verlag, insbesondere den Herren Bender, Einsele und Gomm sowie ihren Mitarbeitern für die ausgezeichnete Zusammenarbeit und für die sorgfältige Ausstattung des Bandes ausgesprochen.

Basel, Oktober 1985 — Dr. E. Jucker

Contents · Inhalt · Sommaire

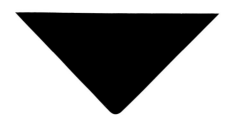

Update of cardiovascular drug interactions

By D. Craig Brater, M. D., and Michael R. Vasko, Ph. D.
Departments of Pharmacology and Internal Medicine, The University of Texas Health Science Center at Dallas, 5323 Harry Hines Blvd, Dallas, TX 75235, and Veterans Administration Medical Center, 4500 Lancaster Road, Dallas, TX 75216

1	Introduction	10
1.1	General principles: Pharmacokinetic interactions	10
1.2	General principles: Pharmacodynamic interactions	12
2	Cardiovascular drug interactions	15
2.1	Drugs interacting at sites of absorption	15
2.2	Cardiovascular drug interactions affecting distribution	17
2.3	Cardiovascular drug interactions that alter elimination	20
2.31	Drug interactions affecting metabolism	20
2.311	Induction of metabolism	21
2.312	Inhibition of metabolism	23
2.312.1	Hepatic metabolism	23
2.312.2	Monoamine oxidase	27
2.32	Drug interactions affecting excretion	27
2.321	Filtration	28
2.322	Active secretion	28
2.323	Reabsorption	30
2.4	Pharmacodynamic drug interactions	31
3	Summary	35
	References	36

1 Introduction

An increasing number of clinical studies address interactions of cardiovascular drugs in terms of both their importance in the treatment of disease and their potential to cause drug-induced toxicity. Several books and reviews are devoted to extensive listings of both observed and theoretical drug interactions [1–4]. Such listings often include extrapolation of animal data, anecdotal case reports, and interactions of questionable importance. Consequently, their literal use could result in withholding therapy or overcomplicating therapeutic decisions. Some drug interactions are critical for optimum patient care, such as combined use of drugs for their additive or synergistic effects, while others, although academically interesting, have little clinical relevance. In this review, it is not our intent to provide extensive listings of cardiovascular drug interactions but to use clinical examples to illustrate pharmacological principles which may then be extrapolated to individual patients [5].

1.1 General principles: Pharmacokinetic interactions

For the purpose of this review, pharmacokinetic drug interactions are those which alter drug absorption, distribution, metabolism, or excretion.

Absorption of a drug has multiple determinants, including physicochemical properties of the drug, gastric pH, site of absorption in the GI tract, rate of gastric emptying, intestinal motility, surface area and mucosal function, and blood flow to the absorption site. When two or more drugs are co-administered, interactions can occur involving any of these factors.

Drug interactions can also occur by alterations in the tissue distribution of drugs. Many drugs bind reversibly to plasma proteins and this binding limits the free drug concentration that is available to tissue sites of action. When drugs are highly bound to plasma proteins, large shifts in free drug concentration can occur if the drug is displaced from binding sites by other drugs. For example, a decrease of warfarin binding from the normal 99 to 96 % may seem like a small decrement. However, the concentration of free drug, that which is available to the site of action, has quadrupled from 1 to 4 %. In most instances an increase in free drug concentration as a result of one drug's displacing another is rapidly compensated for by an increase in metabolism and/or excretion of the free drug such that the increased free concentration is transient [6]. If, however, the patient's drug elimination is compromised by drug interactions of metabolism or by disease, the increased free concentration may be more sustained and serious consequences may result.

Drug interactions of metabolism can result in an increase or decrease in drug clearance. This change in metabolism can occur by multiple mechanisms, including: alterations in hepatic blood flow (affecting drugs for which the limiting step is delivery to metabolic enzymes), competitive inhibition at sites of metabolism, or induction of microsomal enzymes. Since many drugs are metabolized by similar pathways, there is a high probability for competitive inhibition of one or more agents. It is important, therefore, to have an idea of the metabolism of each drug administered to a patient to anticipate the possibility of an interaction's occurring. In many instances, even if drugs compete for metabolic sites, sufficient capacity remains such that the competition is of negligible importance. Since most drugs are metabolized by first order kinetics, the rate of metabolism is dependent on drug concentration, meaning the sites of drug metabolism are unsaturated. All drugs, however, are capable of saturating their metabolic sites and shifting from first order to zero order kinetics. Indeed, with some drugs this shift occurs at very low plasma concentrations. Since the limited metabolic sites are saturated, only a specific amount of drug is metabolized per unit of time. With a majority of drugs, however, the shift from first order to zero-order only occurs at extremely high plasma concentrations that are seldom attained. When two drugs are administered simultaneously, however, they may compete for the same metabolic sites, and conceivably lesser concentrations of each are needed to saturate the metabolic enzymes. Therefore, a drug interaction may result in a disproportional increase in the half-life and decreased clearance of drug if the interaction not only competes for metabolism but also causes a change from a nonsaturated to a saturated state.

Finally, pharmacokinetic drug interactions can occur at sites of drug elimination, mainly the kidney. Drug excretion can be altered by several factors depending on how the individual drug is handled, including blood flow to the kidneys, glomerular filtration rate, physical characteristics of the drug such as molecular size and pKa, urine pH and secretion or reuptake of the drug. Drug interactions can occur via any of these possible mechanisms or their combination.

It is important to note that pharmacokinetic drug interactions can be exaggerated by disease states. For example, disease induced decreases in metabolic capacity of drugs (as may occur in cirrhosis) may increase the likelihood of a drug interaction that shifts metabolism from first to zero-order kinetics. Similar extrapolations can be hypothesized for many disease conditions. Therefore, additional caution should be used when examining drug interactions in patients with various diseases.

1.2 General principles: Pharmacodynamic interactions

Pharmacodynamic drug interactions are defined as those that result in an alteration of the biochemical or physiological effects of a drug. Since a wide variety of dynamic drug interactions can occur, it is important for the clinician to not only be aware of the pharmacokinetics of a drug but also its mechanism of action.

In general, pharmacodynamic drug interactions can be divided into four classes:

(1) Interactions at the drug receptor (pharmacological).
(2) Interactions via different cellular mechanisms acting in concert or in opposition (physiological).
(3) Interactions due to alterations of the cellular environment.
(4) Chemical neutralization in the body.

Most drugs act at specific receptor sites by binding to a receptor (affinity) and that binding triggers a biochemical event or series of events that result in the drug's pharmacological action (intrinsic activity). Such drugs are categorized as agonists (those with both affinity and intrinsic activity), antagonists (having binding affinity without intrinsic activity), and mixed agonist-antagonists (with varying degrees of affinity and intrinsic activity). The overall outcome of drug interactions at receptor sites are dependent on the varying affinities and activities of the different agents involved.

For example, several antagonists are used clinically for their ability to block the actions of an agonist; e. g., β-adrenergic receptor antagonists are used extensively to block the actions of endogenous or exogenous catecholamines. In some instances, however, these interactions are undesirable. For example, patients with congestive heart failure or chronic obstructive pulmonary disease may be compromised by blocking β-adrenergic receptors since such drugs obviate the beneficial effects of endogenous catecholamines. The physician, therefore, needs to be aware of which drugs bind to or act at the same receptor to avoid potential toxicity or loss of therapeutic effect.

Many pharmacodynamic drug interactions occur as a result of drugs acting via different mechanisms to produce similar or opposite effects. An example of a beneficial effect is the use of β-receptor agonists with methylxanthines (theophylline) to relax bronchiolar smooth muscle. An example of a harmful interaction is the effect of aminoglycoside antibiotics to potentiate the blockade of skeletal muscle (particularly respiratory) by the neuromuscular blocking agents [7]. Physiological interactions can also occur by one drug acting in concert or in opposition to another through different mechanisms. An example of the former is use of combinations of agents with different mechanisms of action to have additive effects to lower blood pressure; a converse example is the effect of nonsteroidal anti-inflammatory drugs to attenuate the antihypertensive effects of captopril [8] and β-adrenergic antagonists [9–12] or to decrease the response to loop diuretics [13, 14].

The third type of pharmacodynamic interaction occurs when the action of one drug results in a change in the intra- or extracellular environment that modifies the action of another drug. The best example of this is the interaction between cardiac glycosides and drugs that cause potassium depletion [15, 16]. Another example is the interaction between reserpine and indirectly acting sympathomimetic agents [17, 18]. Reserpine depletes norepinephrine (NE) in nerve terminals [18]. With less NE, there is less response to drugs such as metaraminol that act by releasing the neurotransmitter.

The final type of interaction involves chemical neutralization of one drug by another. Several interactions of this type occur in the GI tract and are discussed in detail in the section on drug absorption. This type of interaction also occurs within the circulation and may be desirable as is the case with the use of protamine to neutralize heparin. They can also be deleterious as is illustrated by the inactivation of gentamicin with carbenicillin, piperacillin, and other penicillins in patients with end-stage renal disease [19–21] due to formation of an inactive complex between the penicillins and the aminoglycoside.

Table 1
Cardiovascular drug interactions at absorption sites.

Mechanism	Causative agent	Drug affected	Result of interaction	Ref.
Formation of complexes, chelation, adsorption	Activated charcoal	Digoxin	↓absorption	[29]
		Phenytoin	↓absorption	[29]
	Antacids	Bishydroxycoumarin	↑absorption	[28, 35]
		Digoxin	↓absorption	[22, 25]
		Propranolol	↓absorption	[23, 26]
	Cholestyramine	Digitoxin	↓absorption, ↑ elimination	[32]
		Digoxin	↓absorption	[31]
		Oral Anticoagulants	↓absorption, ↑ elimination	[27, 33]
	Kaolin-pectin	Digoxin	↓absorption	[22, 30]
	Sucralfate	Warfarin	↓absorption	[34]
Alterations in gastric pH affecting ionization or dissociation		NONE DESCRIBED		
Changes in gastric motility	Amitryptyline	Bishydroxycoumarin	↑absorption due to ↓GI motility	[40]
	Antacids	Phenytoin Propranolol	↓rate of absorption of each of these drugs due to delayed gastric emptying	[37, 38, 23]
	Ethanol	Propranolol	↓rate of absorption	[26]
	Metoclopramide	Cimetidine Digoxin	↓absorption of both due to ↑GI motility	[39]
	Propantheline	Digoxin	↑absorption due to ↓GI motility	[39]
Effects on GI mucosa	Neomycin	Digoxin	↓absorption	[42]
	Phenytoin	Furosemide	↓absorption (possibly due to altered transport)	[44]
	Sulfasalazine	Digoxin	↓absorption	[41]
Effects on GI metabolism	Erythromycin	Digoxin	↑serum concentrations	[43]

2 Cardiovascular drug interactions

2.1 Drugs interacting at sites of absorption

Drug interactions can alter the rate and/or the extent of drug absorption. The rate of absorption determines how fast drug enters the blood and the peak concentrations attained. Extent of absorption affects the total amount systemically available. A decreased extent of absorption can result in a substantial decrease in circulating drug concentrations, thereby compromising therapy. On the other hand, an increase in absorption could subject the patient to drug toxicity. Since oral absorption of drugs is dependent on many factors, it is not surprising that potential drug interactions occur that involve a number of mechanisms. Though the ensuing discussion will focus on oral absorption, the clinician should remember that the same principles apply to IM drug administration. For our purposes, we classify absorption interactions (table 1) as follows:

(1) Formation of drug complexes due to adsorption, chelation, or binding.
(2) Alterations in gastric pH.
(3) Changes in GI motility.
(4) Alterations in GI mucosal function.
(5) Effects on GI metabolism.
(6) Effects on membrane absorption sites.
(7) Alterations in GI perfusion.

Antacids interact with many drugs to alter absorption by formation of complexes with magnesium, aluminum or calcium ions. Usually, the complex formed between drug and ion is less soluble, and/or less absorbable than the parent drug, resulting in a decrease in the extent of absorption [22–26]. In unusual instances, however, the complexes may be more soluble and, thus, antacids may occasionally result in increased absorption. For example, bishydroxycoumarin chelates with magnesium in the stomach to form a more absorbable complex [27, 28]. Thus, patients administered this anticoagulant with antacids may develop higher serum concentrations and a greater anticoagulant effect. Several other interactions occur via a similar mechanism usually to decrease bioavailability. These include binding or adsorption of digoxin and phenytoin by activated charcoal [29], of digoxin by Kaolin-pectin

[30], of a variety of cardiovascular drugs by the hypocholesterolemic agents, cholestyramine and colestipol [27, 31–33], and of warfarin by sucralfate [34]. Cholestyramine binds bile acids in the gut and, thus, decreases absorption and enterohepatic reabsorption of cholesterol. The resin can also bind other compounds in the intestine, thus accounting for decreased absorption of chlorothiazide, cardiac glycosides, and anticoagulants [27, 31–33, 35]. To avoid the interaction rather than compensate for it, patients requiring these drug combinations should be administered the drug (digoxin, digitoxin or warfarin) 1 hour before or 4 hours after ingesting cholestyramine or colestipol.

The interaction between digitoxin or warfarin and cholestyramine can also be used to clinical advantage in patients toxic with either of these agents. Both are excreted in the bile and reabsorbed by enterohepatic circulation. Administration of cholestyramine sequesters drug in the gut, thereby resulting in less reabsorption and decreasing serum concentrations [27, 32].

Antacids can also affect drug absorption by altering gastric pH [35], since drug absorption is, in part, dependent on dissolution of drug and on extent of ionization. In the acidic milieu of the stomach, drugs that are weak acids are less ionized, and thus more rapidly absorbed. The opposite is true for weak bases. Thus, alteration in gastric pH by antacids, and/or cimetidine can affect drug absorption though important interactions by this mechanism have not been described for cardiovascular agents.

Drug interactions can also affect the rate of absorption. Since most drugs are absorbed in the small intestine, changes in gastric emptying can alter delivery to absorption sites and, thereby, alter the rate of absorption. Thus antacids, alcohol and narcotic analgesics [36], which slow gastric emptying, decrease the rate of absorption of several cardiovascular drugs, including phenytoin [37, 38] and propranolol [23, 26].

Drugs that alter intestinal transit time can also affect both rate and extent of drug absorption. The anticholinergic, propantheline, slows GI motility and, thus, increases absorption of poorly soluble drugs such as some of the older preparations of digoxin [39]. Amitryptyline increases absorption of bishydroxycoumarin presumably by the same mechanism [40]. A variety of drugs such as phenothiazines, antidepressants and antihistamines (H_1 receptor antagonists) have anticholinergic side effects and might importantly influence GI transit, emphasizing the importance of understanding the spectrum of a drug's phar-

macologic effects. Conversely, metoclopramide, a drug that increases GI motility, decreases quantitative absorption of older preparations of digoxin by mechanisms analogous to those above [39].
Drug interactions of absorption also occur as a result of one drug affecting the intestinal metabolism and/or transport of another, although there are only limited reports of these interactions. Neomycin and sulphasalazine decrease the absorption and serum concentrations of digoxin, presumably by affecting the integrity of the GI mucosa [41, 42]. In contrast, in approximately 10 % of patients, erythromycin increases digoxin absorption, the presumed mechanism being a decrease in flora that metabolize digoxin in the small intestine [43]. Further studies are needed to determine the importance of these types of interaction.
Theoretically, changes in GI or skeletal muscle perfusion alter absorption. Aspects of this mechanism are discussed subsequently in the section on hepatic metabolism and presystemic elimination.
In summary, many interactions of absorption have been described. However, it is difficult to predict a priori in individual patients whether an interaction will occur and if it will be important. Since the therapeutic margin with most cardiovascular drugs is so narrow, one must pay close attention to the possibility of an interaction occurring and pay scrupulous attention to the medications being administered. A reasonable general rule would be to administer potentially interacting drugs at different times and to be aware that the patient's disease may make him more susceptible to a quantitatively important interaction than one might anticipate based on studies in normal volunteers.

2.2 Cardiovascular drug interactions affecting distribution

Upon entering the circulation, a drug is distributed to various tissues, including its site of action. For many drugs, the most abundant distribution site is binding to serum proteins where an equilibrium is established between bound and free drug. Since free drug can bind to tissue storage sites, bind at sites of action, be metabolized, or be excreted, the equilibrium in the serum constantly changes as free drug is removed. Several drugs can alter serum protein binding and, thus, shift the equilibrium and thereby the amount of free drug in the serum [45].
Displacement of one drug from binding sites by another constitutes an important and often misinterpreted source of drug interactions. The

impact of displacement is dependent on three factors for each drug: concentration, relative binding affinity, and the volume of distribution (V_d). A high concentration of one drug relative to another will shift the binding equilibrium. Thus, assuming equal affinities, addition of large concentrations of one drug to the blood will rapidly decrease binding of a second. Relative binding affinity is the second important factor. In general, only drugs with high binding affinities displace other drugs. These include digitoxin, bishydroxycoumarin, warfarin, diazoxide, phenytoin, clofibrate, valproic acid, salicylates, nonsteroidal anti-inflammatory drugs, sulfonamides, oral sulfonylureas, organic acid diuretics, penicillins and many cephalosporins. Finally, if a drug has a small V_d, then more of the displaced drug remains in the serum and can be delivered to sites of action with a concomitant change in the pharmacologic effect.

Theoretically, all drugs that are highly bound to serum proteins can interact with others to increase free drug concentration. The clinical importance of this displacement is dependent on the therapeutic index of the drugs involved, their rate of elimination, and other available distribution sites. If a drug has a wide therapeutic index such as penicillin, then an increase in free concentration has less impact than drugs with narrow margins of safety.

Most drugs are eliminated by first-order kinetics. Consequently, increasing the free concentration is rapidly compensated for by increased metabolism and/or excretion. After displacement, a new steady state is achieved with the free concentration and pharmacologic effect similar to that before the interaction. Therefore, such interactions are usually transient [6]. If drug concentrations are sufficiently high, or elimination is compromised by disease, then elimination kinetics may shift from first-order to zero-order, prolonging the increase in concentration of free drug and thus creating a greater potential for a toxic drug interaction. Predictably, increased concentrations of free drug are less important when the free drug can be distributed to other tissue storage sites. Therefore, clinically important interactions are less likely with drugs that have a large V_d.

Several points regarding interactions that displace drugs from protein binding sites are important. First, the effect is rapid and transient, since, as mentioned above, it is usually compensated for by increased elimination or distribution to tissues without active sites [6]. Second, the transient increase in free drug concentration results in an increase in drug concentration at the site of action and an enhanced pharmacological response. The duration of this transient increase, a time when a toxic effect might occur, is impossible to predict and depends on the time for distribution and/or elimination to occur. Third, even though elimination is usually enhanced, one cannot predict the overall effect on elimination half-life. Often the V_d is also increased; and since half-life equals 0.693 V_d/Clearance, the overall effect on half-life is dependent on the relative magnitude of change in clearance and in V_d. Finally, after the drug displacement has stabilized and a new steady state has been attained, the resulting total plasma concentration may be below the 'normal' therapeutic range. One should not misinterpret such 'subtherapeutic' values, for they do not adequately reflect the concentration of free drug. The clinician must, therefore, rely on his clinical assessment of drug effect and/or measure free drug concentration.

Table 2
Cardiovascular drug interactions due to displacement from protein binding sites.

Drug displaced	Causative agent	Ref.
Coumarin anticoagulants	Chloral hydrate	[49, 50]
	Clofibrate	[47]
	Diazoxide	[51]
	Ethacrynic acid	[51]
	Mefenamic acid	[51]
	Nalidixic acid	[51]
	Phenylbutazone	[46]
	Phenytoin	[48, 52]
	Salicylates	[48]
Phenytoin	Phenylbutazone	[53]
	Salicylates	[54]
	Tolbutamide	[55]
	Valproic acid	[56, 57]

Clinically the most important drug interactions involving serum protein binding (table 2) occur with coumarin anticoagulants [46–48]. Displacement of anticoagulants from binding sites results in a rapid, yet transient, increase in free drug concentrations and severe bleeding can occur. This clinical effect is enhanced in patients who have a compromised ability to metabolize these drugs due to hepatic disease, or if the displacing drug also blocks metabolism, as phenylbutazone does with warfarin [46]. Even though the increase in free concentrations may be transient, bleeding can occur during this time. It is, therefore, important to determine successive prothrombin times when a potential interaction is suspected. The patient may be able to 'ride through' the interaction and return to the same baseline level of anticoagulation in a few days, but then again, he may not. Close observation is mandatory.

Interactions with phenytoin (table 2) are probably less critical than those occurring with anticoagulants, but may be important under specific circumstances.

In some instances, drug toxicity originally attributed to a displacement interaction is the result of multiple interactions, i. e., displacement plus impairment of metabolism. For example, phenylbutazone displaces warfarin from binding sites on serum proteins but also inhibits its metabolism [46]. The increased concentration of free anticoagulant is not compensated for by increased metabolism, thus increasing the likelihood of bleeding. A similar mechanism accounts for an increased incidence of phenytoin toxicity when tolbutamide is co-administered with the anticonvulsant [55]. Displacement from serum binding sites is usually thought of in terms of increasing a drug's effects. However, the opposite can occur. Phenytoin displaces dicoumarol from serum protein [52]. This increase has little importance in patients chronically administered phenytoin due to the additional effect by phenytoin to induce dicoumarol metabolism. In fact, a decreased overall anticoagulant effect may occur. One must be aware that multiple types of interactions may occur with the same drugs, the clinical consequences of which are difficult to predict in individual patients. A high level of awareness of their potential should minimize their incidence and consequences.

2.3 Cardiovascular drug interactions that alter elimination

Drugs are eliminated by metabolism and/or excretion which, for the most part, occur in the liver and kidney, respectively. Other organ systems eliminate specific drugs such as excretion of paraldehyde by the lungs, glucuronidation of several different agents by the kidney, etc., but clinically important drug interactions have not been described with these pathways.

2.31 Drug interactions affecting metabolism

Interactions via metabolic pathways occur with induction or inhibition of their hepatic metabolism and from inhibition of nonhepatic enzymes (e. g., monoamine oxidase; MAO). A number of drugs, as well as chemicals in our environment, can affect hepatic enzymes, especially the microsomal cytochrome P-450 system. There is a great deal of individual variability in the metabolic capacity of this enzyme system and the degree to which it may be induced or inhibited by drugs [58, 59]. In addition, the time course of drug interaction varies with different drugs and a patient's disease can influence the susceptibility to an interaction. This knowledge, plus the realization that many of the potential inducers and inhibitors have been studied only in animals, makes tenuous any prediction of the extent of an interaction in an individual. Each patient, then, must be followed closely with clinical evaluation of therapeutic endpoints, drug toxicity, and, if possible, with measurement of serum drug concentrations to assess the clinical importance of a drug interaction.

2.311 Induction of metabolism

Induction of hepatic microsomal enzymes results in clinically important drug interactions for cardiovascular drugs, including oral anticoagulants, phenytoin, quinidine, digitoxin and mexiletine [60–64] (table 3). A variety of drugs can increase metabolism of the oral anticoagulants (table 3), increasing the dose required to achieve a therapeutic prolongation of the prothrombin time [48, 61, 65–69]. This type of interaction is not only important during the time of induced metabolism, but also after the inducing agent is withdrawn [66]. When the inducing stimulus is no longer present, the hepatic enzyme activity slowly decreases, increasing serum concentrations and resulting in increased anticoagulation of the patient with the possible consequence of a disastrous bleeding diathesis. When an agent is added to a patient's regimen which induces the metabolism of warfarin, an effect can be seen within one to two days, but usually does not become maximal until approximately one week [66, 70]. Return to baseline after removal of the inducing agent usually requires approximately two weeks. This slow onset and offset of the interaction contrasts with the rapid onset of interaction with drugs that inhibit metabolism (see below).

Phenobarbital is a potent inducer of hepatic microsomal enzymes. Phenobarbital increases the metabolism of phenytoin in some patients [60, 62–64], resulting in subtherapeutic phenytoin concentrations that

Table 3
Induction of metabolism of cardiovascular drugs.

Inducing agent	Induced agent	Ref.
Barbiturates	Digitoxin	[15, 76, 77]
	Oral anticoagulants	[48, 66, 79]
	Phenytoin	[60, 62–64]
	Propranolol	[26]
	Timolol	[80]
Carbamazepine	Oral anticoagulants	[48, 61, 66, 79]
	Phenytoin	[61, 63, 64]
Ethanol (chronic, until hepatic impairment)	Phenytoin	[63, 64, 81–83]
	Propranolol	[26]
	Warfarin	[48, 66]
Ethchlorvynol	Warfarin	[48, 66]
Glutethimide	Warfarin	[48, 66, 79]
Griseofulvin	Warfarin	[48, 66, 79]
Phenylbutazone	Digitoxin	[15, 76, 77]
Phenytoin	Carbamazepine	[63, 64]
	Clonazepam	[63, 64]
	Diazepam	[63, 64]
	Digitoxin	[15, 76, 77]
	Disopyramide	[84]
	Doxycycline	[63, 64]
	Glucocorticoids	[63, 64]
	Meperidine	[63, 64]
	Methadone	[63, 64]
	Metyrapone	[63, 64]
	Oral anticoagulants	[48, 52, 66, 79]
	Oral contraceptives	[63, 64]
	Propranolol	[26]
	Quinidine	[71, 73]
	Theophylline	[85, 86]
	Valproic acid	[63, 64]
Rifampin	Digitoxin	[15, 76, 77, 87]
	Disopyramide	[84]
	Mexiletine	[78]
	Propranolol	[26]
	Quinidine	[75]
	Warfarin	[48, 66, 67–69, 79]

may have clinical relevance depending, in part, on the therapeutic endpoint. If phenytoin is used as an anticonvulsant, the decrease in serum concentrations may be offset by the anticonvulsant effect of phenobarbital. However, if phenytoin is being administered for its antiarrhythmic effect, then phenobarbital could adversely influence therapeusis.

Phenobarbital combined with phenytoin doubles the clearance of quinidine in humans, with a concomitant decrease in the elimination half-life of approximately 50 % [71–73], presumably due to enzyme induction. However, in dogs (74), phenytoin decreased quinidine's elimination half-life by 50 %, but did not affect clearance. The half-life change was caused by a considerable decrease in the volume of distribution indicating a different mechanism of interaction in this experimental model. These data are cited to stress the importance of cautious extrapolation of animal data to man.
Rifampin is also a potent inducer of hepatic metabolizing enzymes and, in so doing, appears to have a 2-fold effect on quinidine handling. It primarily decreases quinidine's elimination half-life with no change in the volume of distribution (i. e., it increases clearance) [75]. From the data, one would predict an approximate 3-fold increase in clearance with an identical decrement in the steady-state serum concentration or area under the curve of serum concentrations vs. time (AUC). The observed change in AUC of quinidine in these subjects, however, was a 6-fold decrease [75]. The probable explanation for this discrepancy is a secondary, concomitant decrease in the bioavailability of quinidine due to an enhanced first pass effect through an induced hepatic enzyme system.
A variety of drugs can induce the metabolism of digitoxin (table 3) resulting in subtherapeutic serum concentrations and recrudescence of symptoms, increased ventricular response in atrial fibrillation, etc. [15, 16, 76, 77]. The importance of this interaction is only minimized by the infrequent use of digitoxin.
Mexiletine is a new oral antiarrhythmic agent with pharmacologic activity similar to lidocaine. Its metabolism can be induced by rifampin with a decrease in elimination half-life from 8.5 ± 0.8 to 5.0 ± 0.4 hr (8 normal volunteers) [78]. This decrease in half-life was solely due to a change in clearance (presumably metabolic). One would predict that other agents with inducing capability (table 3) may also increase metabolism of mexiletine.

2.312 Inhibition of metabolism

2.312.1 Hepatic metabolism

In general, those drugs discussed above which are susceptible to induction of metabolism are also subject to inhibition (table 4). Inhibi-

Table 4
Inhibition of metabolism of cardiovascular drugs.

Inhibiting agent	Inhibited agent	Ref.
Amiodarone	Digixon	[101]
	Quinidine	[101]
	Procainamide	[101]
	Warfarin	[79, 101]
Azapropazone	Phenytoin	[103]
Bishydroxycoumarin	Tolbutamide	[3, 48, 66]
Chloramphenicol	Oral anticoagulants	[7, 48, 66, 104]
	Phenytoin	[7, 63, 64, 104–106]
Chlorpheniramine	Phenytoin	[63, 64]
Chlorpromazine	Phenytoin	[63, 64, 99]
	Propranolol	[26, 108]
Cimetidine	Labetolol	[98]
	Lidocaine	[94, 95, 109]
	Metoprolol	[99]
	Phenytoin	[63, 64, 106, 109–111]
	Propranolol	[26, 99, 100, 109, 112]
	Quinidine	[113]
	Warfarin	[48, 66, 79, 97, 109]
Clofibrate	Oral anticoagulants	[48, 66]
Disopyramide	Warfarin	[48, 66, 79, 114]
Disulfiram	Phenytoin	[63, 64, 106, 115, 116]
	Warfarin	[48, 66, 117]
Ethanol (acute)	Phenytoin	[63, 64, 81–83]
	Warfarin	[48, 66, 81–83]
Imipramine	Phenytoin	[63, 64, 106, 107]
Isoniazid	Phenytoin	[63, 64, 106, 118, 119]
Methsuximide	Phenytoin	[63, 64]
Methylphenidate	Phenytoin	[63, 64, 106, 120]
Metoprolol	Lidocaine	[90]
Metronidazole	Warfarin	[48, 66, 79, 125]
Oral anticoagulants	Phenytoin	[48, 66]
Oral contraceptives	Oral anticoagulants	[48, 66, 122]
Paraminosalicylate	Phenytoin	[63, 64, 118]
Phenylbutazone	Phenytoin	[63, 64, 106]
	Oral anticoagulants	[46, 48, 66, 79, 123]
Phenyramidol	Phenytoin	[63, 64, 118, 124]
Propoxyphene	Phenytoin	[63, 64, 106, 125]
Propranolol	Lidocaine	[26, 88–90]
Ranitidine	Metoprolol	[96]
	Warfarin	[97]
Sulfamethizole	Phenytoin	[7, 63, 64, 126, 127]
	Warfarin	[7, 48, 66, 126, 127]
Sulfaphenazole	Phenytoin	[7, 63, 64, 126]
	Warfarin	[7, 48, 66, 126]
Sulfinpyrazone	Phenytoin	[63, 64]

Sulthiame	Phenytoin	[63, 64, 128]
Thioridazine	Phenytoin	[63, 64]
Tolbutamide	Phenytoin	[55, 63, 64]
Trimethoprim- Sulfamethoxasole	Phenytoin Warfarin	[7, 63, 64, 126, 129] [7, 48, 66, 79, 126, 129–131]
Valproate	Phenytoin	[132]

tors are capable of decreasing the metabolism of any drug which is metabolized by thc same enzyme system. As with induction, whether or not these interactions are clinically important is largely a function of the therapeutic index of the drug whose metabolism is inhibited; cardiovascular agents are of primary importance.

The interaction of β-adrenergic blocking agents with lidocaine is particularly important because the margin for error with lidocaine dosing is so small and the toxicity potentially severe. In healthy volunteers, propranolol decreases lidocaine clearance 40–50 % [26, 88–90] while metroprolol causes a 30 % decrease [90]. Rather than inhibiting metabolic enzymes, this interaction most likely occurs by β-blockade-induced decreases in hepatic blood flow thereby decreasing lidocaine access to the liver [91–93]. Cimetidine has a similar effect on hepatic blood flow, but additionaly, inhibits the hepatic microsomal enzyme system. By so doing, cimetidine decreases the clearance of lidocaine by 25 % or more [94, 95]; however, lidocaine serum concentrations increase 50 % indicating that the volume of distribution also decreases (approximately 10 %). Consequently, when co-administering lidocaine and cimetidine, one should decrease both the loading and maintenance lidocaine doses.

The cimetidine effect illustrates an important clinical strategy for avoiding potentially serious drug interactions. The clinician may be able to use an alternative agent that does not exhibit the drug interaction. For example, another antiarrhythmic could be substituted for lidocaine. On the other hand, ranitidine, another H_2 receptor blocker has been proposed as an alternative to cimetidine. It is similarly effective to decrease gastric acid secretion, yet has less effect to inhibit metabolism of other drugs. Importantly, though, ranitidine clearly can inhibit metabolism of metoprolol [96], warfarin [97], and probably other drugs though its potency in doing so is less than that of cimetidine. Thus, clinicians should realize that individual patients may manifest drug interactions with ranitidine.

All drugs affected by cimetidine (table 4) demonstrate decreased elimination in amounts that should dictate a change in dosing regimen. In addition, in drugs with a large first-pass effect such as labetolol [98], metoprolol [99], and propranolol [26, 99, 100], cimetidine also causes an increase in bioavailability. This effect, coupled with a decrease in clearance due to inhibition of metabolism results in an increase in serum drug concentration greater than would occur from either effect alone.

It is important to note that the effect of cimetidine, as well as other inhibitors of metabolism, is rapid [66, 70]. An effect can be seen with the first dose, and the maximum effect is dependent upon attainment of a new steady-state which requires 4–5 times the new half-life of the drug.

Amiodarone is a new antiarrhythmic that appears to inhibit the elimination of a number of cardiovascular agents [79, 101]. This drug has a very long elimination half-life and the time course of its effects differs from that of other drugs. Its effects on warfarin have been best characterized and have an onset of effect between 1 and 28 days (usually greater than 6). With discontinuation of amiodarone, inhibition of warfarin metabolism lasts up to 16 weeks [101]. Patients receiving warfarin clearly must be followed closely for extended periods of time after exposure to this antiarrhythmic.

Another important interaction which might involve inhibition of hepatic metabolism has been described with furosemide and propranol. In normals, furosemide increases serum propranolol concentrations by approximately 30 % with a concomitant increase in β-blockade [102]. Whether this interaction occurs by decreasing metabolism, increasing bioavailability, or by decreasing the volume of distribution of propranolol, is unknown.

Other factors may influence the clinical relevance of the above interactions. These include age, disease state and genetic factors. Elderly patients have slower metabolism. Consequently, a drug interaction that partially inhibits metabolism in a young healthy volunteer subject may be exaggerated in the elderly. In addition, patients with intrinsic liver disease might manifest an exaggerated effect of agents impairing the metabolic function of remaining hepatocytes. Another factor involves the patient's inherited ability to metabolize drug. Genetic variability of drug metabolism is well documented [133–137]. Phenytoin metabolism, for example, is genetically controlled with certain individuals slowly metabolizing the drug [138, 139]. This genetic variability may

prove to be an additional important determinant of susceptibility to drug interactions.
One should always take the conservative approach and assume that the magnitude of an interaction is likely to be clinically important and err on the side of administering insufficient doses at the beginning of therapy. It is always easier to administer supplemental doses to the patient than to extract what has already been administered. Similarly, with the addition of new drugs to the patient's regimen and with changes in clinical status, one should anticipate quantitative changes in drug handling as well as qualitative changes in response.

2.312.2 Monoamine oxidase

A number of drug interactions occur due to inhibition of monamine oxidase (MAO) [140]. Their significance is only lessened by the infrequent use of these agents. However, procarbazine [141], a drug used in treating Hodgkin's disease, and isoniazid [142, 143] have MAO inhibitory activity. Concomitant use of sympathomimetic agents requires awareness of this potential effect of these two drugs. MAO metabolizes sympathetic neurotransmitters that are retaken into the neuron from the synaptic cleft. During inhibition of this enzyme, administration of catecholamines or agents that release catecholamines can cause fatal hypertensive crises. In addition, some antihypertensive agents like reserpine, guanethidine, methyldopa and clonidine may acutely release endogenous catecholamine stores, paradoxically raising blood pressure to dangerous levels. Conversely, the effects of some antihypertensive agents such as diuretics may be potentiated and prolonged.

2.32 Drug interactions affecting excretion

As noted previously, excretion of a large majority of drugs occurs in the kidney which is responsible for eliminating both the parent drug and metabolites. Some drugs are excreted via the intestinal tract, presumably by biliary secretion. Once secreted, they are often susceptible to being reabsorbed, a process called 'entero-hepatic' circulation. For such drugs, interactions are possible with other agents capable of sequestering the secreted drug in the intestinal tract [144–146]. This phenomenon is entirely analogous to the effect of many of these sequestrants to diminish bioavailability as discussed previously. By this

mechanism, activated charcoal triples the clearance of parenterally administered phenobarbital [144] and doubles that of nadolol [146]. This same mechanism occurs with cholestyramine and digitoxin [32, 147].
The kidney's handling of drugs and drug interactions is easiest to consider in terms of its normal physiologic functions; namely, filtration, active secretion and reabsorption.

2.321 Filtration

Many drugs are eliminated by glomerular filtration; clinically important examples are the aminoglycoside antibiotics and digoxin. Theoretically, changes in glomerular filtration rate (GFR) affect handling of these and other drugs. Critically ill patients are particularly susceptible to such changes because of their dynamic volume status and clinical condition, and administration of other drugs, like dopamine or dobutamine, which are vasoactive and can affect renal perfusion.
Few studies have addressed this important area of drug interactions. By presumably affecting this mechanism, furosemide may increase digoxin's renal clearance [148]; although this is controversial [149, 150]. Unfortunately, furosemide is a poor pharmacologic tool for assessing potential effects on GFR, since its effect on GFR is dependent upon the degree of volume losses and replacement. Consequently, this entire area remains speculative. We caution the clinician to be aware of potentially important drug interactions involving glomerular filtration.

2.322 Active secretion

The kidney is capable of secreting a number of drugs. Active secretion of a variety of agents occurs at the pars recta (straight segment) of the proximal tubule. There are two nonspecific transport systems, one for organic acids and one for bases, in which secreted drugs can compete for transport with another drug within the same group [151–153]. An additional important secretory pathway is that for digoxin which is located in the distal tubule. Here again, competition for secretion occurs (vide infra).
For cardiovascular drugs, the clinically important example of competition for transport of organic acid agents is that which occurs between a variety of drugs and the accumulated endogenous organic acids of

uremia. In mild to moderate renal failure, this mechanism is probably more important for the decreased elimination of a number of organic acids than is the decreased nephron mass. Organic acid diuretics such as furosemide, ethacrynic acid, thiazides, metolazone, etc., reach their site of action by secretion into the renal tubular lumen by the organic acid secretory pathway. Accumulated organic acids in uremia block the access of these diuretics to their active site accounting for the requirement for larger doses of these diuretics needed to attain amounts within the tubular lumen sufficient to cause a diuresis [154–156]. The active transport system for organic bases and its importance in man is less well understood than is that for organic acids. Cardiovascular drugs which are bases and undergo potentially clinically important active secretion include amiloride, mecamylamine, and procainamide [157, 158]. Cimetidine can compete for secretion of these drugs and in 6 normal subjects decreased procainamide clearance by 35 % [159].

The effects of several drugs to decrease the secretion of digoxin is one of the most important drug interactions [160–182]. This phenomenon, described in 1976, is reviewed elsewhere [161, 178]. The interaction occurs in at least 90 % of patients co-administered quinidine and digoxin with, on average, a doubling of the serum digoxin concentration. The magnitude of the effect appears to be dependent on the serum quinidine concentration. The predominant mechanism of the interaction is a decreased tubular secretion of digoxin [183], though some have argued a decreased nonrenal clearance [177]. There is overall a decreased clearance and a need to halve the dose (on average) to maintain the same serum concentration of digoxin. In addition, approximately 2/3 of patients also demonstrate a 10 % or more decrease in digoxin's volume of distribution. The mechanism of this effect is presumably by displacement of digoxin from muscle binding sites by quinidine [174].

Herein lies the most controversial aspect of this interaction. If displacement from muscle occurs, does displacement also occur from cardiac muscle, such that the increased serum concentration does not translate to an increased pharmacologic effect on the heart? If so, the increased concentrations should be of little concern and the interaction merely requires an upward redefinition of the 'therapeutic concentration' of digoxin with no dose adjustment necessary. However, if the increased serum concentrations translate to the heart, dose adjustment is mandatory.

Data on this issue are inconsistent [160, 164, 165, 167, 172, 174–176, 179]. Some animal studies report a decreased cardiac digoxin concentration with an increased concentration in the brain [161]. Although controversial, observations of physiological effects suggest enhanced electrophysiological action on the heart [160, 172, 174] with a decrease in inotropism [164, 167, 174–176, 182]. Other investigators disagree [179, 181]. The former actions may be due to enhanced central nervous system effects on AV nodal conduction. If these effects occur as a result of the interaction, one clearly needs to decrease the dosage of digoxin to maintain the 'therapeutic range' [161, 178]. This decrease may compromise the inotropic actions of the drug. In fact, in those patients who do not require digoxin for effects on conduction, alternative therapy may be appropriate.

Whether or not a similar phenomenon occurs with digitoxin is unclear [184–186]. Since only a minor amount of digitoxin is eliminated by the kidney, if an interaction occurs, it presumably would be via a different mechanism.

It is important to note that this interaction with digoxin is not unique to quinidine. Amiodarone, quinine, spironolactone, triamterene, and verapamil also increase serum digoxin concentrations [161, 187–192]. The mechanism of this effect is presumed to be similar to that with quinidine, though interestingly, quinine (an optical isomer of quinidine) causes a decrease in nonrenal rather than renal clearance of digoxin [189]. In contrast, nifedipine does not affect digoxin elimination [193].

Like quinidine, spironolactone affects both the volume of distribution and the clearance of digoxin [188, 192]. On average, clearance decreases 26 %, however, there is great variability among patients with a range of 0 to 74 %. Similarly, the effect on volume of distribution is highly variable. Therefore, some patients may require dose adjustment while others will not. In general, loading and maintenance doses of digoxin should be decreased by 1/3 in patients receiving spironolactone, realizing some patients may need subsequent upward titration.

2.323 Reabsorption

Some drugs are reabsorbed in the nephron after having gained access to the tubular lumen by filtration or secretion. In animal studies, digoxin clearly has a reabsorptive component presumably in the proximal tubule [194]. Consequently, renal excretion of digoxin involves filtration, reabsorption, and secretion at more distal tubular sites. The mechanism of digoxin's reabsorption is unclear but appears to follow general reabsorptive activity of the proximal tubule. Administration of saline or mannitol decreases proximal tubular reabsorption of sodium and of digoxin while agents acting more distally to increase sodium excretion and urinary volume do not affect digoxin excretion [194]. Whether drug interactions could occur in man via this pathway is unclear.

2.4 Pharmacodynamic drug interactions

Other drugs can alter the response to a drug without affecting its kinetics. Table 5 offers selected examples of pharmacodynamic interactions, to alert the reader to the multiplicity of possible mechanisms. Ensuing will be brief comments concerning caveats or mechanisms of some of the more important and interesting interactions.

Drugs causing changes in acid base and electrolyte status can importantly affect the response to digitalis glycosides [15, 16]. Potassium depletion increases the risk of digoxin toxicity. This interaction is particularly easy to overlook because substantial decreases in intracellular potassium can occur with normokalemia. Hypercalcemia and hypomagnesemia similarly increase the sensitivity to digitalis.

A number of reports have recently discussed the interaction of indomethacin or other nonsteroidal anti-inflammatory agents and loop diurectics. Indomethacin decreases the acute response to furosemide while aspirin has no effect in normal subjects [13]. The indomethacin effect is not a result of a change in the amount of furosemide reaching its intraluminal site of action [13]. Whether this interaction is of clinical importance is unclear. Patients chronically receiving both drugs show no effect on sodium excretion, though the antihypertensive effect of furosemide is blunted by indomethacin [14]. Clinicians should carefully monitor the response to diuretics in patients receiving nonsteroidal anti-inflammatory agents; however, any decreased response can be overcome with an increased dose of diuretics.

Indomethacin administration also decreases the antihypertensive effects of captopril [8] and propranolol [9, 10, 12]. Whether this interaction is related to prostaglandins is unknown. This effect, though, is probably common to all inhibitors of prostaglandin synthesis, for sulindac causes the same effect [11].

Table 5
Examples of pharmacodynamic cardiovascular drug interactions.

Drug or condition Altering response	Drug response altered	Comments	Ref.
Acidemia	Sympathomimetics	Decreased	[228]
β-Adrenergic antagonists	Clonidine	Blood pressure overshoot during withdrawal	[220]
Digitalis	β-Adrenergic antagonists	Profound bradycardia	[223]
Disopyramide	Practolol	Profound bradycardia	[229]
Diuretics	Antihypertensives	Increased response	
Guanidinium antihypertensives	Direct acting α-adrenergic agonists	Increased response	[196–199]
Hypercalcemia	Cardiac glycosides	Toxicity increased	[15, 16]
Inhibitors of prosta-glandin synthesis	Captopril	Decreased response	[8]
	Loop diuretics	Predominantly decreased response	[13]
	Propranolol	Decreased antihypertensive effect	[9–12]
Magnesium depletion	Cardiac glycosides	Toxicity increased	[15, 16]
Methyldopa	Haloperidol	Dementia	[230]
Phenytoin	Lithium	Increased effect	[231]
Potassium depletion	Cardiac glycosides	Toxicity increased	[15, 16]
Reserpine	Indirect acting α-adrenergic agonists	Decreased response	[17, 18]
Tricyclic antidepressants	Direct and indirect acting α-adrenergic agonists	Increased response	[195–199]
	Guanidinium antihypertensives	Decreased response	[195, 196, 200–206]
	Clonidine	Decreased response	[214–216]

Drug interactions with direct and/or indirect effects on the autonomic nervous system can be clinically important. Interactions among drugs primarily used for their effects on the autonomic nervous system include blockade of the β-adrenergic agonist effects of isoproterenol or of endogenous catecholamines by propranolol or of endogenous α-adrenergic agonists by phentolamine, phenoxybenzamine or prazosin. Less commonly anticipated interactions are those occurring with concomitant administration of drugs that have secondary effects on the autonomic nervous system. For example, phenothiazines, tricyclic antidepressants, and butyrophenones have α-adrenergic antagonist properties accounting for the enhanced activity of other α-adrenergic blockers used concomitantly and the rationale for using directly acting α-sympathomimetics to reverse toxic α-adrenergic blockade caused by these drugs.

Tricyclic antidepressants and guanidinium antihypertensives block neuronal uptake of catecholamines at the synaptic cleft [195, 196]. Because reuptake of catecholamines is the major mechanism of attenuation of their effect, the effect of exogenously administered catecholamines may be increased if used with these agents [197]. In studies of normal subjects given imipramine (25 mg TID) for 5 days, the pressor effect of phenylephrine was potentiated by 2 to 3 times, norepinephrine by 4 to 8 times; and epinephrine by 2 to 4 times [198]. In a similar study in subjects administered debrisoquin (a guanidinium antihypertensive agent that blocks the catecholamine reuptake system), the effects of phenylephrine were markedly potentiated and prolonged [199]. Sympathomimetics which are indirectly acting would not be expected to have an enhanced effect with guanethidine-like drugs, and would more likely have a decreased effect, since guanethidine, debrisoquin, and bethanidine deplete endogenous catecholamines.

Another set of important cardiovascular drug interactions involving the catecholamine reuptake mechanism is that occurring between the guanidinium antihypertensives, guanethidine, bethanidine, and debrisoquin and a number of psychoactive agents [196, 200–206]. The former drugs are taken into the synaptosomes by the catecholamine reuptake system where they cause release and depletion of endogenous catecholamine stores. Tricyclic antidepressants reverse their antihypertensive effects by inhibiting their uptake to this site of action [195, 196, 200–206]. Doxepine appears to have less tendency to this effect than other antidepressants [207]. A similar reversal of effect that probably occurs by inhibition of uptake or displacement from the site of action also occurs with amphetamine, ephedrine, methylphenidate, doxepin, phenothiazines, butyrophenones, thiothixene and possibly reserpine [195, 207–212]. This interaction has also been reported with use of a nasal decongestant (Ornade®), containing chlorpheniramine, isopropamide, and phenylpropanolamine [213]. This last interaction is probably not clinically important in most patients but use of over-the-counter 'cold' remedies should probably be avoided in patients receiving guanethidine-like drugs.

Reversal of the antihypertensive effect of clonidine by desipramine is well documented [214–216], implying that caution should be used when other tricyclic antidepressants and psychoactive drugs are administered to patients receiving clonidine.

Drugs, like reserpine, or diseases, like congestive heart failure, that deplete endogenous catecholamine stores can blunt the response to the indirectly acting sympathominetics such as metaraminol and ephedrine, whose major effect depends on release of catecholamines at the nerve ending [17, 18].

Unanticipated but predictable interactions may occur with use of agents that have multiple effects. For example, epinephrine is both an α- and β-adrenergic agonist; the α effect predominates in most instances, causing arteriolar vasoconstriction. Concomitant use of an α-adrenergic antagonist not only will attenuate the vasoconstriction, but also may unveil the vasodilation caused by the β effect. Similarly, the use of propranolol for the treatment of hypertension rarely exacerbates the hypertension by blocking β-induced vasodilation and potentiating preexisting α-mediated vasoconstriction, especially in patients with a pheochromocytoma [217–219]. By a similar mechanism, in patients treated with clonidine, withdrawal of the drug during continued administration of propranolol can cause accentuation of the α-adrenergic effect of the endogenous catecholamines [220].

A number of drugs affect the parasympathetic limb of the autonomic nervous system. These agents may unexpectedly antagonize or potentiate the effects of cardiovascular agents used to affect the parasympathetics. Phenothiazines, antihistamines, and tricyclic antidepressants have clinically important parasympathoplegic effects [221, 222]. Use of digitalis with guanethidine or propranolol can result in profound bradycardia as a result of the vagal activity of digitalis during sympathetic blockade [223].

Another type of drug interaction, that is less clearly defined, is that of psychotherapeutic agents on the myocardium. Phenothiazines and tricyclic antidepressants have quinidine-like effects on conduction and automaticity [224–227]. These effects can be manifested as QRS widening, QT prolongation and arrhythmias. Because of the mechanism of these cardiac effects, concomitant use of quinidine and procainamide may have additive effects. Similarly, treatment of arrhythmias due to this effect of phenothiazines or tricyclic antidepressants requires agents that would not further depress conduction; these include lidocaine, phenytoin, and sympathomimetics. Some arrhythmias caused by these psychoactive agents are due to this quinidine-like effect rather than their parasympathoplegic effect [226]. Inappropriate use of physostigmine for arrhythmias due to the quinidine-like effect could result in worsening of the arrhythmias.

The reader should realize that with increasing knowledge, a number of interactions currently classified as pharmacodynamic may prove to be pharmacokinetic. A good example of this reclassification are studies of the stereoselectivity of the interaction of trimethoprim-sulfamethoxasole, metronidazole, and phenylbutazone with warfarin [121, 130, 131]. Warfarin is administered as a racemate with the S(–)-enantiomer having the predominant anticoagulant effect. The drugs noted above selectively inhibit metabolism of the S(–)-enantiomer causing an increased pharmacologic effect. Assessing the interaction by measuring effects on the racemate had revealed no pharmacokinetic interaction, for the lack of effect on the R(+)-enantiomer 'swamped out' the ability to detect an interaction. The mechanism was elucidated only after separately examining the effects of these compounds on each enantiomer, thereby changing our concepts of the mechanism from pharmacodynamic to pharmacokinetic.

3 Summary

In this review, we have attempted to discuss cardiovascular drug interactions from a mechanistic point of view using specific examples. Several types of disease can potentially alter response when drugs are administered concomitantly. Since patients receiving cardiovascular drugs are often administered multiple medications, it is important that clinicians be aware of the possible complications of disease and drug interactions, even though the literature in this area is limited.

Interactions of drug absorption may be altered by a number of disease conditions including changes in GI motility, hypersecretion of acid, changes in blood flow to the gut, alterations in biliary excretion, changes in liver blood flow altering the first pass effect, and changes in gut mucosa. Interactions involving distribution of drugs can be altered by changes in plasma proteins (hypoalbuminemia, liver disease, renal failure), by altered drug distribution to damaged tissues and by other as yet undefined disease-induced effects.

Obviously, drug metabolism is altered by hepatic disease. Patients with intrinsic compromise of hepatic function are more susceptible to inhibition of hepatic microsomal enzymes by cimetidine [109]. A decreased number of functioning hepatocytes would be exposed to relatively greater amounts of an inhibiting drug and, therefore, a greater effect might result. On the other hand, one might speculate that induction of hepatic metabolism would be less in patients with severe hepatic dysfunction reasoning that the remaining hepatocytes are operating at maximal capacity and are incapable of further induction. Other diseases, like congestive heart failure, could also affect drug metabolism by altering hepatic blood flow. Thus, another possible influence of disease to amplify drug interactions could exist but too few data are available to draw firm conclusions.

Renal disease may critically affect drug interactions. Decreased excretion of drugs will result in accumulation of parent drug and/or metabolites probably increasing the possibilities of interactions. We are aware of only one documented example of the impact of disease on a drug interaction. In vitro mixing of the semisynthetic penicillins, carbenicillin, piperacillin, ticarcillin and others with aminoglycosides results in physicochemical complexing and a loss of aminoglycoside activity. In patients, this only occurs with severe renal impairment in which elimination of both the penicillins and aminoglycosides is impaired thereby allowing sufficient time for the interaction to occur [19–21].

Other similar influences of disease undoubtedly exist and will only be described when noted anecdotally by skilled clinical observers and then more formally explored in prospective clinical studies. Awareness of the potential for their occurrence is the framework on which observation occurs; heightening that awareness is the entire purpose of this review.

References

1 Hansen, P. D.: Drug Interactions. Lea and Febiger, Philadelphia 1979.
2 Morselli, P. L, Garattini, S., and Cohen, S. N.: Drug Interactions. Raven Press, New York 1974.

3 Prescott, L. F.: Pharmacokinetic drug interactions. Lancet *II*, 1239–1243 (1969).
4 Stockley, I.: Drug Interactions. Blackwell Scientific Publications, Oxford 1981.
5 Melmon, K. L., and Nierenberg, D. W.: Drug interactions and the prepared observer. N. Engl. J. Med. *304*, 723–724 (1981) (editorial).
6 Sellers, E. M.: Plasma protein displacement interactions are rarely of clinical significance. Pharmacology *18*, 225–227 (1979).
7 Bint, A. J., and Burtt, I.: Adverse antibiotic drug interactions. Drugs *20*, 57–68 (1980).
8 Moore, T. J., Crantz, T. R., Hollenberg, N. K., Koletsky, R. J., Leboff, M. S., Swartz, S. L., Levine, L., Podolsky, S., Dluhy, R. G., and Williams, G. H.: Contribution of prostaglandins to the antihypertensive action of captopril in essential hypertension. Hypertension *3*, 168–173 (1981).
9 Durao, V., Prata, M. M., and Goncalves, L. M. P.: Modification of antihypertensive effect of β-adrenoceptor-blocking agents by inhibition of endogenous prostaglandin synthesis. Lancet *II*, 1005–1007 (1977).
10 Durao, V., and Rico, J. M. G. T.: Modification by indomethacin of the blood pressure lowering effect of pindolol and propranolol in conscious rabbits. Eur. J. Pharmac. *43*, 377–381 (1977).
11 Easton, P. A., and Koval, A.: Hypertensive reaction with sulindac. Can. Med. Ass. J. *122*, 1273–1274 (1980).
12 Lopez-Ovejero, J. A., Weber, M. A., Drayer, J. I. M., Sealey, J. E., and Laragh, J. H.: Effects of indomethacin alone and during diuretic or β-adrenoreceptorblockade therapy on blood pressure and the renin system in essential hypertension. Clin. Sci. Mol. Med. *55*, 203–205 (1978).
13 Chennavasin, P., Seiwell, R., and Brater, D. C.: Pharmacokinetic-dynamic analysis of the indomethacin-furosemide interaction in man. J. Pharmac. Exp. Ther. *215*, 77–81 (1980).
14 Patak, R. V., Mookerjee, B. K., Bentzel, C. J., Hysert, P. E., Bagej, M., and Lee, J. B.: Antagonism of the effects of furosemide by indomethacin in normal and hypertensive man. Prostaglandins *10*, 649–659 (1975).
15 Binnion, P. F.: Drug interactions with digitalis glycosides. Drugs *15*, 369–380 (1978).
16 Brown, D. D., Spector, R., and Juhl, R. P.: Drug interactions with digoxin. Drugs *20*, 198–206 (1980).
17 Boura, A. L. A., and Green, A. F.: Adrenergic neurone blockade and other acute effects caused by N-benzyl-N′-N″-dimethylguanidine and its orthochloro derivative. Br. J. Pharmac. *20*, 36–55 (1963).
18 Burn, J. H., and Rand, M. J.: The action of sympathomimetic amines in animals treated with reserpine. J. Physiol. *144*, 314–336 (1958).
19 Pickering, I. K., and Rutherford, I.: Effect of concentration and time upon inactivation of tobramycin, gentamicin, netilmicin and amikacin by azlocillin, carbenicillin, mecillinam, mezlocillin and piperacillin. J. Pharmac. Exp. Ther. *217*, 345–349 (1981).
20 Thompson, M. J. B., Russo, M. E., Saxon, B. J., Atkinthor, E., and Matsen, J. M.: Gentamicin inactivation by piperacillin or carbenicillin in patients with end stage renal disease. Antimicrob. Ag. Chemother. *21*, 268–273 (1982).
21 Weibert, R., Keane, W., and Shapiro, F.: Carbenicillin inactivation of aminoglycosides in patients with severe renal failure. Trans. Am. Soc. Artif. Intern. Organs *22*, 439–443 (1976).
22 Brown, D. D., and Juhl, R. P.: Decreased bioavailability of digoxin due to antacids and kaolin pectin. N. Engl. J. Med. *295*, 1034–1037 (1976).
23 Dobbs, J. H., Skoutakis, V. A., Acchardio, S. R., and Dobbs, B. R.: Effects of aluminum hydroxide on the absorption of propranolol. Curr. Ther. Res. *21*, 887–892 (1977).

24 Garty, M., and Hurwitz, A.: Effect of cimetidine and antacids on intestinal absorption of tetracycline. Clin. Pharmac. Ther. *28*, 203–207 (1980).
25 Khalil, S. A. H.: Bioavailability of digoxin in presence of antacids. J. Pharm. Sci. *63*, 1641–1642 (1974) (letter).
26 Wood, A. J. J., and Feely, J.: Pharmacokinetic drug interactions with propranolol. Clin. Pharmacokin. *8*, 253–262 (1983).
27 Robinson, D. S., Benjamin, D. M., and McCormack, J. J.: Interaction of warfarin and nonsystemic gastrointestinal drugs. Clin. Pharmac. Ther. *12*, 491–495 (1971).
28 Ambre, J. J., and Fisher, L. J.: Effect of coadministration of aluminum and magnesium hydroxides on absorption of anticoagulants in man. Clin. Pharmac. Ther. *14*, 231–238 (1973).
29 Neuvonen, P. J., Elfring, S. M., and Elonen, E.: Reduction of absorption of digoxin, phenytoin and aspirin by activated charcoal in man. Eur. J. Clin. Pharmac. *13*, 213–218 (1978).
30 Albert, K. S., Ayres, J. W., DiSanto, A. R., Weidler, D. I., Sakmar, E., Hallmark, M. R., Stoll, R. G., Desante, K. A., and Wagner, J. G.: Influence of kaolin-pectin suspension on digoxin bioavailability. J. Pharm. Sci. *1*, 1582–1586 (1978).
31 Brown, D. D., Juhl, R. P., and Warner, S. L.: Decreased bioavailability of digoxin due to hypocholesterolemia interventions. Circulation *58*, 164–172 (1978).
32 Caldwell, J. H., and Greenberger, N. J.: Interruption of the enterohepatic circulation of digitoxin by cholestyramine. I. Protection against lethal digitoxin intoxication. J. Clin. Invest. *50*, 2626–2637 (1971).
33 Jahnchen, E., Meinertz, T., Gilfrich, H.-J., Kersting, F., and Groth, V.: Enhanced elimination of warfarin during treatment with cholestyramine. Br. J. Clin. Pharmac. *5*, 437–440 (1978).
34 Mungall, D., Talbert, R. L., Phillips, C., Jaffe, D., and Ludden, T. M.: Sucralfate and warfarin. Ann. Intern. Med. *98*, 557 (1983) (letter).
35 Levine, R. R.: Factors affecting gastrointestinal absorption of drugs. Digest Dis. *15*, 171–188 (1970).
36 Nimmo, W. S., Heading, R. C., Wilson, J., Tothill, P., and Prescott, L. F.: Inhibition of gastric emptying and drug absorption by narcotic analgesics. Br. J. Clin. Pharmac. *2*, 509–513 (1975).
37 Garnett, W. R., Carter, B. L., and Bellock, J. M.: Bioavailability of phenytoin administered with antacids. Ther. Drug Monitoring *1*, 435–437 (1979).
38 Kulshrestha, V. K., Thomas, M., Wadsworth, J., and Richens, A.: Interaction of phenytoin and antacids. Br. J. Clin. Pharmac. *6*, 177–179 (1978).
39 Manninen, V., Apajalahti, A., Simonen, H., and Reissel, P.: Effect of propantheline and metoclopramide on absorption of digoxin. Lancet *I*, 398 (1973).
40 Pond, S. M., Graham, G. G., Brikett, D. J., and Wade, D. N.: Effects of tricyclic antidepressants on drug metabolism. Clin. Pharmac. Ther. *18*, 191–199 (1975).
41 Juhl, R. P., Summers, R. W., Guillory, J. K., Blang, S. M., Cheng, F. H., and Brown, D. D.: Effect of sulfasalazine on digoxin bioavailability. Clin. Pharmac. Ther. *20*, 387–394 (1976).
42 Lindenbaum, J., Maulitz, R. M., and Butler, V. P.: Inhibition of digoxin absorption by neomycin. Gastroenterology *71*, 399–404 (1976).
43 Lindenbaum, J., Rund, D. H., Butler, V. P., Tse-Eng, D., and Saha, J. R.: Inactivation of digoxin by the gut flora: Reversal by antibiotic therapy. N. Engl. J. Med. *305*, 789–794 (1981).
44 Fine, A., Henderson, I. S., Morgan, D. R., and Wilstone, W. J.: Malabsorption of furosemide caused by phenytoin. Br. Med. J. *2*, 1061–1062 (1977).
45 Koch-Weser, J., and Sellers, E. M.: Binding of drugs to serum albumin. N. Engl. J. Med. *294*, 311–316, 526–531 (1976).

46 Aggeler, P. M., O'Reilly, R. A., and Leong, L.: Potentiation of anticoagulant effect of warfarin by phenylbutazone. N. Engl. J. Med. *276*, 496–501 (1967).
47 Bjornsson, T. D., Meffin, P. J., Swezey, S., and Blaschke, T. F.: Clofibrate displaces warfarin from plasma proteins in man: An example of a pure displacement interaction. J. Pharmac. Exp. Ther. *210*, 316–321 (1979).
48 MacLeod, S. M., and Sellers, E. M.: Pharmacodynamic and pharmacokinetic drug interactions with coumarin anticoagulants. Drugs *11*, 461–470 (1976).
49 Sellers, E. M., and Koch-Weser, J.: Potentiation of warfarin-induced hypoprothombinemia by chloral hydrate. N. Engl. J. Med. *283*, 827–831 (1970).
50 Udall, J. A.: Warfarin-chloral hydrate interaction. Pharmacological activity and clinical significance. Ann. Intern. Med. *81*, 341–344 (1974).
51 Sellers, E. M., and Koch-Weser, J.: Displacement of warfarin from human albumin by diazoxide and ethacrynic, mefenamic, and nalidixic acids. Clin. Pharmac. Ther. *11*, 524–529 (1970).
52 Hansen, J. M., Siersbaek-Nielsen, K., Kristensen, M., Skousted, L., and Christensen, L. K.: Effects of diphenylhydantoin on the metabolism of dicoumarol in man. Acta Med. Scand. *189*, 15–19 (1971).
53 Neuvonen, P. J., Lehtovaara, R., Bardy, A., and Elonen, E.: Antipyrine analgesics in patients on antiepileptic drug therapy. Eur. J. Clin. Pharmac. *15*, 263–268 (1979).
54 Fraser, D. G., Ludden, T. M., Evens, R. P., and Sutherland, E.W.: Displacement of phenytoin from plasma binding sites by salicylate. Clin. Pharmac. Ther. *27*, 165–169 (1980).
55 Wesseling, H., Mols-Thurkow, I.: Interaction of diphenylhydantoin (DPH) and tolbutamide in man. Eur. J. Clin. Pharmac. *8*, 75–78 (1975).
56 Mattson, R. H., Cramer, J. A., Williamson, P. C., and Novelly, R. A.: Valproic acid in epilepsy: Clinical and pharmacological effects. Ann. Neurol. *3*, 20–25 (1978).
57 Perucca, E., Hebdige, S., Gatti, G., Leccini, S., Frigo, B. M., and Crema, A.: Interaction between phenytoin and valproic acid: Plasma protein binding and metabolic effects. Clin. Pharmac. Ther. *28*, 779–789 (1980).
58 Burns, J. J., and Conney, A. H.: Enzyme stimulation and inhibition in the metabolism of drugs. Proc. R. Soc. Med. *58*, 955–960 (1965).
59 Gelehrter, T. D.: Enzyme induction. N. Engl. J. Med. *294*, 522–526, 589–595, 646–651 (1976).
60 Buchanan, R. A., Heffelfinger, J. C., and Weiss, C. F.: The effect of phenobarbital on diphenylhydantoin metabolism in children. Pediatrics *43*, 114–116 (1969).
61 Hansen, J. M., Siersbaek-Nielsen, K., and Skovsted, L.: Carbamazepine-induced acceleration of diphenylhydantoin and warfarin metabolism in man. Clin. Pharmac. Ther. *12*, 539–543 (1971).
62 Kutt, H., Haynes, J., Verebely, K., and McDowell, F.: The effect of phenobarbital on plasma diphenylhydantoin level and metabolism in man and in rat liver microsomes. Neurology *19*, 611–616 (1969).
63 Perucca, E.: Pharmacokinetic interactions with antiepileptic drugs. Clin. Pharmacokin. *7*, 57–84 (1982).
64 Perucca, E., and Richens, A.: Drug interactions with phenytoin. Drugs *21*, 120–137 (1981).
65 Cucinell, S. A., Conney, A. H., Sansur, M., and Burns, J. J.: Drug interactions in man. I. Lowering effect of phenobarbital on plasma levels of bishydroxycoumarin (Dicumarol) and diphenylhydantoin (Dilantin). Clin. Pharmac. Ther. *6*, 420–429 (1965).
66 Koch-Weser, J., and Sellers, E. M.: Drug interactions with coumarin anticoagulants. N. Engl. J. Med. *285*, 487–498, 547–558 (1971).

67 O'Reilly, R. A.: Interaction of chronic daily warfarin therapy and rifampin. Ann. Intern. Med. *83*, 506–508 (1975).
68 O'Reilly, R. A.: Interaction of sodium warfarin and rifampin. Ann. Intern. Med. *81*, 337–340 (1974).
69 Romankiewicz, J. A., and Ehrman, M.: Rifampin and warfarin: A drug interaction. Ann. Intern. Med. *82*, 224–225 (1975).
70 Dossing, M., Pilsgaard, H., Rasmussen, B., and Poulsen, H. E.: Time course of phenobarbital and cimetidine mediated changes in hepatic drug metabolism. Eur. J. Clin. Pharmac. *25*, 215–222 (1983).
71 Data, J. L., Wilkinson, G. R., and Nies, A. S.: Interaction of quinidine with anticonvulsant drugs. N. Engl. J. Med. *294*, 699–702 (1976).
72 Urbano, A. M.: Phenytoin-quinidine interaction in a patient with recurrent ventricular tachyarrhythmias. N. Engl. J. Med. *308*, 225 (1983) (letter).
73 Kroboth, F. J., Kroboth, P. D., and Logan, T.: Phenytoin-theophylline-quinidine interaction. N. Engl. J. Med. *308*, 725 (1983) (letter).
74 Jaillon, P., and Kates, R. E.: Phenytoin-induced changes in quinidine and 3-hydroxyquinidine pharmacokinetics in conscious dogs. J. Pharmac. Exp. Ther. *213*, 33–37 (1980).
75 Twum-Barima, Y., Carruthers, S. G.: Quinidine-rifampin interaction. N. Engl. J. Med. *304*, 1466–1469 (1981).
76 Solomon, H. M., and Abrams, W. B.: Interactions between digitoxin and other drugs in man. Am. Heart J. *83*, 277–280 (1972).
77 Solomon, H. M., Reich, S., Spirt, N., and Abrams, W. B.: Interactions between digitoxin and other drugs *in vitro* and *in vivo*. Ann. N.Y. Acad. Sci. *179*, 362–370 (1971).
78 Pentikainen, P. J., Koivula, I. H., and Hiltunen, H. A.: Effect of rifampicin treatment on the kinetics of mexiletine. Eur. J. Clin. Pharmac. *23*, 261–266 (1982).
79 Serlin, M. J., and Breckenridge, A. M.: Drug interactions with warfarin. Drugs *25*, 610–620 (1983).
80 Mantyla, R., Mannisto, P., Nykanen, S., Koponen, A., and Lamminsivu, U.: Pharmacokinetic interactions of timolol with vasodilating drugs, food and phenobarbitone in healthy human volunteers. Eur. J. Clin. Pharmac. *24*, 227–230 (1983).
81 Linnoila, M., Mattila, M. J., and Kitchell, B. S.: Drug interactions with alcohol. Drugs *18*, 299–311 (1979).
82 Seixas, F. A.: Alcohol and its drug interactions. Ann. Intern. Med. *83*, 86–92 (1975).
83 Sellers, E. M., and Holloway, M. R.: Drug kinetics and alcohol ingestion. Clin. Pharmacokin. *3*, 440–452 (1978).
84 Karim, A., Nissen, C., and Azarnoff, D. L.: Clinical pharmacokinetics of disopyramide. J. Pharmacokin. Biopharm. *10*, 465–494 (1982).
85 Reed, R. C., and Schwartz, H. J.: Phenytoin-theophylline-quinidine interaction. N. Engl. J. Med. *308*, 724–725 (1983) (letter).
86 Marquis, J.-F., Carruthers, S. G., Spence, J. D., Brownstone, Y. S., and Toogood, J. H.: Phenytoin-theophylline interaction. N. Engl. J. Med. *307*, 1189–1190 (1982).
87 Poor, D. M., Self, T. H., and Davis, H. L.: Interaction of rifampin and digitoxin. Arch. Intern. Med. *143*, 599 (1983).
88 Branch, R. A., Shand, D. G., Wilkinson, G. R., and Nies, A. S.: The reduction of lidocaine clearance by dl-propranolol: An example of hemodynamic drug interaction. J. Pharmac. Exp. Ther. *184*, 515–519 (1973).
89 Ochs, H. R., Carstens, G., and Greenblatt, D. J.: Reduction in lidocaine clearance during continuous infusion and by coadministration of propranolol. N. Engl. J. Med. *303*, 373–377 (1980).
90 Conrad, K. A., Byers, J. M., Finley, P. R., and Burnham, L.: Lidocaine eli-

mination: Effects of metoprolol and of propranolol. Clin. Pharmac. Ther. *33*, 133–138 (1983).
91 Halkin, H., Meffin, P., Melmon, K. L., and Rowland, M.: Influence of congestive heart failure on blood levels of lidocaine and its active monodeethylated metabolite. Clin. Pharmac. Ther. *17*, 669–676 (1975).
92 Stenson, R. E., Constantino, R. T., and Harrison, D. C.: Interrelationships of hepatic blood flow, cardiac output, and blood levels of lidocaine in man. Circulation *43*, 205–211 (1971).
93 Thomson, P. D., Melmon, K. L., Richardson, J. A., Cohn, K., Steinbrunn, W., Cudihee, R., and Rowland, M.: Lidocaine pharmacokinetics in advanced heart failure, liver disease, and renal failure in humans. Ann. Intern. Med. *78*, 499–508 (1973).
94 Feely, J., Wilkinson, G. R., McAllister, C. B., and Wood, A. J. J.: Increased toxicity and reduced clearance of lidocaine by cimetidine. Ann. Intern. Med. *96*, 592–594 (1982).
95 Knapp, A. B., Maguire, W., Keren, G., Karmen, A., Levitt, B., Miura, D. S., and Somberg, J.C.: The cimetidine-lidocaine interaction. Ann. Intern. Med. *98*, 174–177 (1983).
96 Spahn, H., Mutschler, E., Kirch, W., Ohnhaus, E. E., and Janisch, H. D.: Influence of ranitidine on plasma metoprolol and atenolol concentrations. Br. Med. J. *286*, 1546–1547 (1983).
97 Desmond, P. V., Mashford, M. L., Harman, P. J., Morphett, B. J., Breen, K. J., and Wang, Y. M.: Decreased oral warfarin clearance after ranitidine and cimetidine. Clin. Pharmac. Ther. *35*, 338–341 (1984).
98 Daneshmend, T. K., and Roberts, C. J. C.: Cimetidine and bioavailability of labetalol. Lancet *1*, 565 (1981) (letter).
99 Kirch, W., Kohler, H., Spahn, H., and Mutschler, E.: Interaction of cimetidine with metoprolol, propranolol, or atenolol. Lancet *2*, 531–532 (1981) (letter).
100 Reimann, I. W., Klotz, U., and Frolich, J. C.: Effects of cimetidine and ranitidine on steady-state propranolol kinetics and dynamics. Clin. Pharmac. Ther. *32*, 749–757 (1982).
101 Latini, R., Tognoni, G., and Kates, R. E.: Clinical pharmacokinetics of amiodarone. Clin. Pharmacokin. *9*, 136–156 (1984).
102 Chiariello, M., Volpe, M., Rengo, F., Trimarco, B., Violini, R., Ricciardelli, B., and Condorelli, M.: Effect of furosemide on plasma concentration and β-blockade by propranolol. Clin. Pharmac. Ther. *26*, 433–436 (1979).
103 Geaney, D. P., Carver, J. G., Davies, C. L., and Aronson, J. K.: Pharmacokinetic investigation of the interactions of azapropazone with phenytoin. Br. J. Clin. Pharmac. *15*, 727–734 (1983).
104 Christensen, L. K., and Skovsted, L.: Inhibition of drug metabolism by chloramphenicol. Lancet *2*, 1397–1399 (1969).
105 Koup, J. R., Gilbaldi, M., McNamara, P., Hilligoss, D. M., Colburn, W. A., and Bruck, E.: Interaction of chloramphenicol with phenytoin and phenobarbital. Clin. Pharmac. Ther. *24*, 571–575 (1978).
106 Eadie, M. J.: Anticonvulsant drugs: An update. Drugs *27*, 328–363 (1984).
107 Vincent, F. M.: Phenothiazine-induced phenytoin intoxication. Ann. Intern Med. *93*, 56–57 (1980) (letter).
108 Vestal, R. E., Kornhauser, D. M., Hollifield, J. W., and Shand, D. G.: Inhibition of propranolol metabolism by chlorpromazine. Clin. Pharmac. Ther. *25*, 19–24 (1979).
109 Somogyi, A., and Gugler, R.: Drug interactions with cimetidine. Clin. Pharmacokin. *7*, 23–41 (1982).
110 Neuvonen, P. J., Tokola, R. A., and Kaste, M.: Cimetidine-phenytoin interactions: Effect on serum phenytoin concentration and antipyrine test. Eur. J. Clin. Pharmac. *21*, 215–220 (1981).

111 Bartle, W. R., Walker, S. E., and Shapero, T.: Dose-dependent effect of cimetidine on phenytoin kinetics. Clin. Pharmac. Ther. *33*, 649–655 (1983).
112 Feely, J., Wilkinson, G. R., and Wood, A. J. J.: Reduction of liver blood flow and propranolol metabolism by cimetidine. N. Engl. J. Med. *304*, 692–695 (1981).
113 Hardy, B. G., Zador, I. T., Golden, L., Lalka, D., and Schentag, J. J.: Effect of cimetidine on the pharmacokinetics and pharmacodynamics of quinidine. Am. J. Cardiol. *52*, 172–175 (1983).
114 Haworth, E., and Burroughs, A. K.: Disopyramide and warfarin interaction. Br. Med. J. *2*, 866–867 (1977).
115 Kiørboe, E.: Phenytoin intoxication during treatment with Antabuse® (disulfiram). Epilepsia *7*, 246–249 (1966).
116 Olesen, O. V.: Disulfiramum (Antabuse®) as inhibitor of phenytoin metabolism. Acta Pharmac. Tox. *24*, 317–322 (1966).
117 O'Reilly, R. A.: Interaction of sodium warfarin and disulfiram (Antabuse®) in man. Ann. Intern. Med. *78*, 73–76 (1973).
118 Kutt, H., Verebely, K., and McDowell, F.: Inhibition of diphenylhydantoin metabolism in rats and in rat liver microsomes by antitubercular drugs. Neurology *18*, 706–710 (1968).
119 Murray, F. J.: Outbreak of unexpected reactions among epileptics taking isoniazid. Am. Rev. Resp. Dis. *86*, 729–732 (1962).
120 Garrettson, L. K., Perel, J. M., and Dayton, P. G.: Methylphenidate interaction with both anticonvulsants and ethyl biscoumacetate. J. Am. Med. Ass. *207*, 2053–2056 (1969).
121 O'Reilly, R. A.: The stereoselective interaction of warfarin and metronidazole in man. N. Engl. J. Med. *295*, 354–357 (1976).
122 DeTeresa, E., Vera, A., Ortigosa J., Pulpon, L. A., Arus, A. P., and DeArtaza, M.: Interaction between anticoagulants and contraceptives: an unsuspected finding. Br. Med. J. *2*, 1260–1261 (1979).
123 O'Reilly, R. A.: Phenylbutazone and sulfinpyrazone interaction with oral anticoagulant phenprocoumon. Arch. Intern. Med. *142*, 1634–1637 (1982).
124 Solomon, H. M., and Schrogie, J. J.: The effect of phenyramidol on the metabolism of diphenlhydantoin. Clin. Pharmac. Ther. *8*, 554–556 (1967).
125 Abernethy, D. R., Greenblatt, D. J., Steel, K., and Shader, R.I.: Impairment of hepatic drug oxidation by propoxyphene. Ann. Intern. Med. *97*, 223–224 (1982).
126 Kabins, S. A.: Interactions among antibiotics and other drugs. J. Am. Med. Ass. *219*, 206–212 (1972).
127 Lumholtz, B., Siersbaek-Nielsen, K., Skovsted, L., Kampmann, J., and Hansen, J. M.: Sulfamethizole-induced inhibition of diphenylhydantoin, tolbutamide, and warfarin metabolism. Clin. Pharmac. Ther. *17*, 731–734 (1975).
128 Hansen, J. M., Kristensen, M., and Skovsted, L.: Sulthiame (Opsollot®) as inhibitor of diphenylhydantoin metabolism. Epilepsia *9*, 17–22 (1968).
129 Wormser, G. P., Keusch, G. T., and Heel, R. C.: Co-trimoxazole (trimethoprim-sulfamethoxazole). An updated review of its antibacterial activity and clinical efficacy. Drugs *24*, 459–518 (1982).
130 O'Reilly, R. A.: Stereoselective interaction of trimethoprim-sulfamethoxazole with the separated enantiomorphs of racemic warfarin in man. N. Engl. J. Med. *302*, 33–35 (1980).
131 O'Reilly, R. A., and Motley, C. H.: Racemic warfarin and trimethoprim-sulfamethoxazole interaction in humans. Ann. Intern. Med. *91*, 34–36 (1979).
132 Levy, R. H., and Koch, K. M.: Drug interactions with valproic acid. Drugs *24*, 543–556 (1982).
133 LaDu, B. N.: Pharmacogenetics. Med. Clin. North Am. *53*, 839–855 (1969).

134 Vesell, E.: Introduction: Genetic and environment factors affecting drug response in man. Fed. Proc. *31*, 1253–1269 (1972).
135 Weber, W. W.: The relationship of genetic factors to drug reactions. In: Drug-induced diseases, Vol. 4. L. Heyler, N. M. Peck, eds. Excerpta Medica, Amsterdam 1972.
136 Evans, D. A. A., Mantey, K. A., and McKusick, V. A.: Genetic control of isoniazid metabolism in man. Br. Med. J. *2*, 485–491 (1960).
137 Kutt, H., Brennan, R., Dehejia, H., and Verebely, K.: Diphenylhydantoin intoxication. A complication of isoniazid therapy. Am. Rev. Resp. Dis. *101*, 377–384 (1970).
138 Kutt, H., Wolk, M., Scherman, R., and McDowell, F.: Insufficient parahydroxylation as a cause of diphenylhydantoin toxicity. Neurology *14*, 542–548 (1964).
139 Vasko, M. R., Bell, R. D., Daly, D. D., and Pippenger, C. E.: Inheritance of phenytoin hypometabolism: A kinetic study of one family. Clin. Pharmac. Ther. *27*, 96–103 (1980).
140 Sjoqvist, F.: Psychotropic drugs (2). Interaction between monoamine oxidase (MAO) inhibitors and other substances. Proc. R. Soc. Med. *58*, 967–977 (1965).
141 DeVita, V. T., Hahn, M. A., and Oliverio, V. T.: Monoamine oxidase inhibition by a new carcinostatic agent, N-isopropyl-A-(2-methyl-hydrazino)-p-to-luamide (MIH). Proc. Soc. exp. Biol. Med. *120*, 561–565 (1965).
142 Lejonc, J. L., Gusmini, D., and Brochard, P.: Isoniazid and reaction to cheese. Ann. Intern. Med. *91*, 793 (1979) (letter).
143 Smith, C. K., and Durack, D. T.: Isoniazid and reaction to cheese. Ann. Intern. Med. *88*, 520–521 (1978).
144 Berg, M. J., Berlinger, W. G., Goldberg, M. J., Spector, R., and Johnson, G. F.: Acceleration of the body clearance of phenobarbital by oral activated charcoal. N. Engl. J. Med. *307*, 642–644 (1982).
145 Levy, G.: Gastrointestinal clearance of drugs with activated charcoal. N. Engl. J. Med. *307*, 676–678 (1982) (editorial).
146 duSouich, P., Caille, G., and Larochelle, P.: Enhancement of nadolol elimination by activated charcoal and antibiotics. Clin. Pharmac. Ther. *33*, 585–590 (1983).
147 Carruthers, S. G., and Dujovne, C. A.: Cholestyramine and spironolactone and their combination in digitoxin elimination. Clin. Pharmac. Ther. *27*, 184–187 (1980).
148 McAllister, R. G., Howell, S. M., Gomer, M. S., and Selby, J. B.: Effect of intravenous furosemide on the renal excretion of digoxin. J. Clin. Pharmac. *16*, 110–117 (1976).
149 Semple, P., Tilstone, W. J., and Lawson, D. H.: Furosemide and urinary digoxin clearance. N. Engl. J. Med. *293*, 612–613 (1971) (letter).
150 Tilstone, W. J., Semple, P. F., Lawson, D. H., and Boyle, J. A.: Effects of furosemide on glomerular filtration rate and clearance of practolol, digoxin, cephaloridine, and gentamicin. Clin. Pharmac. Ther. *22*, 389–394 (1977).
151 Prescott, L. F.: Mechanisms of renal excretion of drugs. Br. J. Anaesth. *44*, 246–251 (1972).
152 Rennick, B. R.: Renal excretion of drugs: Tubular transport and metabolism. Ann. Rev. Pharmac. *12*, 141–156 (1972).
153 Weiner, I. M., and Mudge, G. J.: Renal tubular mechanisms for excretion of organic acids and bases. Am. J. Med. *36*, 743–762 (1964).
154 Rose, H. J., Pruitt, A. W., and McNay, J. L.: Effect of experimental azotemia on renal clearance of furosemide in the dog. J. Pharmac. Exp. Ther. *196*, 238–247 (1976).
155 Rose, H. J., Pruitt, A. W., Dayton, P. G., and McNay, J. L.: Relationship of urinary furosemide excretion rate to natriuretic effect in experimental azotemia. J. Pharmac. Exp. Ther. *199*, 490–497 (1976).

156 Rose, H. J., O'Malley, K., and Pruitt, A. W.: Depression of renal clearance of furosemide in man by azotemia. Clin. Pharmac. Ther. *21*, 141–146 (1976).
157 Rennick, B. R.: Renal tubule transport of organic cations. Am. J. Physiol. *240*, F83–F89 (1981).
158 McKinney, T. D.: Heterogeneity of organic base secretion by proximal tubules. Am. J. Physiol. *243*, F404–F407 (1982).
159 Somogyi, A., McLean, A., and Heinzow, B.: Cimetidine-procainamide pharmacokinetic interaction in man: Evidence of competition for tubular secretion of basic drugs. Eur. J. Clin. Pharmacol *25*, 339–345 (1983).
160 Belz, G. G., Doering, W., Aust, P. E., Heinz, M., Matthews, J., and Schneider, B.: Quinidine-digoxin interaction. Cardiac efficacy of elevated serum digoxin concentration. Clin. Pharmac. Ther. *31*, 548–554 (1982).
161 Bigger, J. T., and Leahey, E. B.: Quinidine and digoxin. An important interaction. Drugs *24*, 229–239 (1982).
162 Bigger, J. T.: The quinidine-digoxin interaction. What do we know about it? N. Engl. J. Med. *301*, 779–781 (1979) (editorial).
163 Chen, T.-S., and Friedman, H. S.: Alteration of digoxin pharmacokinetics by a single dose of quinidine. J. Am. Med. Ass. *244*, 669–672 (1980).
164 Das, G., Krishnamurthi, S., Barr, C., Carlson, J., and Khalil, S.: Clinical implications of digoxin-quinidine interaction in man. Clin. Res. *29*, 691A (1981) (abstract).
165 Doering, W.: Quinidine-digoxin interaction. Pharmacokinetics, underlying mechanism and clinical implications. N. Engl. J. Med. *301*, 400–404 (1979).
166 Hager, W. D., Fenster, P., Mayersohn, M., Perrier, D., Graves, P., Marcus, F.I., and Goldman, S.: Digoxin-quinidine interaction. Pharmacokinetic evaluation. N. Engl. J. Med. *300*, 1238–1241 (1979).
167 Hirsh, P. D., Weiner, H. J., and North, R. L.: Further insights into digoxin-quinidine interaction: Lack of correlation between serum digoxin concentration and inotropic state of the heart. Am. J. Cardiol. *46*, 863–867 (1980).
168 Holt, D. W., Hayler, A. M., Edmonds, M. E., and Ashford, R. F.: Clinically significant interaction between digoxin and quinidine. Br. Med. J. *2*, 1401 (1979).
169 Leahey, E. B.: Digoxin-quinidine interaction: Current status. Ann. Intern. Med. *93*, 775–776 (1980) (editorial).
170 Leahey, E. B., Rieffel, J. A., Drusin, R. E., Heissenbuttel, R. H., Lovejoy, W. P., and Bigger, J. T.: Interaction between quinidine and digoxin. J. Am. Med. Ass. *240*, 533–534 (1978).
171 Leahey, E. B., Reiffel, J. A., Giardina, E.-J. V., and Bigger, J. T.: The effect of quinidine and other oral antiarrhythmic drugs on serum digoxin. Ann. Intern. Med. *92*, 605–608 (1980).
172 Leahey, E. B., Reiffel, J. A., Heissenbuttel, R. H., Drusin, R. E., Lovejoy, W. P., and Bigger, J. T.: Enhanced cardiac effect of digoxin during quinidine treatment. Arch. Intern. Med. *139*, 519–521 (1979).
173 Mungall, D. R., Robichaux, R. P., Perry, W., Scott, J. W., Robinson, A., Burelle, T., and Hurst, D.: Effects of quinidine on serum digoxin concentration. Ann. Intern. Med. *93*, 689–693 (1980).
174 Schenck-Gustafsson, K., Jogestrand, T., Nordlander, R., and Dahlqvist, R.: Effect of quinidine on digoxin concentration in skeletal muscle and serum in patients with atrial fibrillation. Evidence for reduced binding of digoxin in muscle. N. Engl. J. Med. *305*, 209–211 (1981).
175 Steiness, E., Waldorff, S., Hansen, P. B., Kjaergard, H., Buch, J., and Egeblad, H.: Reduction of digoxin-induced inotropism during quinidine administration. Clin. Pharmac. Ther. *27*, 791–795 (1980).
176 Williams, J. F., and Mathew, B.: Effect of quinidine on positive inotropic action of digoxin. Am. J. Cardiol. *47*, 1052–1055 (1981).
177 Fenster, P. E., Hager, W. D., Perrier, D., Powell, J. R., Graves, P. E., and

Michael, U. F.: Digoxin-quinidine interaction in patients with chronic renal failure. Circulation *66*, 1277–1279 (1982).
178 Fichtl, B., and Doering, W.: The quinidine-digoxin interaction in perspective. Clin. Pharmacokin. *8*, 137–154 (1983).
179 Belz, G. G., Doering, W., Munkes, R., and Matthews, J.: Interaction between digoxin and calcium antagonists and antiarrhythmic drugs. Clin. Pharmac. Ther. *33*, 410–417 (1983).
180 Walker, A. M., Cody, R. J., Greenblatt, D. J., and Jick, H.: Drug toxicity in patients receiving digoxin and quinidine. Am. Heart J. *105*, 1025–1028 (1983).
181 Schenck-Gustafsson, K., Jogestrand, T., Brodin, L.-A., Nordlander, R., and Dahlqvist, R.: Cardiac effects of treatment with quinidine and digoxin, alone and in combination. Am. J. Cardiol. *51*, 777–782 (1983).
182 Das, G., Barr, C. E., and Carlson, J.: Reduction of digoxin effect during the digoxin-quinidine interaction. Clin. Pharmac. Ther. *35*, 317–321 (1984).
183 Gibson, T. P., and Quintanilla, A.: Effect of quinidine on the renal handling of digoxin. J. Lab. Clin. Med. *96*, 1062–1070 (1980).
184 Fenster, P. E., Powell, J. R., Graves, P. E., Conrad, K. A., Hager, W. D., Goldman, S., and Marcus, F. I.: Digitoxin-quinidine interaction: Pharmacokinetic evaluation. Ann. Intern. Med. *93*, 698–701 (1980).
185 Garty, M., Sood, P., and Rollins, D. E.: Digitoxin elimination reduced during quinidine therapy. Ann. Intern. Med. *94*, 35–37 (1981).
186 Ochs, H. R., Pabst, J., Greenblatt, D. J., and Dengler, H. J.: Noninteraction of digitoxin and quinidine. N. Engl. J. Med. *303*, 672–674 (1980).
187 Schwartz, J. B., Keefe, D., Kates, R. E., and Harrison, D. C.: Verapamil and digoxin: Another drug-drug interaction. Clin. Res. *29*, 501A (1981) (abstract).
188 Waldorff, S., Andersen, J. D., Heebøll-Neilsen, N., Nielsen, O. G., Moltke, E., Sørensen, U., and Steiness, E.: Spironolactone-induced changes in digoxin kinetics. Clin. Pharmac. Ther. *24*, 162–167 (1978).
189 Wandell, M., Powell, J. R., Hager, W. D., Fenster, P. E., Graves, P. E., Conrad, K. A., and Goldman, S.: Effect of quinine on digoxin kinetics. Clin. Pharmac. Ther. *28*, 425–430 (1980).
190 Pedersen, K. E., Christiansen, B. D., Kjaer, K., Klitgaard, N. A., and Nielsen-Kudsk, F.: Verapamil-induced changes in digoxin kinetics and intraerythrocytic sodium concentration. Clin. Pharmac. Ther. *34*, 8–13 (1983).
191 Klein, H. O., Lang, R., Weiss, E., Segni, E. D., Libhaber, C., Guerrero, J., and Kaplinsky, E.: The influence of verapamil on serum digoxin concentration. Circulation *65*, 998–1003 (1982).
192 Waldorff, S., Hansen, P. B., Egeblad, H., Berning, J., Buch, J., Kjaergard, H., and Steiness, E.: Interactions between digoxin and potassiumsparing diuretics. Clin. Pharmac. Ther. *33*, 418–423 (1983).
193 Pedersen, K. E., Dorph-Pedersen, A., Hvidt, S., Klitgaard, N. A., Kjaer, K., and Nielsen-Kudsk, F.: Effect of nifedipine on digoxin kinetics in healthy subjects. Clin. Pharmac. Ther. *32*, 562–565 (1982).
194 Gibson, T. P., and Quintanilla, A. P.: Effect of volume expansion and furosemide diuresis on the renal clearance of digoxin. J. Pharmac. Exp. Ther. *219*, 54–59 (1981).
195 Boullin, D. J.: The action of antidepressants on the effects of other drugs. Primary Care *2*, 669–688 (1975).
196 Stafford, J. R., and Fann, W. E.: Drug interactions with guanidinium antihypertensives. Drugs *13*, 57–64 (1977).
197 Cocco, G., and Ague, C.: Interactions between cardioactive drugs and antidepressants. Eur. J. Clin. Pharmac. *11*, 389–393 (1977).
198 Boakes, A. J., Laurence, D. R., Teoh, P. C., Barar, F. S. K., Benedikter,

L. T., and Prichard, B. N. C.: Interactions between sympathomimetic amines and antidepressant agents in man. Br. Med. J. *1*, 311–315 (1973).

199 Allum, W., Aminu, J., Bloomfield, T. H., Davies, C., Scales, A. H., and Vere, D. W.: Interaction between debrisoquin and phenylephrine in man. Br. J. Clin. Pharmac. *1*, 51–57 (1974).

200 Hanahoe, T. H. P., Ireson, J. D., and Large, B. J.: Interactions between guanethidine and inhibitors of noradrenaline uptake. Arch. Int. Pharmacodyn. *182*, 349–353 (1969).

201 Leishman, A. W. D., Matthews, H. L., and Smith, A. J.: Antagonism of guanethidine by imipramine. Lancet *1*, 112 (1963).

202 Mitchell, J. R., Arias, L., and Oates, J. A.: Antagonism of the antihypertensive action of guanethidine sulfate by desipramine hydrochloride. J. Am. Med. Ass. *202*, 973–976 (1967).

203 Mitchell, J. R., Cavanaugh, J. H., Arias, L., and Oates, J. A.: Guanethidine and related agents. III. Antagonism by drugs which inhibit the norepinephrine pump in man. J. Clin. Invest. *49*, 1596–1604 (1970).

204 Skinner, C., Coull, D. C., and Johnston, A. W.: Antagonism of the hypotensive action of bethanidine and debrisoquin by tricyclic antidepressants. Lancet *2*, 564–566 (1969).

205 Stone, C. A., Porter, C. C., Stavorski, J. M., Ludden, C. T., and Totaro, J. A.: Antagonism of catecholamine-depleting agents by antidepressant and related drugs. J. Pharmac. *144*, 196–204 (1964).

206 Gokhale, S. D., Gulati, O. D., and Udwadia, B. P.: Antagonism of the adrenergic neurone blocking action of guanethidine by certain antidepressant and antihistamine drugs. Arch. Int. Pharmacodyn. *160*, 321–329 (1966).

207 Fann, W. E., Cavanaugh, J. H., and Kaufmann, J. S.: Doxepin: Effects on transport of biogenic amines in man. Psychopharmacologia *22*, 111–125 (1972).

208 Chang, C. C., Costa, E., and Brodie, B. B.: Reserpine-induced release of drugs from sympathetic nerve endings. Life Sci. *3*, 839–844 (1964).

209 Day, M. D.: Effect of sympathomimetic amines on the blocking action of guanethidine, bretylium, xylocholine. Br. J. Pharmac. *18*, 421–439 (1962).

210 Day, M. D., and Rand, M. J.: Antagonism of guanethidine by dexamphetamine and other related sympathomimetic amines. J. Pharm. Sci. *14*, 541–549 (1962).

211 Day, M. D., and Rand, M. J.: Evidence for a competitive antagonism of guanethidine by dexamphetamine. Br. J. Pharmac. *20*, 17–28 (1963).

212 Janowsky, D. S., El-Yousef, M. K., Davis, J. M., and Fann, W. E.: Antagonism of guanethidine by chlorpromazine. Am. J. Psychiat. *130*, 808–812 (1973).

213 Misage, J. R., and McDonald, R. H.: Antagonism of hypotensive action of bethanidine by 'common cold' remedy. Br. Med. J. *2*, 1–3 (1970).

214 Briant, R. H., Reid, J. L, and Dollery, C. T.: Interaction between clonidine and desipramine in man. Br. Med. J. *1*, 522–523 (1973).

215 Hoobler, S. W., and Sagastume, E.: Clonidine hydrochloride in the treatment of hypertension. Am. J. Cardiol. *28*, 67–83 (1971).

216 vanZwieten, P. A.: The reversal of clonidine-induced hypotension by protiptyline and desipramine. Pharmacology *14*, 227–231 (1976).

217 McMurtry, R. J.: Propranolol, hypoglycemia, and hypertensive crisis. Ann. Intern. Med. *80*, 669–670 (1974).

218 Nies, A. S., and Shand, D. G.: Hypertensive response to propranolol in a patient treated with methyl dopa – a proposed mechanism. Clin. Pharmac. Ther. *14*, 823–826 (1973).

219 Prichard, B. N. C., and Ross, E. J.: Use of propranolol in conjunction with alpha receptor blocking drugs in pheochromocytoma. Am. J. Cardiol. *18*, 394–398 (1966).

220 Bailey, R. R., and Neale, T. J.: Rapid clonidine withdrawal with blood pressure overshoot exaggerated by beta-blockade. Br. Med. J. *1*, 942–943 (1976).
221 Newton, R. W.: Physostigmine salicylate in the treatment of tricyclic antidepressant overdosage. J. Am. Med. Ass. *231*, 941–944 (1974).
222 Noble, J., and Matthew, H.: Acute poisoning by antidepressants: Clinical features and management of 100 patients. Clin. Toxicol *2*, 403–421 (1969).
223 Roberts, J., Ito, R., Reilly, J., and Carioli, V. J.: Influence of reserpine and beta TM 10 on digitalis induced ventricular arrhythmia. Circ. Res. *13*, 149–158 (1963).
224 Arita, M., and Surawicz, B.: Electrophysiologic effects of phenothiazines on canine cardiac fibers. J. Pharmac. Exp. Ther. *184*, 619–630 (1973).
225 Davis, J. M., Bartlett, E., and Termini, B. S.: Overdosage of psychotropic drugs. A review. Dis. Nerv. Syst. *29*, 157–164 and 246–256 (1968).
226 Fowler, N. O., McCall, D., Chou, T., Holmes, J. C., and Hanenson, I. B.: Electrocardiographic changes and cardiac arrhythmias in patients receiving psychotropic drugs. Am. J. Cardiol. *37*, 223–230 (1976).
227 Williams, R. B., and Sherter, C.: Cardiac complications of tricyclic antidepressant therapy. Ann. Intern. Med. *74*, 395–398 (1971).
228 Nash, C. W., and Heath, C.: Vascular responses to catecholamines during respiratory changes in pH. Am. J. Physiol. *200*, 755–782 (1961).
229 Cumming, A. D., and Robertson, C.: Interaction between dispoyramide and practolol. Br. Med. J. *2*, 1264 (1979).
230 Thornton, W. E.: Dementia induced by methyldopa with haloperidol. N. Engl. J. Med. *294*, 1222 (1976).
231 MacCallum, W. A. G.: Interaction of lithium and phenytoin. Br. Med. J. *1*, 610–611 (1980).

Platelets and Atherosclerosis

By Robert N. Saunders
Department of Platelet Research, Sandoz Research Institute, East Hanover, New Jersey, USA

1	Introduction	50
2	History	50
3	The platelet	51
3.1	Platelet morphology	51
3.2	Platelet physiology	51
3.3	Platelet function	52
3.4	Platelet dysfunction	52
3.5	Platelet adhesion	52
4	Atherosclerosis	53
4.1	Atherogenesis	53
4.2	Response-to-injury hypothesis	53
4.3	Verification of the response-to-injury hypothesis	54
4.4	Loss of endothelium examined	55
4.5	Endothelium permeability concept	55
4.6	Current response-to-injury hypothesis	56
5	The critical role of the platelet	56
5.1	Vascular smooth muscle cell migration and proliferation	56
5.2	Endothelial cell and platelet inhibition of intimal proliferation	57
5.3	Animal models and clinical conditions which suggest platelets are involved in atherogenesis	58
5.4	The effect of antiplatelet agents in animal models and human disease	60
6	Potential pharmacological sites of intervention	61
7	Conclusion	62

1 Introduction

The concept that platelets play a role in the genesis of atherosclerosis has assumed a level of general acceptance in the past ten years. The magnitude of that role has fluctuated repeatedly during that time interval from that of a minor component to a major factor. The purpose of this review, although written by a somewhat biased individual, will be to present an overview of the concept and hopefully place platelets in their proper perspective within that concept. The etiology of atherosclerosis is multifaceted with a lengthy list of associated risk factors characterized by epidemiological studies and succinctly summarized by the US National Heart, Lung and Blood Institute Working Group on Arteriosclerosis [1]. The intention of this review will be to focus on those aspects of atherogenesis where platelets are predicted to be involved with comments on what remains to be answered.

2 History

Atherosclerosis is not limited to modern environments since its presence in Egyptian mummies has been described [2]. Atherosclerosis, differentiated from arteriolar sclerosis in 1904 [3], was associated with the clinical syndrome of myocardial infarction in 1912 [4]. Improved clinical diagnosis and definition of cause of death, associated with a reduction in other disease-related fatalities, suggested that an apparant rise in cardiovascular disease occurred during the first half of the twentieth century. An underlying aspect or inducer of cardiovascular disease is atherosclerosis. With this facet in mind, an appreciation of the enormous impact atherosclerosis has had on the health and financial well being of developed nations is obvious. During the year 1977, atherosclerosis caused almost half of all deaths in the United States and cost an estimated $ 39 billion in health expenditures and lost productivity [5].

The decline in cardiovascular disease in the United States over the past twenty years has been analyzed with great interest, especially since a similar decline has not occurred in most other western countries [6]. The source of this decline is assumed to be related to a reduction of risk factors, increased exercise and better primary health care, although the exact reason is undetermined [7, 8].

3 The platelet

Slightly more than 100 years ago Bizzozero suggested the name blood plates for the then recently discovered cell which differed morphologically from both the red and white corpuscle and possessed properties important to coagulation [9]. Although the role of these blood platelets, which they became to be called, was quickly recognized in blood coagulation, they did not become the focus of extensive research until the middle of the twentieth century. Perhaps the greatest stimulus to platelet research was the introduction of the aggregometer by Born [10].

3.1 Platelet morphology

Platelets were among the first cells utilized in early electron microscopic investigations [11]. Discoidal and anuclear, platelets are the smallest circulating blood cell. They possess discrete granules and an open canalicular system which connects to the exterior of the cell. Just below the membrane around the circumference of poles of the discoid cell lie microfilaments composed of actin and myosin. The middle gel zone contains granules and mitochondria surrounded by actin in an unpolymerized form [12] with an open channel system throughout. The granules are divided into the dense and alpha by their electron opacity. The dense granules contain primarily adenosine di- and triphosphate and serotonin. The alpha granules contain numerous proteins including platelet factor 4, β-thromboglobulin and platelet derived growth factor (PDGF) [12]. Platelet lysosomes contain the expected acid hydrolases [12] and perhaps a specific heparitinase which is suggested to play a role in the prevention of atherosclerosis [13].

3.2 Platelet physiology

As dynamic storehouses and primary sentinels of the cardiovascular system, platelets have the potential to adhere to surfaces and each other, change shape from discoid to sphere to flat, and to release numerous potent biochemical agents [14]. These responses are controlled by interaction of structural elements of the vessel wall or chemical signals with platelet membrane receptors [15]. The most relevant agents to this discussion are collagen, thrombin, ADP, thromboxane A2, platelet activating factor and epinephrine. With the exception of epinephrine, the platelet responds to mediator-receptor interaction by changing to a spherical shape. If the mediator-receptor interaction is of sufficient quantity, phospholipases C and A2 are activated to initiate the phosphoinositol pathway and arachidonic acid cascade [15]. These metabolic steps lead to the formation of pseudopods and granule secretion.

3.3 Platelet function

Without platelets, our vascular system would soon lose its integrity [16]. Minor gaps in the endothelial lining of the blood vessels would lead to the loss of plasma proteins and fluid [17]. The platelet senses these breaks in the vascular tree and quickly plugs the gap, acting as a pseudoendothelium for a short-time interval. When a blood vessel is severed, platelets form a hemostatic plug preventing further blood loss. Platelet granule contents will aid in the plug-forming process or in the healing-repair process that follows [16].

3.4 Platelet dysfunction

Dysfunction of platelets is usually observed as an increase in bleeding time [18] or the presence of platelet thrombi [19]. Genetic disorders are known in which platelets lack alpha and dense granule constituents [20–22] and contractile protein [22]. Genetic disorders also have been described in which the platelet function is normal but plasma factors required for adhesion or clot formation are reduced or absent [22]. These required proteins are von Willebrand factor and fibrinogen. Idiopathic platelet dysfunction occurs when individuals take drugs which enhance or reduce platelet response mechanisms [22].

3.5 Platelet adhesion

Rapid platelet adhesion occurs to the subendothelial connective tissue matrix when endothelial cells are removed from the vessel wall. This subendothelium-platelet adhesion is different from platelet-platelet adhesion since aspirin and other cycooxygenase inhibitors prevent the latter but not the former [23]. The adherent platelets spread over the denuded vessel surface forming a pseudoendothelium [24]. During the process of adhesion and spreading, the platelets degranulate and release the granule contents extracellularly. The presence of alpha granule proteins within the vessel wall after platelet adhesion and degranulation has been observed [25].

4 Atherosclerosis

Arteriosclerosis is a generic term applied to conditions in which a permanent reduction in vessel lumen size occurs. Atherosclerosis is a form of arteriosclerosis in which intimal neomuscular lesions with lipid deposition occurs. These lesions or plaques more often appear at branch sites in arteries where turbulent blood flow is prominent. Atherosclerosis may be slow in formation as part of the aging process or relatively rapid when vascular surgery or other forms of arterial damage have occurred.

4.1 Atherogenesis

In the mid-nineteenth century, Virchow introduced the concept that atherosclerosis was a degenerative process that induced lipid-laden lesions after vessel injury [26]. Some fifty years later the cell growth-promoting properties of serum were observed [27]. This observation eventually led to the discovery that platelets contain a cell growth factor which might be released when platelets responded to an injury of the vessel wall [28]. This concept formed the basis of the 'response-to-injury hypothesis' of atherosclerosis.

4.2 Response-to-injury hypothesis

As first proposed, this hypothesis assumed that an insult to the endothelial tissue results in endothelial desquamation followed by platelet adhesion [28]. PDGF released by the adherent platelets induces the medial smooth muscle cells to migrate to the luminal surface and divide. As originally proposed, low density lipoprotein uptake by these smooth muscle cells would then lead to foam cells. This lesion would be recovered with endothelial cells and nearly reabsorbed with the passage of time unless another insult to the vessel wall occurred before the resorptive process was complete. Repeated injuries at the same site would result in accumulative lesions with low local tissue oxygenation. Severe lesions with necrotic centers, large lipid and calcium deposits could ensue [28]. Modification of this theory occurred over the past ten years as additional experimental evidence was obtained [29].

4.3 Verification of the response-to-injury hypothesis

Physical injury of several forms had been used to induce atherosclerotic lesions in animals prior to Ross' and Glomset's hypothesis. The most popular of these vessel injury techniques was introduced by Baumgartner in 1963 [30]. This model uses an inflated Fogarty balloon catheter to strip the endothelium as it is drawn through the vessel. Platelets adhere immediately in several layers to the denuded surface forming small platelet thrombi [31]. These aggregates quickly dispense but leave a monolayer of spread platelets covering the subendothelium. By the fourth day after the injury, the adherent platelets have dispersed back to the circulation leaving a nonthrombogenic subendothelium surface. In the injured rat aorta smooth muscle cells begin to appear and by the tenth day, the intima has thickened from the normal one cell thickness to an average of twelve cells by migration and proliferation [32]. Re-endothelialization occurrs quickly in the rat but is much slower in the rabbit where regeneration of the endothelium may take 12 months [33]. Human re-endothelialization rates are assumed to be more like the rabbit than the rat. Repeated injuries over time with the balloon catheter technique will increase the neointima area but rarely leads to major lipid accumulation or vessel stenosis [34].
Using a different injury technique, an indwelling catheter in the rabbit aorta, Moore [35] induced lipid-laden, raised lesions similar to those seen in humans. When the platelet count was maintained at very low levels by use of antiplatelet serum, the development of raised lesions in this model was greatly reduced or prevented [36]. This experiment was repeated using the balloon-catheter technique with similar results [37]. Ohnishi and co-workers [38] took this concept one step farther. They platelet-depleted rats with antiplatelet serum one day prior to de-endothelialization of the right common carotid artery by the air-drying technique previously described by Clowes and Clowes [39]. They maintained the platelets in reduced numbers with additional antiplatelet serum and injected the rats daily with a crude isolate of PDGF [38]. Proliferation was minimal in the platelet-depleted rats without the PDGF supplementation and normal for this model in those rats with

the PDGF supplementation. These platelet-depletion experiments are often cited as the prima-facie evidence for the involvement of platelets in atherosclerosis although the toxicity associated with repeated use of antiplatelet serum has led at least one group to question the validity of these conclusions [40].

4.4 Loss of endothelium examined

Large vessel denudation was not observed in early dietary-induced atherosclerotic animal models [41, 42]. It was still conceivable that small areas of denudation which might be rapidly re-endothelialized might lead to focal sites of neointimal proliferation. To address this question, investigators produced a narrow (3 to 5 cells wide) endothelial denudation in rabbit [43] and rat [44] aortas. Platelets adhered to the exposed subendothelium but neointimal proliferation did not occur [44]. This suggests that re-endothelialization must prevent the smooth muscle cell response (migration and proliferation). Denudation as the first event in atherogenesis seemed to be less plausible. Other investigators found that in denuded rabbit aorta experiments, the re-endothelialized intima was more prone to lipid accumulation than were the areas which remained endothelial cell free [45]. This indicated that the presence of endothelial cells would be more likely to lead to lipid-laden lesions.

4.5 Endothelium permeability concept

An intact endothelium appears to be more likely to produce a lipid-laden lesion whereas a denuded vessel would be more likely to form a proliferation lesion. The early lesion in dietary atherogenesis in animal models was found to consist of subendothelial foam cells, therefore, one would assume the endothelial cells were present but somehow lost their normal barrier function. Endothelial cell permeability to plasma proteins was an early observation in cholesterol-fed rabbits [46, 47]. These permeability changes were attributed to histamine release and prevented by antihistaminic drugs [8]. Platelet activating factor, an inflammatory mediator, has also been shown to induce the loss of plasma proteins from the fluid vascular space and increase the extracellular matrix of guinea-pig aortas [49]. In these studies no platelet adherence to the intact endothelium nor significant neointimal proliferation was observed [49]. Thus, lipid infiltration or permeation to the subendothelial space may occur in the absence of prominent endothelial cell injury or platelet involvement.

4.6 Current response-to-injury hypothesis

To clarify the early atherogenic events, Ross and co-workers evaluated changes in the vascular morphology over time in pigtail monkeys fed a high fat/high cholesterol diet [50]. An early event was observed to be the attachment of monocytes to the endothelium which migrate to the subendothelial space, convert to macrophages and accumulate lipids. Within four weeks, fatty streaks under an intact endothelial cell cover was observed. The macrophages were apparently intending to scavenger the intimal lipids and egress to the plasma compartment [50]. The lipid laden macrophages were several layers deep by 2 to 3 months, producing a weakened endothelial cover with irregular surface characteristics. During the third month of the study endothelial desquamation with platelet adherence was observed and proliferative lesions in these anatomical sites occurred one to 2 months later [51].

5 The critical role of the platelet

The modified 'response-to-injury' hypothesis suggests that platelets are involved in intermediate stages of atherogenesis. They are given the role of inducing proliferation lesions after patchy endothelial cell desquamation has occurred. One must clearly distinguish between the proliferative and thrombotic aspects of platelets within atherogenesis. In the proliferative phase of atherogenesis platelets would be envisioned to attach, degranulate and detach, rejoining the circulation. Human atherosclerotic lesions do not demonstrate the presence of platelets by the use of specific antiplatelet serum until an organized, focal fibrotic plaque develops [52]. At late stages of atherogenesis, thrombus formation on necrotic lesions would add quickly to the lesion and perhaps produce vessel occlusion.

5.1 Vascular smooth muscle cell migration and proliferation

Intimal proliferation of medial smooth muscle cells involves dedifferentiation followed by migration to the vessel lumen. PDGF is the only purified growth factor known to induce smooth muscle cell migration [29, 53]. Platelets can produce 12-L-hydroxy-5,8,10,14-eiosatetraenoic acid (12-HETE) which will also stimulate smooth muscle cell migration in vitro [54]. The presence of macrophages in the early lesions would suggest that macrophage-derived growth factor (MDGF) may also be present. The effect of MDGF on smooth muscle cell migration is undefined at present. MDGF is a potent mitogen [55] and would be expected to contribute to the neointimal proliferative lesion with or without PDGF assuming the presence of premigrated smooth muscle cells in the intimal area. Low density lipoprotein (LDL) was also observed to induce the proliferation of smooth muscle cells from arterial explants [56, 57]. Most of these investigations utilized serum from hyperlipidemic animals and were criticized because platelet factors could also be present in the serum. The same mitogenic effect of hyperlipedemic serum was however noted when it was obtained from platelet-poor plasma [58] suggesting that the mitogenic effect of LDL is independent of platelet releasate. An endothelial cell-derived growth factor which will induce smooth muscle cell mitogenesis has also been described [59, 60]. Serotonin was recently reported to be mitogenic to cultured vascular smooth muscle cells and synergistic with PDGF in this response [61]. Degranulation of platelets at subendothelial attachment sites would be expected to release both PDGF and serotonin. Several smooth muscle cell mitogenic factors therefore exist within the vascular compartment and the exclusive requirement for PDGF in this process may be questioned. The suggestion that PDGF (or 12-HETE) is required for smooth muscle cell migration from the medial layer to the intima [28] implies that platelet degranulation is critical for the initial step in neointimal proliferation. Once smooth muscle cells are present in the intimal area, other growth promoters may enhance the lesion development.

5.2 Endothelial cell and platelet inhibition of intimal proliferation

An intact endothelium prevents platelet attachment and degranulation but also produces other anti-atherogenic factors. Heparin is attached to the endothelial cell surface and will prevent smooth muscle cell proliferation either *in vitro* [62] or *in vivo* [63]. The antimitogenic effect does not depend upon calcium chelation by heparin [64] nor upon the anticoagulant property [65] and remains effective when the heparin molecule is cleaved to subunits of six or greater saccaharides in length [66]. An enzyme capable of fractionating heparin into such subunits is located within the lysosomal fraction of platelets [62]. Therefore, platelets which degranulate and secrete near endothelial cells with a heparin coat would potentially produce a potent antiproliferative agent which could prevent the formation or progression of neointimal lesions.

Endothelial cells are also a source of prostacyclin (PGI_2), the most potent endogenous platelet anti-aggregatory compound [67]. PGI_2 at very high levels is capable of reducing platelet adhesion to subendothelial surfaces [68, 69] and preventing platelet degranulation [69]. These aspects have led several investigators to suggest that atherosclerosis is the result of a deficiency in PGI_2 production [70]. The converse arguments are that plasma levels of PGI_2 in healthy individuals are far below the effective concentration required to alter platelet behavior [71], PGI_2 potentiates the inflammatory effects of other mediators such as histamine [72] and bradykinin [73] and PGI_2 has no effect on fibroblast proliferation in culture [74]. A recent study has suggested that PGI_2 biosynthesis is actually elevated in patients with severe atherosclerosis [75]. This study would suggest that the platelet-vascular interaction is a stimulus to PGI_2 production. Two of the platelet released mediators, PDGF [76] and serotonin [77] act synergistically to stimulate PGI_2 synthesis from cultured vascular smooth muscle cells. These two platelet mediators also act synergistically to induce vascular smooth muscle cell proliferation [61].

The balance between the anti-atherosclerotic agents produced by the vascular wall and the atherogenic agents released by the platelet must be tipped in favor of neointimal proliferation primarily when endothelial cells are absent.

5.3 Animal models and clinical conditions which suggest platelets are involved in atherogenesis

The reduction of atherosclerotic lesions in the de-endothelialized aorta of rabbits whose platelet count was severely reduced was previously mentioned [37]. Platelet adhesion to the de-endothelialized subendothelium is not accompanied by clotting mechanisms whereas platelet accumulation on the neointimal lesion surface is in concert with activation of the coagulation system and thrombus formation [78]. Once again, we are reminded of the two different aspects of platelets in atherogenesis, proliferation and thrombus formation. Prevention of platelet adhesion to the vessel wall in this model might be expected to yield results similar to platelet count suppression. Platelet adhesion to the vessel wall requires the presence of von Willebrand factor [79]. In humans [80] and animals [81], individuals with varying severity of the syndrome of reduced von Willebrand factor disease are known. In pigs with severe homozygous von Willebrand's disease, a high cholesterol diet produced only moderate plaque formation which consisted of flat fatty lesions [81]. Severe aortic endothelial damage was apparent but intimal proliferation was absent. Heterozygous von Willebrand pigs with moderate expression of the syndrome are susceptible to atherogenesis similar to that exhibited by normal pigs [81]. The European Thrombosis Research Organization has initiated the Rokitansky-Duguid project to determine if patients with severe von Willebrand's disease have a reduced risk of atherosclerosis [82]. An anticipated complication of this study is the observation that the majority of these severe patients will require transfusions or Factor VIII therapy which will return the platelet adhesive properties of these patients to near normal [82].

The formation of atherosclerotic fibromuscular intimal thickening following arterial de-endothelialization of hypophysectomized rats with normal platelet counts is nearly absent [83]. This reduced neointimal proliferation may be related to the reduced somatomedin-C levels in these animals. Somatomedin-C addition to hypopituitary human patient plasma allowed normal fibroblast proliferation in culture after exposure to PDGF [84].

Other human conditions or animal models where platelet involvement in atherogenesis is suggested will be briefly cited. Patients with homocystinemia experience accelerated atherosclerosis and reduced platelet half-life [85]. Baboons with experimental homocystinemia develop arteriosclerotic lesions that are attributed to endothelial cell desquamation and platelet attachment [86]. Pulmonary arteries of dogs infected with heartworms for 30 days demonstrate endothelial cell loss, platelet-leukocyte adhesion and the development of myoproliferative lesions [87]. Numerous surgical procedures such as heart transplants [88], aorto-coronary bypass [89] and transluminal angioplasty [90] result in accelerated atherosclerosis and have been considered as ideal clinical studies for the evaluation of the role of platelets in atherogenesis.

5.4 The effect of antiplatelet agents in animal models and human disease

A definition of the term antiplatelet agent must be made since this is important to the concept of their use in atherosclerosis. Most currently marketed antiplatelet agents are antithrombotic since they prevent platelet-platelet interaction. To effectively intervene with the proliferative aspect of the platelet in the 'response-to-injury hypothesis', the agent would need to prevent platelet adhesion to the subendothelial surface, prevent the release of stored granule contents or modify the response of the medial smooth muscle cells to PDGF. Trapidil, a coronary vasodilator [91], is reported to have minimal effects on platelet-platelet interaction but inhibit the effect of PDGF on smooth muscle cell proliferation [92, 93]. Trapidil was likewise shown to inhibit myointimal thickening induced by air injury of the rat carotid artery [37] and balloon catheter de-endothelialization of the rat aorta [93]. Whether this compound will ever be used clinically for this indication is uncertain but it has provided an example of an antiproliferative agent which can reduce lesion formation.

Another type of antiproliferative agent is the non-anticoagulant heparin fraction [64, 94]. Heparin and heparin fragments can inhibit PDGF induced vascular smooth muscle cell proliferation in culture [95] and neointimal proliferation induced by air-drying injury in the rat carotid artery [96]. A potential difficulty with these preparations of heparin is low oral absorptivity, requiring that they be used by systemic administration.

Two classical antiplatelet agents, dipyridamole and aspirin, have been repeatedly evaluated in atherosclerotic models with conflicting results [97–100]. The concept of antithrombotic versus antiproliferative must again be mentioned since these two agents have assumed the role of accepted therapy post coronary bypass operations [101–105]. A reduction of the quantity of platelets adhering to aortocoronary bypass grafts in dogs [106] and peripheral arterial bypass grafts in humans [107] has been noted. An assessment of the quantity of platelets may well relate to the platelet-platelet interaction rather than the platelet-vessel wall adhesion assumed to be more critical for neointimal proliferation. Aspirin or dipyridamole given individually to humans did not alter the subsequent adhesion of their platelets to de-endothelialized rabbit aortas ex vivo [108]. Aspirin ingestion does reduce the ex vivo platelet adhesion when the blood is anticoagulated with citrate [108] or heparin [109] whereas dipyridamole under the same conditions had no effect [108, 109]. The combination of dipyridamole plus aspirin had a significant effect on the early thrombotic-type occlusions of aortocoronary bypass grafts in dogs and man but had only a slight effect on the slower atherosclerotic type of graft occlusion [88]. Minimal effects were also observed with subendothelial proliferation of veno-venous allografts in canine femoral veins when treated with either aspirin or dipyridamole separately or in combination [110].

Prostacyclin has been infused into patients during aortocoronary bypass surgery [111] with advanced arteriosclerosis obliterans [112] and vascular prosthetic grafts [113]. The infusions were short term with numerous reversible side effects. The beneficial effects observed were an increase in the platelet count, decrease in platelet adhesion [111, 114] relief of pain and healing of ischemic ulcers [115]. Effects on atherogenesis have yet to be accomplished and may require the preparation of a chemically stable PGI_2 analogue with less hypotensive effect.

Calcium is required for platelet secretion [116] and plays a central role in PDGF-induced cell proliferation [117]. One of the initial events after the exposure of diploid human fibroblasts in culture to a mitogen (PDGF, EGF or serum) is a rise in free intercellular calcium levels [118] which requires the presence of calcium in the medium. An increase in calcium concentration in the medium will enhance the response of human fetal lung fibroblasts to PDGF [119]. An increase in arterial calcium content may occur in diet-induced atherogenesis [120]. Several calcium channel blockers have been shown to reduce the extent of arterial lipid and calcium deposition in animal models of atherosclerosis [121], but other investigations with these agents have questioned the validity of these observations [122, 123]. The usefulness of this approach will require additional evaluation.

6 Potential pharmacological sites of intervention

Several sites for preventing the platelet aspect of atherogenesis may be considered. These include: inhibition of platelet adhesion, alpha granule (PDGF) release, PDGF synthesis, PDGF receptor binding, vascular smooth muscle cell migration or proliferation. In most cases the success or failure of each intervention site in the in vivo state remains to be determined. Completely blocking the adhesion of platelets to subendothelium would be expected to place the patient at risk to bleeding tendencies similar to those observed with von Willebrand's disease. Therapy which would inhibit PDGF synthesis or receptor binding would be of high interest but agents with this activity have not been described. The role PDGF may have in wound repair and tissue regeneration may limit the continued use of such agents.

The observation that trapidil and heparin act as inhibitors of smooth muscle cell proliferation both in vitro and in vivo suggest that such an approach is feasible. These agents, although perhaps not ideal for clinical evaluation, may improve our understanding of atherogenesis in at least experimental models.

7 Conclusion

The 'response-to-injury hypothesis' of atherogenesis has gained wide acceptance in recent years. According to this hypothesis, platelets play a critical role in the evolution of serious occusive arterial plaques at two phases of the lesion development. Thrombocytopenia and limited antiproliferative drug investigations have supported the need for the platelet releasate PDGF at least in neointimal proliferative lesions. The presence of platelets and fibrin in later stages or advanced atherosclerotic plaques indicates that thrombosis may also play a part of atherogenesis. The etiology of atherosclerosis is varied and the degree of proliferative lesions and thrombotic additions would also be expected to vary. Evidence does not support the need for platelet involvement in arterial lipid deposits. Early events in atherogenesis may be more focused on plasma lipoprotein (LDL) levels, endothelial cell permeability and foam cell production. Considering antiproliferative therapy for the slowly evolving atherosclerosis may not be advised because of the requirement for extended drug therapy. Perhaps the better clinical target would be the accelerated atherogenesis associated with vascular surgical procedures.

References

1 Working Group on Arteriosclerosis: Arteriosclerosis *2,* 52 (1981).
2 M. A. Ruffer: J. Path. Bact. *15,* 453 (1911).
3 E. V. Cowdry (ed.): Arteriosclerosis. A Survey of the Problem. MacMillan Company, New York 1933.
4 J. B. Herrick: J. Am. Med. Ass. *23,* 2015 (1912).
5 Working Group on Arteriosclerosis: Arteriosclerosis *1,* 1 (1981).
6 Ibid. p. 37.
7 W. B. Kannel: JAMA *247,* 877 (1982).
8 J. Stamler and R. Stamler: Am. J. Med. *76,* 13 (1984).
9 Editoral, Sci. Am. *250,* 9 (1984).
10 G. V. R. Born: Nature *194,* 927 (1962).
11 J. G. White: Am. J. Path. *83,* 589 (1976).
12 K. L. Kaplan, in: Platelets in Biology and Pathology. Ed. J. L. Gordon, Elsevier/North-Holland Biomedical Press, Amsterdam 1981.
13 G. M. Oosta, L. V. Favreau, D. L. Beeler and R. D. Rosenberg: J. biol. Chem. *257,* 11249 (1982).
14 M. B. Zucker: Sci. Am. *242,* 86 (1980).
15 E. G. Lapetina: Life Sci. *32,* 2069 (1983).
16 R. G. Mason and H. I. Saba: Am. J. Path. *92,* 775 (1978).
17 P. D'Amore: Microvasc. Res. *15,* 137 (1978).
18 H. J. Weiss: Semin. Hematol. *17,* 228 (1980).
19 K. K. Wu and J. C. Hoak: Lancet *ii,* 924 (1974).
20 H. J. Weiss, L. D. Witte, K. L. Kaplan, B. A. Lages, A. Chernoff, H. L. Nossel, D. S. Goodman and H. R. Baumgartner: Blood *54,* 1296 (1979).
21 S. Levey-Toledano, J. P. Caen, J. Breton-Gorius, F. Rendu, C. Cywiner-Golenzer, E. Dupuy, Y. Legrand and J. MacLouf: J. Lab. clin. Med. *98,* 831 (1981).
22 M. J. Stuart: Semin. Hematol. *12,* 233 (1975).
23 H. R. Baumgartner, R. Muggli, T. B. Tschopp and V. T. Turitto: Thromb. Haemostasis *35,* 124 (1976).
24 H. R. Baumgartner and A. Studer, in: Atherosclerosis IV. Eds. G. Schettler, Y. Goto, Y. Hata and G. Klose. Springer Verlag, Berlin 1976.
25 I. D. Goldberg, M. B. Stemerman and R. I. Handin: Science *209,* 611 (1980).
26 R. Virchow: Phlogose und Thrombose in Gefässystem, Gesammelte Abhandlungen zur wissenschaftlichen Medicin. Meidinger Sohn, Frankfurt am Main 1856.
27 A. Carrel: J. exp. Med. *15,* 516 (1912).
28 R. Ross and J. Glomset: Science *180,* 1332 (1973).
29 R. Ross, in: Ciba Fdn. Symp. *100,* 198 (1983).
30 H. R. Baumgartner and A. Studer: Pathol. Microbiol., Basel *26,* 129 (1963).
31 H. R. Baumgartner: Thromb. Haemost. *51,* suppl., 161 (1972).
32 M. L. Tiell, M. B. Stemerman and T. H. Spaet: Circulation Res. *42,* 644 (1978).
33 S. Moore: Diabetes *30,* suppl. 2, 8 (1981).
34 M. Richardson, I. O Ihnatowycz and S. Moore: Lab. Invest. *43,* 509 (1980).
35 S. Moore: Lab. Invest. *29,* 478 (1973).
36 S. Moore, R. J. Friedman, D. P. Singal, J. Gauldie, M. A. Blajchman and R. S. Roberts: Thromb. Haemostasis *35,* 70 (1976).
37 R. J. Friedman, M. B. Stemerman, B. Weinz, S. Moore, J. Gauldie, M. Gent, M. L. Tiell and T. H. Spaet: J. clin. Invest. *60,* 1191 (1977).
38 H. Ohnishi, K. Yamaguchi, S. Shimada, M. Sato, H. Funato, Y. Katsuki, T. Dabasaki, Y. Suzuki, Y. Saitoh and A. Kumagai: Life Sci. *31,* 2595 (1982).
39 A. W. Clowes and M. M. Clowes: Lab. Invest. *43,* 535 (1980).
40 I. Joris, P. W. Braunstein, Jr., L. Pechet and M. Majno: Exp. molec. Path. *33,* 283 (1980).

41 K. Taylor, S. Glagor, J. Lamberti, D. Vesselinovitch and T. Schaffner: Scanning Electron Microscope *2,* 449 (1978).
42 M. A. Reidy and S. M. Schwartz: Fed. Proc. *39,* 1109 (1980).
43 M. A. Reidy and S. M. Schwartz: Lab. Invest. *44,* 301 (1981).
44 E. Z. Hirsch and A. L. Robertson: Atherosclerosis *28,* 271 (1977).
45 C. R. Minick, M. Stemerman, W. Insull, Jr.: Proc. natn. Acad. Sci. USA *74,* 1724 (1977).
46 D. Harmon: Circulation Res. *11,* 277 (1962).
47 G. K. Owens and T. M. Hollis: Atherosclerosis *34,* 365 (1979).
48 W. Hollander, D. M. Kramsch, C. Franzblau, J. Paddock and M. A. Colombo: Circulation Res. *34,* suppl., I–131 (1974).
49 D. A. Handley, M. L. Lee and R. N. Saunders, in: Platelet-Activating Factor. Eds. J. Benneviste and B. Arnoux, Elsevier Science, Amsterdam 1983.
50 A. Faggiotto, R. Ross and L. Harker: Arteriosclerosis *4,* 323 (1984).
51 A. Faggiotto and R. Ross: Arteriosclerosis *4,* 341 (1984).
52 N. Woolf and K. C. Carstairs: Am. J. Path. *51,* 373 (1967).
53 L. R. Bernstein, H. Antoniades and B. R. Zetter: J. Cell. Sci. *56,* 71 (1982).
54 J. Nakao, T. Ooyama, W.-C. Chang, S. Murota and H. Orimo: Atherosclerosis *43,* 143 (1982).
55 K. C. Glenn and R. Ross: Cell *25,* 603 (1981).
56 K. Fischer-Dzoga, R. Fraser and R. W. Wissler: Exp. molec. Path. *24,* 346 (1976).
57 K. Fischer-Dzoga and R. W. Wissler: Atherosclerosis *24,* 515 (1976).
58 K. Fischer-Dzoga, Y.-F. Kuo and R. W. Wissler: Atherosclerosis *47,* 35 (1983).
59 C. Gajdusek, P. Dicorleto, R. Ross and S. M. Schwartz: J. Cell Biol. *85,* 467 (1980).
60 C.-H. Wang, E. E. Largis and S. A. Schaffer: Artery *9,* 358 (1981).
61 S. R. Coughlin, G. M. Nemecek, D. A. Handley and M. A. Moskowitz: Platelets, Prostaglandins and the Cardiovascular System. Florence, Italy, 1984.
62 G. M. Oosta, L. V. Favreau, O. L. Beeler and R. D. Rosenberg: J. biol. Chem. *257,* 11249 (1982).
63 A. Clowes and M. Karnovsky: Nature *265,* 625 (1977).
64 R. N. Saunders and G. M. Nemecek: Pharmacologist *26,* 157 (1984).
65 J. R. Guyton, R. D. Rosenberg, A. W. Clowes and M. J. Karnovsky: Circulation Res. *46,* 625 (1980).
66 M. J. Karnovsky: Am. J. Path. *105,* 200 (1981).
67 R. J. Gryglewski, S. Bunting, S. Moncada and J. R. Flower: Prostaglandins *12,* 685 (1976).
68 H. J. Weiss and V. T. Turitto: Blood *53,* 244 (1979).
69 B. Adelman, M. B. Stemerman, D. Mennell and R. I. Handin: Blood *58,* 198 (1981).
70 R. J. Gryglewski, A. Dembinska-Kiec, A. Chytkowski and T. Grylewska: Atherosclerosis *31,* 385 (1978).
71 I. A. Blair, S. E. Barrow, K. A. Waddell, P. J. Lewis and C. T. Dollery: Prostaglandins *23,* 579 (1982).
72 E. M. Davidson, A. W. Ford-Hutchinson, M. J. H. Smith and J. R. Walker: Br. J. Pharmac. *64,* 437P (1978).
73 J. Morley, C. P. Page, W. Paul, N. Mongelli, R. Ceserani and C. Gandolfi: Prostagland. Leuk. Med. *8,* 239 (1982).
74 G. M. Nemecek, D. A. Handley and R. N. Saunders: Platelets, Prostaglandins and Cardiovascular System. Florence, Italy, 1984.
75 G. A. Fitzgerald, B. Smith, A. K. Pedersen and A. R. Brash: New Engl. J. Med. *310,* 1065 (1984).
76 S. R. Coughlin, M. A. Moskowitz, B. R. Zetter, H. N. Antoniades and L. Levine: Nature *288,* 600 (1980).

77 S. R. Coughlin, M. A. Moskowitz, H. N. Antoniades and L. Levine: Proc. natn. Acad. Sci. USA *78,* 7134 (1981).
78 H. M. Groves, R. L. Kinlough-Rathbone, M. Richardson, L. Jorgensen, S. Moore and J. F. Mustard: Lab. Invest. *46,* 605 (1982).
79 T. Tschopp, H. J. Weiss and H. R. Baumgartner: J. Lab. clin. Med. *83,* 296 (1974).
80 A. L. Bloom: Semin. Hematol. *17,* 215 (1980).
81 V. Fuster, E. J. W. Bowie, J. C. Lewis, D. N. Fass, C. A. Owen and A. L. Brown: J. clin. Invest. *61,* 722 (1978).
82 P. M. Mannucci, A. L. Bloom, M. J. Larrieu, I. M. Nilsson and R. R. West: Br. J. Haemat. *57,* 163 (1984).
83 M. L. Tiell, M. B. Stemerman and T. H. Spaet: Circulation Res. *42,* 644 (1978).
84 D. R. Clemmons and J. J. Van Wyk: J. cell. Physiol. *106,* 361 (1981).
85 L. A. Harker, S. J. Slichter, C. R. Scott and R. Ross: New Engl. J. Med. *291,* 537 (1974).
86 L. A. Harker, R. Ross, S. J. Slichter and C. R. Scott: J. clin. Invest. *58,* 731 (1976).
87 R. G. Schaub, C. A. Rawlings and J. C. Keith: Am. J. Path. *104,* 13 (1981).
88 R. B. Griepp, E. B. Stinson, C. P. Bieber, B. A. Reitz, J. C. Copeland, P. E. Oyer and N. E. Shumway: Surgery *81,* 262 (1977).
89 V. Fuster and J. H. Chesebro: Platelets, Prostaglandins and Cardiovascular System. Florence, Italy, 1984.
90 D. P. Faxon, V. J. Weber, C. Haudenschild, S. B. Gottsman, W. A. McGovern and T. J. Ryan: Arteriosclerosis *2,* 125 (1982).
91 H. Fuller, F. Hauschild, D. Modersohn and E. Thomas: Pharmazie *26,* 554 (1971).
92 H. Ohnishi, K. Yamaguchi, S. Shimada, Y. Suzuki and A. Kamagai: Life Sci. *28,* 1641 (1981).
93 M. L. Tiell, I. I. Sussman, P. B. Gordon and R. N. Saunders: Artery *12,* 33 (1983).
94 A. W. Clowes, R. D. Rosenberg and M. M. Clowes: Surg. Forum *34,* 357 (1983).
95 J. J. Castellot, Jr., L. V. Favreau, M. J. Karnovsky and R. D. Rosenberg: J. biol. Chem. *257,* 11256 (1982).
96 A. W. Clowes and M. J. Karnovsky: Nature *265,* 625 (1977).
97 R. N. Saunders: Ann. Rev. Pharmacol. Toxicol. *22,* 279 (1982).
98 P. Clopath: Br. J. exp. Path. *61,* 440 (1980).
99 V. Fuster, J. H. Chesebro, M. K. Dewanjee, M. P. Kaye, M. Josa and J. M. Byrne: Thromb. Haemostasis *42,* 404 (1979).
100 R. G. Schaub, C. A. Rawlings and J. C. Keith, Jr.: Thromb. Haemostasis *46,* 680 (1981).
101 B. G. Brown, R. A. Cukingnan, L. Goede, M. Wong, H. Fee, J. Roth, J. Wittig and J. Carey: Am. J. Cardiol. *47,* 494 (1981).
102 J. H. Chesebro, I. P. Clements, V. Fuster, L. R. Elveback, H. C. Smith, W. T. Bardsley, R. L. Frye, D. R. Holmes, R. E. Vlietstra, J. R. Pluth, R. B. Wallace, F. J. Puga, T. A. Orszulak, J. M. Piehler, H. V. Schaff and G. K. Danielson: New Engl. J. Med. *307,* 73 (1982).
103 J. H. Chesebro, V. Fuster, L. R. Elveback, I. P. Clements, H. C. Smith, D. R. Holms, Jr., W. T. Bardsley, J. R. Pluth, R. B. Wallace, F. J. Puga, T. A. Orszulak, J. M. Piehler, G. K. Danielson, H. V. Schaff and R. L. Frye: New Engl. J. Med. *310,* 209 (1984).
104 R. L. Lorenz, M. Weber, J. Kotzur, K. Theisen, C. V. Schacky, W. Meister, B. Reichardt and P. C. Weber: Lancet *i,* 1261 (1984).
105 J. Mehta: JAMA *249,* 2818 (1983).
106 M. K. Dewanjee, M. Tago, M. Josa, V. Fuster and M. P. Kaye: Circulation *69,* 350 (1984).

107 C. W. Pumphrey, J. H. Chesebro, M. K. Dewanjee, H. W. Wahner, L. H. Hollier, P. C. Pairolero and V. Fuster: Am. J. Cardiol. *51,* 796 (1983).
108 H. J. Weiss, V. T. Turitto, W. J. Vicic and H. R. Baumgartner: Thromb. Haemostasis *45,* 136 (1981).
109 C. N. McCollum, M. J. Crow, S. M. Rajah and R. C. Kester: Surgery *87,* 668 (1980).
110 V. A. Gaudiani, D. C. Miller, J. C. Kosek, J. Berg and S. W. Jamieson: J. Surg. Res. *34,* 263 (1983).
111 D. Heinrich, E. Schlenssner, W. L. Wagner, R. Sellmann-Richter and F. W. Hehrlein: Thromb. Res. *32,* 409 (1983).
112 A. Szczeklik, R. Nizankowski, S. Skawinska, J. Szczeklik, P. Gluszko and R. J. Gryglewski: Lancet *i,* 1111 (1979).
113 H. Sinzinger, J. O'Grady, M. Cromwell and R. Hofer: Lancet *i,* 1275 (1983).
114 H. Sinzinger and P. Fitscha: Lancet *i,* 905 (1984).
115 R. J. Gryglewski and A. Szczeklik: Adv. exp. Med. Biol. *164,* 211 (1984).
116 H. J. Weiss: Platelets, Pathophysiology and Antiplatelet Drug Therapy. Alan R. Liss, New York, N. Y., 1982.
117 W. T. Shier, D. J. Dubourdieu and L. A. Hull: Ions, Cell Proliferation, Cancer. Academic, New York, N. Y., 1982.
118 W. H. Moolenaar, L. G. J. Tertoolen and S. W. de Laat: J. biol. Chem. *259,* 8066 (1984).
119 T. Ohno, in: Proceedings of the Seventh Asia and Oceania Congress of Endocrinology. Eds. K. Shizume, H. Imura and N. Shimizh. Exerpta Medica, Amsterdam 1982.
120 W. Hollander, J. Paddock, S. Nagraj, M. Colombo and B. Kirkpatrick: Atherosclerosis *33,* 111 (1979).
121 F. V. de Feudis: Life Sci. *32,* 557 (1983).
122 M. Naito, F. Kuzuya, K. Asai, K. Shibata and N. Yoshimine: Atherosclerosis *51,* 343 (1984).
123 S. Stender, I. Stender, B. Nordestgaard and K. Kjeldsen: Arteriosclerosis *4,* 389 (1984).

Structures, properties and disposition of drugs

By Eric J. Lien
Section of Biomedicinal Chemistry, School of Pharmacy, University of Southern California, Los Angeles, Calif. 90033, USA

1 Introduction ... 68
2 A general mathematical model ... 68
3 Percutaneous absorption ... 70
3.1 Chemicals for enhancing percutaneous penetration of drugs ... 71
4 Transperitoneal membrane absorption and intraperitoneal chemotherapy ... 74
5 Passage of drugs and chemicals from blood into the uterine fluids and fetus ... 75
6 Permeability of drugs across the blood testis barrier of the rat ... 76
7 Permeation of drugs from blood to pancreas ... 77
8 Accumulation and penetration of some antibiotics in rat lungs ... 78
9 Pharmacokinetic parameters – Volume of distribution and extent of absorption ... 80
9.1 Effect of age on pharmacokinetic parameters ... 80
10 Targeting of drugs by liposomes ... 81
11 Multiple membrane interaction by polymer-bound anthracyclines ... 82
12 β-Sympathomimetic agents – Agonists and antagonists (β-Blockers) 83
12.1 β-Sympathomimetic agonists ... 83
12.2 β-Adrenergic blockers (antagonists) ... 84
13 Calcium blockers and accumulation of drugs by cardiac tissue ... 84
14 Drug metabolism ... 91
15 Conclusion ... 92
References ... 92

1 Introduction

Many of the quantitative structure-activity relationships (QSARs) published over the last twenty years are, in fact, determined by some common rate-limiting pharmacokinetic steps, such as permeation of the drugs through lipoprotein membranes [1, 2] or renal elimination following administration. Although it is conceptually instructive to distinguish among various phases involved by a drug in exerting the pharmacological or toxicological action (i. e., pharmaceutical, pharmacokinetic, and pharmacodynamic phases) [3, 4], in real testing system it may be difficult to separate out the pharmacokinetic phase from the pharmacodynamic phase. This is shown schematically in figure 1. Even if one is studying a purified enzyme system, one can not eliminate the diffusion step and the adsorption-desorption between the drug (substrate or inhibitor) and the macromolecule (enzyme). A drug molecule will undoubtedly collide with nonspecific binding sites many times before it can reach the active site. On the other end of figure 1, as in the case of percutaneous absorption or absorption through corneal membrane, there is always a possibility that the drugs being studied can affect either the structure and/or the function of the membrane, so some pharmacodynamic effect may be mixed with the kinetic parameters being measured (e. g., absorption constant k) [5]. A couple of good examples of this type of complication are the ocular absorption of catecholamines [6] and β-adrenergic blockers [5, 7]. This necessitates the development of a versatile mathematical model, which will be applicable to both kinetically (k) controlled as well as dynamically (K) controlled processes.

2 A general mathematical model

Following the parabolic model of Hansch et al. [8–10], a more general extended mathematical model was first introduced in 1974 [1]. This model has since been extensively used in correlating data obtained from various biological systems, ranging from in vitro to in vivo testing

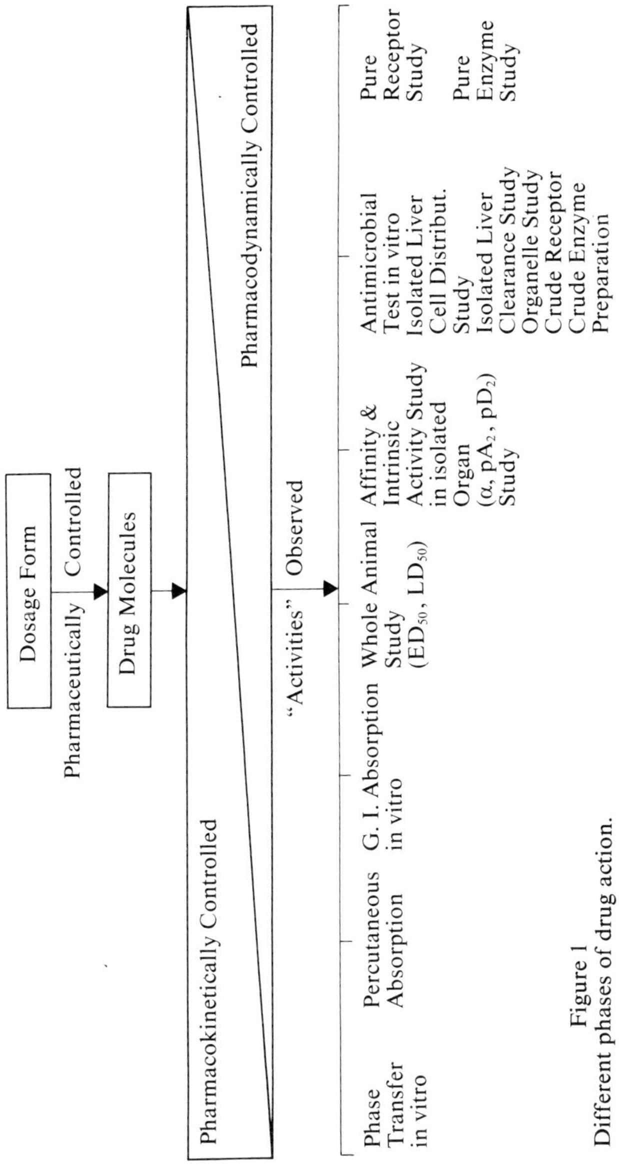

Figure 1
Different phases of drug action.

[1, 2, 11–20]. The underlying principles for the terms used in this model are explained as follows:

$$\text{Eq. 1.} \log A = \underbrace{-a(\log P)^2 + b \cdot \log P}_{\text{from the parabolic model of Hansch et al.}} + \underbrace{c \cdot \log MW}_{\text{from the Sutherland-Einstein diffusion equation}} + \underbrace{d \cdot \mu}_{\text{for dipolar interactions}} + \underbrace{e \cdot \log (U/D)}_{\text{from the pH-partition hypothesis}} + \underbrace{f}_{\text{constant}}$$

where A is either absorption (k or % Abs) or activity (1/C). The coefficients a, b . . . f can be obtained from multiple regression analysis. Additional parameters may be needed for some specific drug-receptor interactions:

H-bonding – n_{OH}, n_H e. g., Anticholinergics (Lien et al., 1976), α-Adrenergic blockers (Hansch and Lien, 1968).

Group symmetry factor – I e. g., Dopamine receptor stimulating agents (Lien and Nilsson, 1983), Pesticides and toxic substances (Cohen et al., 1974).

In many cases, not all the properties are varied wide enough in a series of compounds studied, therefore, some of the terms in eq. (1) may be combined with the constant term f. This will result in simpler forms of equation.

Among the different parameters in eq. (1), the log MW term had been neglected for quite some time, mainly because in most studies of series of congeners, the molecular weight ranges were usually fairly narrow. Furthermore, if only nonpolar side chain or the number of halogen atoms were modified, the use of log P or chain length n would give similar correlation as log MW, since there would be high degree of colinearity among these parameters. If, on the other hand, a wide range of molecular weight is involved (say more than one log unit) as well as the inclusion of both polar and nonpolar groups then, the role of log MW may become apparent [13, 18]. This is shown in figure 2. These apparent correlations with log MW should not be construed to mean that all high molecular weight compounds will be more active than lower molecular weight compounds. They may simply reflect the same rate-limiting step for the series of drugs examined, namely, renal clearence after the drugs are administered.

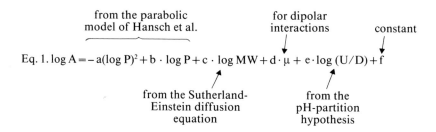

3 Percutaneous absorption

The structure and biophysics of skin as well as methods of measuring skin permeability have been thoroughly analyzed by Scheuplein and Black [21], various factors affecting percutaneous absorption have been reported by Idson [22]. Excellent reviews are also available on the pharmaceutical, pharmacokinetic, and biopharmaceutical aspects of topical drug products [23, 24]. From multiple regression analysis, it has been shown by Lien and Tong [25] that for many series of neutral drugs, percutaneous absorption is highly dependent upon the lipophilic character (log P), electronic and steric terms appear to play only minor roles. Molecular weight has also been shown to play a role [26], especially if the range is wide enough (one to two log units).

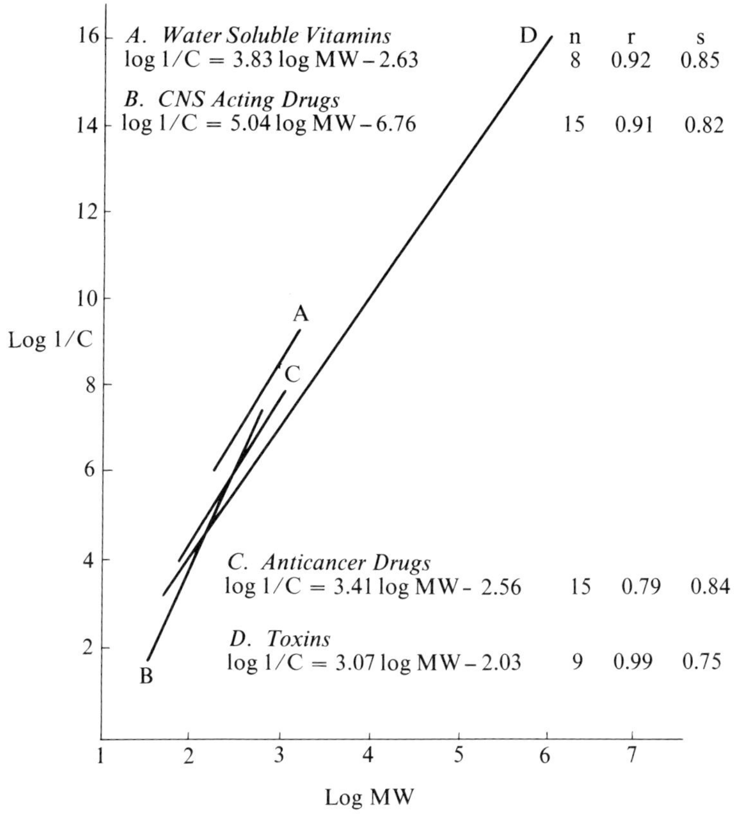

Figure 2
Apparent correlation between biological activity and log MW.

3.1 Chemicals for enhancing percutaneous penetration of drugs

It has been known for over twenty years that simple organic molecules like dimethylsulfoxide (DMSO), dimethylacetamide (DMAC), and dimethylformamide (DMFA) can enhance both percutaneous absorption of various drugs and retention of drugs in the stratum corneum [27–30]. In addition to these small open chain semipolar compounds, cyclic lactam derivatives like N-methyl-2-pyrrolidone [31] and more recently 1-dodecylazacycloheptan-2-one have been shown to increase the permeability of skin and enhance the penetration of antibiotics, glucocorticoids, and 5-fluorouracil [32, 33]. Table 1 summarizes the physicochemical properties of the dipolar molecules capable of enhancing percutaneous penetration of drugs. It is interesting to note that the single common denominator among all these agents is the relatively narrow range of electric dipole moment (4.0–4.5 Debyes), while the octanol/water partition coefficient ranges from –1.4 to 5.7 and molecular weight ranges from 78 to 282.

Table 1
Physicochemical properties of dipolar molecules capable of enhancing percutaneous penetration of various drugs.

Agent	MW	Dipole moment (D)[a]	log P[b] (oct/w)	Drugs for which increased percutaneous penetration has been shown
DMSO $CH_3-\underset{\underset{O}{\downarrow}}{S}-CH_3$	78.1	4.50 ± 0.1	– 1.35	quaternary ammonium drugs (hexopyrronium bromide) naphazoline fluocinolone acetonide [28], griseofulvin, hydrocortisone [29]
DMAC $CH_3-\underset{\underset{O}{\|}}{C}-N(CH_3)_2$	87.1	3.96 ± 0.03	– 0.77	griseofulvin [29], penecillin G, erythromycin [30], hydrocortisone [29], thiabendazole, tolnaftate [30]
DMF $\underset{\underset{O}{\|}}{HC}-N(CH_3)_2$	73.1	3.95 ± 0.02	– 1.01	griseofulvin [29], hydrocortisone [29]

N-Methyl-2-pyrrolidone 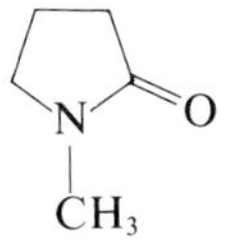	99.1	3.75a) 4.09 ± 0.04c)	− 0.71d)	clindamycin into the comedo in the treatment of Acne vulgaris [31]
1-Dodecylazacyclo-heptan-2-one 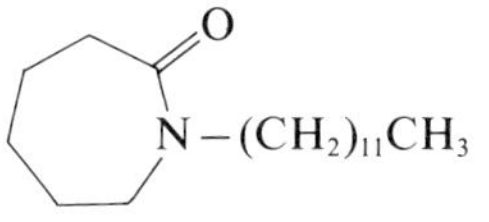	281.5	4.23e)	5.65d)	antibiotics, 5-FU glucocorticoids [32, 33]

a) From A. L. McClellan: Tables of Experimental Dipole Moments. Rahara Enterprises, El Cerrito 1974.
b) From C. Hansch and A. J. Leo: Substituent Constants for Correlation Analysis in Chemistry and Biology. Wiley, New York 1979.
c) From E. Fisher: J. Chem. Soc., 1382–1383 (1955).
d) Calculated from the homologues, C. Elison, E. J. Lien, A. P. Zinger, M. Hussain, G. L. Tong and M. Golden: J. Pharm. Sci. *60,* 1058–1062 (1971).
e) The value of the N-methyl derivative is used, since the extension of the nonpolar N-alkyl group does not change the permanent dipole moment of the lactam ring.

4 Transperitoneal membrane absorption and intraperitoneal chemotherapy

Torres et al. [34] have investigated the absorption of various drugs into the systemic circulation following intraperitoneal administration of large volumes of drug solutions. It was shown that the peritoneal membrane in the rat behaved like other biological membranes. In general, unionized lipid-soluble compounds were absorbed to a greater extent than ionized lipid-insoluble compounds. Neutral compounds were absorbed in reserve relationship to their molecular weights [34]. In an attempt to devise more effective strategies for the treatment of ovarian cancer, Myers et al., at the National Cancer Institute, further studied intraperitoneal administration of anticancer drugs [35]. Table 2 summarizes the absorption data and the physicochemical parameters of eleven anticancer drugs and nine miscellaneous compounds reported by these two groups. In addition to log % absorption, log k_a values were also included in the regression analysis. This was suggested by Seydel and Schaper [36].

Table 2
Intraperitoneal absorption and the physicochemical properties of anticancer drugs and miscellaneous compounds.

log % Abs	log k_{Abs}[a]	log MW	log $P_{heptane/w}$[b]	Drug
0.95[b]	– 1.05	5.12	– 1.00	Asparaginase
1.00[b]	– 1.00	2.73	–	Doxorubicin
1.09[b]	– 0.89	3.15	– 2.70	Bleomycin
1.18[b]	– 0.80	2.67	– 3.00	Methotrexate
1.32[b]	– 0.62	3.10	– 0.64	Actinomycin D
1.39[b]	– 0.55	2.48	–	*cis*-DDP
1.40[b]	– 0.54	2.51	– 1.00	Melphalan
1.45[b]	– 0.48	2.11	– 1.05	5-FU
1.47[b]	– 0.46	2.39	– 2.22	Ara-C
1.87[b]	0.13	2.27	– 0.68	ThioEPA
1.96[b]	0.40	2.32	1.05	Hexamethylmelamine
1.84[c]	0.07	1.25		H_2O
1.62[c]	– 0.27	1.78		Urea
1.78[c]	– 0.04	1.88		Propyleneglycol
1.36[c]	– 0.59	2.05		Creatinine
1.48[c]	– 0.43	2.26		Mannitol
1.40[c]	– 0.54	2.52		Sucrose
1.11[c]	– 0.85	2.95[d]		PEG 800–1000
0.99[c]	– 1.00	3.70		Inulin
0.34[c]	– 1.70	6.30		Dextran blue

a) $k_{Abs} = \dfrac{-\ln[1-(\%\,Abs/100)]}{t\,(1\,hr)}$.

b) From [35].

c) From [34].

d) The valuc of PEG 900 was used.

Eq. (2) and (3) correlate the absorption data with the log MW term for all twenty compounds:

	n	*r*	*s*	
$\log \%\,Abs = -0.282 \log MW + 2.132$	20	0.85	0.21	(2)
$\log k_{Abs} = -0.332 \log MW + 0.361$	20	0.80	0.30	(3)

About 72% and 64% of the variance in the data can be accounted for by these two equations, respectively. From these two equations it appears that log % Abs gives slightly better correlation than log k_{Abs}. Since k_{Abs} was calculated from the % absorption data, it is not expected to give any additional information than % absorption. As long as the first order kinetics is obeyed, if log k_{Abs} gives a meaningful correlation, so will log % Abs. This author does not see the use of log % Abs in correlation as a 'pitfall', as it was claimed by Seydel and Schaper [36]. For the subset of nine anticancer drugs for which the heptane/water partition coefficients are available, eq. (4)–(8) are obtained:

	n	r	s	
$\log \% \text{Abs} = -0.254 \log \text{MW} + 2.133$	9	0.70	0.26	(4)
$\log \% \text{Abs} = 0.177 \log \text{P} + 1.631$	9	0.66	0.27	(5)
$\log \% \text{Abs} = -0.236 \log \text{MW} + 0.163 \log \text{P} + 2.286$	*9*	*0.92*	*0.15*	(6)
$F_{1,6} = 14.8; F_{1,6_{0.99}} = 13.7$				
$\log k_{\text{Abs}} = -0.323 \log \text{MW} + 0.440$	9	0.64	0.39	(7)
$\log k_{\text{Abs}} = 0.266 \log \text{P} - 0.146$	9	0.70	0.36	(8)
$\log k_{\text{Abs}} = -0.296 \log \text{MW} + 0.248 \log \text{P} + 0.674$	*9*	*0.91*	*0.23*	(9)
$F_{1,6} = 11.8; F_{1,6_{0.975}} = 8.81$				

Both log MW and log P are about equally important in determining the peritoneal absorption of these anticancer drugs. The addition of the log P term in eq. (6) is statistically significant at 99 percentile level, while the addition of log MW term in eq. (9) is significant at 97.5 percentile level, as indicated by an F-test. Over 82% ($r^2 > 0.82$) of the variance in the data can be explained by these two parameter equations. Addition of $(\log P)^2$ term does not result in significant improvement in correlation. The coefficients associated with the log MW term is about -0.24 to -0.30, not very different from that derived from the Sutherland-Einstein equation of diffusion [12, 18, 37].

$$D = \frac{RT}{6\pi N\eta} \sqrt[3]{\frac{4\pi N}{3\nu \text{MW}}} \qquad (10)$$

where ν = partial specific volume in cm^3/g of solute, η = viscosity, R = gas constant. When the temperature T and η are held constant, one obtains the following equation:

$$\log D = \text{constant} - 0.33 \log \text{MW} - 0.33 \log \nu. \qquad (11)$$

5 Passage of drugs and chemicals from blood into the uterine fluids and fetus

The potential hazard of having drugs or chemicals passing from the maternal blood to the fetus is of concern to all health professionals as well as the pregnant woman. Fabro [38] has studied the distribution into uterine of a series of C-14 labeled drugs and chemicals following the intravenous administration to rabbits. It has been concluded from this animal study that there is an inverse relationship between the molecular weight and its uterine fluid/plasma radioactivity ratio ($R_{u/p}$). Using the logarithmic transformation of the data (see table 3), the following correlation is obtained:

$$\log R_{u/p} = -0.552 \log MW + 1.318 \qquad n = 5 \quad r = 0.98 \quad s = 0.14 \qquad (12)$$
$$F_{1,3} = 79.6; \; F_{1,3\;0.995} = 55.6$$

It has been suggested by Levy and Hayton [39] that the fetal/maternal drug concentration ratios must be determined over a considerable period of time to attain a constant value if the maternal tissues and/or the fetus are 'deep' compartments (i. e., slowly penetrable). Unfortunately, very scarce data of this nature are available for quantitative analysis. From the correlation obtained above, it is obvious that all drugs with high molecular weights (> 1000) will be relatively impermeable [40].

Table 3
Passage of compounds from blood into uterine in rabbits.

$\log R_{u/p}$ a)	log MW	Compound	
0.08	2.27	N-Methyl-C^{14}-antipyrine	
−0.17	2.55	Phenyl-C^{14}-DDT	
−0.55	3.74	Carboxy-C^{14}-inulin	
−1.10	4.20 b)	Carboxy-C^{14}-dextran	15,000–16,000
−1.40	4.88 b)	Carboxy-C^{14}-dextran	60,000–90,000

a) From [38].
b) The median value was used.

6 Permeability of drugs across the blood testis barrier of the rat

Johnson and Setchell [41] demonstrated that the Blood-Testis Barrier (BTB) excluded high molecular weight immunoglobulins from the seminiferous tubules. It was further demonstrated that iodinated albumin, inulin, and even some small molecules were excluded from the seminiferous tubules and suggested the existence of a protective barrier [42].
Okumura et al. [43] have examined the permeability of selected drugs and chemicals and found that the permeability of the BTB to nonelectrolytes was dependent upon their molecular size, suggesting bulk flow through water filled pores. Permeability of acidic drugs, on the other

hand, depends on their apparent partition coefficients ($CHCl_3$/buffer), similar to transport from blood to cerebrospinal fluid. From the transfer data and physicochemical constants listed in table 4, eq. (13)–(16) are obtained:

Table 4
Transfer rate and physicochemical properties of drugs.

log k (min^{-1})[a]	log MW	log P'[a] ($CHCl_3$/buffer $pH_{7.4}$)	log U/D[b]	Drug
−0.97	2.38	2.01	0.20	Thiopental
−1.55	2.35	1.32	0.70	Pentobarbital
−1.66	2.26	0.30	0.10	Barbital
−1.43	2.45	0.20	−0.70	Sulfamethoxy-pyridazine
−1.70	2.24	−1.57	3.00	Sulfanilamide
−2.28	2.37	−2.74 (a base)	−4.70	Sulfaguanidine
−2.39	2.14	−3.52	−4.40	Salicyclic acid

a) From [43]; P' = apparent partition coefficient.
b) Log U/D = (pK_a − pH) for acids, log U/D = (pH − pK_a) for bases.

	n	*r*	*s*	
$\log k = 0.219 \log P'_{CHCl_3/B} - 1.586$	*7*	*0.93*	*0.19*	(13)
$\log k = 0.214 \log P'_{CHCl_3/B} - 0.194 \log MW - 2.037$	7	0.93	0.22	(14)
$\log k = 0.190 \log P'_{CHCl_3/B} + 0.036 \log U/D - 1.573$	7	0.95	0.20	(15)
$\log k = -0.005 (\log P'_{CHCl_3/B})^2 + 0.212 \log P - 1.571$	7	0.93	0.22	(16)

For the limited number of compounds tested (n = 7), the chloroform/buffer partition coefficient appears to be the most important parameter in determining the transfer rate constant through BTB. Additions of log MW, log U/D, or $(\log P)^2$ term does not result in significantly improved correlations. It has been suggested that the BTB is a complex multicellular system composed of membranes surrounding seminiferous tubules and the several layers of spermatogenic cells within the tubules. Thus, the BTB restricts the permeability in male germ cells of many foreign compounds, just like the blood-brain barrier (BBB).

7 Permeation of drugs from blood to pancreas

Hori et al. [44] have studied penetration of drugs in the rabbit pancreas by cannulation into the pancreatic duct and by collecting pancreatic juice. From this study, it has been shown that drugs with high lipid solubility distribute within the pancreas. Permeability of drugs through the pancreas has been shown to be dependent on their molecular size and lipophilicity. For the nine data points reported (see table 5), eq. (17)–(19) are obtained by regression analysis. Eq. (17) correlates the (pancreas/plasma free drug) ratios with chloroform/buffer partition coefficients. About 90% ($r^2 = 0.90$) of the variance in the data can be explained by the lipophilicity. Addition of log MW or $(\log P')^2$ term does not result in improved correlations (r = 0.95).

	n	*r*	*s*	
$\log Panc/P_f = 0.139 \log P'_{CHCl3/B} + 0.112$	*9*	*0.95*	*0.07*	(17)
$\log J/P_f = -1.977 \log MW + 3.919$	9	0.57	0.45	(18)
$\log J/P_f = 0.121 \log P'_{CHCl3/B} - 0.328$	9	0.34	0.52	(19)
$\log J/P_f = 0.250 \log P'_{CHCl3/B} - 3.015 \log MW + 6.767$	*9*	*0.86*	*0.31*	(20)

$F_{1,6} = 9.4; F_{1,6\ 0.975} = 8.8$

Eq. (18)–(20) correlate the pancreatic juice/plasma ratio of the drugs with both log MW and log $P'_{CHCl3/B}$. Here the log MW term appears to be more important than log P′ term. The log P term in eq. (20) is statistically significant at 97.5 percentile level as indicated by the F-test. Eq. (20) is in agreement with the conclusion of Hori et al. [44].

Table 5
Distribution of drugs in pancreas and permeation from blood into pancreatic juice.

log Panc/P_f a)	log J/P_f b)	log MW	log $P'_{CHCl3/buffer}$ c)	Drug
0.07	−1.00	2.37	−0.57	Procainamide
0.00	−0.70	2.44	−0.80	Sulfisomidine
−0.01	−0.77	2.41	−1.34	Sulfathiazole
−0.02	0.02	2.24	−1.59	Sulfanilamide
−0.14	−0.04	2.09	−1.34	Isonicotinamide
−0.14	0.26	2.11	−1.66	Dimethadione
−0.24	−0.92	2.43	−1.62	Sulfisoxazole
−0.37	−0.72	2.09	−3.82	Isonicotinic acid
−0.60	−1.22	2.24	−5.00	Sulfanilic acid

a) Panc/P_f = the concentration ratio of the drug in the pancreas to unbound drug in the plasma.
b) The concentration ratio of the drug in the pancreatic juice to unbound drug in the plasma.
c) Partition coefficient between $CHCl_3$/phosphate buffer pH 7.4 at 37° C, from [26].

8 Accumulation and penetration of some antibiotics in rat lungs

Leucomycin A_3 and erythromycin (see fig. 3), which are widely used for the treatment of pulmonary infections, have been shown to accumulate well in the isolated blood-perfused rat lungs [45], while tetracycline (log $P_{CHCH3/B}$ = − 1.22, log $P_{oct/w}$ = − 1.37) and chloramphenicol (log $P_{oct/w}$ = 1.14) did not show specific accumulation. The pulmonary accumulation of leucomycin A_3 (log $P_{CHCl3/B}$ = 3.65) and erythromycin (log $P_{CHCl3/B}$ = 2.18, log $P_{oct/w}$ = 2.48) has been attributed to the presence of a basic amino group and a strong lipophilic group, as reflected by the partition coefficients in either chloroform/buffer or octanol/water. The transport of these antibiotics from alveoli to the blood following intratracheal administration in the perfused lungs has been shown to be dependent on their lipid solubility [45].

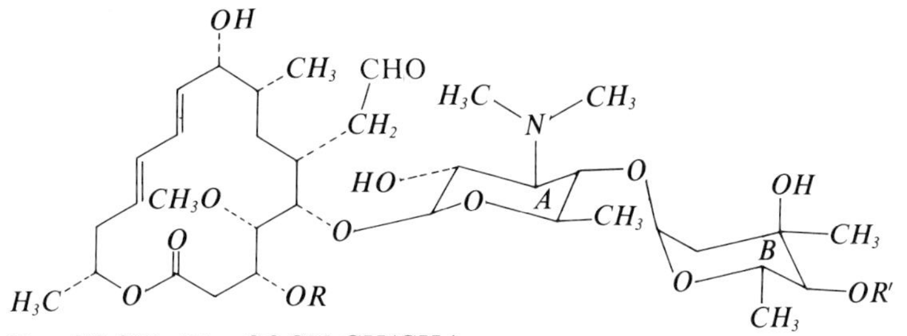

R = $COCH_3$, R′ = $COCH_2CH(CH_3)_2$

Leucomycin A_3 $C_{42}H_{69}NO_{15}$
MW = 828.0
log $P_{CHCl3/B}$ = 3.65

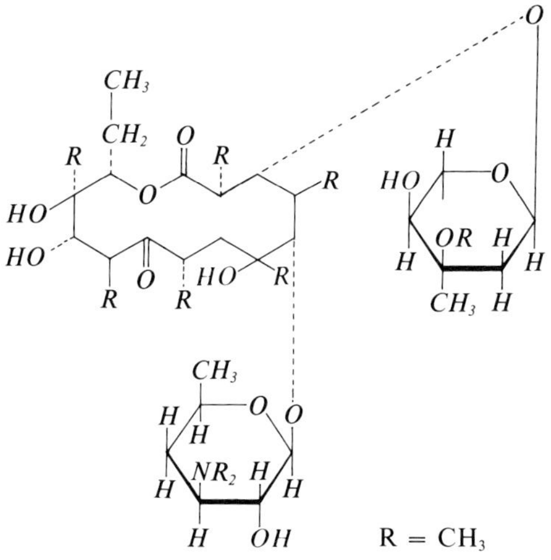

Erythromycin $C_{37}H_{67}NO_{13}$
MW = 733.9
log $P_{CHCl3/B}$ = 2.18 [45]
log $P_{oct/w}$ = 2.48 [46]

Figure 3
Molecular structures and properties of leucomycin A_3 and erythromycin.

9 Pharmacokinetic parameters – Volume of distribution and extent of absorption

Using both the literature and experimental data of the volume of distribution (Δ')[1]) of 125 different drugs, Ritschel and Hammer have derived equations for the prediction of the blood level of a drug knowing the extent of protein binding (EPB), apparent partition coefficient (APC), and certain pharmacokinetic rate constants [47].
The Δ'-APC-EPB relationship will enable one to estimate a drug's volume of distribution in a patient from in vitro data, and estimate the extent of absorption when intravenous data cannot be obtained.
Watanabe and Kozaki [48] have reported that the apparent volumes of distribution (V'_d)[2]) for basic drugs with low apparent partition coefficient ($P' < 0.16$) had an almost constant value (1.47 l/kg, 30 times greater than plasma volume 0.05 l/kg), while the V'_d value of basic drugs with high apparent partition coefficient ($P' = 3.23$ to 670) were related to the blood plasma volume, the lipid space, and other nonlipid space [49]. Other factors affecting the fluctuation of the V'_d values were also noted [49].

9.1 Effect of age on pharmacokinetic parameters

Many physiological and pathophysiological changes in elderly patients may affect the pharmacokinetic parameters, and consequently, the therapeutic and side effects as well. For example, delayed gastric emptying, higher gastric pH may affect drug absorption. Changes in total body water and the ratio of lean to fatty tissue influence drug distribution. Drug metabolism may be altered because of changes in the hepatic microsomal enzymes. Reduced hepatic blood flow will lead to a decreased clearance. Compromised renal function (in both renal blood flow and glomerular filtration) is a major contributing factor to drug toxicity in geriatric patients. Precautions for preventing overdosing of elderly patients have been suggested [50].

1) $\Delta' = \frac{V_d}{\text{Body wt}}$ (ml/g) $V_d = \frac{\text{Dose}}{C^\circ_p}$ (ml), where C°_p = drug concentration upon i.v. administration at time 0.

2) V'_d estimated from blood plasma concentration of a drug is a proportionality constant that relates the plasma concentration of a drug to the total amount in the body. For a detailed description, see [49].

10 Targeting of drugs by liposomes

Liposomes are either multilamellar large vesicles (MLV) or small (2–500 Å) or large (0.1–1 μm) unilamellar vesicles (SUV or LUV), made from phospholipids with other components like sterols, glycolipids, organic bases or acids, membrane proteins, artificial polymers, and additives like mannitol and α-tocopherol. Liposomes have been extensively investigated over the last decade as drug delivery systems [51, 52].

Considerable progress has been made in recent years in terms of encapsulation efficiency of not only drugs but radioactive metal ions [53, 54] as a result of a better understanding of the chemical structure of metal-ligand complexes in the lipid bilayer [55].

Liposomes are predominantly taken up by the reticulo-endothelial system. However, Fidler et al. [56] have successfully shown that negatively charged MLV liposomes [phosphatidylserine (PS): phosphatidylcholine = 3:7 molar ratio] localize efficiently in the lung (up to 25% of injected dose). It has also been shown that macrophage-activating agents (lymphokine preparations) can effectively activate lung macrophage in situ [57]. Schroit and Fidler [57] have further reported that activation of macrophage to become tumoricidal against syngeneic tumor cells with liposome-encapsulated muramyldipeptide (MDP) was superior in both extent and duration when MLV composed of distearoylphosphatidylcholine:phosphatidylserine (7:3 molar ratio) were used.

Abra et al. [58] have found that a strong relationship between liposome size and lung accumulation using a variety of extrusion and dialysis of liposomes containing phosphatidyl choline. A maximum lung accumulation of 30.9% of the administered dose was achieved without detectable gross pathological lung lesion up to 24 hours after dosing.

Anionic liposomes, using a double packing technique for doxorubicin encapsulation, have been prepared by Forssen and Tökes [60] for the reduction of the drug's chronic cardiotoxicity. Anionic vesicles (800–1,000 Å) containing doxorubicin-phosphatidylcholine complexes entrapped in phosphatidylcholine, phosphatidylserine and cholesterol at a molar ratio of 0:6:0.2:0:3 per mole of drug were tested for chemotherapeutic potential against L-1210 and P-388 murine leukemias, Lewis Lung and Sarcoma-180 as well as for cardiotoxicity. The authors reported that these vesicles possess increased antileukemic activity, reduced cardiotoxicity, skin toxicity, reduced immunosuppressive activity, and eliminate the growth inhibition observed in young mice treated with doxorubicin [60].

11 Multiple membrane interaction by polymer-bound anthracyclines

Tökes et al. [61] reported membrane directed action of adriamycin-coupled polyglutaraldehyde microspheres (4,500 Å) and their cytostatic activity. In drug resistant human and murine leukemic cell lines a 10 times increase in sensitivity was observed. The drug polymer complexes remained chemically stable for three days during the incubation with cells. Scanning and transmission electron micrographic data revealed extensive cell surface modification and a lack of internalization of the drug-polymer complexes. They proposed that this mode of drug delivery may provide multiple and repetitious sites for drug-cell surface interactions. Furthermore, drug resistance due to decreased drug binding at the cell surface or due to increased intracellular drug-degradation may be overcome by this type of drug-polymer complexes [61–63]. A 1,000-times increase in the cytotoxic activity of adriamycin against a highly resistant rat liver cell line (RLC) has been achieved by adriamycin-polyglutaraldehyde microspheres [62]. Even an inactive anthracycline analog, 4-demethoxy-7,9-di-epidaunorubicin, has been imparted significant cytostatic activity against doxorubicin resistant and sensitive murine L-1210 leukemia cells [63]. This raises the question whether DNA binding is mandatory for anticancer activity of anthracyclines. Several observations suggest that DNA binding may not be the sole mechanism for their toxicity, for example, N-trifluoroacetyl adriamycin-14-valerate is an Adr analog which has greater cytotoxicity than Adr, but does not bind to DNA. However, the doxorubicin-polymer complexes may represent a novel mode of cytotoxicity which may not be available to the free form of the drug [63]. Since these polymeric forms of anthracyclines increase the drug's effectiveness, they represent an ideal candidate in combination with organ specific monoclonal antibodies for targeted drug delivery.

12 β-Sympathomimetic agents – Agonists and antagonists (β-blockers)

In 1970, Collin et al. reported the β-stimulant activities of saligenin analogs of sympathomimetic catechol amines [64]. SAR analysis confirmed the hypothesis that interactions of the phenylethanolamine side chain with adenyl cyclase bound ATP resulted in pyrophosphate fission and activation of β-adrenergic pathways via cyclic AMP.

12.1 β-Sympathomimetic agonists

Enhancement of β-sympathomimetic activity by specific branched alkyl and aryl groups is in agreement with the corresponding catecholamines [65, 66]. One of these active compounds is now used clinically as a bronchodilator under the generic name of albuterol (the free base as inhaler and the sulfate as tablets).
Terbutaline sulfate, a resorcinol derivative is administered orally and is not metabolized by catechol-O-methyltransferase. It is used clinically in the treatment of bronchial asthma [67].
Ritodrine hydrochloride, another β-sympathomimetic agent with greater affinity for β_2-receptor on the uterus, has been approved by FDA as a class 1-A drug for the management of preterm labor [68].

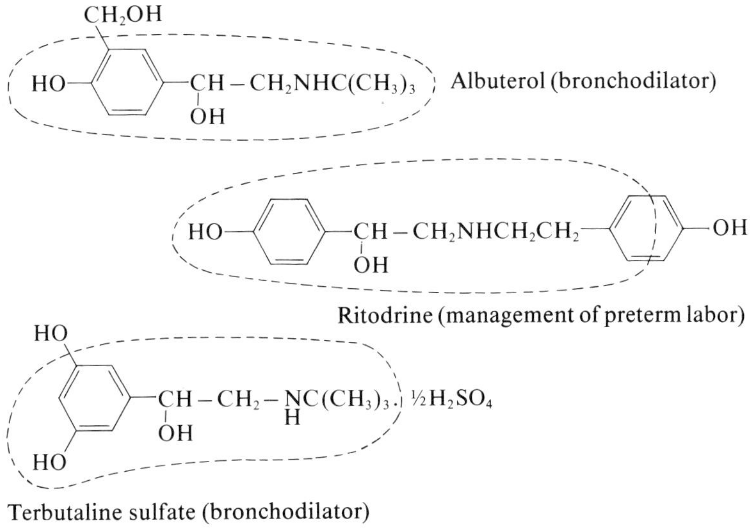

Figure 4
The pharmacophore of the β-receptor agonists.

12.2 β-Adrenergic blockers (antagonists)

During the last few years, several β-blockers have been marketed in Europe and in the US. This class of drugs is useful for a wide variety of diseases, including hypertension, sinus tachycardia, atrial flutter and atrial fibrillation, ventricular tachycardia and fibrillation, migraine headache, angina pectoris, idiopathic hypertrophic subaortic stenosis (IHSS), open angle glaucoma, hyperthyroidism, myocardial infarction, essential tremor, and others [69].

Table 6 summarizes the pharmacophore, pharmacodynamic, pharmacokinetic, and physicochemical properties of ten β-adrenergic blockers [70–78]. It is interesting to note that the β-blockers which are absorbed to greater than 90 % all have fairly high true octanol/water partition coefficients for the unionized form (log P = 1.61 to 3.18), while the less lipophilic drugs have lower degrees of absorption. Similar ranking order (although not a one-to-one correlation) also exists for protein binding. For example, the most lipophilic drug, propranolol (log P = 3.1, 3.3), has 95 % protein binding, while the least lipophilic drug, sotalol (log P = 0.08), has 0 % protein binding. Binding of various drugs to plasma, muscle, and different tissues has been reviewed [79]. Lipid solubility appears to be important in binding of drugs to muscle tissue [79]. Two compounds with β-blockade potency ratio of 6.0 have, on the other hand, relatively narrow log P range of 1.61–1.91 [70].

In recent studies of the corneal penetration behavior of β-blockers [70, 80, 81], Huang et al. have found linear dependence of the log (epithelial permeability) and log (endothelial permeability) on log P, and nonlinear dependence of log (permeability through intact cornea) on log P with an ideal log P_0 of around 4.3 for maximum permeation [70, 81]. Negative dependence on log (degree of ionization) and log MW has also been observed [71] as it has been seen from many examples of passive diffusion [1, 12].

13 **Calcium blockers and accumulation of drugs by cardiac tissue**

The important roles of calcium ions as charge carriers in excitable membranes and cardiac tissue have been known for some time [82, 83]. However, because of the ubiquitous distribution of calcium ions in the various organs and tissues of the body [83, 84], only selectivity achieved by distribution (affinity) and/or high intrinsic activity toward specific receptors can lead to potential therapeutic applications. Verapamil, a papaverine derivative, was introduced in Germany in 1962 as a coronary vasodilator. Later, it was shown to have antiarrhythmic properties and has been used first in Europe and then in the US as an agent in the treatment of re-entrant AV junctional tachyarrhythmias.

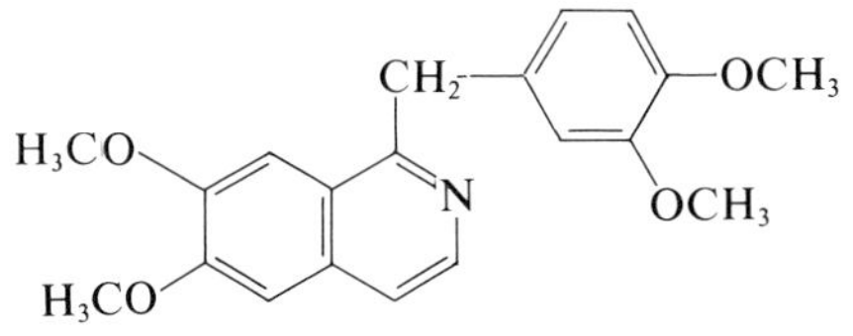

Papaverine

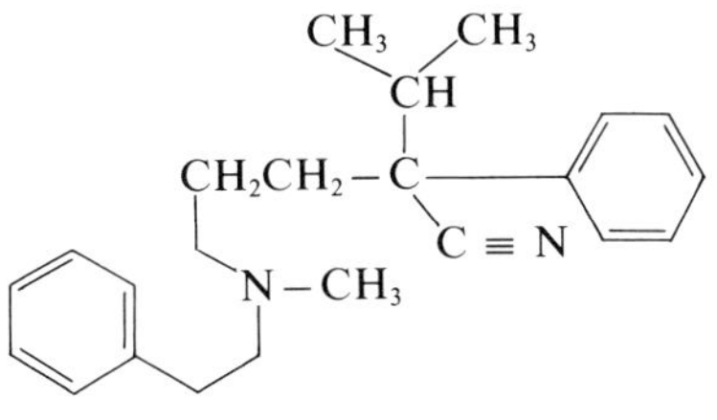

Verapamil

Verapamil inhibits the calcium ion (and possibly sodium ion) influx through slow channels into conductile and contractile myocardial cells and smooth muscle cells. These lead to a slowing down of AV conduction and the sinus rate, and a negative inotropic effect on myocardial tissue.

Lüllmann et al. [85] have studied the accumulation of verapamil with other neutral, anionic, and cationic drugs in resting and 2-Hz-stimulated isolated left auricles of the guinea-pig. Electrical stimulation accelerated the uptake process. Table 7 summarizes the equilibrium tissue/medium ratio (T/M) in guinea-pig left auricles stimulated with 2-Hz for verapamil and various drugs and their octanol/water partition coefficients (log P).

Regression analysis shows that a linear equation of log P gives a correlation coefficient (r) of 0.93 with log T/M, about 87% ($r^2 = 0.87$) of the variance in the data can be accounted for by this linear equation. Addition of the + $(\log P)^2$ term further improved the correlation ($r = 0.96$, $F_{1,8} = 7.2$, $F_{1,8_{0.95}} = 5.3$). The exact meaning of such a hyperbolic equation is not clear, it may be due to the lower accumulation by two depressant drugs which happened to have medium log P values, namely carticaine (a local anesthetic) and phenobarbital (a depressant) (see fig. 4). Addition of the log MW term to either eq. (21) or (22) does not result in significant improvement in correlation. This is probably due to the relatively narrow range of log MW involved.

Table 6
The pharmacophore, pharmacodynamic, pharmacokinetic, and physicochemical properties of β-adrenergic blockers.

Agonist (stimulant)	MW	log P	log P′ (log D)	pK_a	β-Blockade potency ratio[a] propranolol = 1.0	Extent of absorption[a] (% of dose)	Protein binding[a] (%)
Acebutolol 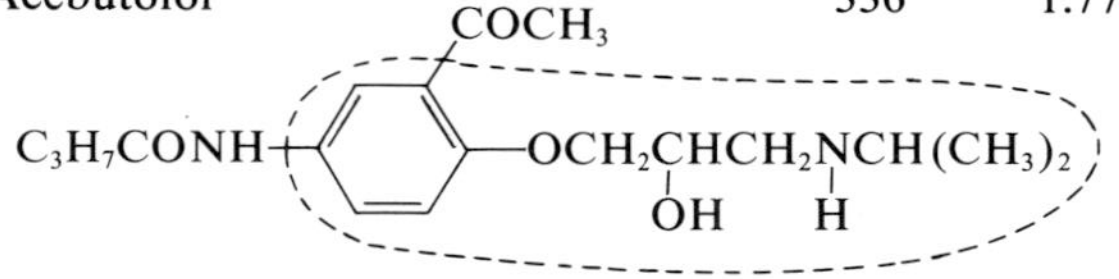	336	1.77[b]	−0.17[c]	9.20[b]	0.3	≃ 70	30–40
Atenolol 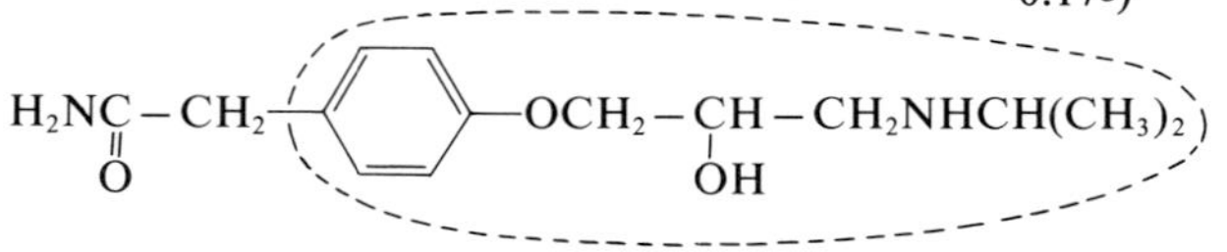	266	0.43[e] 0.17[g]	−1.82[c] −1.62[e] −1.94[g]	9.6[g]	1.0	≃ 50	< 5
Labetalol 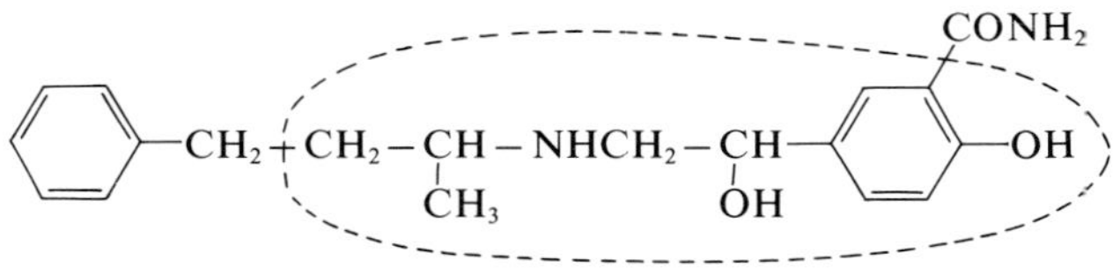	328	3.18[e]	1.13[e] 1.06[c]	9.45[e]	0.3	> 90	≃ 50
Metoprolol 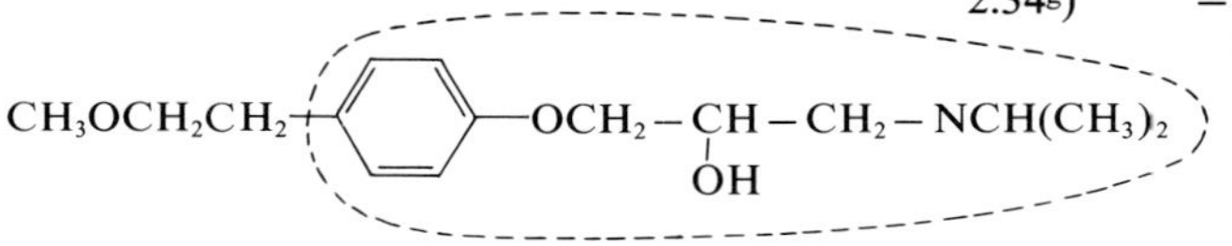	267	2.04[e] 2.34[g]	−0.01[c] −0.25[e] 0.04[g]	9.60[g] 9.68[e]	1.0	> 90	12

Table 6 (continued)

Agonist (stimulant)	MW	log P	log P′ (log D)	pK_a	β-Blockade potency ratio[a]) propranolol = 1.0	Extent of absorption[a]) (% of dose)	Protein binding[a]) (%)
Nadolol 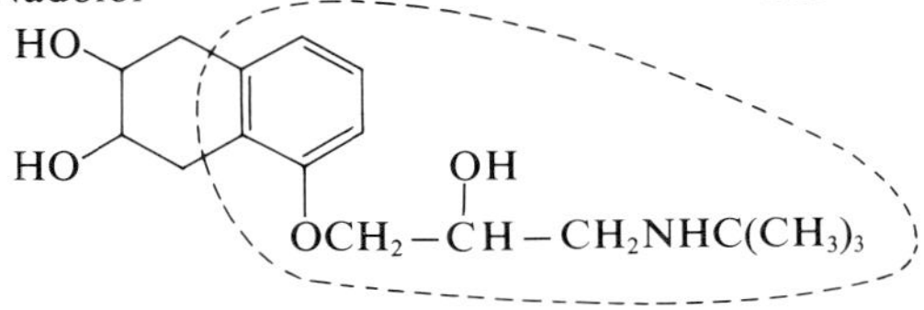	309	1.09[f]) 0.93[b])	−1.18[c])	9.67[h]) 9.39[b])	1.0	≃ 30	≃ 30
Oxprenolol 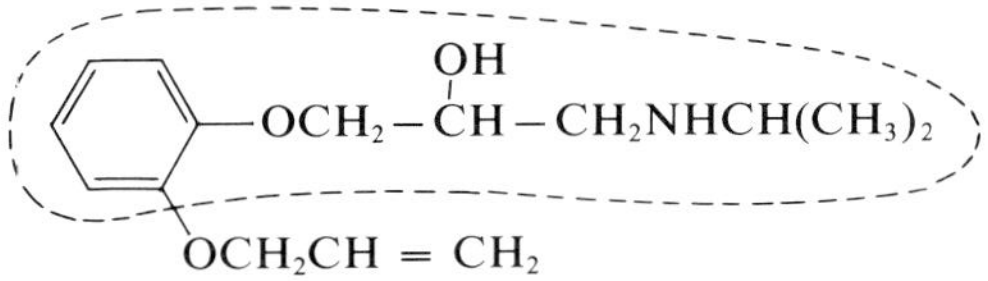	265	2.37[b])	−0.37[d]) 0.36[c])	9.32[b])	0.5–1.0	≃ 90	80
Pindolol 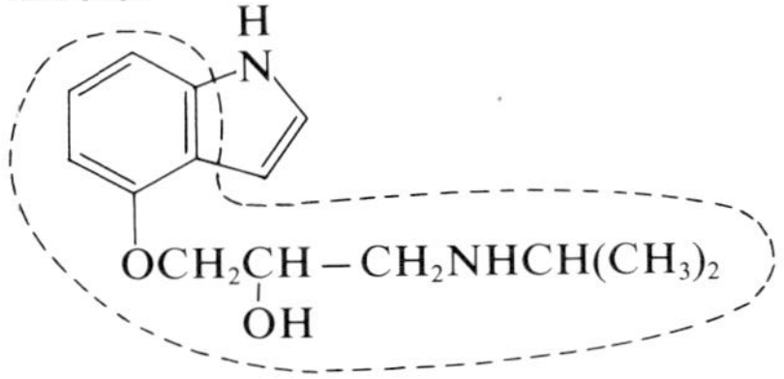	248	1.61[f])	−0.09[c])	8.81	6.0	> 90	57
Propranolol 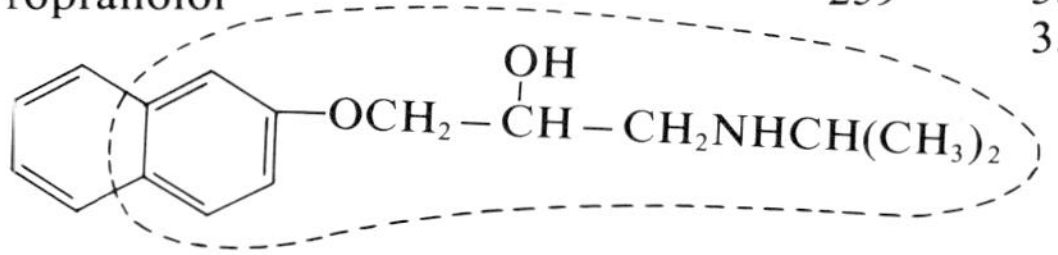	259	3.29[d]) 3.14[g])	0.73[d]) 1.30[c]) 1.24[e]) 1.41[g])	9.60[g]) 9.45[e])	1.0	> 90	93

Table 6 (continued)

Agonist (stimulant)	MW	log P	log P′ (log D)	pK_a	β-Blockade potency ratio[a)] propranolol = 1.0	Extent of absorption[a)] (% of dose)	Protein binding[a)] (%)
Sotalol	272	0.08[f)] 0.26[d)]	− 1.96[d)] − 1.40[c)]	9.05[j)]	0.3	≃ 70	0
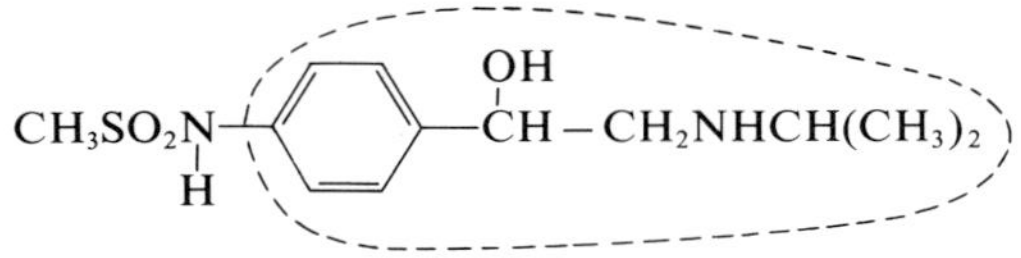							
Timolol	316	1.91[b)]	0.06[c)]	9.21[b)]	6.0	> 90	≃ 10
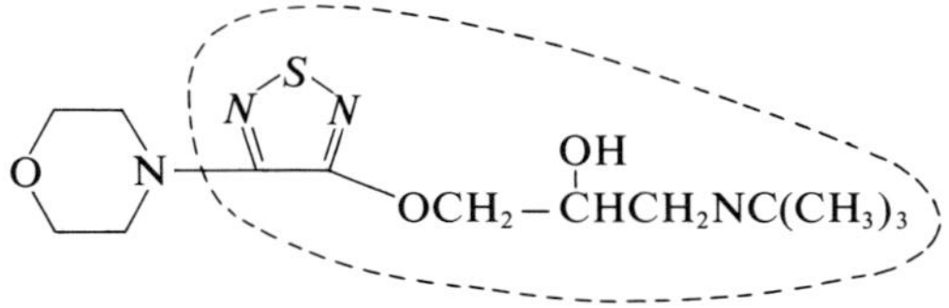							

a) From [70]
b) From [71].
c) From [72], n-octanol-phosphate buffer pH 7.4, not ion-corrected.
d) From [73], n-octanol-phosphate buffer pH 7.4, ion-corrected.
e) From [74], n-octanol-phosphate buffer pH 7.4, not ion-corrected.
f) Calculated from $\log P = \log P' - \log(1-\alpha)$, where $\alpha = 1/(1 + \text{antilog}[pH - pK_a])$.
g) From [75].
h) From [76].
i) From [77].
j) From [78].

	n	r	s	
$\log T/M = 0.34 \log P + 0.46$	11	0.93	0.37	(21)
$\log T/M = 0.06 (\log P)^2 + 0.35 \log P + 0.01$	11	0.96	0.29	(22)

It is interesting to note that verapamil and alprenolol (a β-blocker) have more favorable cardiac tissue accumulation than digitoxin, because of their higher log P values (2.51 and 3.10 vs. 1.76, respectively). From the clinical point of view, verapamil is contraindicated in the treatment of patients receiving β-adrenergic blockers. Combination of the two drugs may cause a profound depression of ventricular function.

Table 7
The cardiac tissue accumulation data, the octanol/water partition coefficients, and the molecular weights of verapamil and miscellaneous drugs.

Drug	log p[a]	log MW	log T/M obsd.[a]	calcd.[b]	calcd.[c]
Sucrose	−3.70	2.53	−0.52	−0.81	−0.44
Glucuronic acid	−2.57	2.29	−0.52	−0.42	−0.48
Acrecaidine ethylester methiodide	−2.34	2.23	−0.30	−0.34	−0.47
Phenobarbital	1.43	2.37	0.34	0.96	0.65
Carticaine	2.40	2.45	0.65	1.29	1.23
Dexamethasone	1.74	2.59	0.95	1.06	0.82
Digitoxin	1.76	2.88	1.18	1.07	0.83
Verapamil	2.51	2.66	1.48	1.33	1.29
Alprenolol	3.10	2.40	1.67	1.53	1.70
Phenylbutazone	3.04	2.49	1.78	1.51	1.66
Phenprocoumon	3.62	2.45	2.18	1.71	2.11

a) From [85].
b) Calculated from eq. (20).
c) Calculated from eq. (21).

Mannhold et al. [86] have reported the structure-activity relationships of verapamil derivatives. For a series of seven compounds with different substituents (R) on the benzene ring, eq. (23) has been obtained:

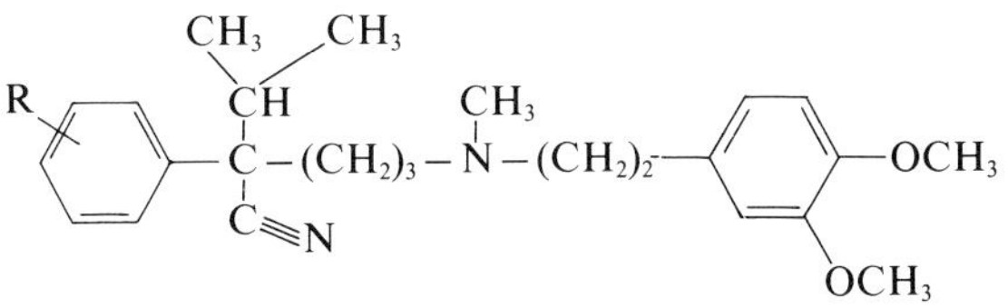

Group I

	n	r	s	
$\log 1/ED_{50} = 0.96\sigma + 0.63\ MV + \text{constant}$	7	0.988	0.06	(23)

where σ is the Hammett substituent constant and MV is the molar volume. This equation suggests that bulky electron-withdrawing group (R) on the ring enhances the negative inotropoc activity.

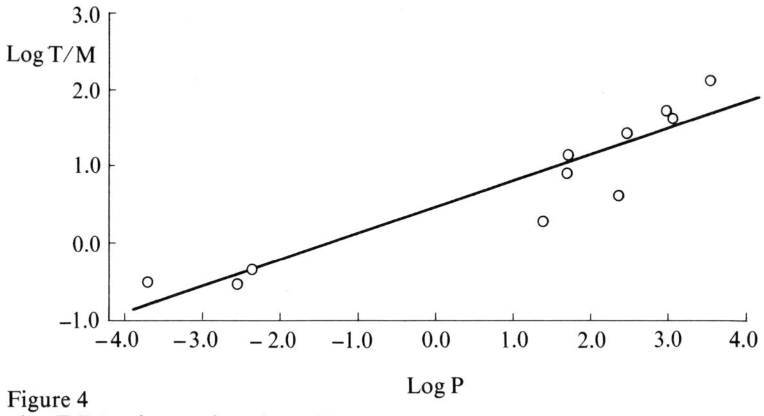

Figure 4
Equilibrium log T/M ratio as a function of log $P_{oct/w}$.

In a later study involving 9 to 13 compounds, Mannhold et al. reported the following correlations with the combination of inductive constant F and Taft's E_S, Hansch's π, and molar refraction (MR) [87]:

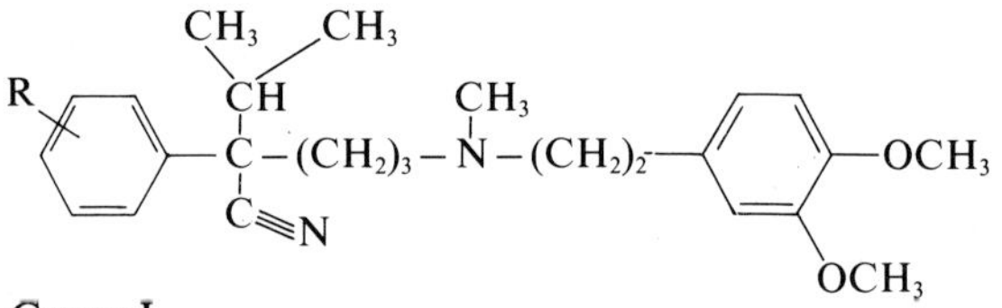

Group I

	n	r	s	
$\log 1/ED_{50} = 0.96\,F - 0.41\,E_S + \text{constant}$	9	0.89	0.27	(24)
$\log 1/ED_{50} = 0.81\,F - 0.30\,\pi + \text{constant}$	9	0.88	0.28	(25)

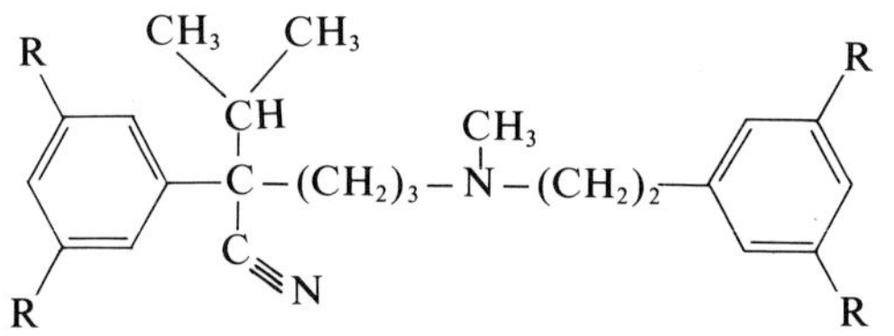

Group II

Group I + Group II

	n	r	s	
$\log 1/ED_{50} + 0.93\,F - 0.59\,MR + \text{constant}$	13	0.82	0.35	(26)

Partial structure activity relationships of 1,4-dihydropyridines as calcium channel antagonists have been reported [88]. From the limited data available, it appears that the nonpolar substituent at the 4-position of the 1,4-dihydropyridine ring increases activity. Ortho substituent (e.g., NO_2, F) in the 4-phenyl ring increases the activity while the para substituted compounds are much less active. Non-coplanarity of the phenyl ring and the 1,4-dihydropyridine ring has been suggested as a requirement for high activity [88]:

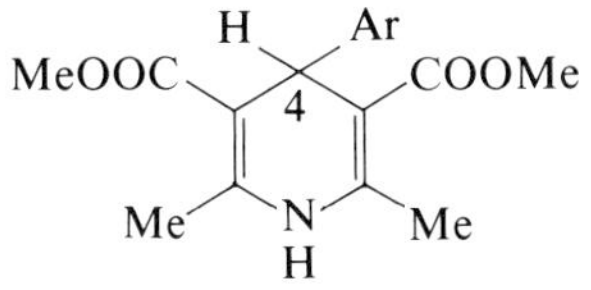

high activity observed for
Ar = 2-NO_2-Ph-
2-F-Ph-

1,4-Dihydropyridines as calcium channel antagonists

Apparent correlation between the biological activity (as measured by inhibition of guinea-pig ileal longitudinal smooth muscle mechanical response) and deviation of planarity (torsion angle around C^4) of 1,4-dihydropyridine ring in three derivatives has been reported by Triggle et al. [89, 90]. More data points are needed to provide sufficient degrees of freedom, and to avoid possible chance correlation.

14 Drug metabolism

While a large volume of data exists in the literature on metabolism of individual drugs, relatively few sets of data are available for quantitative analysis. Hansch et al. have reported on the correlations of oxidative deamination of different types of amines by different liver preparations [91]. Parabolic equations of log $P_{oct/w}$ with ideal log P_0 of 1.7 to 2.7 have been reported for primary and secondary amines. Eq. (27) and (28) were derived from the glucuronide and hippuric acid formation data of substituted benzoic acids [91]. Although different enzymes were involved, the log P_0 values of 2.3 and 2.4 are quite similar to those for the oxidative deamination of amines, suggesting that similar barriers (lipoprotein membranes) exist between the enzyme and the substrate.

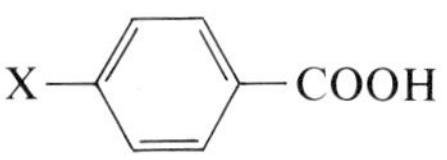

where X = CN, NO_2, I, Cl,
F, Me,
H, $MeCONH_2$

Glucuronide formation in rabbit — n 8, r 0.83, s 0.41

$$\log k_g = -1.03(\log P)^2 + 4.81 \log P - 6.62 \qquad n = 8 \quad r = 0.83 \quad s = 0.41$$
$$\log P_0 = 2.3 \qquad (27)$$

Hippuric acid formation — n 8, r 0.92, s 0.19

$$\log \%\, HA = -0.67(\log P)^2 + 3.15 \log P - 1.76 \qquad n = 8 \quad r = 0.92 \quad s = 0.19$$
$$\log P_0 = 2.4 \qquad (28)$$

Higher log P_0 value (5.69) has been found, on the other hand, for the N-oxidation of tertiary amines by a purified microsomal mixed function oxidase system [92, 93]. Many linear equations of log P have been reported for limited series of congeners [91, 93].

15 Conclusion

It is gratifying to see that the general mathematical model derived originally for in vitro data can be extended to in vivo systems as well. It is further demonstrated that the combination of the concept of *pharacophore,* together with the technique of *QSAR analysis* enables one to interpret large amount of data and to extract useful generalizations, which cannot be easily obtained by other means.

Several important areas of recent development, like drug delivery systems by liposomes, polymer-bound anthracyclines, β-adrenergic agonists, β-blockers, and calcium blockers have also been included in this review.

It is expected that more useful specific drug entities and preparations can be materialized in the next few years, as more experience is being accumulated in linking the disposition and the biological activities of specific drug molecules or the carrier systems to the optimum structural requirements of the specific pharmacological receptors.

References

1 E. J. Lien, in: Medicinal Chemistry IV, Proceedings of the 4th International Symposium on Medicinal Chemistry, p. 319. J. Maas (ed.). Elsevier, Amsterdam 1974.
2 E. J. Lien, in: Drug Design, vol. 5, p. 81. E. J. Ariëns (ed.). Academic Press, New York 1976.
3 E. J. Ariëns and A. M. Simonis, in: Towards Better Safety of Drugs and Pharmaceutical Products. D. D. Breimer (ed.). Elsevier/North Holland Press, Amsterdam 1980.
4 E. J. Ariëns, in: Drug Design, vol. IX, p. 1. E. J. Ariëns (ed.). Academic Press, New York 1980.
5 E. J. Lien, A. A. Alhaider and V. H. L. Lee: Parenteral Sci. Technol. *36,* 86 (1982).
6 K. Asghar and S. Riegelman: Arch. int. Pharmacodyn. *194,* 18 (1971).
7 F. E. Rose, H. C. Innemee and P. A. van Zweiten: Documenta Ophthal. *48,* 291 (1979).
8 C. Hansch, A. R. Steward, S. M. Anderson and D. Bentley: J. Med. Chem. *11,* 1 (1968).
9 E. J. Lien, C. Hansch and S. M. Anderson: J. Med. Chem. *11,* 430 (1968).

10 J. T. Penniston, L. Beckett, D. L. Bentley and C. Hansch: Molec. Pharmacol. *5,* 333 (1969).
11 E. J. Lien and P. H. Wang: J. Pharm. Sci. *69,* 648 (1980).
12 E. J. Lien: Annu. Rev. Pharmacol. Toxicol. *21,* 31 (1981).
13 E. J. Lien: J. clin. Hosp. Pharm. *7,* 101 (1982).
14 E. J. Lien, E. J. Ariëns and A. J. Beld: Europ. J. Pharmacol. *35,* 245 (1976).
15 C. Hansch and E. J. Lien: Biochem. Pharmac. *17,* 709 (1968).
16 E. J. Lien and J. L. G. Nilsson: Acta Pharm. Suec. *20,* 271 (1983).
17 J. L. Cohen, v. Lee and E. J. Lien: J. Pharm. Sci. *63,* 1068 (1974).
18 E. J. Lien: Abstract from the Annual Meeting of the Society of Environmental Toxicology and Chemistry, Arlington, November, 1983, p. 101.
19 A. T'ang and E. J. Lien: Acta Pharm. Jugosl. *32,* 87 (1982).
20 C. L. Chan, A. T'ang and E. J. Lien: Acta Pharm. Jugosl. *33,* 193 (1983).
21 R. J. Scheuplein and I. H. Blank: Physiol. Rev. *51,* 702 (1971).
22 B. Idson: J. Pharm. Sci. *64,* 901 (1975).
23a M. Katz: Chapter 4, in: Drug Design IV, p. 93, E. J. Ariëns (ed.). Academic Press, New York 1973.
23b B. J. Poulson: Chapter 5, in: Drug Design IV, p. 149. E. J. Ariëns (ed.). Academic Press, New York 1973.
24 T. J. Franz: Int. J. Dermatol. *22,* 499 (1983).
25 E. J. Lien and G. L. Tong: J. Soc. Cosmet. Chem. *24,* 371 (1973).
26 A. W. Tai and E. J. Lien: Acta Pharm. Jugosl. *30,* 171 (1980).
27 R. B. Stoughton and W. C. Fritsch: Arch. Derm. *90,* 512 (1964).
28 R. B. Stoughton: Arch. Derm. *92,* 675 (1965).
29 D. D. Munro and R. B. Stoughton: Arch. Derm. *92,* 585 (1965).
30 R. B. Stoughton: Arch. Derm. *101,* 160 (1970).
31 W. Resh and R. S. Stoughton: Arch. Derm. *112,* 182 (1976).
32 R. S. Stoughton: Arch. Derm. *118,* 474 (1982).
33 A. J. Lewis, R. J. Capetola and J. A. Mezick: Chapter 19, in: Ann. Rept. Med. Chem. *18,* p. 181, edit. H. J. Hess. Academic Press, New York 1983.
34 I. J. Torres, C. L. Litterst and A. M. Guarino: Pharmacology *17,* 330 (1978).
35 C. E. Myers and J. M. Collins: Cancer Invest. *1,* 394 (1983).
36 J. K. Seydel and K. J. Schaper: Pharmacol. Ther. *15,* 131 (1982).
37 C. D. Selassie, P. H. Wang and E. J. Lien: Acta Pharm. Jugosl. *30,* 135 (1980).
38 S. Fabro, in: Fetal Pharmacology, p. 443. L. Boreus (ed.). Ravens Press, New York 1973.
39 G. Levy and W. L. Hayton, in: Fetal Pharmacology, p. 29. L. Boreus (ed.). Ravens Press, New York 1973.
40 B. L. Mirkin, in: Fetal Pharmacology, p. 1. L. Boreus (ed.). Ravens Press. New York 1973.
41 M. H. Johnson and B. P. Setchell: J. Reprod. Fert. *17,* 403 (1968).
42 B. P. Setchell, J. K. Voglmayr and G. M. H. Wattes: J. Physiol., Lond. *200,* 73 (1969).
43 K. Okumura, I. P. Lee and R. L. Dixon: J. Pharmac. exp. Ther. *194,* 89 (1975).
44 R. Hori, M. Arakawa and K. Okumura: Chem. Pharm. Bull. *26,* 1135 (1978).
45 R. Hori, H. Yoshida and K. Okumura: Chem. Pharm. Bull. *27,* 1321 (1979).
46 C. Hansch and A. J. Leo, Substituent Constants for Correlation Analysis in Chemistry and Biology. Wiley, New York 1979.
47 W. A. Ritschel and G. V. Hammer: J. clin. Pharmac. Ther. Toxicol. *18,* 298 (1980).
48 J. Watanabe and A. Kozaki: Chem. Pharm. Bull. *26,* 665 (1978).
49 J. Watanabe and A. Kozaki: Chem. Pharm. Bull. *26,* 3463 (1978).
50 P. P. Lamy: J. Am. Geriat. Soc. Suppl. *30,* S11 (1982).

51 R. L. Juliano: Drug Delivery Systems. Oxford University Press, New York, Oxford 1980.
52 G. Gregoriadis, J. Senior and A. Trouet: Targeting of Drugs. Plenum Press, New York 1981.
53 K. J. Hwang, J. T. Merriam, P. L. Beaumier and K. F. S. Luk: Biochem. biophys. Acta *716,* 101 (1982).
54 P. L. Beaumier and K. J. Hwang: J. Nucl. Med. *23,* 810 (1982).
55 K. J. Hwang: J. Nucl. Med. *19* (1978).
56 I. J. Fidler, A. Raz, W. E. Fogler, R. Kirsh, P. Bugelski and G. Poste: Cancer Res. *40,* 4460 (1980).
57 A. J. Schroit and I. J. Fidler: Cancer Res. *42,* 161 (1982).
58 R. M. Abra, C. A. Hunt and D. T. Lau: J. Pharm. Sci. *73,* 203 (1984).
59 K. Mehta, R. L. Juliano and G. Lopez-Berestein: Immunology *51,* 517 (1984).
60 E. A. Forssen and Z. A. Tökes: Cancer Res. *43,* 546 (1983).
61 Z. A. Tökes, K. E. Rogers and A. Rembaum: Proc. natn. Acad. Sci. USA *79,* 2026 (1982).
62 K. E. Rogers, B. I. Carr and Z. A. Tökes: Cancer Res. *43,* 2741 (1983).
63 K. E. Rogers and Z. A. Tökes: Biochem. Pharmac. *33,* 605 (1984).
64 T. Collin, D. Hartley, D. Jack, L. H. C. Lunts, J. C. Press, A. C. Ritchie and P. Toon: J. Med. Chem. *13,* 674 (1970).
65 H. D. Mold, J. van Dijk and H. Niewind: Rec. Trav. Chim. Pays-Bas *77,* 273 (1958).
66 J. H. Biel, E. G. Schwarz, E. P. Sprengeler, H. A. Leiser and H. L. Friedman: J. Am. Chem. Soc. *76,* 3149 (1954).
67 R. F. Doerge (ed.): Wilson and Gisvold's Textbook of Organic and Pharmaceutical Chemistry, 8th ed., p. 411, 1982.
68 D. A. Hussar: Am. Pharm. *21,* 37 (1981).
69 M. Zarakov: California Pharmacist *28,* 34 (1981).
70 W. H. Frishman, in: Clinical Pharmacology of the β-Adrenoreceptor Blocking of Drugs, p. 13. W. H. Frishman (ed.). Appleton-Century-Crofts, 1984.
71 R. D. Schoenwald, H. H. Huang and J. L. Lach: J. Pharm. Sci. *72,* 1273 (1983).
72 P. B. Woods and M. L. Robinson: J. Pharm. Pharmac. *33,* 172 (1981).
73 D. Hellenbrecht, B. Lemmerm, G. Wiethold and H. Grobecker: Naunyn-Schmiedebergs Arch. Pharmac. *277,* 211 (1973).
74 P. H. Wang and E. J. Lien: J. Pharm. Sci. *69,* 662 (1980).
75 H. Lüllmann and M. Wehling: Biochem. Pharmac. *28,* 3409 (1979).
76 M. Windholz (ed.), The Merck Index, Merck Sharp & Dohme Research Laboratories. Rahway, New Jersey 1983.
77 G. S. Avery (ed.): Drug Treatment, p. 890. Publishing Sciences Group, Inc., Littletown, MA, 1976.
78 C. Hansch and A. Leo: The Log P Database, from The Pomona College Medicinal Chemistry Project. Technical Database Services, Inc., New York 1983.
79 H. Kurz and B. Fichtl: Drug Metab. Rev. *14,* 467 (1983).
80 H. S. Huang, R. D. Schoenwald and J. L. Lach: J. Pharm. Sci. 72, 1272 (1983).
81 H. S. Huang, R. D. Schoenwald and J. L. Lach: J. Pharm. Sci. 72, 1279 (1983).
82 P. F. Baker and H. Reuter, Calcium Movement in Excitable Cells. Pergamon Press, Oxford 1975.
83 G. B. Weiss, Calcium in Drug Action. Plenum Press, New York 1978.
84 S. L. Flaim and R. Zelis: Calcium Blockers. Urban & Schwarzenburg, Baltimore–Munich 1982.

85 H. Lüllmann, P.B.M.W.M. Timmermans and A. Ziegler: Europ. J. Pharmacol. *60,* 277 (1979).
86 R. Mannhold, R. Steiner, W. Haas and R. Kaufman: Naunyn-Schmiedebergs Arch. Pharmac. *302,* 217 (1978).
87 R. Mannhold, P. Zierden, R. Bayer, R. Dodenkirchen and R. Steiner: Arzneimittel-Forsch. *31,* 773 (1981).
88 L. Rosenberger and D. J. Triggle, in: Calcium in Drug Action, p. 3. G. B. Weiss (ed.). Plenum Press, New York 1978.
89 A. M. Triggle, E. Shefter and D. J. Triggle: J. Med. Chem. *23,* 1442 (1980).
90 D. J. Triggle, in: New Perspectives on Calcium Antagonists, p. 1. G. B. Weiss (ed.). American Physiological Society, Bethesda 1981.
91 C. Hansch, E. J. Lien and F. Helmer: Arch. Biochem. Biophys. *128,* 319 (1968).
92 D. M. Ziegler, C. H. Mitchell and D. Jollow: Microsomes and Drug Oxidation, p. 173. Academic Press, New York 1969.
93 G. L. Tong and E. J. Lien: J. Pharm. Sci. *65,* 1651 (1976).

Medicinal research: Retrospectives and perspectives

George deStevens
Department of Chemistry, Drew University, Madison, N. J., USA

1	Introduction	98
2	Retrospectives	98
2.1	Early years	98
2.2	CIBA: 1955–1969	99
2.3	CIBA-GEIGY merger	103
2.4	The next phase: 1971–1979	106
3	Perspectives	109
3.1	Central nervous system	109
3.2	Cardiovascular-renal system	112
3.3	Rheumatoid arthritis	115
3.4	Prostaglandins and cytoprotection	116
3.5	Recombinant DNA, gene expression and gene structure	117
	References	119

1 Introduction

At the early part of last year the editor of this well-known series invited me to write a review chapter for an up-coming volume. In his invitation he allowed a generous latitude in the selection of topics and inferred that a review which would encompass my experience of a quarter of a century in the pharmaceutical industry, all of which was spent with CIBA and CIBA-GEIGY, would be most desirable. To this end and following several exchanges of letters with the editor, I have decided to present a review partly in the form of an autobiographical essay as it relates to the various influences in my background and training associated with the personalities and events which shaped some of the significant developments in medicinal research with which I was fortunate to be involved. My years in management will also permit me to comment retrospectively on the events of those times.

2 Retrospectives

2.1 Early years

From my earliest recollection I had always wanted to be a chemist. I recall my mother encouraging this desire by presenting me with a chemistry set at the age of 12. In 1936 these kinds of kits were first becoming popular. Many an hour was spent mixing various substances, noting changes in colors, formation of precipitates and the bubbling over of gaseous materials. Much of these phenomena, though not understood, left memorable impressions and gave substance to one's youthful imagination. This was manifest one quiet summer afternoon when I mixed several chemicals (their identity which has long been forgotten) and to my astonishment, a white milk-like solution was obtained. As chance would have it, the family cat entered at the time and it occurred to me that my discovery could be incorporated in the cat's milk bowl. This was done. The cat voraciously drank the milk with its additional contents, then hesitated for a moment, its back went up in typical defensive form, gave a howl and raced out of the house. My first pharmacological experiment and I knew not what I had wrought! Fortunately, our cat returned a few hours later, a bit haggard but feeling better and prepared to live to a ripe old age.

The ensuing adolescent years followed uneventfully only to be met by the harsh and maturing reality of ten months of combat duty in France, Germany and Austria. The only association with chemistry at that time was in the early spring of 1945 in the battle for the city of Munich. I was with a group that came upon the destroyed remains of the Technische Hochschule of Munich. It was only much later that I learned that this was the laboratory of Professor Hans Fischer, Nobel Laureate for his brilliant research on the structure of hemin and chlorophyll. Obviously this devastation of his life's work was instrumental in his untimely demise.

My continuing interest in a career in chemistry eventually led me to pursue my doctorate with Professor F. F. Nord at Fordham University. Nord was one of the many staff members of the Kaiser Wilhelm Institut at Berlin-Dahlem, who was forced to leave Germany after 1933 and subsequently he had established a strong school of Organic Chemistry and Enzymology at Fordham. I joined a large group of graduate students and post-doctorals engaged in diverse areas of research in bioorganic chemistry, synthesis, hydrogenation, cryoscopy, natural products from fungi, enzymic reactions and lignin structure. This was chemistry in the European tradition, large multifaceted research programs involving bio-chemical approaches to their solutions. This type of environment extended the scope of one's education and exposure to the significant research problems and personalities of the day. This surely was so in my case. My research on the enzymic effects on bagasse lignin led to contacts over a three-year period with a number of eminent scientists; namely, Sumner, Freudenberg, Kratzl, Brauns and Erdtman.

2.2 CIBA: 1955–1969

Having completed my requirements for the Ph. D., I was anxious to join a pharmaceutical house. It was 1955 and suddenly the golden age of drug discovery was upon us. Professor Emil Schlittler of CIBA, Basel, had recently arrived in Summit, New Jersey, to head up the Chemical Research Department of CIBA Pharmaceutical Company. He engaged me to join the newly formed synthetic organic chemistry group. These were momentous times within the CIBA Research organization. Schlittler and his collaborator Muller had isolated reserpine from *Rauwolfia serpentina* and Professor Hugo Bein had clearly defined its pharmacological activity and its potential clinical utility. The structural elucidation of reserpine by Schlittler and his colleagues in Summit was widely acknowledged as an important research achievement. Reserpine was an immediate success as a neuroleptic agent and in addition, as noted from the animal pharmacology, it significantly lowered blood pressure in hypertensive patients. This coupled with the earlier synthesis of hydralazine by Jean Druey and P. Ringier, and the discovery of its potent antihypertensive effect by Albert Plummer and F. F. Yonkman in Summit established CIBA as a pioneer in cardiovascular research. Such a development was an important change for the company since it was obvious to most observers, both within and outside the firm, that it was losing its pre-eminent position as 'the Steroid-House'. This is not to say that important steroid research was not in progress at the CIBA Basel Research Center. Certainly the total synthesis of aldosterone by Albert Wettstein and co-workers, was a *tour* de *force* in steroid synthesis; however one considers the academic excellence of this research, its commercial significance was nil. In the meantime prednisone, prednisolone, triamcinolone and dexamethasone were discovered and brought to the market place as therapeutically useful anti-arthritic drugs by Schering, Squibb, Lederle and Merck. In addition Wettstein and his collaborators had isolated a steroid which they had identified as SEF (sodium excreting factor); pharmacological evaluation did not result in pronounced saluresis. However, this group was responsible for the discovery of one of the first anabolic drugs, methandrostenolone.

On the other hand the quality and style of research within CIBA remained undaunted. The management had great confidence in research and supported it vigorously. Within my own experience in Summit, Schlittler's philosophy was to select the best scientists available and permit a free flow and exchange of ideas. The Basel organization functioned the same way. In 1958 I was fortunate to spend seven months in the Basel laboratories. It was on this occasoin that I had the opportunity to learn of the Swiss connection in research. The older, more conservative and staid organic chemists were in the twilight of their careers and were giving way, although reluctantly, to a new wave of very bright and more modern chemists. These included Karl Schenker who had worked with Prelog at the ETH on transannular reactions and then with R. B. Woodward at Harvard on the total synthesis strychnine, Hans Bickel from the University of Zurich and post-doctoral with

Woodward on the synthesis of reserpine; George Huber, a carbohydrate chemist from the ETH and Erwin Jenny, Paul Schmidt and Karl Heusler from the University of Basel. Ernst Vischer also from Basel, joined the CIBA organization after his classic research with Chargaff on the nucleotide composition of DNA and now was in charge of microbiological research on steroids and antibiotics. It was also at this time that I established relations with Albrecht Hüni who had been assistant to Hans Fischer of Munich prior to joining CIBA. These contacts were immensely helpful to me in the years that followed. The CIBA worldwide research environment was not only highly stimulating but the collegial competition between the two research centers was a beneficial force. The consequence was that in the years 1956 to 1966 several important drugs were discovered and introduced into clinical medicine (see table).

My experience in the discovery of hydrochlorothiazide bears brief mention. Chlorothiazide had been discovered by Novello and Sprague at Merck Sharp and Dohme and my exchange of information with Paul Schmidt of our Basel Laboratories led us to different approaches to potentially useful diuretics. Working with my collaborator Lincoln Werner and two able assistants, S. Ricca and A. Halamandaris, we quickly came upon hydrochlorothiazide and testing by J. Chart and co-workers in biology gave us the important information to proceed further. Hydrochlorothiazide, ten times more potent than chlorothiazide, was introduced in early 1959 and within a short time became the drug of choice in mild hypertension. Over a period of 3 years, our small group prepared over 400 derivatives of this class which was the subject of several patents and publications. Although several derivatives with more potent properties than hydrochlorothiazide were prepared (e. g., cyclopenthiazide is 1000 times more potent than hydrochlorothiazide), the prototype remains to this day as the most widely prescribed drug, alone and in combination, for the treatment of hypertentsion.

In the meantime my next door laboratory associate, Robert Mull, had discovered guanethidine and Charles Huebner, the CIBA chemist who had elucidated the stereochemistry of reserpine, had synthesized a potent antihistamine, methindene. William Beneze also clarified the amphenone story and synthesized metyrapone which was shown by Robert Gaunt and co-workers to he a diagnostic aid for pituitary function.

New drug entities discovered at CIBA: 1956–1966

Drug	Indication	Year
chlorisondamine	antihypertensive	1956
deserpidine	antihypertensive	1956
glutethimide	hypnotic	1956
methylphenidate	stimulant	1957
hydrochlorothiazide	diuretic, antihypertensive	1958
methindene	antihistamine	1959
methandrostenolone	anabolic	1959
guanethidine	antihypertensive	1960
cyclopenthiazide	diuretic, antihypertensive	1960
oxprenolol	antiyhypertensive	1963
niridazine	anti-schistosomal	1964
sulfaphenazole	anti-infective	1964
maprotoline	anti-depressant	1966

The medicinal chemistry group at CIBA Basel was also quite active during this time. The section headed by Schmidt capitalized on the basic findings of Schindler at GEIGY on tricyclic substances (imipramine) for the treatment of endogenous depression. Schmidt and Wilhelm prepared a tetracyclic drug, maprotoline, which has been used as an anti-depressant since 1968. The same group also noted the important developments from ICI (Black et al) on adrenergic antagonists and brought forward oxprenolol. The Schmidt group in addition was responsible for the discovery of the anti-schistosomal drug, niridazole and the anti-infective sulfaphenazole.

Although CIBA had been involved from the beginning with the cephalosporins, its contribution in terms of a significant drug in this area has been minimal to date primarily due to a lack of concerted effort to prepare necessary amounts of the key intermediate 7–ACA for scale up in the preparation of cephacetrile. This led to interminable delays in the development process, a consequence of which was that Eli Lilly and others were able to establish strong priority marketing positions for cephalosporins, especially in the US. Notwithstanding, Vischer and co-workers in a joint effort with Lepetit did pioneering research in the discovery and development of rifampicin, an effective antibiotic for the treatment of tuberculosis and some specific gram-negative, bacterial infections.

In 1967 Schlittler retired and I succeeded him as head of research and development of CIBA, US. We were in a state of change especially with impending NAS–NRC review of all drugs marketed prior to 1962.

In addition I accelerated Schlittlers's initial efforts to strengthen biochemistry and drug metabolism. The alkaloid research group which for 12 years had searched, in vain, for another reserpine was disbanded and their research was directed toward synthetic programs. Research was consolidated in four therapeutic areas: cardiovascular, central nervous system, inflammation and anti-infectives. It was 1968 and as the end of this eventful decade was approaching, the rumors of merger began to emerge.

2.3 CIBA-GEIGY merger

Nowadays mergers and acquisitions are a way of life and are taken for granted. But a merger, the likes of which was being considered in 1968 in Basel, Switzerland, still staggers the imagination. Two major chemical firms of equal size, CIBA and GEIGY, with different management and research styles and philosophies, different business traditions, opposite types of personalities in middle and upper managements, both headquartered in Basel, a city in which family, social and cultural relationships are closely allied with one's company affiliation and position; a merger which would immediately create redundancy in almost every important position. Although some within each company saw the serious consequences of such a move, the decision to merge was taken. Whether or not this merger will ever be considered truly successful remains to be seen. The losses due to years of readjustment, diminished productivity, disillusioned talented people and their traumatized families, unrealized potential products, and the demoralizing impact on human motivation, innovation and creativity can never be measured or quantified.

Be that as it may, one of the most difficult tasks on a world-wide basis was consolidation and integration of the vast CIBA-GEIGY research and development organization. As the two chemistry groups in Basel came together in discussions and exchange of programs, the relative import of Ruzicka's parody gave one reason to pause. The Swiss research group was consolidated and all scientists were retained. In the US, this could not be done due to the loss of products and revenue associated with the divestiture. As Executive Vice President and Director of Research of the CIBA-GEIGY Pharmaceuticals Division in the US, the task fell upon me to plan a reduction in staff of almost 200 scientists. There were some in upper management who urged that this be done quickly and thus the benefits to the bottom line would be immediate. Instead, I proposed that this reduction be accomplished over an 18-month period by means of transfers, attrition, resignations, etc. It was my contention that the strong image of CIBA and GEIGY as research oriented companies was due not only to the quality of its product line, but more importantly to the high caliber scientists they attracted and this in turn resulted in excellence in research. To suddenly terminate 200 scientists in the US for a short-term gain would tarnish that image and reputation which had taken over a quarter of a century to establish, I argued. Reason and good judgment finally prevailed and our reduction in staff was accomplished uneventfully. Shortly thereafter, Max Tishler, retired President of Merck, Sharp and Dohme Laboratories and at the time President of the American Chemical Society wrote to me on this matter (see the accompanying letter).

The terms of the agreement with the Justice Department in order to permit the merger in the US required the divestiture of several products. Management finally agreed to negotiate with USV (the pharmaceutical division of Revlon) on the exchange of products. The details of these negotiations are well outlined in *The Basel Marriage* by Paul Erni and in the Charles Revson biography, *Fire and Ice* by Andrew Tobias and will not be discussed herein. Eventually CIBA-GEIGY gave up chlorthalidone and the combination chlorothalidone-reserpine and received in exchange phenformin. Phenformin was already under close scrutiny by the FDA because of its questionable efficacy and safety record as noted in the UGDP report. These questions were brought to the attention of the CIBA-GEIGY negotiating team by the Research Department prior to consummation of the deal. However, the forward thrust of the negotiators could not be thwarted. Too much time and energy had been committed.

As expected, phenformin proved to be an albatross around the neck of the Research and Medical Department. Over the first three years CIBA-GEIGY increased the annual sales of phenformin from $ 14,000,000 to $ 28,000,000. Correspondingly, more and more cases of lactic acidosis appeared and were reported to the FDA.

Massive efforts were made by Drug Metabolism, Biochemistry and Medical to get a handle on the problem and control it. Teams of scientists were put on the problem to study the cause of lactic acidosis, the role of phenformin in this metabolic aberration, and what measures could be taken to limit its occurrence. After several years of excessive

American Chemical Society

OFFICE OF THE
PRESIDENT

Max Tishler, *President*

Department of Chemistry
Wesleyan University
Middletown, Connecticut 06457
Telephone (203) 347-9411 Ext. 777 or 778

December 29, 1972

RECEIVED
JAN 5 1973
GdeS

George deStevens, Ph. D.
Executive Vice President and Director
of Research
Pharmaceuticals Division
Ciba-Geigy Corporation
Summit, New Jersey 07901

Dear George,

I remember very well our conversation over a year ago when you told me about the problem Ciba-Geigy faced due to the Merger. At the time when the employment situation throughout the country was very poor and mass terminations occurred in many substantial companies, I thought you and your associates were very far-sighted in planning to resolve your problem of excess personnel in a more human and gentler manner. I think you have shown wonderful statesmenship and you deserve thanks and kudos from the scientific community. I wish that your story could be told since I believe that other companies could take wisdom and courage for what you did.

One of my last letters as President of the American Chemical Society is to congratulate Ciba-Geigy for the way a difficult problem was handled and to extend the sincere thanks of the American Chemical Society for having given your scientific people reason to hold their heads high and to be proud of their science heritage.

Good luck -- it was beautifully done.

Sincerely,

Max

Max Tishler

MT:jbp
cc: R. Cairns
A. Nixon
H. Block

support at the basic level, the recommendation from Research with the advice of outside consultants, was that phenformin had a specific and limited use in the treatment of diabetes and more stringent efforts should be made to reduce the level of promotion, advertising and detailing of this drug. The changes made were not quite up to the proposals. In 1978 FDA ordered the removal of phenformin from the market place. The lesson here is simple: it is better to have lower sales and profit over a long period of time, rather than none at all.

2.4 The next phase: 1971–1979

The selection of research programs and the emphasis or priority to be given to these by a corporation are the consequences of a continual dialogue between Management, Marketing and Research. The final decision is one which Management must make but requiring full input from the scientists. In all of these deliberations it is quite clear that *Research cannot predict the rate or sequence of discovery* and although one may hear on occasion 'Research is too important to be left up to the scientists', any other alternative is folly and reveals a naivete and lack of understanding of the complexities of the discovery process.

Two of the main pillars of research for CIBA-GEIGY have been cardiovascular and inflammatory diseases. Although extensive strength and depth were available internally in each of these areas, Professor Bein, head of CIBA-GEIGY Pharmaceutical Research in Basel, anticipated the desirability of extending this base, especially in the cardiovascular area. Consequently, he consummated a joint research agreement with Hässle, Sweden on β-adrenergic antagonists. This was especially important in the US because propranolol was the only β-blocker available, a situation which contrasted sharply with the state of affairs in the UK and Europe where several drugs of this type had been available for several years for the treatment of angina, arrythmias and hypertension. Bein and I were impressed with the research excellence of the Hässle cardiovascular section headed by Professor Oblad. Their emphasis on cardioselectivity was clearly the direction for second generation β-blockers. One of the substances, H 93/26, later to be known as metoprolol, was selected for world-wide development. All toxicology, considerable amounts of pharmaceutical development work and some phase I and phase II clinical trials were done in the US, Hässle carried out most of the pharmacology, biochemistry, metabolism and

phase III clinical trials. The objective in all of this effort was to introduce metoprolol in as many markets as possible as soon as possible, with the target that metoprolol should be the next β-blocker to be introduced in the US. Several discussions between FDA and Hässle and CIBA-GEIGY scientists led to clear guidelines for pre-clinical and clinical requirements for approval of the New Drug Application. The NDA approval for metoprolol (Lopressor®) for the treatment of hypertension was obtained in the summer of 1978, approximately 4 years after our collaboration with Hässle on this project began. Metoprolol was the second β-blocker to be approved in the US and it was the first of these agents to be approved by FDA as a cardioselective agent. This development has been discussed in some detail to illustrate how good product selection, planning, coordination, communication and execution by major research organizations on two continents can expedite new product development. Obviously this could not have been achieved without the motivation, commitment and just plain hard work by talented scientists in both companies.
These years saw a further strengthening of the cardiovascular franchise with the introduction of Apresazide® and Slow-K®. The unique-feature of the latter drug was that for the first time it permitted the safe administration of a potassium chloride supplement to hypertensive and edemetous patients on long term diuretic therapy. This was accomplished by means of the slow-release of potassium chloride from a sugar-coated wax matrix. The consequence was that the salt was not dumped out in one large portion in the stomach or in the small intestine (enteric coated form) thus preventing mucous inflammation, ulceration and stenosis. Slow-K became an immediate success and emphasized the importance of drug delivery systems. Discussions with Dr. George Ferguson, Corporate Vice President for Research Services, in 1974 resulted in a joint program between the corporate polymer group and the CIBA-GEIGY pharmaceutical scientists for the discovery of new and patentable delivery systems. Through the years this became an integrated group and its viability was further enhanced in 1977 by the strong ties developed by CIBA-GEIGY with Alza.
After the significant discovery of the corticosteroids, their widespread use revealed that though they were effective for the treatment of arthritis, their many side effects made them less than desirable for the long-term treatment of this crippling disease. The Merck, Sharp and Dohme research management recognized that alternative approaches, such as

non-steroidal anti-inflammatory agents, could be more fruitful. After several years T. Y. Shen and co-workers discovered indomethacin, an indole acetic acid derivative, which from 1965 on has gained wide use as an effective anti-inflammatory drug. Shortly thereafter, Boots introduced the phenylpropionic acid derivative, ibuprofen, in the UK, while Upjohn, under license from Boots was proceeding with the same substance in the US. Indomethacin proved to be a much more potent compound than ibuprofen, although the latter appeared to cause less gastrointestinal effects. It was quickly ascertained that the acetic acid portion was the common feature in both of these molecules. As a consequence a virtual explosion of research ensued in the synthesis of acetic and propionic acid derivatives for evaluation of their anti-inflammatory properties.

At merger time, CIBA-GEIGY had three such compounds ready for clinical evaluation. Due to the importance of this therapeutic area the international research management agreed to move forward with all three clinical candidates. Within a short-time Ba-47,210 dropped out because of carcinogenic effects in rats. Diclofenac in early clinical trials in Europe proved to be effective although gastrointestinal side effects were of some concern. When we in the US sought to submit an IND for diclofenac, all sorts of obstacles were encountered with the FDA. The animal toxicity in rodents, dogs and monkeys was quite severe at all dose levels. The FDA imposed such excessive demands for additional animal work that Management decided to withdraw pursuit of this substance in the US and to seek to register it only in selective markets in Europe to ascertain the response. It was only after 3–4 years of successful clinical usage in Europe and Japan along with much basic biochemical work, it was learned that the pharmacokinetic profile of diclofenac is quantitatively different in animals than in humans. As a result, it appeared that the alarming animal toxicity did not carry over into humans. On the basis of these findings, the IND for diclofe-

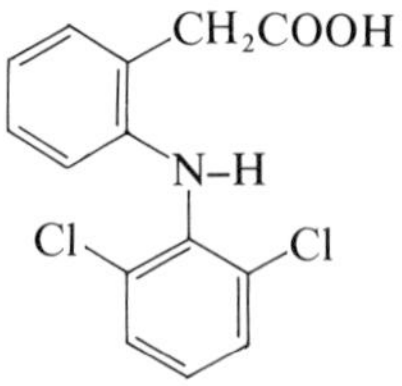

Diclofenac

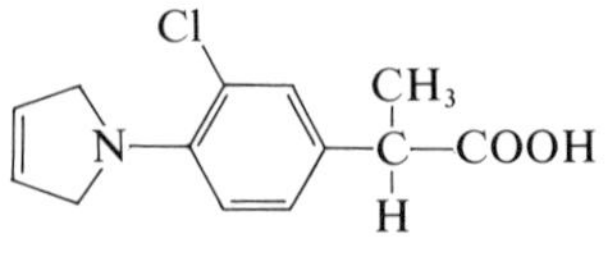

Pirprofen

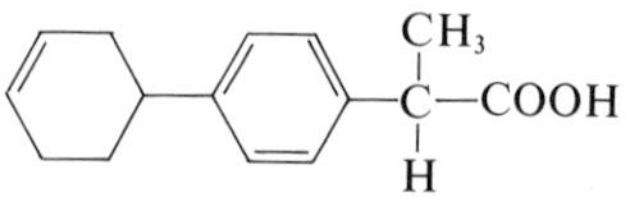

Ba-47,210

nac was submitted to FDA by CIBA-GEIGY in the winter of 1978. In the meantime pirprofen was registered in several European countries and its NDA was recommended for approval by the Anti-inflammatory Advisory Board.

During these years, the Research and Development Department which consisted of all operations essential in taking a chemical substance from discovery through NDA approval had adjusted to the changes brought about by the merger and began to function effectively as an integrated unit. From the beginning it had been my conviction that the whole enterprise of drug discovery and development to product approval was a continous process and thus could and should be executed most efficiently under the overall direction of one research management. Others in the corporate milieu, for whatever reasons one may posit, sought to separate the responsibility for research from that of the development phase (including medical and regulatory operations). This dialogue and altercation went on for some 6 years. Finally, in 1979 rather than preside over the demise of the organization which I had devoted so much time and effort to build, I chose instead to pursue a new career in academia.

My association with Drew University began quite unexpectedly on the golf course at Morris County Golf Club. On opening day of 1976 I was paired in a foursome with Dr. Paul Hardin the newly appointed President of Drew. Several weeks later at a luncheon I suggested to Hardin the idea of establishing an annual research conference on the Drew campus. His response was immediate and positive and thus resulted the annual CIBA-GEIGY-DREW Symposium on Frontiers in Biomedical Research which has attained both national and international prominence.

My transition to academia occured quite smoothly in September 1979 with my appointment to Research Professor of Chemistry at Drew University.

3 Perspectives

This section of the review will be concerned with my impressions of selected areas of medicinal research which will be of utmost importance in the closing years of this century.

3.1 Central nervous system

The ultimate aim of medicinal research is the discovery of drugs to cure disease so that sick people can return to normal health. Presently many drugs do not cure disease but merely are used to treat and alleviate symptoms. For example, although much is known about the biochemical changes that occur in the brain leading to schizophrenia, the drugs presently in use are not curative for this condition. This is not to say that these drugs are not of value. In fact, they are of great value since they permit many people to lead productive lives and to function effectively in society. In the case of schizophrenia, the available neuroleptic agents are universally acting on the dopamine system of the brain, primarily as receptor blocking substances. Dopaminergic neurons are widely spread in the brain and consequently drug treatment has lacked specificity resulting in a variety of adverse reactions. A side effect of antipsychotic drugs of principal concern has been extrapyramidal symptoms which are usually manifested as Parkinsonianism or persistent dyskinesia. Until recently, all neuroleptic agents, because of their non-selective blockade of post-synaptic dopamine receptors, have caused this serious and disconcerting side reaction. However, first with clozapine, and now more recently with fluperlapine [1], it appears that a class of drugs will be available which has good antipsychotic action but is practically devoid of persistent dyskinesia. This is a major step forward and emphasizes that even more selectively active substances will become available in the future.

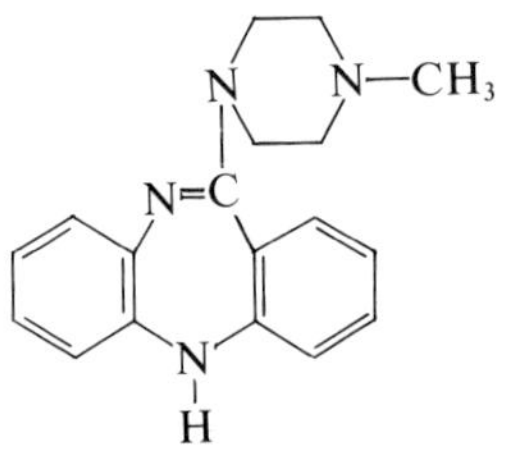

Clozapine

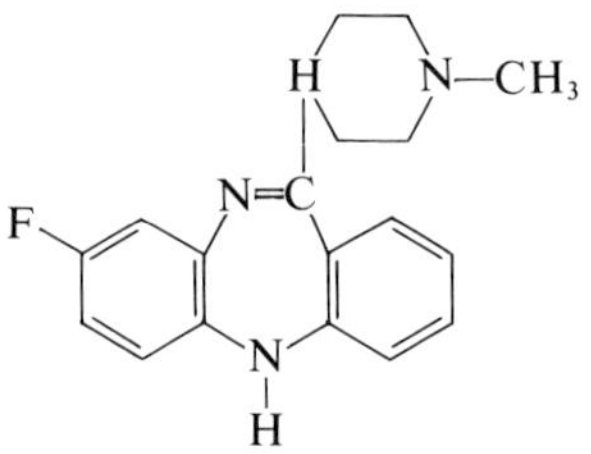

Fluperlapine

Alzheimer's disease, a dementing disorder for which the diagnosis is unequivocal only postmortem, the therapy almost nonexistent, and the prognosis grim, represents a major challenge for CNS research. This disease causes loss of memory followed by confusion, disorientation, loss of affect and intellect, and eventual physical deterioration in over 100,000 Americans between the ages of 45 to 65 and in an estimated 1.5 million persons over the age of 65. On a world-wide basis, Alzheimer's can be considered a mental health problem of major proportions. The work of Davies and co-workers [2] at Albert Einstein College of Medicine has shown that this condition is associated with a marked deficiency of acetylcholine in the CNS. They have discovered that the activities of the neurotransmitter's synthesizing and degrading enzymes, namely, cholineacetyl transferase and acetylcholine esterase, respectively, were greatly reduced in the cerebral cortex of persons with Alzheimer's disease.

On the other hand, McKinley and Prusiner [3] have found unique proteinaceous particles called prions which may be associated with Alzheimer's. Prions resist inactivation by procedures that modify nucleic acids. These substances have been found in the brains of hamsters infected with scrapie, a degenerative, fatal, brain disease of sheep and goats.

Prions are known to contain little or no DNA or RNA and are smaller than viruses. McKinley and Prusiner have also been able to stain prion rods with Congo red dye and show that they elicit green birefringence under polarization microscopy. Amyloid, a substance found in the brains of persons who have died of Alzheimer's disease, also shows the same effect when so treated. It is thus suggested that amyloid plaques may be composed of prions which could be the cause of this disease. This has led most major research-based pharmaceutical companies to establish programs in this area. As of now, a variety of drugs has been tried and found wanting. Nimodipine, a calcium channel blocker with good cerebrovascular effects, is in clinical trials as is pentoxifylline, a vasodilator with possible effects on cerebral metabolism.

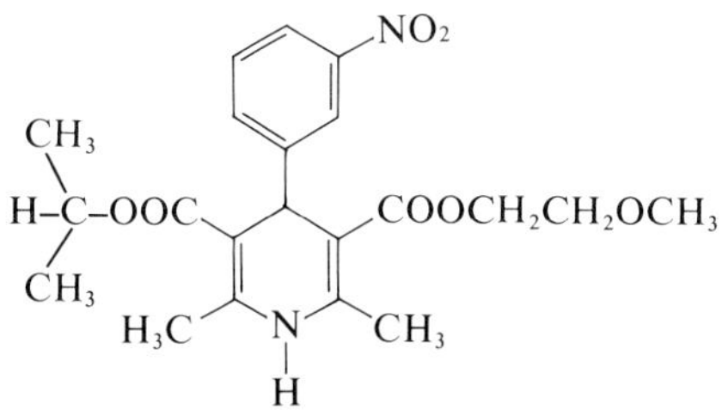

Nimodipine

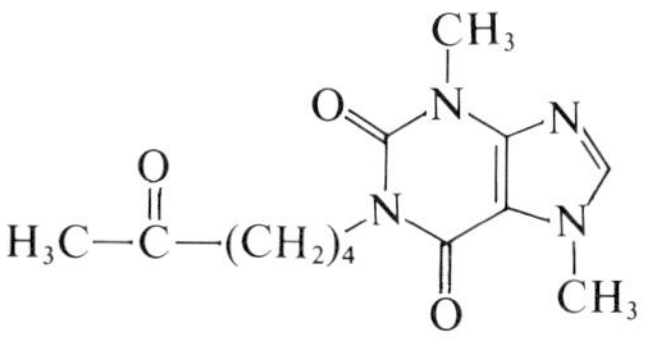

Pentoxifylline

Probably the pyrrolidinone derivatives have been most extensively studied. Piracetam and pramiracetam [4] are amongst those most widely studied. Although some improvement in the patient's memory response has been reported, it is far from certain if efficacy can be proven. The clinical problems associated with these studies are quite difficult. First of all there are inherent problems in conducting controlled studies among demented persons; also it is difficult for old and sick patients to stop all other medication while they are taking part in a trial. These factors are already compromised by the selection of the substance for clinical evaluation; that is, *there is really no satisfactory animal model for senile dementia.* Nevertheless, extensive research continues in this field since the medical needs and social benefits are immense.

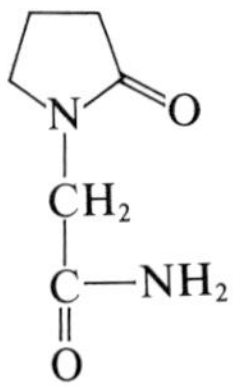

Piracetam

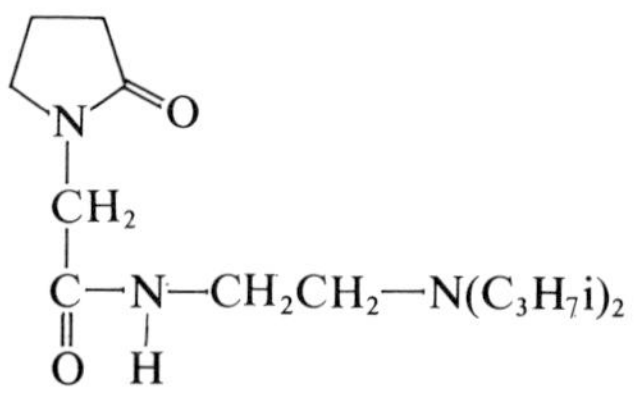

Pramiracetam

3.2 Cardiovascular-renal system

The cardiovascular-renal disease area has had considerable attention focused on it for almost 30 years. Initially research was directed toward extensive screening of compounds in animals in which experimental high blood pressure had been induced. However, as more knowledge was discovered about the central and autonomic nervous systems, their neurotransmitters, the $\alpha$and β receptors includingα_1 α_2 and β_1 β_2 subsets, and the relationship of these to the control of blood pressure, a rationale for drug design has evolved. Equally important has been the research on the renin-angiotensin system, its role in the kidney and how changes in this system can lead to hypertension. Within the past decade, calcium ion has been found to exercise a strong influence on the normal functioning of the heart and the vascular smooth musle. In fact it has been suggested that calcium should really be considered the ubiquitous second messenger instead of cAMP. Most recently natriuretic atrial factor has been identified and shown to be released from the heart in order to facilitate a diuretic action in the kidneys and also to relax the smooth muscles of blood vessel walls[1]. The consequence of these discoveries has been a plethora of effective drugs for treating hypertension and angina; β-adrenergic blockers, α_1-postsynaptic antagonists, α_2-presynaptic agonists, angiotensin converting enzyme inhibitors and calcium channel blockers. The calcium channel blockers are gaining, by far, the greatest amount of attention and it is anticipated that they will become the drugs of choice in the next decade for the treatment of hypertension. Fleckenstein [6] in Freiburg im Breisgau has done most of the basic work on the pharmacology of these substances. Others have expanded on this work. Recently, Bühler and co-workers [7] in Basel have studied free-calcium levels in blood-platelets, which have many features in common with vascular smooth muscle cells. These studies were carried out in hypertensive patients who were treated with β-adrenergic blockers, calcium channel blockers or a diuretic. Treatment resulted in a reduction in intracellular calcium and this correlated with the fall in blood pressure. According to the authors, 'calcium entry blockers probably reduce intracellular concentrations by reducing calcium influx, β-adrenoceptor blockers may act through inhibition of renin and angiotensin II, and diuretics may act by decreasing intracellular sodium levels'. They also consider it possible that increased intracellular calcium levels may be a consequence rather than a cause of elevated blood pressure. Further research will elucidate the role of calcium in the renal system and the effect on blood pressure.

That the kidney is intrinsically involved in hypertension, there can be little doubt. The paper by Curtis, Lake and Dustan [8] may prove to be a landmark. Six patients in whom essential hyptertension led to nephrosclerosis and kidney failure received kidney transplants from normotensive donors. After an average follow-up of 4.5 years, all were normotensive and had evidence of reversal of hypertensive damage to the heart and retinal vessels. These patients were then matched with

1 Matsuo and co-workers [5] have cloned the gene from the human heart which is responsible for the synthesis of the precursor peptide. This will now permit sequencing of the DNA and this information means that substantial amounts of the peptide now can be produced by recombinant DNA techniques. Although several peptides are involved in regulating the blood pressure, the detailed study of the mechanism of action of these substances will considerably enhance our knowledge of blood pressure control and also will be an important step in the discovery of more effective drugs for treating hypertension.

six control subjects (age, sex and race) and observed for 11 days for changes in blood pressure and their responses to salt deprivation and salt loading. There were no differences between these patients who had had essential hypertension and the control group. That is, mean arterial pressure among the patients who had had essential hypertension was similar to that of the normal controls and both groups had similar responses to salt deprivation and salt loading. Thus, as shown in the spontaneously hypertensive rat, essential hypertension in humans can be corrected by transplantation of a kidney from a normotensive donor. This observation lends strong support to the significant role the kidney plays in causing hypertension.

Another aspect of the cardiovascular-renal system which has been of much interest for many years is the search for positive inotropic drugs to improve myocardial contractility in congestive heart failure. At present the only oral positive inotropic agents clinically available are the digitalis glycosides. However, these have a narrow therapeutic range and a high incidence of cardiac toxicity (8–35 %).

Recently important discoveries have been made on substances which appear to offer marked advantages over the digitalis class of drugs. The pyridone derivatives, amrinone and milrinone, reported by Winthrop Laboratories, [9] appear to be the furthest advanced.

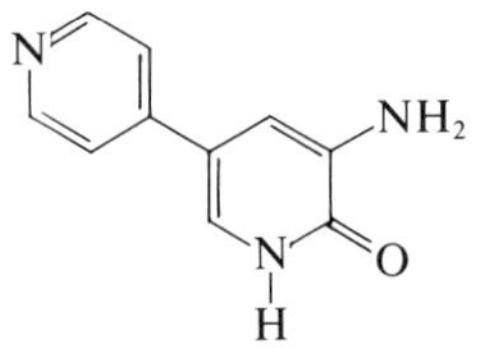

Amrinone

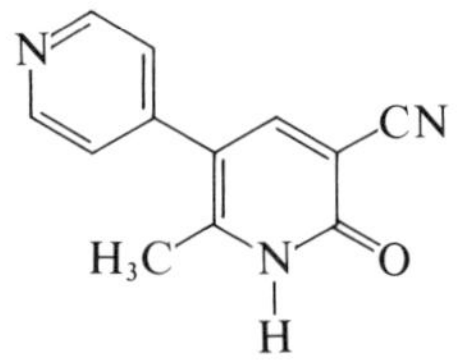

Milrinone

Amrinone has been approved in the US for parenteral administration under closely monitored circumstances, due to some toxicity associated with the drug. However, milrinone is a far superior derivative which is 10–30 times more potent than amrinone and has a therapeutic index of about 100. Extensive clinical trials are now underway. The positive inotropic effect in humans has been clearly manifested with little toxicity.

Another group of substances showing promise in the treatment ofheart failure is the imidazolones prepared in the Merrell-Dow Laboratories. The clinical efficacy of MDL-17,043 has been most encouraging in patients with congested heart failure. The papers by Uretsky et al. [10] and Crawford et al. [11] show the effects of repeated IV doses on various hemodynamic parameters. Crawford studied the 0.5 mg/–3.0

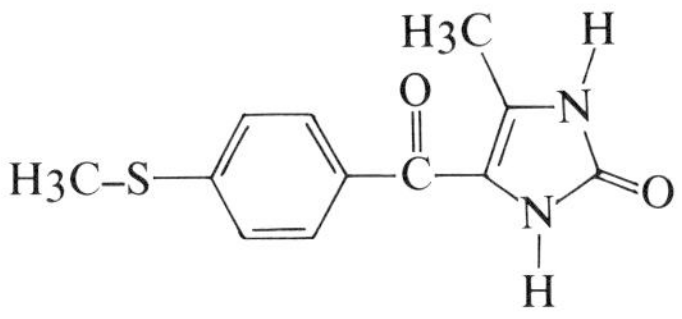

MDL–17,043

mg/kg dose range while Uretsky increased the total cumulative dose from 0.5 to 10.5 mg/kg. Both groups observed a marked improvement in cardiac output and pulmonary artery wedge pressure. Uretsky noted that the change in cardiac index persisted at 6 hours. Arbogast and co-workers [12] also have studied the effectiveness of IV MDL 17,043, but also examined its pharmacokinetic parameters. In eight patients, a comparison was made with sodium nitroprusside. Both agents increased cardiac output and decreased wedge pressure. However, MDL 17,043 increased left ventricular stroke work wheras nitroprusside did not. They suggest that MDL 17,043 produces its effects by both a direct positive inotropic effect along with peripheral vasodilation.
It is well established that dopaminergic substances can also exert cardiotonic effects, although the protoype, dopamine, lacks oral activity. Its beneficial hemodynamic actions in patients with heart failure have been attributed not only to a potent positive inotropic effect that is mediated by activation of the β_1-adrenergic receptor, but also to its agonist activity at the dopamine vascular receptor. Activation of the dopamine receptor at the renal vascular bed appears to be responsible for the strong diuresis observed with administration of this substance parenterally. Goldberg et al. [12] have reported on the effect of orally administered 1-dihydroxyphenylalanine (levodopa, 1–2 g) to 10 patients with severe congestive heart failure. Levodopa, of itself inactive, is metabolized to dopamine in sufficient quantities to cause a significant positive inotropic effect which persisted for a least 6 hours. Half of these patients were then put on long-term therapy (3 to 4 months)without ill effects. Further studies are underway to confirm these findings. The authors have concluded that the beneficial hemodynamic responses oberserved can be attributed to the activation of the β_1-adrenergic, dopamine$_1$, and dopamine$_2$ receptors by dopamine derived from levodopa.

3.3 Rheumatoid arthritis

Rheumatoid arthritis is a chronic, relapsing inflammatory condition usually affecting multiple diarthrodial joints with varying degree of systemic involvement. The primary tissues affected are the joints and their surrounding structures including tendons, bursae and periarticular subcutaneous tissues. Within the joint, the primary pathological findings appear within the synovium. The resulting pathology involves exudation, cellular infiltration and the formation of granulation tissue. Congestion and edema are greatest at the internal surface of the synovium, close to cartilage margins. Polymorphonuclear leukocytes emigrate into the synovial fluid. Small lymphocytes also infiltrate within the synovial lining.
The etiology of rheumatoid arthritis remains an enigma. Current theories center about an immunologic response in a genetically predisposed host to an enviornmental stimulus or infectious agent; e. g. virus, that is promoted or allowed to perpetuate by altered immunoregulatory mechanisms. Once the inciting agent has gained access to joint structures, subsequent immunologic and inflammatory events exacerbate the reaction in the synovium.
The present treatment modalities, namely non-steroidal anti-inflammatory drugs, do much to alleviate the pain and inflammation associated with this condition. However, they do not appear to alter the progress of the disease. It has been reported that some of these drugs inhibit monocyte migration and perhaps even lymphocytes. The inference is that they may affect the immune system. Goodwin [14] has suggested that piroxicam may be acting as such on latex titers before and after tratment. Further clinical studies are presently in progress to elucidate these findings. Thus, improvement in drug therapy in this area is highly desired. The goal is a substance which is orally effective, causing little or no gastric disturbances, no adverse effects on the liver or kidney, and at the same time capable of arresting cartilege and bone damage (the reversal of the damage would probably require other treatment modalities).

3.4 Prostaglandins and cytoprotection

It has been over 25 years since the intensive efforts on the prostaglandins were begun by Bergstrom. Although no drug of this class has resulted from this extensive research, the new knowledge that has been acquired has greatly enriched our understanding of fundamental biological mechanisms in the cells and membranes of living systems. Following the work of Bergstrom, the contributions of Vane and Samuellson have been basic and extensive; the chemical synthesis contributions of Corey at Harvard have been outstanding.

Major efforts are now directed toward controlling platelet aggregation through the synthesis of thromboxane A_2inhibitors, or the preparation of more stable and longer acting prostacyclin derivatives. Several substances are under clinical evaluation and within the next few years effective anti-platelet drugs for treating atherosclerosis may be forthcoming.

Another group of prostaglandin and prostacyclin derivatives of great interest are those exhibiting cytoprotective effects. These would appear to have an anti-ulcer action in a natural way, that is, through the protection of the mucous membrane in the gastrointestinal track rather than by controlling histamine release and subsequent acid concentrations in the stomach, (e. g., H_2 = histamine inhibitors). Some compounds of this class are:

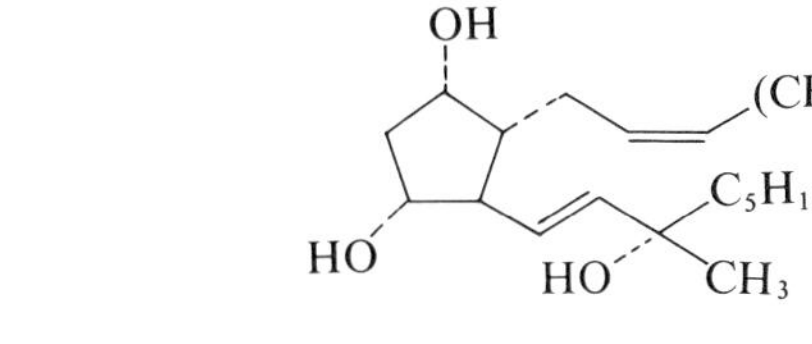

R021–6937

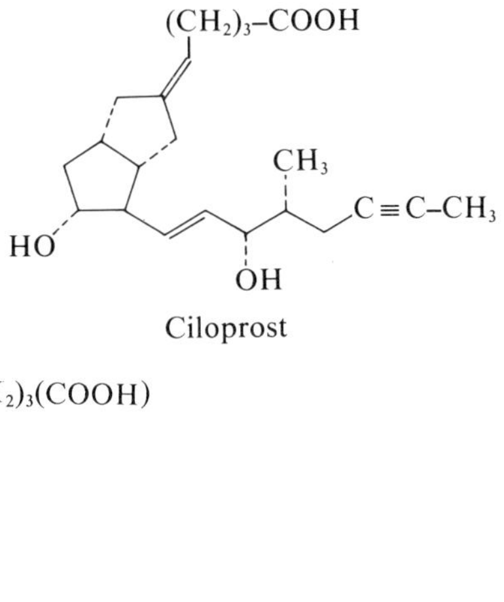

Ciloprost

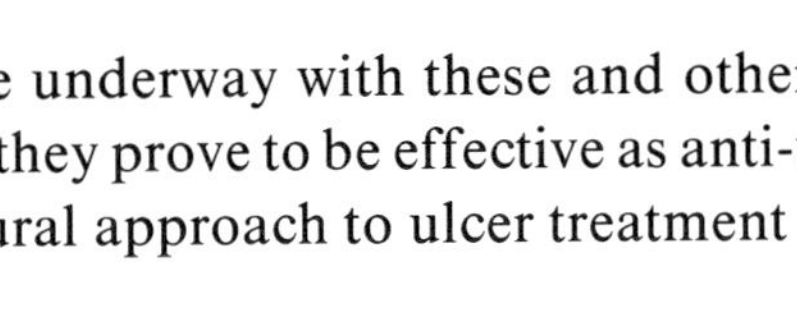

Carbaprost

Clinical trials are underway with these and other compounds of this class and should they prove to be effective as anti-ulcer drugs, they will offer a more natural approach to ulcer treatment and prevention.

3.5 Recombinant DNA, gene expression and gene structure

Within the past 15 years, the evolution of new knowledge in the biological sciences associated with recombinant DNA, gene expression and gene structure has had a profound effect on how medicinal research will be carried out in the future. Certainly synthesis and biological testing for evaluation of new lead compounds will continue to be the bedrock of drug research for a number of years to come. But slowly and inexorably the approach is changing.

Several years ago some laboratories were involved in the synthesis of natural macromolecules, such as insulin, and indeed were successful in achieving its total synthesis. However, the methods were laborious, time-consuming and prohibitively expensive. The discovery of recombinant DNA techniques now permit the facile transfer of the appropriate gene (in this case the insulin gene) into a microbial cell system resulting in the preparation of large quantities of the gene product on a commercial scale. Accordingly, genetic engineering will make available for the first time large quantities of naturally occuring macromolecules (somatostatin, interferons, interleukin-2, plasminogen activators and a large number of life-essential enzymes). In addition, specific antigens for use in the manufacture of new vaccines, can be produced via this technique. Most recently, Merck, Sharp and Dohme has successfully introduced into clinical studies the first gene-spliced vaccine against hepatitis B. In this case the Merck scientists led by Scolnick [15] joined the gene which produces hepatits B antigen to yeast cells. The multiplication of the yeast cells then leads to millions of clones of the gene which in turn produce quantities of antihepatitis antigen or vaccine. This approach to vaccine production thus eliminates the sometime cumbersome problem of starting with infected plasma.

At this writing human tests are in progress with recombinant interleukin-2 similar to that produced by white blood cells in the human body. Since this substance is known to stimulate the growth of cells that control and regulate the immune system, it is believed that interleukin-2 may be important in treating diseases caused by malfunctioning of the immune system, such as multiple sclerosis, rheumatoid arthritis and acquired immune deficiency syndrome (AIDS).

The great promise of the recombinant interferons (α, β, γ) for the treatment of cancer has now been somewhat moderated. Extensive clinical trials have shown that α-interferon does show a measure of efficacy toward a few neoplasms (Kaposi's sarcoma, renal cell cancer, and chronic myeloid leukemia), but it has failed versus lung cancer and bone marrow cancer. Its efficacy as an anti-viral agent, and specifically for prophylaxis against the rhino virus (common cold) appears clinically firm; however, except under unusual circumstances, its widespread use for this indication seems doubtful.

The explosive growth of knowledge on gene structure due to extensive application of recombinant DNA technology has espanded many times our understanding of how genes evolved and normal gene structure. Consequently, the sequence of events by which information in a gene is normally decoded into a precursor message, followed by conversion to a mature message and then translated into protein is now understood. This information will in turn enable us to discern what happens in either genetic or induced diseases where gene processing is interrupted. Our knowledge of specific disease processes will be understood at the molecular level which in turn will lead to effective cures of the diseases whether by correcting DNA sequences that give faulty transcription, or by replacing a defective gene with a normal one by cloning.

DNA sequencing will be of immense value in expanding our knowledge base in rational drug design. As the structures of enzymes and proteins containing receptor sites are elucidated in detail, then chemists with the use of the three-dimensional computer graphics systems will be able to analyze the dimensions, polarity and receptivity of the site and thus plan the synthesis of organic structures which would have the best chance of eliciting the desired biological activity. This approach is presently underway in many research-based pharmaceutical companies.

References

1 E. Eichenberger: Arzneimittel-Forschung 34 (1), 109 (1984).
2 P. Davies: Med. Res. Rev. 3 (3) 221 (1983).
3 M. P. McKinley and S. B. Prusiner: J. Am. Med. Ass. 251 (14) 1806 (1984).
4 T. M. Itil, S. Mukerjee, G. Dayican, D. M. Shapiro, A. M. Freedman and L. A. Borgan: Psychopharm. Bull. 19 (4) 709 (1983).
5 S. Oikawa, M. Imai, A. Ueno, S. Tanaka, T. Noguchi, H. Nakayata, K. Kangawa, A. Fukuda and H. Matuso: Nature 309 (5970), 724 (1984).
6 A. Fleckenstein: Calcium Antagonism in Heart and Smooth Muscle. John Wiley and Sons, New York 1983.
7 P. Erne, P. Bolli, E. Burgisser and F. R. Buhler: New Engl. J. Med. 310 (17), 1084 (1984).

8 J. J. Curtis, R. G. Luke and H. P. Dustan: New Engl. J. Med. 309 (16), 1009 (1983).
9 D. S. Baim, A. V. McDowell, J. Cherniles, E. S. Monrad, J. A. Parker, J. Edelson, E. Braunwald and E. Grossman: New Engl. J. Med. 309 (13), 748 (1983).
10 B. F. Uretsky, T. Generalovitch, P. S. Reddy, R. P. Spangenberg and W. P. Follansbee: Circulation 67, 823 (1983).
11 M. H. Crawford, S. G. Sorensen, K. L. Richards, M. T. Sodums: Clin. Res. 30, 866 a (1982); Circulation 68 (4), 372 (1983).
12 R. Arbogast, C. Brandt, K. D. Haegeli, J. L. Fincker and P. J. Schechter: J. Cardiovasc. Pharmacol. 5, 998 (1983).
13 S. I. Rajfer, A. H. Anton, J. D. Rossen, and L. I. Goldberg: New Engl. J. Med. 310 (17), 1357 (1984).
14 J. W. Goodwin: Mod. Med. 130 Feb. (1984).
15 E. M. Scolnick, A. A. McLean, D. J. West, W. J. McAleer, W. J. Miller, and E. B. Buynak: J. Am. Med. Ass. 251 (21), 2812 (1984).

A review of advances in prescreening for teratogenic hazards

By E. Marshall Johnson, Ph. D.
Daniel Baugh Institute, Jefferson Medical College, Thomas Jefferson University,
1020 Locust Street, Philadelphia, PA 19107

1 Introduction

During the past several years there has been a marked increase of interest in, and attention to, reproductive and developmental toxicology. Knowledge that reproductive capacity and embryonic development can be altered by environmental and other factors is not new though it has expanded considerably, perhaps due in part to increased public awareness and expanded research efforts. Coupled with this awareness is the fact that women are entering the workplace at an increasing rate and in a wider variety of settings. In contemporary society the fact of pregnancy does not necessarily preclude gainful employment any more than it has ever precluded the considerable work of the homemaker. Taken together, these factors tend to focus attention on considerations of reproductive effects. All parties are aware that both the primary and secondary organs of reproduction may be vulnerable to the toxic action of specific agents just as may any other organ or organ system.

Somewhat more slowly recognized has been the fact that the population consists of not just two groups – males and females – but encompasses also a third and totally unawares resident – the conceptus. Just as the adult male or female reproductive organs may be the target of some agents, so, too, are there substances with a unique predilection to disrupt the embryo/fetus. The classic example of this is the drug thalidomide. The remarkable thing about this chemical was not that it caused congenital malformations, but that it did so at dosages so far below those acutely toxic to the mothers. The ability to selectively disrupt development without producing overt signs of toxicity in the mother is unusual but not totally unique to thalidomide. Agents such as this must be considered as primary developmental hazards because the 'target' most vulnerable to their action is the embryo. Many other agents may also interfere with embryonic development but the majority do so only at dosage or exposure levels high enough to also make the mother acutely ill. These become developmental hazards if the pregnant female is exposed to levels high enough to affect herself. The manifestations of toxicity she experiences may be tolerable or transient and disappear as the offending agent is eliminated, but the conceptus may have suffered irreparable damage to its carefully balanced developmental sequences. Such agents are coaffective teratogens in that they affect both the adult and the conceptus at generally similar dosages.

2 Need for prescreening

In the mid-1960s, a series of toxicological safety evaluation protocols was developed in an effort to improve and standardize the means whereby adverse effects on reproduction and development could be detected. There were three experiments and, with but minor modification, they are still the standard experimental developmental toxicity safety evaluation tests. Propagation of the species is divided into three parts or segments by these protocols so that adverse effects in each portion of procreation can be examined. The first or Segment I study provides for administration of the test agent during the time of gamete formation in both and/or either the male and female test animals (usually rats) and entails evaluation to reveal adverse effects such as reduced fertility. The second standard experiment is the Segment II evaluation which necessitates administration of the test agent to pregnant females (usually rats and/or rabbits) throughout the period of embryonic development. Treatment begins in each species at the time of blastocyst implantation into the uterus just as the three primary germ layers of the embryo become established. It terminates at the end of major organogenesis. In rats this is generally from about day 7 through day 17 of their 22-day gestation period. The dams are killed just before expected delivery and their young are examined for external and internal gross anatomical normalcy of both soft tissues and viscera, as well as osseos and occasionally cartilaginous skeleton. The last or Segment III evaluation calls for treatment to begin after the major organs have formed but as the fetal stages of development begin (about day 17 in rats). Exposure continues through and even beyond natural delivery and may include all or part of lactation. This is the experiment most frequently employed to test for postnatal behavioral or other functional effects.
The three-segment series of studies, supplemented in special circumstances by a multigenerational experiment, is the current and state-of-the-art means for using animal surrogates to detect potential hazard to human reproduction and development. The Segment II protocol is the most frequently executed of these experiments and is generally considered as a powerful tool in that all, or virtually all, agents known to adversely affect human embryos also adversely affect animal embryos exposed at these stages of development. One might then logically ask, why seek another test system? There are several reasons for this and

among them are that the use of standard test assays in pregnant animals are quite expensive and demand extensive facilities and expertise for their adequate performance. Also, there are well in excess of 50,000 chemicals already in commerce and at least several hundred new ones are added each year. In contrast, there are perhaps less than 100 Segment II evaluations made in the world each year and it would be impractical to significantly expand this effort. We are, therefore, falling increasingly behind in our efforts to protect the conceptus from adverse effects due to exogenous agents.

3 Goals of prescreening

A sometimes stated goal for a developmental toxicity prescreen is to serve as an alternative to testing in animals. This goal is unlikely to be attained, though some testing, at least in pregnant animals, could be obviated by a prescreen. An essential part of standard toxicologic testing in animals is to determine the no-observed-effect level (NOEL) defined as that exposure level that produces no dose-related signs of adverse effect in a sensitive animal species examined by contemporary means. The NOEL is markedly influenced not just by the animal species examined, but also by the route and duration of dosage within the species. In the case of the conceptus one must be aware that differences in maternal absorption, transport, metabolism, elimination and placental character, as well as conceptus genotype and developmental stage will all influence the nature of effects seen in the embryo. It is doubted that a streamlined short-term test executed in vitro can account for all of these variables and reliably predict a whole-animal NOEL. Though it is highly unlikely that an artificial system will be capable of encompassing this many variables, the goal of *pre*screening can still be met. This is because determination of a whole-animal NOEL is not the goal of a prescreen. It is the goal of the standard three segment evaluations using pregnant animal surrogates. Their NOEL is determined and a safety factor is applied to establish the permissible human exposure level. Present safety evaluation methods have proven notably capable of predicting human developmental toxicity. Statements to the effect that they are inadequate are just not correct [101]. Unfortunately, it is not possible to examine all of the new formulations, much less the substances already in commerce by these elaborate means alone.

What is needed is a *pre*screening system that will serve as a means to prioritize chemicals so that only those needing detailed developmental toxicity testing are actually evaluated in pregnant animals. Some explanation of the problem is in order. Figure 1 illustrates the types of

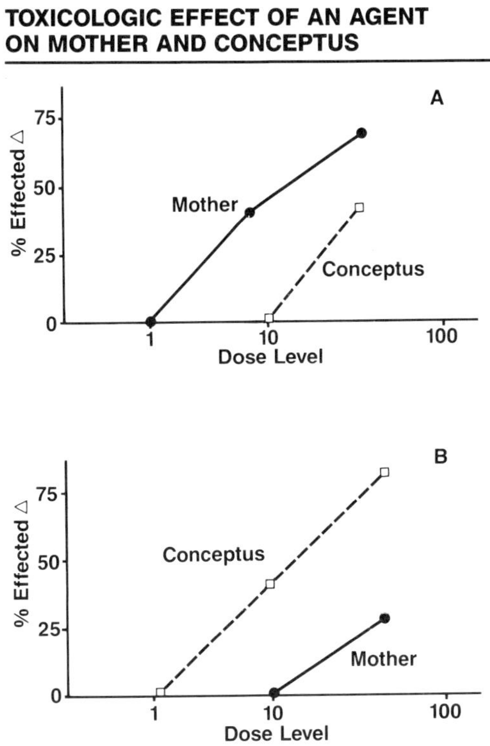

Figure 1

A pair of hypothetical examples of the types of relationships known to exist between dosage needed to adversely affect adult and embryonic animals examined in standard developmental toxicity safety evaluations. See text for detailed explanation.

data necessary in order to determine whether or not a particular agent is more dangerous to the mother or to the embryo. In the instance depicted in figura 1 A, the mother is more vulnerable to the test agent than is the conceptus. If her NOEL is not exceeded, the threshold of embryonic toxicity is not either. Note, however, that should maternal exposure become excessive, the embryo may be affected. In figure 1 B, which is more akin to the situation with thalidomide and other primary developmental hazards, the embryo is the more sensitive target. The basis for establishing a safe exposure level must, in this situation, be based on development of the embryo. Figure 1 A shows the relationship between adult and developmental toxicity most frequently encountered when data from Segment II-type experiments are plotted. If a prescreen could identify the type A (coaffective teratogens) and type B (primary developmental hazards) agents, it would have immediate and significant utility. This is a valid and presently attainable developmental toxicity prescreening goal. It ignores the NOEL so essential for risk estimation and concentrates instead, and much more realistically, on the nature of the test agent itself, i. e., its hazard potential.

The difference between hazard and risk are considerable. Hazard is the intrinsic character of an agent as regards its ability to produce an adverse effect. Risk is the probability that an effect will occur from a particular exposure. Though use of the terms differs, it is of paramount importance to keep the concepts clearly delineated when discussing prescreening systems for developmental toxicity. The goal of risk estimation is to calculate the probability of an effect occurring at a particular dosage. The starting point for quantification of risk is the no-effect, or more pragmatically, the no-observed-effect level. The complexities of the intact mammalian maternal/placental/developmental unit far exceed what can be anticipated of simplified systems. Without these complexities, however, NOEL cannot be determined, so a prescreen is of scant use for risk estimation. From such methods a NOEL relevant to another species and treatment protocol is difficult to imagine. It follows that NOEL determinations cannot be the goal of a prescreen for developmental effects. Prioritization of agents based on their effect or no-effect level is unlikely. Some excellent and important writings miss this point [112] and, while discussing in vitro systems, concentrate too much on intact systems and effect levels that are considerations apropos of live animal testing such as is achieved in the Segment II evaluations. A simple prescreen performed in vitro cannot determine risk but it can determine degrees of hazard.

Hazard is the intrinsic nature of the agent and need not involve consideration of route of administration or placental permeability, for instance. Attempts to assess the developmental hazard of chemicals have been attempted along two lines. The first is identification of substances as either teratogenic or nonteratogenic. Such designations are difficult validation goals for an in vitro assay. Even if a reference is found indicating that a particular test agent has or has not been shown to produce live abnormal offspring, another may be found or may be published later showing the opposite. Species, routes of treatment, timing and duration of exposure so markedly affect outcomes that most any substance must be considered capable of somehow affecting development [53] yet if exposure is low or brief enough, the conceptus may suffer no ill effects. The second involves determination of the magnitude of the difference between the dosage of an agent needed to produce adverse signs in the mother and that needed to disrupt embryonic development (compare fig. 1 A and 1 B).

4 Whole-embryo methods as prescreening assays

4.1 Mammalian embryos

The in vitro cultivation of rodent embryos began over a half century ago and was firmly established as a somewhat routine technique for study of mammalian development within the past decades [85, 86] after having been fully described in 1966 [84]. In the basic technique [86, 90], timed pregnancies are produced in either rats or mice and the dams are killed during week two of gestation. The embryos are collected from the uterine horns either with or without their associated yolk sacs. Rodents have a functioning yolk sac circulation, so these embryos have a dual 'placental' circulation. The chorioallantoic circulation is established by day 11 in the rat (day of finding evidence on mating taken as day 0 of pregnancy) and becomes the dominant avenue of maternal/fetal exchange by the end of week two. Prior to this time the embryo relies to some extent on the somewhat unique circulation passing to and from its visceral yolk sac which incidentally is also subject to alteration by a teratogenic agent [50, 51]. The rodents' visceral yolk sac is covered by an elaborate columnar epithelium supported by a loose mesoderm containing an elaborate set of embryo-derived blood vessels. The entire structure is of embryonic origin and in conjunction with the chorioallontoic placenta, achieves either a complementary or supplementary exchange directly with the contents of the uterine lumenal contents. Its retention for in vitro studies may aid in gas exchange in vitro and thereby overcome some of the problems associated with oxygenation of these rather large embryos [81, 82].

Freshly harvested embryos are placed in a partially defined media containing, in addition to the quantitatively-known supportive constituents, serum and appropriate antibiotics. The culture vessels' gas phase is oxygen, nitrogen and carbon dioxide at a ratio of about 1:18:1 and the culture apparatus is maintained at 37°C while being rotated constantly at 20–30 rpm to facilitate gas exchange to and from the liquid medium.

In addition to prescreening trials where correlation between in vitro and in vivo outcomes are reported [91], this system has provided excellent material for the study of in vitro metabolism of xenobiotics achieved through addition of activated hepatic microsomes and the necessary cofactors. Mirkes et al. [76] used it to examine the likely products of cyclophosphamide metabolism for their effects on embryonic rat development in vitro. The results obtained were different from those previously reported [72] from embryonic mouse limb buds in culture in some regards, but were analyzed in a manner illustrating the considerable value of this technique. Also by this means, it was demonstrated [30] that cyclophosphamide did indeed require maternal metabolic activation to produce its teratogenic action. Furthermore, through measurements of the embryos' total protein, size and developmental stage (somite number), as well as types of anatomical developmental abnormalities, the authors developed a convincing group of data regarding the relevance of one purported teratogenic end product of cyclophosphamide metabolism.

The nature of the media used in each study merits close attention, as its composition may itself affect the outcomes observed [89], and growth in homologous serum usually is taken as the normal or bascline level for comparisons in these studies. The embryos are scored by numerous means such as external morphology or chemical composition and a recently developed comprehensive scoring system by Brown and Fabro [19] may facilitate future interlaboratory comparisons of data. These and other investigators, e. g., Grennaway et al. [33] have used embryo culture to advantage in a variety of ways. Of particular utility is their demonstrated ability to permit investigators to closely observe the sequence in pathogenesis of specific patterns of malformations due to selected agents.

Though it is often stated, and correctly so, that the maternal organism is an essential part of developmental toxicity and that her status markedly influences the conceptus [57], elimination of maternal variables can prove advantageous. Kitchen et al. [61] illustrated this facet of embryo-culture not only for examination of potential mechanisms of action of a heavy metal and the similarities of in vivo and in vitro outcomes, but also to initiate studies of protection. In the absence of the maternal/placental contribution they were able to directly add glutathione to the culture medium and at least partially prevent the expected effects of mercuric chloride addition. The comparative examination of in vitro and in vivo effects has been exploited also on numerous occasions to evaluate the effects of specific agents [77, 78, 105, 106] on embryonic morphology as well as composition [65, 66, 92]. The system is not restricted to testing only water soluble agents, as means for solvent use are available [60].

In an intriguing series of experiments, rat embryos were cultured in blood serum from patients treated with specific pharmacologics [21, 22] by Klein and associates. The outstanding aspect of this approach is that it may be capable of detecting specific exposures [64] or individuals [63] at risk for adverse developmental outcomes. Though whole embryo culture is a complex technique and may not serve as a prescreen [62], it merits vigorous investigation as a means for examining mechanisms of abnormal development. This would be a basis for continued study even if there were not the additional potential of examining effects of human sera of varying composition and from different settings. Because it may be capable of such, the system is unusually interesting.

4.2 Preimplantation embryos

Preimplantation embryos are markedly amenable to culture and in vitro manipulation and their application to problems of safety evaluations has been discussed in some detail [17]. Embryos of this age consist of relatively few cells which are of somewhat restricted diversity, though possessing tremendous potential. Injury or even death of one or a few of these cells is rapidly overcome by replacemant with newly divided cells and, perhaps because of phenomena such as this, congenital abnormalities are difficult to produce from agents acting this early in gestation. It is possible, of course, to severely injure an early embryo and once its repair capabilities are exceeded, it dies. On the other hand, these highly responsive organisms [15, 16] could eventually become part of a general reproductive toxicity testing system to examine for agents interfering with the developmental stages between fertilization and implantation [29].

4.3 Chicken embryos

The most widely used embryo has been that of the chicken. This is a classic system. Its in ovo development is readily available for treatment and study when either explanted, via 'widowed' eggs, or by means of injection into the subgerminal yolk or chorioallantoic vessels. Quantification of the relevant dosage is a major limiting factor in this system [58] but embryonated chicken eggs have been used extensively for mechanistic studies [54, 70, 71, 79] and have proven highly responsive. Actually, their responsiveness is perhaps their greatest difficulty as a prescreen. Embryonated eggs react positively to most substances. The toxicologic responses of avian embryos are markedly dose-dependent [26] but those treatment levels needed to produce abnormal development bear no clear relationship to those levels overtly toxic and lethal. An excellent study by Verrett et al. [107] clearly illustrates this point. In an exhaustive study of many chemicals it was established that the treatment dose, route and the developmental stage all markedly influenced the outcome. Based on their results the authors conclude that the chicken embryo reacts in a specific manner to test agents, and, though their data provide several instances of specific responses, it is difficult to relate these to those of other developing systems. Jelinek has also carefully studied the chicken embryo as a potential prescreening method [40]. His experiments demonstrate specific effects and no-effect levels correlated in a dose-related manner also correctly dependent on the embryonic stage treated.

Some doubt has been expressed [109] regarding this system, but Dr. Jelinek [39] holds that the reasons stated for not using the chick egg for teratology testing are either incorrect or overcome by recent refinements of technique. His point is well made and it does appear that two of the three reasons advanced for not seriously considering this system are really off target. They were based on the assumption that the chick egg could be used for NOEL determinations; a highly unlikely occurrence for such a closed system. The third reason is still germane however. The chicken egg markedly overestimates the hazard potential of test substances. As presently proposed it raises too many alarms and would, perhaps, dilate attention from the relatively fewer exogenous substances actually hazardous to embryonic development.

The Chick Embryotoxicity Screening Test (CHEST) is interesting and worthy of additional study. Worthwhile avenues for such perhaps would produce: (1) a uniform test method capable of testing even in the absence of pre-existing toxicity data regarding the test agent, (2) a definable means for determining the limits of dosage to be tested, and (3) a fixed relationship to data from pregnant animals actually predictive of what is known to occur due to the action of well-tested agents. In contemporary usage it is a largely neglected tool for studies of teratogenic mechanisms where it can be markedly useful [52]. Its applicability for prescreening and hazard prioritization remains to be published in a form allowing one to become confident of utility. Such comparative attempts [74] have been reported in some detail but whether the experimental outcomes and doses employed [40] to achieve them have practical utility is difficult to perceive.

4.4 Other whole embryos

The intact embryos of other species also have been examined for their potential as a developmental toxicity prescreen. Streisinger [103] used zebra fish. Their potential for examining effects on behavior and reproductive capacity in clones of homozygous fish is particularly striking, but the system needs further study before definitive statements are possible regarding its potential as a prescreening test.
Dumont and Epler [28] examined numerous substances for their effect on frog embryo development. They use a teratogenic index or TI to rank test agents. The TI is the EC50 (concentration of test agent causing 50 % of the embryos to be abnormal) divided by the LC50. Substances with large TIs are considered as teratogenic hazards while a TI near unity is taken as a coaffective teratogen [41]. The assay has only been validated on the basis of whether each agent was a 'teratogen' or a 'nonteratogen', which is less than useful. It, nevertheless, has marked potential of being able to predict the A/D ratio of mammals. If this is possible, it would have applicability as the primary developmental toxicity prescreen in that it may also have some predictive ability regarding the organs or organ systems most vulnerable to test agents.

5 Organ culture methods as prescreening assays

5.1 Limb buds

The most widely used organ culture system has been that of limb buds. This method shares many of the attributes of whole embryo culture but does have some advantages of its own. A little discussed but perhaps valid consideration is that the smaller and less thick limb bud preparation probably is better oxygenated than some of the more deeply positioned organs of an intact embryo. The extent to which this is overcome by retention of the yolk sac in whole-embryo culture has yet to be determined. The use of cultured limb bud cells certainly overcomes this problem and such use has been reported [38] for cells of prechondrogenic limb bud mesenchymal cells.

The basic method of limb bud culture involves production of timed pregnancies in mice or rats. The dams are killed early in the second week of pregnancy and the embryos are retrieved. Just as in embryo culture, their external morphology can be used to determine their developmental stage and status prior to use in an experiment [1]. They are cultured in media of various types to which the substance of interest is added directly. Instead of floating free as is common in embryo culture, limb buds are more typically cultured on a raft or platform at the gas/media interface [3, 72]. At the end of the desired period of exposure the developing limb can be removed for analysis and study or perhaps continued in fresh media lacking the test agent. A variation on this theme is treatment of the dam prior to removal of the embryo [5, 31] or its limb buds for in vitro culture. One possible advantage of this approach could be metabolic activation of the test agent by the intact mother with thc explant being used to study subsequent development in the more readily accessible in vitro situation.

The technique has been used in numerous excellent studies of teratogenesis and events related to the abnormal biochemical [95, 104], and morphologic [68, 69, 84], developmental sequences produced of specific agents [4] both with [73] and without metabolic activation [67].

6 Summary of embryo and organ culture methods

It is difficult to perceive the use of either whole embryos or organ culture of limbs as a prescreening method. There are a great many differences between in vivo and in vitro circumstances which seemingly preclude economical evaluation of the relationship between exposure levels toxic to adults and embryos. These differences will render NOEL determinations unreliable if they are determined only in vitro. There may, however, be considerable potential for their evolving into specific predictive assays of a more detailed and specific nature such as are being explored by Klein and associates. An extension of this may lead to exploration of the effect of specific blood levels and outcomes of sera used either for embryo or for organ culture. Either by this means or through direct addition of a test substance to the media or embryo [105] it may be possible to equate specific levels with specific outcomes and correlate these back to human experience. This latter has been attempted by Brown et al. [18] who sought to determine levels needed to achieve effects. This was achieved with some success even if effects seen in humans were somewhat different from those produced in vitro. Embryo and organ culture may also be of marked utility for production of syndromes from those agents that apparently are more restricted in their action and apparently are limited to affecting only particular organs. Studies such as these could prove of considerable value when constructing human epidemiological studies by serving as an indicator of adverse outcomes possibly related to an otherwise untested agent. This is not to denigrate use of in vitro systems for more broadly acting agents, since elegant studies of developmental pathogenesis are possible with them also [2]. Last, but not least, is the possibility that tests for weak teratogens [6] that rely on specific organ development in vivo might be markedly expanded through application of organ culture techniques.

7 Cell culture methods as prescreening assays

7.1 Binding to lectin-coated surface

Use of individual cells as a prescreening technique would be highly attractive because of its potential low cost. Braun et al., developed this technique which was used [11–14] to examine for a correlation between teratogenic chemicals and the response (attachment) of cells to lectin-coated surfaces. In this technique, mouse tumor cells were incubated in the presence of lectin-coated plastic disks. A selected concentration of a test chemical was added to the medium and incubated for a short period. Since the tumor cells were radioactively tagged, the numbers which attached to the disk could be counted at the end of the incubation period. The authors reported a striking correlation between substances which inhibit cell attachment and agents which they feel are reported to be teratogenic in mammals.

This is an intriguing system perhaps capable of forming a part of a prescreening matrix if it could be validated and the system merits careful validation vis-à-vis a documented mammalian developmental toxicity data base. The compounds tested to date often lack such documentation and most of the designations of substances as teratogens or nonteratogens are based on secondary, compendia-type references. These are inadequate because they are not intended or analyzed for validity but only to convey the conclusion of the original author regardless of his/her studies' thoroughness. A second area of attention is represented by EDTA. It is reported as a noninhibitory drug in the assay in that it does not inhibit tumor cell attachment. It is, therefore, classed as a nonteratogen accurately identified. The reference cited for EDTA not being a teratogen [94] is of an excellent study involving EDTA where no abnormal offspring were encountered at the doses and administration route tested. Another report, however, using different methods and doses of EDTA, demonstrates that the compound is indeed capable of producing structurally abnormal offspring in rodents [59]. On the other side of the coin, substances which are listed as being inhibitory of cell attachment and, therefore potential teratogens, are referenced to techniques where the means by which the compounds have produced abnormal development is in chicken eggs or when injected directly into mammalian embryos. When these considerations are examined carefully and added to the cytotoxic drugs selectively excluded from the calculations, the technique looks as though it may be somewhat better than 50 % accurate. That is, it is capable of detecting agents which have already been tested in pregnant mammals and shown, at some dose, route, etc., to interfere with development of some mammalian system. If the technique could be applied at some standard dosing regimen and then correlated with a fully and carefully evaluated literature, it theoretically at least should have the capability of detecting agents which act through cell/cell interactions mediated via membrane configuration.

7.2 Human embryonic cells

Human mesenchymal cells considered as the undifferentiated fibroblast-like cells derived form the unfused palatal shelves of a human abortus were grown on a standard medium containing fetal calf serum and antibiotics [88]. The cells were plated at a fixed density and after 24 hours the medium was changed and the cells continued in culture for an additional 72 hours in the presence of graded concentrations of test agents after which they were harvested and counted. With increas-

ing concentrations of test agent, fewer cells were counted after exposure. The assay provided excellent dose response relationships between test agent concentrations and cell number. The test agents were ranked according to their toxicity and they clearly differed from one another in this parameter over several orders of dosage magnitude.
The assay is proposed as a complement to the cell binding assay of Braun et al. [11, 14] to test agents considered to be cytotoxic and a major effort was begun last year by the National Toxicology Program (US) to evaluate this system jointly in two different contracts (RFP–NIH–ES–83–50018). Determining that one substance is more toxic than another is a valuable tool when applied to embryonic or other cells. The relationship of this parameter to teratogenicity may become evident when the ongoing studies are completed.

7.3 Neuroblastoma differentiation

Neuroblastoma cells of murine derivation, grown in a standard tissue culture medium supplemented with 7.5 % fetal calf serum were evaluated for their predictive value by Mummery et al. [80]. The cells were plated into wells to which test concentrations of 62 agents were added. Control or untreated cultures in serum-containing medium grow to confluence by day 7 and 1.5 % show evidence of neurite induction. The assay seeks to take advantage of this and of the fact that removal of the serum from the medium induces a larger proportion of the cells to differentiate. A positive response was indicated by an increased percentage of the cells forming neurites on addition of the test agent to serum-containing medium or, alternately, the agent preventing neurite formation from occurring when tested in the absence of fetal calf serum in the medium.
Thirty-nine or forty of the agents tested were identified by the authors as teratogens, 18 were considered as nonteratogens and four or five were not recorded. The basis for selecting the test agents examined and for designating them as either teratogens or nonteratogens was reported as being drawn from lists such as those of Schardein [93], Shepard [100], Council for Environmental Quality [27] and Braun [11]. The criterion for a positive response by this system was interference with differentiation *expression*. This could be evident in either of two ways in this system as indicated above. The first is induction of neurite formation by a chemical in the presence of fetal calf serum and the second is

its prevention by the agent in the absence of calf serum. All experiments were repeated three times. Variability in the affective doses from one experiment to the next apparently did not hinder interpretation of the results. An unexplained order of magnitude difference sometimes existed between the dose interfering with differentiation expression evaluated as either the induction or suppression of induction by agents tested with and without fetal calf present.

The authors compare their results with those of Braun et al. [11, 14] and conclude that though the neuroblastoma cell test identified some agents as 'teratogens' or 'nonteratogens' differently from the lectin binding assay, the final percentage of false positives and false negatives compared favorably. The basis for considering these two assays together is interesting because each suggests that it is founded on membrane-based phenomena, yet each evaluates somewhat different aspects of this embryologically significant [87] organelle. The combination of studies is intriguing and could be exploited for some types of mechanistic studies, and the extent of their value in prescreening may become evident when compared with a more precisely analyzed toxicologic data base.

7.4 Neural crest cells

Neural crest cells were obtained from explanted stage 9 chick embryo neural tubes and evaluated as a potential prescreening method. This technique [110] was evaluated by testing a number of chemicals [34] at exposure levels based on serum levels known to occur in animals in the presence of teratogenesis or, in the absence of such data, at a series of preselected concentrations. Neural crest cells are known to undergo elaborate differentiation in vitro and have been used productively even for studies of specific syntheses [35, 36]. When these cells were cultured with specific agents, a variety of morphological effects were observed over a wide range of exposure levels. The author concludes that this system is a relevant model of practical use in studying mechanisms of teratogenesis.

8 Summary of cell culture methods

These are potentially useful methods warranting evaluation in comparison to a well-analyzed mammalian data base, as there may be a number of agents that disrupt embryonic development by means of membrane-mediated factors. Correlations reported to date from these assays are frought with uncertainties resulting from less than full use and analysis of existing mammalian data.

9 Subvertebrates as prescreening assays

9.1 Drosophila

The embryonic cells of the extensively studied fly *Drosophila* have been examined as a potential prescreening assay for several years [9]. Bournias-Vardiabasis [10] isolated embryonic *Drosophila* cells which were plated and after they had attached to the culture vessel, were exposed to a medium containing test agents at selected concentrations. Dose levels were based on the adult fly LD50 dosage ('fed to adult females') in an attempt ot detect developmental hazard as distinct from coaffective teratogens. When left undisturbed, such preparations normally differentiate myotubes and neurons within about 24 hours. Within a particular test dish, control cultures contained from 200 to 400 ganglion and 100 to 500 myotubes based on calculations from fixed microscopic fields of view. Test agent concentrations of from 10^{-3} to 10^{-6} M were scored according to both the number of myotubes and number of neurons differentiated in their presence. If the number of either was statistically significantly different from controls, the test agent was considered as a teratogenic agent. This is a system being virogously explored and numerous related issues have been carefully investigated by these authors.
The animal strains employed for most of the testing were Oregon R, Canton S109 and Canton S. This may be an advantage for examination of genetic and perhaps biochemical differences associated with differing mechanisms of abnormal development. For prescreening purposes, however, one notes that the three strains have strikingly different responses to some agents over a series of doses. Some responded quite differently from others to an extent markedly influencing and confusing the issue of which classification was appropriate. These authors have very carefully avoided cell lethality from becoming read as a positive response by testing at, and for, various transient periods during the total test duration. This is potentially an important means to explore mechanisms of action if the exact developmental events taking place in the cells at specific times could be determined. The authors in-

dicate that they have examined the original literature for the agents selected and, therefore, report as teratogens only those substances actually reported by others to have demonstrated capability to produce clear gross anatomical abnormalities. The list of references for this, however, is drawn primarily from secondary sources [93, 100] instead of from carefully evaluated primary references. Examination of the system's degree of validity, therefore, requires the reader to make an original literature search of each substance tested. A significant advancement of the test is that it examines for effects on differentiation at a fraction of the adult LD50 dose level. Should the test prove useful, this would reduce markedly the incidence of false positive designations. The means for determining the LD50 in the adult fly, ingesting, e. g., sugars or poorly absorbed substances, warrants further explanation, however.

Another means for employing *Drosophila* as a test system was described by Schuler et al. [96]. These investigators fed larvae of a standard wild-type strain *Drosophila* on the test chemicals of interest. After the larvae matured into adults, the flies were collected as they emerged and were examined for gross external abnormalities. The responses of the developing larvae, manifest as adult fly morphology, were notably elaborate and the authors raise a series of valid cautions in the utility of this test which may or may not preclude its actual use as a prescreen. The very straight-forward nature of the exposure is an advantage and permits transient or stage-specific treatments. The major difficulty in this method for prescreening may be related more to relevance of dosage between larvae immersed in and ingesting a test agent and levels needed to disrupt mammalian development. It is too early in the evaluation stage to even initially estimate the system's potential utility. Perhaps experiments such as this study will serve as a stimulus for others to examine other and diverse animal genera by similar means.

9.2 Planaria

These relatively simple aquatic flatworms have been studied and reported on extensively by those interested in both behavior and development. When cut in two, the absent segment is regenerated. The sequence of this regeneration is a fixed pattern normally but it is readily disrupted by experimental manipulation. Systematic interference with, and manipulation of, this organism has long been a valuable research method, particularly in developmental biology. During the course of such studies and partially as a result of studies initiated to explore its applicability to problems of toxicity [7], planaria have been exposed to a variety of chemical substances of interest to experimental teratologists.

When amputated (or intact) planaria were treated with graded concentrations [8] of selected chemicals, a concentration of each was found that would adversely affect the animals. Reactions to various substances appeared to have a degree of substance-specificity which, if documented further, would markedly enhance the method's probability of eventual applicability. This valuable animal has already proven of great utility for studies of normal developmental mechanisms. It could be that it may serve equally well for studies of abnormal development and its systematic evaluation as a possible developmental toxicity prescreening assay should be a high priority item. An early thought in this task might be that it could provide the adult/regeneration-toxic dose relationships of specific agents. The ability to detect agents disruptive of development at a fraction of the dosage toxic to adults may or may not be possible with planaria. It may be that its remarkable sensitivity to insult constitutes an impediment to this goal. The data presently available give some indication that both the adult and regenerating animals are always affected at nearly the same exposure levels. Even if this were to prove true, it is possible that some other means could be divised or discovered for it to predict responses of pregnant mammals.

9.3 Hydra

Hydra attenuata is a fresh water coelenterate largely uncontaminated by confounding algae and fungi. When adults are dissociated into their individual cells, the cells can be randomly repacked or associated into small pellets by gentle centrifugation. Adult hydra consists of two basic classes of cells, fully differentiated or committed cells, and undifferentiated or pluripotent cells, so these reaggregated clusters consist of a random mixture of these types. If left undisturbed, these preparations undergo total and whole-body regeneration of new adult hydra in a few days. In doing so they employ most of the developmental mechanisms utilized by any ontogenic system. The early stages of cellular spatial orientation, migration, recognition of neighbors, and pattern formation apparently are achieved by the fully differentiated cells, whereas later developmental events such as induction/response may rely more on the pluripotent cells. Because of their complex differentiation, these reaggregates are called artificial 'embryos' [47]. When they and intact adults are exposed to test chemicals, the lowest concentrations capable of injuring adults and 'embryos' can both be determined in a dose-related and stage-specific manner [46].

From chemicals already adequately evaluated in standard rodent and rabbit Segment II safety evaluations, as diverse a group of chemicals as possible was evaluated [48] in the hydra assay (US patent 4346070, Thomas Jefferson University) and the minimal affective concentrations (MAC) for both adult and developmental toxicity were determinded for each substance. The MAC was taken as that lowest concentration of test agent producing a predetermined toxicity endpoint assay. The concentrations affective in the hydra assay generally bore no relationship to those affecting embryonic development or adult homeostasis in the more common laboratory animals. However, their relationship to one another did. The adult MAC divided by the developmental MAC provides an A/D ratio. The larger the ratio, the greater is the chemical's propensity for disrupting development in the absence of effects on the adult. An A/D ratio close to unity indicated that the test substance was capable of interfering with development but only at exposures high enough to also injure adults. These ratios were compared with those of vigorous Segment II evaluations of rodents and rabbits published in the open literature and were remarkably similar. The actual doses differed greatly just as they may when the same substance is evaluated in multiple species or by various routes of administration, but the ratios did not. If one uses the endpoint assays of developmental and adult toxicity common to the Segment II protocol, each substance tends to have an A/D ratio largely independent of species and route of administration.

The A/D ratio, therefore, is indicative of the chemical's nature or, in this regard, developmental hazard potential. Though most any substance may be capable of interfering with development, very few (the primary developmental hazards) do so at a dosage markedly below those toxic to adults. Those with smaller A/D ratios may injure the embryo but they do so only as an adult toxic exposure level is approached. These are the coaffective teratogens which are substances not primarily targeted on development, yet capable of injuring it at very high dosage.

The hydra assay is of limited utility. For the one goal of separating the developmental hazards from the coaffective teratogens in a quantified and ranked means it is effective. Its ability to test water insoluble agents is modest [45] though perhaps significant, and its use for substances requiring metabolic activities is undocumented. If there are substances producing developmental toxicity only as a direct result of a change in normal maternal physiology [20], hydra (and probably all in vitro systems) would be anticipated to be an inadequate test. Just as most potentially useful systems, the assay has marked long-term potential for studies of mechanisms [24, 25] and interactions between otherwise unrelated agents and has already been used to evaluate critical periods and events of development vulnerable to specific agents.

10 Summary of subvertebrate forms

For detection of hazards these systems are already established in one instance and highly likely of being developed in others. Their utility resides on the commonality held to exist among embryos of all species. Embryos do things on a large scale that adults don't. For example, pattern formation is crucial for organogenesis of every organ though it is perhaps most evident in limb bud growth. Initiation of pattern formation is not an obvious phenomenon of adults. Embryos and adults do, however, share more basic biologic factors than not. Cell division or maintenance of cellular respiration, for instance, are expected at both stages of life and a substance capable of interfering at one stage will interfere at another. Substances of this type may prove to be the coaffective teratogens while those more selectively altering developmental phenomena may be the primary developmental hazards. The basis for the latter being more hazardous (or at least causing adverse effects in embryos at lower dosage than is needed to affect adults) to embryos is possibly because some poorly understood yet crucial developmental event is uniquely vulnerable to their action.

11 An in vivo assay

Though not strictly prescreening assays, there are two other systems potentially avialable for use that may have a place in contemporary developmental toxicity safety evaluations. They are mentioned here because of their utility.

11.1 Chernoff-Kavlock assay

This is a test that is still evolving, but basically it is an abbreviated Segment II evaluation. At the present time the protocol entails chronic exposure of pregnant rodents (mice) to a test agent at the maternal maximum tolerated dose (MTD) which in this instance is defined as no more than 10 % maternal deaths. After natural delivery and at a predetermined postnatal age, the young are counted and weighed [23]. The assay is the subject of ongoing validation studies but already has a significant data base generated some time ago. During the '50s and early '60s a large number of agents was evaluated in mice treated at or near their MTD by several investigators. From results of these studies it is anticipated that at its present state of evolution this assay will produce a large number of positives. There are a great many substances, e. g., table salt and glucose, that interfere with mammalian development at the maternal MTD and this test would accurately predict this , and if it did not, then the discrepancy from published studies would merit examination. When evaluated at a single dose, all such substances will be in the same category or group without prioritization and the group so identified will be very large.

The Chernoff-Kavlock test could be modified slightly in either of two ways and each would be of use. It could be performed at several dose levels as a more economical, albeit somewhat less precise, form of the Segment II evaluation. The larger the difference between the adult and the developmental toxic dosage levels, the greater the hazard potential of the agent. Those interfering with development only at or very near to the MTD could be considered low priority items for further developmental toxicity testing unless potential exposure or other considerations move them up the priority ladder. A second way to modify the test could be to adjust its sensitivity. Instead of testing at the maternal MTD, it could be structured to test instead at a selected fraction of this level. Setting its sensitivity at a predetermined level would reduce the number of positives and simultaneously provide a ranking proportional to each agent's developmental hazard index. For example, if an agent were tested, for instance, at one-fifth of the adult MTD and produced no developmental effect, its A/D ratio could be considered as less than 5. If an effect were seen in the offspring, the agents's teratogenicity hazard index is larger than 5, indicating the substance to be somewhat of a primary developmental hazard.

The abbreviated Segment II form of the test can be used as an alternative to a fullblown Segment II evaluation. Its precision is less than a full study and the safety factor for cross-species extrapolation may need to be larger in this case, but the test could be put to actual use [42]. Setting the sensitivity to a fraction of the MTD allows consideration of the test for prescreening application. The practical applicability of this is reduced markedly by cost since the essential application of good laboratory practice and documentation thereof make this somewhat expensive of both money and expertise.

12 Structure-activity relationships

SAR apparently have some general relevance for predicting hormonal activity and perhaps mutagenic potential of substances before they are tested in vivo. Though it is generally true that mutagens are teratogens, the reverse does not obtain. The majority of substances demonstrated to overtly affect mammalian embryos are not known mutagens. Both Schumacher [97] and Zimmerman [113] have examined published literature to determine whether or not it is possible to relate chemical structure to developmental toxicity. It has proven unpredictive. Actually, it may be that SAR applied to developmental toxicity is a reverse discriminator using up funds that might be more productively used for developmental toxicity safety evaluations of other agents. For example, it is sometimes stated that since a particular substance is a mutagen or highly suspected of being such, it should be tested in a Segment II evaluation. Actually the reverse is probably more accurate. Knowing that it disrupts DNA is sufficient to identify the substance as also injurious to the conceptus. However, experience also indicates that developmental toxicity by such agents is achieved only at or near treatment levels also injurious of the mother. Knowing this makes the agent a low rather than a high priority item for developmental toxicity testing, unless, of course, there are considerations such as exposure or usage involved.

13 Discussion and conclusions

13.1 Nature of the problem

The largely standardized toxicologic evaluations for developmental toxicity are powerful tests capable of detecting agents hazardous to development of the conceptus. The only possible exception to this generality may be coumadin. Through application of rather modest safety factors one has great confidence that is is possible, in theory at least, to provide the conceptus with a degree of safety equivalent to that provided to avoid overt adult toxicity. Unfortunately, some substances known to adversely affect development in these standard animal tests, e. g., vitamin A and valproic acid, are in human use at levels apparently producing abnormal young. The A/D ratio for excess vitamin A, for instance, is about 2 regardless of species or route of administration when both the 'A' and the 'D' are calculated from the same Segment II evaluations. A prescreen test must be accurate on this point. If it is, then potential exposure can either be controlled below this level or at least warnings of outcome to be expected can be provided. This concept could be applied to information written for package inserts. If used properly it would provide a better prospective than current practice which may precipitate unnecessary sequelae. A prescreen also must provide an alert to agents with large A/D ratios because substances such as thalidomide must be detected as having a unique propensity for disrupting development.

Standard in vivo tests cannot be applied to all the old and new substances in commerce because they are too expensive and require resources of expertise and facilities beyond present capabilities. Interestingly, when tested in the standard laboratory animals, the majority of chemicals prove capable of interfering with development only at or very near to exposure levels high enough to acutely perturb the adult. In other words, maternal homeostasis is their target, not development. If a means were available to determine this ahead of time, most need not have been tested initially in pregnant animals. Testing could have concentrated more productively on signs of adult toxicity and as long as this level of exposure was avoided the conceptus would not be at risk. Similarly, if those substances with the developmental events of an embryo as their target or most sensitive point of attack could be identified ahead of time, they could be prioritized into a tier system [111] going even beyond the presently standardized tests in pregnant rodents and rabbits.

13.2 A solution

There is need for continued and even broader use of presently available technology to prioritize substances according to their developmental toxicity hazard potential [42]. The hydra assay achieves this already and *Drosophila* and possibly planaria and frog embryos have potentials either supplemental, complementary or preemptory in nature. Other assays such as cell binding tests merit careful evaluation because, for a range of potentially hazardous agents, they eventually may be shown to have utility.

Both the limb bud and whole embryo techniques can serve at this time for some applications. If one needed to know which of two or more isomeres or substituion forms of a chemical was more, or less, toxic to a developing system, both of these assays achieve the answer to this question. In addition, they have the marked advantage of detecting patterns of effects due to selected agents. Though the techniques involved are not simple, the information provided also could become part of the planning phase for human epidemiologic studies.

13.3 Points of confusion regarding validation

There are four classes of perturbed in utero development: death of the products of conception, birth of structurally abnormal young, developmentally delayed or stunted young and production of functionally impaired offspring. There is no absolute consistency in the class or classes of outcome that a particular test agent may produce in a particular test species but it is generally accepted that at some dosage some animal species will have at least one type of effect from most every agent. It may be that the only type of agent incapable of producing any sign of developmental toxicity is one where death of the mother occurs before the agent can reach the embryo and affect a change in one or more of its myriad developmental sequences.

Perhaps the main thing unique about a 'human teratogen' is that it is a substance with a large A/D ratio (thalidomide) or, if it has a small ratio, is used near the adult toxic dosage (vitamin A, aspirin?).

It is less than realistic to seek validation of a prescreen by comparing its results with reports of structurally abnormal young having been reported for a test agent. Live gross-anatomically malformed young are but one of the four signs of disrupted development. Actually, absence of terata in a report could be an example [49] of embryonic death of the severely abnormal embryos, though more frequently it reflects treatment only at very low dosage. Even more subtle is the consideration that the Segment II study is not an efficient means for producing live structurally abnormal young [44]. Most published studies of developmental toxicity are Segment II evaluations. There are very severe and, from the viewpoint of the embryo, lifetime exposures more apt to cause death and resorption than birth of structurally abnormal live fetuses at term.

It avails but little to evaluate the predictive power of a prescreening systems against inadequate data bases, though this continues to occur. Second hand references generally are not written as analyzed, peer reviewed, or quality controlled statements regarding the published reports of others. They tend to be references to the literature providing a summary of the views of the original authors but lacking 'confidence intervals'. Investigators seeking to determine whether or not a potential prescreen is valid must find original papers clearly testing and determining the potential of agents administered to a particular species by a particular route. When confidence is achieved that the effects reported in mammalian studies are valid, the prescreen must recapitulate the results seen in those pregnant animals. The journals: Journal of the American College of Toxicology, Toxicology and Applied Pharmacology, and Fundamental and Applied Toxicology of the past five or six years contain ample Segment II evaluations of diverse substances. These and some of the original papers referenced in the list of smith et al. [102], provide ample starting places for development of valid mammalian data bases. Investigators contemplating efforts at prescreening or investigating truly innovative systems [32, 55, 108] might remember the old adage that a day in the library may save a month at the bench.

The bottom line on the utility of a prescreening method in the real world of safety evaluations is the degree to which it recapitulates some useful datum already documented in standard tests. If it can recapitulate known valid data, then it can be used in a predictive context.

Toward this end, a list of 48 substances considered of some utility for prescreen validation was published recently [102]. This well-intended effort, however, proves counter-productive of careful validation studies and is inaccurate. This group of 48 chemicals provides the chemical name, CAS number, LD50 as determined by various routes of exposure in various species, and a dose, route, and species where developmental toxicity was or was not reported. This latter is referenced to the original papers. Substances are listed as + or '−' with a '+' indicating that the article referenced reports either death, malformation or growth retardation of the offspring in the *absence* of maternal toxicity. Examination of the actual papers referenced reveal that some listed als '+' contain no data or information regarding maternal condition [98] and others so listed report severe maternal toxicity and even maternal death [56]. This same publication is cited for aspirin as a '+' but the actual publication mentions adult condition only in passing, does not indicate the means of ascertainment, and contains no tabular data on the topic. Some substances, however (e. g., formaldehyde), are correctly referenced to valid and fully documented studies [75].

The best correlation to be made from this list is that many of the substances listed as positive were tested at dose levels near the adult toxic range and many listed as negative were tested only at a fraction of the adult toxic dose.

It cannot be overstressed that the goal of a prescreen is not to determine the nature of outcomes at or in close proximity to the adult toxic dose. If a prescreening system were to exactly recapitulate the '+'s and '−'s of this report it would have demonstrated its lack of validity.

A second major type of validation difficulty is highlighted also in the tabular data of this otherwise well-written paper. The maternal toxicity data consists of only the LD50 and this is often by a route not represented in the data of embryo toxicity. The LD50 is a notoriously inaccurate means to evaluate toxicologic potential and one reason for this is that it cannot provide information regarding the slope of the dose-response curve. Agents with identical LD50 values may have very different no-effect levels. The only valid way to compare adult and developmental toxicity is when both are determined in the same study so that route, duration, etc., are identical. The standard endpoint assays of developmental toxicity required in national and international guidelines is determination of fetal weight, sex and gross anatomical inspection of fetal soft tissues and skeleton. The most commonly reported parameter of adult toxicity is maternal weight gain. If other maternal parameters are measured [37] they will tend to somewhat lower the 'A' of the A/D ratio, thereby encouraging more detailed evaluation of animals used for these safety evaluations. Genetic differences do not markedly alter the relationship between adult and developmentally toxic dose levels. They influence the relevant doses, threshold and incidences above background markedly, however [99].

13.4 A useful outcome

The relationship of adult and developmental toxicity is a valuable tool capable of prioritizing substances for testing [42] in conjunction with other new methods. Additional and hopefully even better means will eventually be developed. Those available now should be used for both prescreening and for workplace safety [43].

References

1 N. D. Agnish and D. M. Kochhar: The role of somites in the growth and early development of mouse limb buds. Dev. Biol. 56, 174–183 (1977).

R. C. Armstrong and J. J. Elias: Development of embryonic rat eyes in organ culture. II. An in vitro approach to teratogenic mechanisms. J. Embryol. exp. Morph. 19 (3), 407–414 (1968).

3 M. B. Aydelotte and D. M. Kochhar: Development of mouse limb buds in organ culture; chondrogenesis in the presence of proline analog, L-azetidine-2-carboxylic acid. Dev. Biol. 28, 191–201 (1972).

4 M. B. Aydelotte and D. M. Kochhar: Influence of 6-diazo-5-oxo-L-norleucine (DON), a glutamine analog, on cartilaginous differentiation in mouse limb buds in vitro. Differentiation 4, 73–80 (1975).

5 A. R. Beaudoin and D. L. Fisher: An in vivo/in vitro evaluation of teratogenic action. Teratology 23, 57–61 (1981).

6 C. L. Berry and C. D. Nickols: The effects of aspirin on the development of the mouse third molar. A potential screening system for weak teratogens. Arch. Toxicol. 42, 185–190 (1979).

7 J. B. Best, M. Morita and B. Abbotts: Acute toxic responses of the freshwater planarian, Dugesia dorotocephala, to chlordane. Bull. env. Contam. Toxicol. 26, 502–507 (1981).

8 J. B. Best and M. Morita: Planarians as a model system for in vitro teratogenesis studies. Teratogen. Carcinogen. Mutagen. 2, 277–291 (1982).

9 N. Bournias-Vardiabasis and R. L. Teplitz: Use of **Drosophila** embryo cell cultures as an in vitro teratogen assay. Teratogen. Carcinogen. Mutagen. 2, 333–341 (1982).

10 N. Bournias-Vardiabasis, R. L. Teplitz, G. F. Chernoff and R. L. Seecof: Detection of teratogens in the **Drosophila** embryonic cell culture test: Assay of 100 chemicals. Teratology 28, 109–122 (1983).

11 A. G. Braun, D. J. Emerson and B. B. Nichinson: Teratogenic drugs inhibit tumour cell attachment to lectin-coated surfaces. Nature 282 (5738), 507–509 (1979).

12 A. G. Braun and J. P. Dailey: Thalidomide metabolite inhibits tumor cell attachment to conconavalin A coated surfaces. Biochem. biophys. Res. Commun. 98, 1029–1034 (1981).

13 A. G. Braun, C. A. Bruckner, D. J. Emerson and B. B. Nichinson: Qualitative correspondence between in vivo and in vitro activity of teratogenic agents. Proc. natn. Acad. Sci. USA 79, 2056–2060 (1982 a).

14 A. G. Braun, B. B. Nichinson and P. B. Horowicz: Inhibition of tumor cell attachment to concanavalin A-coated surfaces as an assay for teratogenic agents: Approaches to validation. Teratogen. Carcinogen. Mutagen. 2, 343–354 (1982b).
15 R. L. Brinster: Studies on the development of mouse embryos in vitro. I. The effect of osmolarity and hydrogen ion concentration. J. exp. Zool. 158, 49–58 (1965).
16 R. L. Brinster: Uptake and incorporation of amino acids by the preimplantation mouse embryo. J. Reprod. Fertil. 27, 329–338 (1971).
17 R. L. Brinster: Teratogen testing using preimplantation mammalian embryos, p. 113–124. In: Methods for Detection of Environmental Agents that Produce Congenital Defects. Eds. T. H. Shepard, J. R. Miller and M. Marois. North-Holland, Amsterdam (1975).
18 N. A. Brown, E. H. Goulding and S. Fabro: Ethanol embryotoxicity: Direct effects on mammalian embryos in vitro. Science 206, 573–575 (1979).
19 N. A. Brown and S. Fabro: Quantitation of rat embryonic development in vitro: a morphological scoring system. Teratology 24, 65–78 (1981).
20 J. Buekle-Sam, R. A. Byrd and C. J. Nelson: Blood flow during pregnancy in the rat: III. Alterations following mirex treatment. Teratology 27, 401–410 (1983).
21 C. L. Chatot, N. W. Klein, J. Piatek and L. J. Pierro: Successful culture of rat embryos on human serum: Use in the detection of teratogens. Science 207, 1471–1473 (1980).
22 C. L. Chatot and N. W. Klein: Teratogenic activity of serum from human epileptic subjects studied by rat embryo cultures. Teratology 23, 30A (1981).
23 N. Chernoff and R. J. Kavlock: An in vivo teratology screen utilizing pregnant mice. J. Toxicol. envir. Health 10, 541–550 (1982).
24 Y. H. Chun, E. M. Johnson and B. E. G. Gabel: Relationship of developmental stage to effects of vinblatine on the artificial 'embryo' of hydra. Teratology 27, 95–100 (1983a).
25 Y. H. Chun, E. M. Johnson, B. E. G. Gabel and A. S. A. Cadogan: Effect of vinblastine sulfate on the growth and histologic development of reaggregated hydra. Teratology 27, 89–94 (1983b).
26 N. C. Coon, P. H. Albers and R. C. Szaro: No. 2, fuel oil decreases embryonic survival of great black-backed gulls. Bull. env. Contam. Toxicol. 21, 152–156 (1979).
27 Council on Environmental Quality. Chemical hazards to human reproduction. US Government Printing Office, Washington 1981.
28 J. N. Dumont and R. G. Epler: Validation studies on the FETAX teratogenesis assay (frog embryos). Teratology 29, 27A (1984).
29 H.-G. Eibs, H. Spielmann, U. Jacob-Müller and J. Klose: Teratogenic effects of cyproterone acetate and medroxyprogesterone treatment during the pre- and postimplantation period of mouse embryos. II. Cyproterone acetate and medroxyprogesterone acetate treatment before implantation in vivo and in vitro. Teratology 25, 291–299 (1982).
30 A. G. Fantel, J. C. Greenaway, M. R. Juchau and T. H. Shepard: Teratogenic bioactivation of cyclophosphamide in vitro. Life Sci. 25, 67–72 (1979).
31 D. L. Fisher: Accumulation of DNA, RNA, and protein by cultured rat embryos following maternal administration of a teratogenic dose of trypan blue. J. exp. Zool. 216, 415–422 (1981).
32 E. Freese: Use of cultured cells in the identification of potential teratogens. Teratogen. Carcinogen. Mutagen. 2, 355–360 (1982).
33 J. C. Greenaway, A. G. Fantel, T. H. Shepard and M. R. Juchau: The in vitro teratogenicity of cyclophosphamide in rat embryos. Teratology 25, 335–343 (1982).

34 J. H. Greenberg: Detection of teratogens by differentiating embryonic neural crest cells in culture: Evaluation as a screening system. Teratogen. Carcinogen. Mutagen. 2, 319–323 (1982).
35 J. H. Greenberg and R. M. Pratt: Glycosaminoglycan and glycoprotein synthesis by cranial neural crest cells in vitro. Cell Differ. 6, 119–132 (1977).
36 J. H. Greenberg, J. M. Foidart and R. M. Greene: Collagen synthesis in cultures of differentiating neural crest cells. Cell Differ. 9, 153–163 (1980).
37 B. D. Hardin, G. P. Bond, M. R. Sikov, F. D. Andrew, R. P. Beliles and R. W. Niemeier: Testing of selected workplace chemicals for teratogenic potential. Scand. J. Work envir. Health 7 (4), 66–75 (1981).
38 J. R. Hassell and E. A. Horigan: Chondrogenesis: A model developmental system for measuring teratogenic potential of compounds. Teratogen. Carcinogen. Mutagen. 2, 325–331 (1982).
39 R. Jelinek: Use of chick embryo in screening for embryotoxicity. Teratogen. Carcinogen. Mutagen. 2, 255–261 (1982).
40 R. Jelinek and Z. Rychter: Morphogenetic systems and screening for embryotoxicity. Arch. Toxicol., suppl. 4, 267–273 (1980).
41 E. M. Johnson: Screening for Teratogenic Hazards: Nature of the Problems. Ann. Rev. Pharmacol. Toxicol. 21, 417–429 (1981).
42 E. M. Johnson: A prioritization and biological decision tree for developmental toxicity safety evaluations. J. Am. Coll. Toxicol. 3 (2), 141–147 (1984 a).
43 E. M. Johnson: Mechanisms of teratogenesis: The extrapolation of the results of animal studies to man. pp 135–151, *Piegnont Women at Work,* Ed: G. Chamberlain, The Royal Soc. Med. and Macmillan Pien. (1984 b).
44 E. M. Johnson and M. S. Christian: When is a teratology study not an evaluation of teratogenicity? J. Am. Coll. Tox. 3 (6), 431–434. (1984 c).
45 E. M. Johnson, B. E. G. Gabel, M. S. Christian and E. W. Sica: The developmental toxicity of xylene and xylene isomers. Teratology 29, 38A (1984 d).
46 E. M. Johnson and B. E. G. Gabel: Application of thc hydra assay for rapid detection of developmental hazards. J. Am. Coll. Tox. 1 (3), 57–71 (1982 a).
47 E. M. Johnson, R. M. Gorman, B. E. G. Gabel and M. E. George: The *Hydra attenuata* system for detection of teratogenic hazards. Teratogen. Carcinogen. Mutagen. 2, 263–276 (1982 b).
48 E. M. Johnson and B. E. G. Gabel: An artificial 'embryo' for detection of abnormal developmental biology. Fund. appl. Toxicol. 3, 243–249 (1983).
49 E. M. Johnson, M. M. Nelson and I. W. Monie: Effects of transitory pteroylglutadic acid. (PGA) deficiency on embryonic and placental development in the rat. Anat. Rec. 146, 215–224 (1963).
50 E. M. Johnson and R. Spinuzzi: Enzymic differentiation of rat yolk-sac placenta as affected by a teratogenic agent. J. Embryol. exp. Morph. 16, 271–288 (1966).
51 E. M. Johnson and R. Spinuzzi: Differentiation of alkaline phosphatase and glucose-6-phosphate dehydrogenase in rat yolk-sac. J. Embryol. exp. Morph. 19, 137–143 (1968).
52 S. M. Kaldare and L. Mulherkar: Alleviation of the toxicity of actinomycin D by uridine and thymidine on the morphogenesis of chick embryos cultivated in vitro. Experientia 28 (6) 690–692 (1971).
53 D. A. Karnofsky: Mechanisms of action of certain growth-inhibiting drugs. In: Teratology: Principles and Techniques, p. 185–194. Eds. J. G. Wilson and J. Warkany. University of Chicago Press., Chicago. (1965 a).
54 D. A. Karnofsky: The chick embryo in drug screening; survey of teratological effects observed in the 4-day chick embryo. In: Teratology: Principles and Techniques, p. 194–213. Eds. J. G. Wilson and J. Warkany. University of Chicago Press. Chicago (1965 b).

55 S. J. Keller and M. K. Smith: Animal virus screens for potential teratogens. I. Poxvirus Morphogenesis. Teratogen. Carcinogen. Mutagen. 2, 361–374 (1982).
56 K. S. Khera: Teratogenicity studies with methotrexate, aminopterin and acetylsalicylic acid in domestic cats. Teratology 14, 21–28 (1976).
57 K. S. Khera: Maternal toxicity – A possible factor in fetal malformations in mice. Teratology 29, 411–416 (1984).
58 K. S. Khera and D. A. Lyon: Chick and duck embryos in evaluation of pesticide toxicity. Toxicol. appl. Pharmacol. 13, 1–15 (1968).
59 C. A. Kimmel: Effect of route of administration on the toxicity and teratogenicity of EDTA in the rat. Toxicol. appl. Pharmacol. 40, 299–306 (1977).
60 K. T. Kitchin and M. T. Ebron: Further development of rodent whole embryo culture: solvent toxicity and water insoluble compound delivery system. Toxicology 30, 45–57 (1984).
61 K. T. Kitchin, M. T. Ebron and D. Svendsgaard: In vitro study of embryotoxic and dysmorphogenic effects of mercuric chloride and methylmercury chloride in the rat. Fd. Chem. Toxic. 22 (1), 31–37 (1984).
62 N. W. Klein and L. J. Pierro: Whole embryos in culture. Handbook exp. Pharm. 65, 315–333 (1983).
63 N. W. Klein, J. D. Plenefisch, S. W. Carey, W. T. Fredrickson, G. P. Sackett, T. M. Burbabacher and R. M. Parker: Serum from monkeys with histories of fetal wastage causes abnormalities in cultured rat embryos. Science 215, 66–69 (1982).
64 N. W. Klein, M. A. Vogler, C. L. Chatot and L. J. Pierro: The use fof cultured rat embryos to evaluate the teratogenic activity of serum: cadmium and cyclophosphamide. Teratology 21, 199–208 (1980).
65 D. M. Kochhar: Assessment of teratogenic response in cultured postimplantation mouse embryos: effects of hydroxyurea. In: New Approaches to the Evaluation of Abnormal Embryonic Development, p. 250–277. Eds. D. Neubert and H.-J. Merker. G. Thieme, Stuttgart 1975.
66 D. M. Kochhar: In vitro testing of teratogenic agents using mammalian embryos. Teratogen. Carcinogen. Mutagen. 1, 63–74 (1980).
67 D. M. Kochhar: Embryonic organs in culture. Handbook exp. Pharm. 65, 301–314 (1983).
68 D. M. Kochhar and M. B. Aydelotte: Susceptible stages and abnormal morphogenesis in the developing mouse limb, analyzed in organ culture after transplacental exposure to vitamin A (retinoic acid). J. Embryol. exp. Morph. 31, 721–734 (1974).
69 D. M. Kochhar: Embryonic limb bud organ culture in assessment of teratogenicity of environmental agents. Teratogen. Carcinogen. Mutagen. 2, 303–312 (1982).
70 W. Landauer: Insulin-induced abnormalities of beak, extremities, and eyes in chickens. J. exp. Zool. 105, 145 (1947).
71 W. Landauer: Genetic and environmental factors in the teratogenic effects of boric acid on chicken embryos. Genetics 38, 216 (1953).
72 J. M. Manson and C. C. Smith: Influence of cyclophosphamide and 4-ketocyclo-phosphamide on mouse limb development. Teratology 15, 291–300 (1977).
73 J. M. Manson and R. Simons: In vitro metabolism of cyclophosphamide in limb bud culture. Teratology 19, 149–158 (1979).
74 O. Marhan and R. Jelinek: Efficiency of embryotoxicity testing procedures. II. Comparison between the official, mest and chest methods. Toxicol. Letts. 4, 389–392 (1979).
75 T. A. Marks, W. C. Worthy and R. E. Staples: Influence of formaldehyde and sonacide© (potentiated acid glutaraldehyde) on embryo and fetal development in mice. Teratology 22, 51–58 (1980).

76 P. E. Mirkes, A. G. Fantel, J. C. Greenaway and T. H. Shepard: Teratogenicity of cyclophosphamide metabolites: phosphoramide mustard, acrolein, and 4-ketocyclophosphamide in rat embryos cultured in vitro. Toxicol. appl. Pharmacol. 58, 322–330 (1981).
77 G. M. Morriss and C. E. Steele: The effect of excess vitamin A on the development of rat embryos in culture. J. Embryol. exp. Morph. 32, 505–514 (1974).
78 G. M. Morriss and C. E. Steele: Comparison of the effects of retinol and retinoic acid on postimplantation rat embryos in vitro. Teratology 15, 109–120 (1977).
79 M. H. Moscona and D. A. Karnofsky: Cortisone-induced modifications in the development of the chick embryo. Endocrinology 66, 533 (1960).
80 C. L. Mummery, C. E. van den Brink, P. T. van der Saag and S. W. deLaat: A short-term screening test for teratogens using differentiating neuroblastoma cells in vitro. Teratology 29, 271–279 (1984).
81 M. L. Netzloff, K. P. Chepenik, E. M. Johnson and S. Kaplan: Respiration of rat embryos in culture. Life Sci. 7, 401–405 (1968).
82 M. Netzloff, E. M. Johnson and S. Kaplan: Effects of a teratogen (9-methyl pteroylglutamic acid) on oxygen consumption by dispersed embryonic rat cells. Teratology 5, 19–22 (1972).
83 D. Neubert, N. Hinz, I. Baumann, H. J. Barrach and K. Schmidt: Attempt upon a quantitative evaluation of the degree of differentiation of or the degree of interference with development in organ culture. In: Role of Pharmacokinetics in Prenatal and Perinatal Toxicology, p. 337–349. Eds. D. Neubert, H. J. Merker, H. Nau and J. Langman. G. Thieme, Stuttgart 1978.
84 D. A. T. New: The Culture of Vertebrate Embryos. Logos Press, London 1966.
85 D.A.T. New: Development of explanted rat embryos in circulating medium. J. Embryol. exp. Morph. 17, 513–525 (1967).
86 D. A. T. New: Whole embryo culture of mammalian embryos during organogenesis. Biol. Rev. 53, 81–122 (1978).
87 V. P. Patel and H. F. Lodish: Loss of adhesion of murine erythroleukemia cells to fibronectin during erythroid differentiation. Science 224, 996–998 (1984).
88 R. M. Pratt, R. I. Grove and W. D. Willis: Prescreening for environmental teratogens using cultured mesenchymal cells from the human embryonic palate. Teratogen. Carcinogen. Mutagen. 2, 313–318 (1982).
89 L. L. Reti, F. Beck and S. Bulman: Culture of 9 ½ day rat embryos in human serum supplemented and unsupplemented with rat serum. J. exp. Zool. 223, 197–199 (1982).
90 T. W. Sadler: Culture of early somite mouse embryos during organogenesis. J. Embryol. exp. Morph. 49, 17–25 (1979).
91 T. W. Sadler, W. E. Horton and C. W. Warner: Whole embryo culture: A screening technique for teratogens? Teratogen. Carcinogen. Mutagen. 2, 243–253 (1982).
92 T. W. Sadler and D. M. Kochhar: Biosynthesis of DNA, RNA and proteins by mouse embryos cultured in the presence of a teratogenic dose of chlorambucil. J. Embryol. exp. Morph. 36, 273–281 (1976).
93 J. L. Schardein: Drugs as teratogens. CRC Press, Florida, 1976.
94 J. L. Schardein, R. Sakowski, J. Petrere and R. R. Humphrey: Teratogenesis studies with EDTA and its salts in rats. Toxicol. appl. Pharmacol. 61, 423–428 (1981).
95 R. R. Schmidt, P. K. Abbott and J. M. Cotler: In vitro effects of the teratogen and folic acid antagonist, 9-methyl pteroylglutamic acid, on glycosaminoglycan accumulation in fetal rat limbs. Teratology 26, 53–58 (1982).

96 R. L. Schuler, B. D. Hardin and R. W. Niemeier: Drosophila as a tool for the rapid assessment of chemicals for teratogenicity. Teratogen. Carcinogen. Mutagen. 2, 293–301 (1982).
97 H. J. Schumacher: Chemical structure and teratogenic properties. In: Methods for Detection of Environmental Agents That Produce Congenital Defects, p. 65–77. Eds. T. H. Shepard, J. R. Miller and M. Marois, North-Holland, Amsterdam 1975.
98 H. J. Schumacher, J. Terapane, R. L. Jordan and J. G. Wilson: The teratogenic activity of a thalidomide analogue, EM_{12}, in rabbits, rats, and monkeys. Teratology 5, 223–240 (1972).
99 M. J. Seller, K. J. Perkins and M. Adinolfi: Differential response of heterozygous curly-tail mouse embryos to vitamin A teratogenesis depending on maternal genotype. Teratology 28, 123–129 (1983).
100 T. H. Shepard: Catalog of Teratogenic Agents. Johns Hopkins University Press, Baltimore 1982.
101 T. H. Shepard, J. R. Miller and M. Marois (ed.): Methods for Detection of Environmental Agents that Produce Congenital Defects, p. 249. Proceedings of the Guadeloupe Conference sponsored by l'Institut de la Vie. North-Holland Publishing Company, Amsterdam, Oxford, and American Elsevier Publishing Company, Inc., New York 1975.
102 M. K. Smith, G. L. Kimmel, D. M. Kochhar, T. H. Shepard, S. P. Spielberg and J. G. Wilson: A selection of candidate compounds for in vitro teratogenesis test validation. Teratogen. Carcinogen. Mutagen. 3, 461–480 (1983).
103 G. Streisinger: On the possible use of zebra fish for the screening of teratogens. In: Methods for Detection of Environmental Agents that Produce Congenital Defects, p. 59–61. Eds. T. H. Shepard, J. R. Miller and M. Marois. North-Holland, Amsterdam 1975.
104 S. P. Sugrue and J. M. Desesso: Altered glycosaminoglycan composition of rat forelimb-buds during hydroxyurea teratogenesis: An indication of repair. Teratology 26, 71–84 (1982).
105 M. M. Turbow: Trypan blue induced teratogenesis of rat embryos cultivated in vitro. J. Embryol. exp. Morph. 15, 387–395 (1966).
106 M. M. Turbow and J. G. Chamberlain: Direct effects of 6-aminonicotinamide on the developing rat embryo in vitro and in vivo. Teratology 1, 103–108 (1968).
107 M. J. Verrett, W. F. Scott, E. F. Reynaldo, E. K. Alterman and C. A. Thomas: Toxicity and teratogenicity of food additive chemicals in the developing chicken embryo. Toxicol. appl. Pharmacol. 56, 265–273 (1980).
108 B. T. Walton: Chemical impurity produces extra compound eyes and heads in crickets. Science 212, 51–53 (1981).
109 WHO: Principles for testing of drugs for teratogenicity. Techn. Rep. Ser. 364. Geneva 1967.
110 A. L. Wilk, J. H. Greenberg, E. A. Horigan, R. M. Pratt and G. R. Martin: Detection of teratogenic compounds using differentiating embryonic cells in culture. In Vitro 16, 269–276 (1980).
111 J. G. Wilson: Critique of current methods for teratogenicity testing in animals and suggestions for their improvement. In: Methods for Detection of Environmental Agents That Produce Congenital Defects, p. 29–48. Proceedings of the Guadeloupe Conference sponsored by l'Institut de la Vie. Eds. T. H. Shepard, J. R. Miller and M. Marois. North-Holland Publishing Company, Amsterdam, and American Elsevier Publishing Company Inc., New York 1975.
112 J. G. Wilson: Survey of in vitro systems: Their potential use in teratogenicity screening. In: Handbook of Teratology 4, Research Procedures and Data

Analysis, p. 135–153. Eds. J. G. Wilson and F. C. Fraser. Plenum Press, New York 1978.

113 E. F. Zimmerman: Chemical structure and teratogenic mechanism of action. In: Methods for Detection of Environmental Agents That Produce Congenital Defects, p. 79–88. Eds. T. H. Shepard, J. R. Miller and M. Marois. North-Holland, Amsterdam 1975.

Carcinogenicity testing of drugs

G. M. Williams and J. H. Weisburger
Naylor Dana Institute for Disease Prevention, American Health Foundation, Valhalla, NY 10595–1599

1 Introduction

The dramatic advances in medicine in the last fifty years have stemmed in large part from extensive research to discover new effective medications for controlling and treating many types of diseases that afflict mankind. In the development of new drugs, an increasingly important aspect is to predict and assess any adverse effects that the products might have. Potential undesirable effects depend on the specific properties of the medication, including its biotransformation in the human body, the dose regimen, the age and sex of the patient and the general condition of the patient, especially in regard to nutritional or immunological status. The demonstration of drug safety involves assurance that major acute toxic effects will be avoided and evidence that long-term use will not cause chronic diseases, including cancer.

Pharmaceuticals, as with other chemicals, have caused cancer in humans [1–6]. A classic example is bladder cancer induced by chlornaphazine, a derivative of the known human carcinogen 2-naphthylamine, which was used in the treatment of polycythemia vera [7].

The ability of drugs with chemically reactive constituents to cause cancer is exemplified by pharmaceuticals used in cancer chemotherapy, which have been found to lead to second cancers. In many cases, the new cancers were lymphomas or leukemias [3, 7a]. These may have occurred because the anticancer drugs, often alkylating agents, are toxic to the bone marrow and lymphoid system and are thereby immunosuppressive. Lymphomas or leukemias can arise under conditions of an altered immune status.

Other types of drugs presumably acting through their pharmacologic effects have caused cancer. The synthetic hormone diethylstilbestrol (DES), used primarily for the maintenance of pregnancy, is an example. Under conditions where very sizable dosages of up to 125 mg/day were prescribed during fetal embryogenesis, a small fraction of the female offspring of treated women presented at puberty with vaginal adenosis and in some instances vaginal adenocarcinoma [8].

In other instances, normal use of a medication had no adverse effect, whereas excess intake was deleterious. Phenacetin is safe when used alone on a routine basis or together with mixtures of other drugs such as aspirin and caffeine as an analgesic at prescribed doses. Nevertheless, some case histories, mainly from Sweden and Switzerland, showed that the addictive abuse of phenacetin involving total intakes over a period of years of more than 10 kg led to cancer of the bladder [9]. In animal studies, moderate level dosing of rats failed to disclose any adverse effect, but very high-level dosing did induce bladder cancer, mimicking what was observed in humans.

In all, 12 pharmaceuticals or treatments have been linked by working groups of the International Agency for Research on Cancer with human cancer (table 1). These represent approximately 50 % of chemicals recognized as human carcinogens by the Agency as of 1982 [6]. Another 26 drugs are listed as probable carcinogens. The nature of the hazard to humans of these agents, however, requires detailed evaluation.

In view of these findings, methods for assessing drugs for possible adverse effects during the process of their development are of great public health importance. This review of contemporary approaches to carcinogen bioassay of pharmaceuticals, using in vitro and in vivo methodologies, will focus on testing of human medications. Nevertheless, similar approaches have been described for the testing of veterinary drugs [10] or, indeed, of any chemical [11, 12].

2 Development of carcinogen bioassay of pharmaceuticals

The procedures for bioassay for carcinogens in animal models were first standardized by scientists at the United States Food and Drug Administration (FDA) in the 1950s, especially Lehman, Fitzhugh, and Nelson, who were concerned with procedures for the safety assessment of food additives and drugs. Their studies involved mainly rats, dogs, and occasionally mice. Some of the results were published in the pharmacological or toxicological literature, but the detailed procedures and the rationale related thereto were published in specialized, monographs of limited circulation printed by the Association of Food and Drug Officials of the United States [13]. In 1982, the FDA released the draft of a manual embodying current procedures for safety assessment [14].

In 1961, the National Cancer Institute of the United States created a special unit, the Carcinogen Screening Section to develop procedures for the detection and quantification of carcinogenic risks of various chemicals [14a]. This group evaluated chronic bioassay methods, refined and standardized them, and applied them to the screening of many chemicals, including some pharmaceuticals [15, 16]. This program of routine carcinogen testing was transferred to the National

Table 1
Pharmacological agents judged by International Agency for Research on Cancer to be associated with cancer in humans.

Causally associated	Azathioprine
	N,N-Bis(2-chloroethyl)-2-naphthylamine (Chlornaphazine)
	1,4-Butanediol dimethanesulphonate (myleran)
	Certain combined chemotherapy for lymphomas (including MOPP)
	Chlorambucil
	Conjugated oestrogens
	Cyclophosphamide
	Diethylstilboestrol
	Melphalan
	Methoxsalen with ultraviolet A therapy (PUVA)
	Phenacetin-containing analgesic mixtures
	Treosulphon
Probably carcinogenic	Actinomycin D
	Adriamycin
	Bis(chloroethyl)-nitrosourea (BCNU)
	Chloramphenicol
	1-(2-Chloroethyl)-3-cyclohexyl-1-nitrosourea (CCNU)
	Chloroform
	Cisplatin
	Dacarbazine
	Dienoestrol
	Ethinyloestradiol
	Mestranol
	Metronidazole
	Nitrogen mustard
	Norethisterone
	Oestradiol-17 B
	Oestrone
	Oral contraceptives, combined and sequential
	Oxymetholone
	Phenacetin
	Phenazopyridine
	Phenytoin
	Procarbazine
	Progesterone
	Propylthiouracil
	Tris(aziridinyl)-para-benzoquione (triaziquone)
	Tris(1-aziridinyl)phosphine sulphide (thiotepa)
	Uracil mustard

Taken from [6]

Toxicology Program in 1981. The procedures used in such standard bioassays and the related science have recently been reviewed [17, 18]. Parallel to these activities in the United States, guidelines for bioassay were developed by several other countries [19, 20]. Also, international groups have been involved in assessing bioassay methodology. A number of workshops on carcinogen testing were organized by the Union Internationale Contre le Cancer (UICC) [21, 22]. The International Agency for Research on Cancer regularly reviews carcinogenicity data and in 1980 produced a monograph on carcinogen testing [23].
In 1969, the World Health Organization published recommendations specifically for testing of drugs for carcinogenicity [24]. Unfortunately, these have not been updated.
In the development of a new therapeutic agent, the requirements of regulatory agencies for safety testing must ultimately be complied with in order to obtain marketing approval. The guidelines for testing to meet these requirements, as described in the above documents, are not directed toward providing the greatest efficiency in the process of drug development. Consequently, additional specific strategies have been proposed for integration of toxicology testing into the development of new drugs [25]. Also, only a few recommendations involve deliberate efforts to establish the mechanisms by which a chemical may elicit a neoplastic response in exposed animals [20]. The present review will attempt to demonstrate how new advances in the field of chemical carcinogenesis can be applied to a systematic approach to safety testing of pharmaceuticals, which incorporates studies that provide information on the underlying mechanism(s) of any carcinogenic effects.

3 Chemical Carcinogenesis

A chemical carcinogen is defined operationally by its ability to increase the occurrence of neoplasms. Four types of response have been generally accepted [24] as evidence of an increase of neoplasms: (1) the development of types of neoplasms not seen in controls, (2) an increased incidence of the types of neoplasms occurring in controls, (3) the occurrence of neoplasms earlier than in controls, and (4) an increased multiplicity of neoplasms in individual animals. The neoplasms may be of either epithelial or mesenchymal origin. In almost all instances, the neoplasms produced by chemicals include both benign and malignant types. Conceptually, both are of significance, but some authorities recommend that only in an increase in malignancies is acceptable evidence of carcinogenicity. This issue is discussed further in section 7.92.

A highly diverse collection of chemical substances including organic and inorganic chemicals, solid-state materials, hormones, and immunosuppressants has produced increases in neoplasms in experimental animals, and therefore has been designated as carcinogens [26–29]. The dominant view has been that all agents producing such effects, i. e. carcinogens, are fundamentally alike. In contrast, some scientists have differentiated between carcinogens with different properties [30–34]. Such distinctions rest in part on the recognition that the production of cancer in animals and also in humans by chemicals is the end result of a complex series of individual reactions each of which can be affected by a chemical to influence the final result [35].

The process of chemical carcinogenesis can be divided into two logically necessary and mechanistically distinct sequences, one in which the normal cell is converted to a neoplastic cell and a second in which the neoplastic cell develops into an overt neoplasm (figure 1).

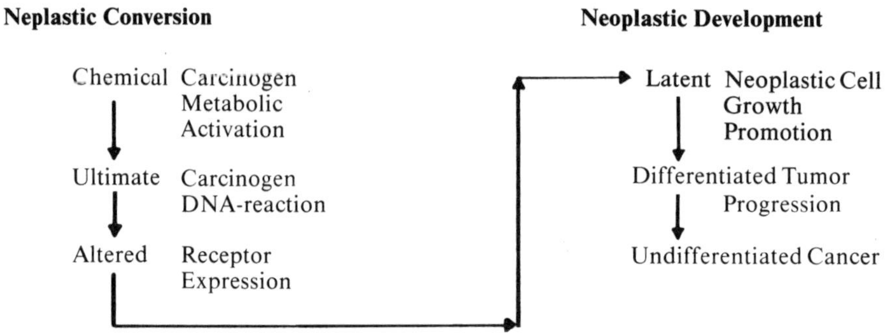

Sequence of complex events during chemical carcinogenesis: The effectiveness of a given agent under specific circumstances hinges in part on the ratio of metabolic activation/detoxification reactions, yielding the ultimate carcinogen, an electrophilic reactant or radical cation, able to react with cellular macromolecules including DNA unless the abnormal DNA is repaired. This step constitutes the mutational event that in turn leads to transformation through further biochemical reactions such as translocations in relation to the expression involving cell duplication. The preceding steps constitute the overall genotoxic events related to neoplastic conversion. The latent tumor cell obtained in the foregoing sequence undergoes development and progression that can be mediated by other agents that are nongenotoxic and operate by epigenetic mechanisms. This sequence of steps involves distinct controling events that can play a major role in the eventual appearance of clinical neoplastic disease in humans or in bioassay systems. There are a number of modifying factors and their action is often tissue-specific, reversible, and highly dose-dependent with a threshold no-effect level.

3.1 Neoplastic conversion

(a) Biotransformation by host enzyme systems. Numerous different enzyme systems function in the detoxification and elimination of endogenous substances and xenobiotics. Unfortunately, as a result of the action of these same systems, certain carcinogens undergo bioactivation to a reactive ultimate carcinogen, usually an electrophilic species [35a]. This step is heavily influenced by a number of agents, including sex hormones [36,37].

A small number of carcinogens, mostly industrial intermediates and chemotherapeutic drugs are reactive in their parent form and, therefore, do not require bioactivation.

(b) Interaction of the ultimate carcinogen with critical cellular and molecular receptors. Carcinogens that form reactive species participate in covalent reactions with a variety cellular macromolecules, including DNA [38]. Abundant evidence now supports the concept that reaction with DNA is a critical event [38,39], although other interactions, such as with the mitotic apparatus, may also be important. The region of DNA containing the damage is subject to removal and restoration by repair enzyme systems, which, however, vary in their efficiency in different species and tissues.

(c) Conversion of carcinogen damage to a permanent alteration. When a cell replicates while DNA damage is present, permanent alterations in the genome are produced in several ways, including the mispairing of bases resulting in point mutations, errors in replication yielding frame-shift mutations, transpositions leading to codon rearrangement, and perhaps combinations of these mutagenic effects. These various genetic alterations may involve gene sequences known as cellular oncogenes [40,41], which are emerging as critical regions of the genome for neoplastic conversion. In the case of interactions with the mitotic apparatus, chromosomal mutations and aneuploidy could result. All these alterations in the genome generate a permanently abnormal cell. Apparently, if gene functions involved in growth control are altered, the cell aquires neoplastic potential.

3.2 Neoplastic development and progression

(a) Multiplication of the altered cells. Genetically altered abnormal cells may be held in check by tissue homeostatic factors or if the conditions of carcinogen exposure and the abnormalities introduced in the cells permit, the abnormal cells may engage in limited proliferation to form 'preneoplastic' lesions. During replication of these cells, further alterations in DNA as a result of transpositions and other error-prone processes are possible.

(b) Progressive growth to neoplasm formation. Cells with the requisite abnormalities may have the capacity to proliferate under permissive conditions beyond tissue constraints to form a neoplasm. This step is also facilitated by agents referred to as neoplasm promoters, which may be endogenous substances or xenobiotics [42].

(c) Progression. Neoplasms can undergo qualitative changes in their phenotypic properties, possibly including transition from benign to malignant behavior [43]. This probably reflects the selection during growth of a population with a genotype coding for advantageous phenotypic properties. New genotypes could arise in neoplasms through errors in DNA replication [44] or alterations in chromosome constitution.

4 Mechanisms of carcinogenesis

The neoplastic state of somatic cells is generally passed on to their progency generated by cell duplication. For this reason, the process whereby a carcinogen converts a normal cell to a neoplastic one (sect. 3) is considered to entail an alteration in the genome.

Theories on the mechanisms of action of chemical carcinogens have usually sought to provide a unifying account for this action of all agents. The team of James and Elizabeth Miller introduced the concept that many structural types of carcinogens act through ultimate electrophilic reactants that interact with cellular macromolecules and have proposed a major generalization that states that the ultimate carcinogenic forms of chemical carcinogens 'usually, if not always, are electrophilic reactants' [45]. This property underlies the ability of some carcinogens to interact with DNA and is the basis for the further generalization, as demonstrated by Malling [46], Ames and co-workers [47], as well as others [48, 49], that many carcinogens are mutagens. These generalizations on the action of carcinogens, although accurate for some carcinogens, have not been proven to apply to all chemical carcinogens, especially many pharmaceuticals that have produced neoplasms in experimental animals.

As noted in section 3, chemicals that induce cancer comprise an extremely diverse group of agents, and thus it would be truly remarkable if all of them acted on cells in exactly the same manner. In fact, there are a number of carcinogens such as saccharin, nitrilotriacetic acid, immunosuppressants, and hormones that have not been documented to react with DNA, or to be gene mutagens [50], in contrast to the properties of typical organic chemical carcinogens. Thus, it seems likely that a variety of modes of action is involved in the overall carcinogenic effects of different chemicals. To facilitate consideration of this possibility, a mechanistic classification of agents involved in the carcinogenic process, separating them into two major categories, genotoxic and epigenetic, has been proposed [30, 32].

Carcinogens that interact with and alter DNA are classified as DNA-reactive or genotoxic. This category contains principally the widely studied organic carcinogens that are electrophilic reactants either in their parent form or after biotransformation. Carcinogens in this category can be identified by the biochemical demonstration of DNA damage or by the finding of genotoxic effects in short-term tests, as described in section 7.2. It seems highly likely that DNA alteration is the key event in the carcinogenicity of these compounds. These agents operate primarily in the sequence of neoplastic conversion, but may also promote the carcinogenic process through reactive or other types of metabolites.

Carcinogenic agents for which no evidence of DNA-reactivity by the chemical or its derivatives has been found by appropriate, reliable studies and that also produce another biologic effect that could be the basis for their involvement in the carcinogenic process are classified as epigenetic carcinogenic agents. Possible mechanisms for epigenetic effects include cytotoxicity and chronic tissue injury, hormonal imbalance, immunologic effects, or promotional activity on cells that are either genetically abnormal or have been independently altered by genotoxic carcinogens. These mechanisms include indirect processes of genetic alteration such as production of active oxygen species [51–54], aberrant methylation [55, 56], or chromosomal alterations [57–59], which could lead to neoplastic conversion. Others, such as hormonal effects, immunosuppression and neoplasm promotion probably operate primarily in the sequence of neoplastic development.

Categorization of carcinogenic chemicals along these lines has been applied by several investigators [30–34] and a working group of the International Agency for Research on Cancer has supported the basic concept that different types of agents appear to be involved in carcinogenesis [60].

If the concept of mechanistically distinct carcinogens becomes established, the categorization may eventually have to be expanded to include other types of agents. As one possibility, types of chemicals that are not DNA-reactive as such, but may indirectly produce DNA damage might be classified as indirect genotoxins, if these effects can be reasonably established to be mechanisms of carcinogenesis. This would give recognition to the fact that a preceding biological effect, i.e., toxicity, must be produced in order for the genetic alteration to occur. Also, if the production of chromosomal effects by agents that do not damage DNA can be shown to occur in vivo in target tissues and to be related to carcinogenesis, agents producing this effect could be separately categorized, perhaps as karyotoxic, to distinguish them from DNA-reactive carcinogens.

5 Classes of chemical carcinogens

The two categories of chemicals involved in the carcinogenic process, DNA-reactive or genotoxic and epigenetic, have been separated into eight classes based upon their chemical and biological properties (table 2).

The DNA-reactive category contains the agents that function as electrophilic reactants, as originally postulated by the Millers [45], or as radical cations. Also, because some inorganic chemicals have displayed such effects [61], they have been tentatively placed in this category. Additionally, the studies of Loeb and colleagues [62] on effects of inorganic chemicals on the fidelity of DNA polymerases suggest that carcinogenic metals might yield abnormal DNA by a distinct mechanism involving alteration of the fidelity of DNA polymerases. In any event, such inorganic carcinogenic chemicals can be construed to yield cells with altered DNA.

DNA-reactive carcinogens, probably because of their effects on genetic material are occasionally effective after a single exposure, frequently are carcinogenic at subtoxic doses, act in a cumulative manner, and act together with other DNA-reactive carcinogens having the same organotropism. They usually produce neoplasms in more than one target organ and with a short latent period.

The second broad category, designated as epigenetic agents comprises those chemicals for which no evidence exists of direct interaction with genetic material, but which produce another biological effect. This category contains cytotoxic agents, solid-state carcinogens, hormones, immunosuppressants, peroxisome proliferators and promoters. Some agents may yield genetically altered cells by indirect processes. A possible, but not yet proven, example is the hypolipidemic drugs such as clofibrate and fenofibrate which are peroxisome-proliferators. Reddy and co-workers [63] have postulated and provided considerable supporting evidence that these drugs have the potential to yield DNA-damaging reactive oxygen species as a consequence of elevated levels of cytoplasmic hydrogen peroxide released from increased numbers of peroxisomes. This mechanism is further supported by evidence that certain of these drugs reduce the activity of glutathione peroxidase [64, 64a], a cytoplasmic scavenger of hydrogen peroxide whose reduction would exacerbate the consequences of release of hydrogen peroxide.
Epigenetic agents comprise a highly diverse group about which generalizations on their carcinogenic effects are difficult. Nevertheless, epigenetic agents usually affect only select organ systems where their physiologic perturbations are produced and usually require a long latent period for their effects. With some classes, such as substances that affect the hormone system and promoters, carcinogenic effects occur mainly with high and sustained levels of exposure which produce prolonged physiologic abnormalities.
A considerable number of carcinogens has been assigned to these different classes (table 2). Sufficient information for classification is not available on many carcinogens, however, and these remain presently unclassified. One objective of the testing approach to be described is to generate the data needed for classification, should the drug prove to be carcinogenic.

6 Carcinogen testing

The standard conventional bioassays [17] for the detection of chemical carcinogens involve the use of male and female rats, mice, and occasionally hamsters of strains selected for their sensitivity to certain experimental carcinogens. The standard test involves determination of a maximally tolerated dose (MTD) of the chemical, based upon which, groups of 50 male and female animals are given the MTD and ½ or ⅓

Table 2
Classes of chemicals involved in the carcinogenic process.

Type	Mode of action	Example
	DNA-reactive carcinogens	
(1) Activation-independent	Electrophile, organic compound, genotoxic, interacts with DNA	bis(chlorethyl)nitrosourea
(2) Activation-dependent	Requires conversion through metabolic activation by host to electrophile	chlornaphazine, procarbazine
(3) Inorganic carcinogen	Some may be DNA-reactive, but others appear to lead to changes in DNA by selective alteration in fidelity of DNA replication	nickel, chromium, cis-platin
	Epigenetic carcinogens	
(4) Promoter	Enhances development of pre-existing transformed cells into neoplasms	phenobarbital
(5) Endocrine modifier	Mainly alters endocrine system balance and differentiation; often acts as promoter	estradiol, diethylstilbestrol, propylthiouracil
(6) Immunosupressor	Appears to permit expression of "virally induced" transformed cells	azathioprine, 6-mercaptopurine
(7) Cytotoxin	Chronic cell killing which leads to increased proliferation may be involved	nitrilotriacetic acid
(8) Peroxisome proliferator	Generate intracellular reactive oxygen species	clofibrate, fenofibrate
(9) Solid-state carcinogen	Physical form vital; may involve cytotoxicity	iron polymers, prosthetic implants
	Unclassified	
(10) Miscellaneous	Not established	methapyrilene

the MTD in a two-year test and, in some cases, a lifetime test. Such testing is still the definitive way to assess the carcinogenicity of a chemical.

Chronic bioassay involves large resources in time and in money and requires scarce specialty skills such as veterinary medicine and pathology for reliable execution. Moreover, chronic bioassays not infrequently yield borderline results that present difficulties in interpretation. In addition, chronic bioassay alone does not provide the mechanistic information needed for proper hazard assessment. Thus, it is now being recognized that a systematic approach to carcinogen testing is required to minimize the testing required for hazard identification, and at the same time, to provide information on the mechanism of action of a test chemical found to increase neoplasia. These considerations are especially relevant for the testing of drugs which are by their very nature agents that produce pharmacological effects and consequently may produce an increase in neoplasms under the extreme conditions of bioassay which are not comparable to normal therapeutic use.

7 The decision point approach

Various approaches to safety testing of chemicals have been developed [23, 25, 65, 66]. As a model for a systematic approach to the evaluation of carcinogenicity, the 'Decision Point Approach' developed by Weisburger and Williams in 1978 [67] and later amplified [11, 12, 69] will be presented. This approach was formulated to incorporate into safety assessment the concept that chemicals could lead to an increase in the incidence of neoplasms in exposed animals by several distinct mechanisms each having different theoretical and practical implications.

The decision point approach takes the two main categories of DNA-reactive and epigenetic carcinogens into account in two ways: (1) a battery of short-term tests is constructed based upon an effort to identify DNA-reactive carcinogens at an early stage and to include systems that may respond to epigenetic agents; (2) testing is structured with the recognition that all forms of subchronic testing may fail to detect chemicals that can produce neoplasms in animals under specific conditions involving chronic administration. An outline of the decision point approach is given in table 3. As shown, it consists of a series of sequential steps of increasing scope. A critical evaluation of the information obtained and its significance in relation to the testing objective is performed at the completion of each phase. Based on this, a decision is made as to whether the data generated are sufficient to reach a definitive conclusion on whether a higher level of testing is required. Attention is paid to both qualitative and quantitative effects. The decision point approach involves five stages of testing which will be discussed as they would apply to evaluation of pharmaceuticals.

Table 3
Decision point approach to carcinogen testing.

Stage A.	Structure of chemical	
Stage B.	Short-term tests in vitro	(1) Mammalian cell DNA repair
		(2) Bacterial mutagenesis
		(3) Mammalian mutagenesis
		(4) Chromosome tests
		(5) Cell transformation
Decision point 1: Evaluation of all tests conducted in stages A and B		
Stage C.	Tests for promoters	(1) In vitro
		(2) In vivo
Decision point 2: Evaluation of results from stages A through C		
Stage D.	Limited in vivo bioassays	(1) Altered foci induction in rodent liver
		(2) Pulmonary neoplasm induction in mice
		(3) Breast cancer induction in female Sprague-Dawley rats
		(4) Skin neoplasm induction in mice
Decision point 3: Evaluation of results from stages A, B and C and the appropriate tests in stage D		
Stage E.	Long-term bioassay	
Decision point 4: Final evaluation of all the results and application to hazard analysis. This evaluation must include data from stages A, B, C to provide basis for mechanistic considerations		

7.1 Structure of drug and biochemical properties

For a number of reasons, the evaluation begins with a consideration of structure, which is a familiar undertaking, of course, for pharmacologists. As for many pharmacological effects, reasonable predictions as to whether or not a given chemical might be carcinogenic can be made for certain structural classes of chemicals [70–72]. This is particularly true in the case of a drug that is structurally related to known DNA-reactive carcinogens, as with dapsone which is an aromatic amine. Among pharmaceuticals, cancer chemotherapeutic agents such as nitrogen mustard and that class of drugs generally, BCNU, and CCNU are DNA-reactive carcinogens [72a]. Any similarly structured alkylating agent would be almost certainly a carcinogen.
Another consideration is the potential of the drug to liberate a DNA-reactive or otherwise carcinogenic molecule upon metabolism. Several

examples can be cited. Chlornaphazine, which formerly was used in the treatment of polycythemia vera, yields reactive of derivatives 2-naphthylamine, a known bladder carcinogen. It is probably on this basis that chlornaphazine produced bladder cancer in treated humans [7].

Some local anesthetics contain aniline derivatives incorporated by an amide bond which is readily hydrolyzed, thereby releasing the aniline derivative. Thus, lidocaine liberates 2,6-dimethylaniline [73], which is mutagenic and likely carcinogenic, since the closely related 2,4,6-trimethylaniline (mesidine) is a carcinogen.

DNA-reactive derivatives can be formed from drugs in other ways. Many drugs contain available amino groups, often tertiary, which can be nitrosated to form nitroso derivatives having the essential structures of carcinogenic nitrosamines. In the case of secondary amines, the more acidic the amino group is, the more readily the reaction takes place. Examples of drugs that can be nitrosated are piperazine, cimetidine, ranitidine, phenacetin, oxytetracycline [74]. Such nitrosation might be the basis for the carcinogenicity of the H_2-receptor antagonist tiotidine [74a].

The assessment of possible DNA-reactive metabolites is greatly assisted if information is available from pharmacodynamic studies. In particular, if radiolabeled material is available, DNA-binding studies should be performed [75].

Information on structure and potential metabolites permits predictions of carcinogenicity and also provides a guide to the selection of appropriate test systems at later stages.

7.2 In vitro short-term tests

A variety of in vitro assays for mutagens and carcinogens is available [23, 76–81]. The utility of several of these in identifying DNA-reactive drugs has been described [82–85]. Extensive application of these tests, however, has shown that no single system has detected all carcinogens tested. Therefore, multiple tests are essential. Moreover, some types of tests are not appropriate for certain agents, for example, bacterial mutagenicity tests in the case of antibiotics.

The two critical elements of a short-term test are the endpoint and the metabolic parameters. The endpoint should be reliable and of definite biological significance. Most in vitro tests identify genetic effects, and thus, would detect only carcinogens of the type classified as DNA-reactive. If epigenetic agents are to be detected, additional tests will have to be developed. One type is discussed in section 7.41 dealing with assays for promoters.

Most tests require a mammalian enzyme preparation to provide for metabolism of procarcinogens and this factor is often the key limiting element of a test series. Thus, a test that employs an enzyme preparation for bioactivation does not expand the scope of a battery in which other detection systems are dependent upon similar preparations.

The rationale for test selection in the decision point approach has been presented in detail [69]. Briefly, the battery includes tests for the three principal genetic endpoints in mammalian cells, DNA damage, mutagenesis, and chromosome effects, in addition to bacterial mutagenesis. Transformation is included as an optional test to supplement equivocal data and because it may detect epigenetic agents. For each of the primary tests, generally available systems have been selected. Cumulative data support the sensitivity and specificity of this battery, but other equally useful tests such as mutagenicity in fungi or *Drosophila melanogaster* could be substituted if these are available.

This battery meets the current requirements of most regulatory agencies [86].

7.21 DNA damage

DNA damage can be measured in numerous ways. The three major types of DNA damage produced by chemicals, base damage, cross linkage and strand breakage all elicit DNA repair synthesis [38, 87]. Therefore, DNA repair has been recognized as a general test for DNA damage [23, 88]. A variety of systems have been used for this purpose [23], but the hepatocyte primary culture/DNA repair test of Williams [89] is now widely recommended [14, 17, 23, 90].

The hepatocyte/DNA repair test uses freshly isolated, non-dividing liver cells which can be derived from various species, including rat, mouse, hamster and rabbit. Cultured hepatocytes metabolize carcinogens and respond to DNA damage with DNA repair synthesis, which is measured autoradiographically or biochemically. The former is recommended because of its greater specificity. This assay has demonstrated substantial sensitivity and reliability with a wide variety of activation-dependent procarcinogens [91–93]. In the battery, it offers the specific advantage of providing whole cell metabolism. Accruing evidence indicates that chemicals active in this system, and also in bacterial mutagenicity tests, are almost certain to be carcinogenic [91, 93]. New approaches under development include the detection of DNA fragmentation [93a] and SOS repair responses to damage of bacterial or viral DNA by genotoxic agents [94–96]. These remain to be validated by a systematic study of known carcinogens. Their advantage would be speed of evaluation.

7.22 Bacterial genotoxicity and mutagenesis

Valuable bacterial screening tests have been developed in the laboratories of Ames [97, 98] and Rosenkranz [99, 100]. The Ames test measures back mutation to histidine independence of histidine mutants of *Salmonella typhimurium* and can be conducted with strains that are also repair deficient, that possess abnormalities in the cell wall to make them permeable to carcinogens, or that carry an R factor enhancing mutagenesis. Hence, these organisms are highly susceptible to mutagenesis making them sensitive indicators. Recently, strains that are mutated by reactive oxygen have been introduced [101].

The test developed by Rosenkranz and associates [99, 100] utilizes DNA repair-deficient *Escherichia coli* and measures their enhanced susceptibility to cell killing by genotoxins. In this system, a chemical that interacts with DNA is more toxic to the repair-deficient mutants than to wild type organisms because the mutant strain cannot repair the DNA damage. Thus, by measurement of relative toxicity, an indication of DNA interaction is obtained.

The bacteria used in these tests possess some bioactivation capability, but generally are dependent upon mammalian enzyme preparations for metabolism of carcinogens. For certain carcinogens, preincubation of the compound and the biochemical activation system with the test organism has enhanced detection [102, 103].

7.23 Mammalian mutagenicity tests

Mammalian mutagenicity assays are a valuable addition to the bacterial assays to corroborate gene effects in the more complex mammalian genome and for the testing of bacteriocidal agents such as antibiotics. Also, mammalian cell mutagenesis can be quantified according to the incidence of mutants in the cell population. A variety of techniques for identifying mutants has been described [104–106].

Three mutational assays in mammalian cells that have been widely used for carcinogen screening are mutagenesis at the hypoxanthine-guanine phosphoribosyltransferase (HGPRT) locus [107–109], thymidine kinase (TK) locus [110], or plasma membrane adenosine triphosphatase (ATPase) locus [111, 112]. In the HGPRT mutagenesis assay, mutants lacking the purine salvage pathway enzyme are identified by their resistance to toxic purine analogs such as 8-azaguanine or 6-thioguanine, which kill wild type cells that can utilize the analogs. This assay has the advantage over ATPase mutagenesis, measured as ouabain resistance, in that it involves a nonessential function, and for this reason, there are no lethal mutants. Its advantage over the measurement of TK deficient mutants by resistance to thymidine analogs is that the gene for HGPRT is on the X-chromosome rather than a somatic chromosome, as with TK. Hence, there is only one functional copy in each cell and consequently mutations in wild type cells can be determined, whereas a heterozygous mutant is required for measurable mutation to homozygous TK deficiency. A potentially useful aspect of mutagenesis at the TK locus is the possibility that both gene and chromosomal mutants can be identified [113], but the utility of this aspect in carcinogen detection has not yet been established. In all of these forward mutation assays, all types of mutations, e. g. point, frameshift, deletion, would be detected.

The target cells used in purine analog resistance assays have almost all been fibroblast-like, such as the V79 and CHO lines. Fibroblast lines have displayed little ability to activate carcinogens, other than polycyclic aromatic hydrocarbons. This deficiency has been overcome in several systems by providing exogenous metabolism mediated by either enzyme preparations [108, 109] or cocultivated cells [107, 114–116]. The former again provides no extension in metabolic capability over that used for bacterial systems. In contrast, the use of freshly isolated hepatocytes as a feeder system [114, 116] offers additional possibilities because the metabolism of hepatocytes has been shown to be different than that of liver enzyme preparations [116–119]. Organ specificity can sometimes be reproduced in whole cell systems [116]. HGPRT mutagenesis can also be measured in human cells, which can be combined with hepatocytes for biotransformation [119a]. Such systems provide clear evidence of effects on the human genome, although thus far, mutability of human cells has not differed qualitatively from that in other animal cells. Also, liver epithelial cell lines with intrinsic activation capability have been used in mutagenicity studies [120].

7.24 Chromosome tests

Chromosome tests are of conceptual importance because they reveal effects at a higher level of genetic organization than do mutagenesis assays. Also, some chemicals may be chromosome mutagens but not gene mutagens. This could be of particular importance in light of the concordance between the breakpoints involved in chromosome translocations and oncogene localization [120a].

A variety of chromosome aberrations can be measured [81]. Some alterations, however, are of uncertain significance. The basis for gaps (achromatic regions) is unknown and these should be distinguished from true breaks. Breakage itself may reflect only cytotoxicity unless heritable aberrations are demonstrated.

A variety of cell types is used to measure chromosome aberrations. The use of permanent stable lines seems to be preferable to freshly isolated cells, such as human peripheral leukocytes, because the genome of cells in primary culture may be artificially fragile during their adaptation to a new environment. Moreover, only in continuous lines can it be demonstrated that aberrations are heritable. The correct determination of chromosome aberrations requires an expert cytogeneticist and the necessary karyotype analysis can be very time-consuming.

As an alternative, measurement of sister chromatid exchanges (SCE) overcomes these problems and has shown sensitivity to carcinogens not readily detected in other in vitro assays [120b–122]. Moreover, a valuable correlation with potential human effects can be made, since SCEs can be monitored in peripheral lymphocytes of exposed populations [122a]. Chemically-induced SCEs are usually assessed in CHO or V79 cells [123] because of the simplicity of their karyotype. These cells require supplementation with an exogenous metabolizing system for biotransformation of most carcinogens. Human liver preparations have been used for this purpose [124]. Whole cell systems can also be used to mediate SCE induction in co-cultured target cells, including human cells [125, 126]. In addition, liver cell systems with intrinsic activation capability are available [127–129]. The determination of an increase in SCEs must be made cautiously; the background appears to be a consequence of the procedures used for differentiation of chromatids and changes in the components of the culture medium and perhaps their utilization by cells can increase the background [130]. Therefore, anything less than a doubling of SCEs should be regarded as an equivocal result.

7.25 Cell transformation

The hamster fibroblast system introduced by Sachs and associates [131] for transformation studies has been developed into a reliable screening test [132–134]. Other transformation assays using fibroblasts are also available [135, 136]. In addition, because human cancers usually involve epithelial tissues, transformation in epithelial systems is actively being pursued [137–140]. The correlation between induction of transformation and carcinogenicity of chemicals appears to be good in several systems. Transformation assays, unfortunately, are not presently as reproducible or as widely available as other tests and, therefore, we recommend them only to supplement results from the preceding four systems. Nevertheless, a potential advantage of transformation is that it may provide a general indication of the carcinogenicity of chemicals, either through genotoxic or epigenetic mechanisms. As examples of the latter, certain naturally occurring hormones and sterols that have not been shown to be DNA-reactive [141] have led to cell transformation [142–143].

7.3 *Decision point 1*

Performance of the two steps, evaluation of structure and genetic toxicology testing, provide a basis for decision making at this stage. If clear-cut evidence of genotoxicity in three or more tests has been obtained, the chemical is virtually certain to be a carcinogen and regardless, represents a clear genetic hazard. As an example of the latter, one form of drug toxicity that may have as its basis an effect on DNA is drug-induced systemic lupus erythematosus. This adverse reaction is seen predominently in slow acetylator individuals exposed to drugs such as isoniazid, hydralazine, and procainamide which are substrates for acetylation [144]. Evidence has been provided that slow acetylation can permit DNA reaction by hydralazine and it has been suggested that this might result in a low level of DNA alteration rendering the DNA immunogenic [145].

Positive results in two tests makes the agent highly suspect. In particular, positive results in the Ames test and that of Williams are a strong indication of potential carcinogenicity [93].
Positive results in only one test must be evaluated with caution. In particular, several types of chemicals such as intercalating agents are mutagenic to bacteria, but not reliably carcinogenic [146].
A statistical approach to the evaluation of data from short-term tests has been developed by Rosenkranz and colleagues [147]. This approach, known as the Carcinogenicity Prediction and Battery Selection, used Bayes's theorem to compute the predictivities of individual assays from which the overall predictivity of a battery can be calculated. This procedure can be readily applied to data from the decision point approach battery.
If the results from in vitro tests are equivocal, there is a variety of in vivo tests that could be used to gain information on genotoxicity. These include the dominant lethal test, specific locus test, heritable translocation test, host-mediated mutagenicity, chromosomal damage, testicular DNA synthesis inhibition, sperm abnormality, sebaceous gland suppression, granuloma pouch assay and DNA fragmentation or repair in various organs [148]. Chemicals negative in all the in vitro genotoxicity tests are rather unlikely to be positive in any one of these in vivo tests, except for chemicals that are activated to DNA-reactine metabolites by host bacteria, for example, cycasin and certain single ring nitroaromatic compounds [149, 150]. To allow for this, the in vitro systems can be modified to detect agents requiring bacterial metabolism by incorporating bacterial enzymes [149]. The in vivo tests are more complex and costly than in vitro tests and, therefore, they are not recommended as primary screening tests. Nevertheless, they may help to clarify the data base on drugs that are negative or equivocal in the in vitro tests. In addition, the in vivo genotoxicity tests may serve the addition purpose of providing evidence for DNA damage in specific target organs [148].
The in vivo genotoxicity tests, however, do not provide definitive evidence of carcinogenicity. In contrast, there is a variety of limited in vivo bioassays that accomplish this (sect. 7.6). Drugs with structures that suggest possible sites of activation may reveal their carcinogenicity in limited in vivo bioassays.
If the in vitro test systems yield no indication of genotoxicity, the strategy for further testing depends mainly upon the structure and

known pharmacological properties of the drug. Hormonally-active substances, immunosuppressants and neoplasm promoters which are not DNA-reactive [50, 141, 150a, 150b] can nevertheless produce neoplasms in animals through complex and as yet poorly understood mechanisms. For these, it is unlikely that the limited in vivo bioassays would yield positive results when such materials are tested alone. The next step that is recommended for such chemicals is specific mechanistic studies or an assay for promoting activity, depending upon structure.

Compounds other than those of the strict androgen and estrogen type can nevertheless exert effects on endocrine glands. Such chemicals can lead to cancer in animals given high levels mainly because they alter normal hormone levels and balances [151]. It is known, for example, that certain drugs lead to release of prolactin, thyroid stimulating hormone or other hormones from the pituitary gland. Chronic intake of such drugs causing higher blood and tissue hormone levels, would, in turn, alter the levels of other hormones. At this time, any substance with such properties needs to undergo a chronic bioassay with carefully selected doses to determine whether neoplasms would be produced in endocrine-sensitive tissues. To circumvent this time consuming effort, more research on methods to effectively test for such properties is required.

A recent example of another type of possibly endocrine-mediated neoplasm induction is the finding that the antisecretory ulcer medication omeprazole produced carcinoid tumors in rat stomach [151a]. The enterochromaffin-like cell from which carcinoids arise are stimulated by gastin which is elevated by suppression of gastric acid production.

Several types of agents lacking DNA-reactivity have produced primarily or exclusively liver neoplasms in rodents [152, 153]. Among these, the potential of several barbiturates [154–156] and hormones [157–160] to act as enhancers of liver neoplasm development has been described. As yet, the structural requirements for promoting activity are poorly understood, outside the class of phorbol esters. Two properties that have been suggested to be of importance are induction of the cytochrome P450 system [161] and alteration of the cell membrane [162]. Evidence of these properties from pharmacological studies would suggest the need for a promotion assay.

Thus, where possible promoting potential might be suspected from several types of evidence such as chemical relationship to known neoplasm promoters, an effect on organ growth, hormone-like properties, and such, the chemical can be tested in the assays for neoplasm promoting agents described in section 7.4.

7.4 Assays for neoplasm promoting agents

The tests recommended at this stage are primarily for nongenotoxic agents. Among carcinogens that operate through epigenetic mechanisms, those that enhance the effects of genotoxic carcinogens or facilitate the development of neoplasms comprise a major group, including a number of drugs and hormones [35]. Both in vitro and in vivo assays for promotion are available. The in vitro assays have not been as extensively validated as the in vivo ones, but their potential seems extremely promising. Also, since they are relatively simple, they are recommended as a first approach.

7.41 In vitro tests for neoplasm promoting agents

Numerous studies on cells have been done with skin neoplasm promoters, 12-0-tetradecanoylphorbol-13-acetate (TPA) and related phorbol esters, and teleocidin B, using in vitro systems to delineate effects [162, 163]. Responses related to interactions with specific membrane receptors have been noted [163, 164]. Such effects, however, have not been found yet with other types of promoters.
The transformation culture systems described in section 7.25 can be modified and used to test for promoting substances if a limited amount of a carcinogen is applied followed by the test agent [165–168]. Other approaches that have been proposed include the determination of sister chromatid exchanges [169] and aneuploidy [170].
Based on the concept that important informational molecules are exchanged between cells in contact through gap junctions, the groups of Trosko [171,172] and of Murray and Fitzgerald [173]have developed systems to detect promoting potential by the specific property of chemicals to block the intercellular communication that occurs through the gap junctions. A similar approach using liver cells [174,175] provides the adavantage of intact cell metabolism.
The inhibition of intercellular molecular exchange in vivo would cause cells with an abnormal genome to be isolated from the growth controlling elements provided by neighboring normal cells and there-

by release them for progressive growth. Thus, this effect may actually be one basis for neoplasm promotion.

7.42 In vivo promoting assays

Promoters of neoplastic development often display a high degree of organ specificity. Consequently, in vivo assays involve the administration of a DNA-reactive agent that affects the organ in which the promoting activity of the test substance is to be examined. It must be remembered that DNA-reactive agents can enhance carcinogenicity when given in sequence with another DNA-reactive carcinogen [176], probably as a result of summation of DNA damage [177]. In fact, this has been proposed as a sensitive means of identifying DNA-reactivity of a test agent given before or after a genotoxin [177]. Therefore, enhancement of carcinogenesis in sequential exposure is not proof of promotion, unless other properties of the chemical are established [35].
To test for promoting activity by a chemical in liver carcinogenesis, rapid in vivo bioassay tests have been designed in which a few doses of a limited amount of an appropriate hepatocarcinogen such as diethylnitrosamine or N-2-fluorenylacetamide is followed by the test chemical [178–181]. Using histochemical markers such as gamma glutamyl transpeptidase [179,180] or iron exclusion [181] enhancement of carcinogen-induced altered foci can be identified within 6 to 12 weeks as evidence of promotion. Since the appearance of gamma-glutamyl transpeptidase in foci is also associated with elevated serum levels [182], the latter is a convenient way of monitoring promoting activity or indeed hepatocarcinogenicity, as will be discussed.
Drugs that are enzyme inducers, in particular those that in a chronic bioassay have given some evidence of liver tumor induction in mice and less often in rats, have shown promoting activity in such systems. These include phenobarbital [154], barbital [155], pentobarbital and amobarbital [156]. Testing of a series of benzodiazepines in this manner revealed them to be inactive [182]. Interestingly, acetaminophen, although it is hepatotoxic, did not promote liver carcinogenesis, but did enhance the incidence of kidney adenomas after N-ethyl-N-hydroxyethylnitrosamine [182a].
Promoting activity can be tested in almost any tissue. Mouse skin exposed to small doses of benzo(a)pyrene or 7,12-dimethylbenz(a)anthracene has been a popular model for tobacco smoke components

and related materials [183,184]. A substance exhibiting endocrine properties or one affecting endocrine balances in general, likewise may show an effect in modifying breast or endometrial cancer induction in animals given limited amounts of methylnitrosourea as an initiating dose, or utilizing mice with or without the mammary tumor virus [185,186]. Similarly, promoters for urinary bladder or colon cancer may be assessed by administration following pretreatment with limited amounts of a carcinogen for these organs [187–190]. As will be discussed, such schemes, when designed with a number of dose levels including the possible therapeutic levels, would provide the necessary background information for health hazard analysis and the establishment of a threshold level.

7.5 *Decision Point* 2

A positive response in an in vitro test for inhibition of intercellular molecular exchange is highly suggestive of potential promoting activity. Several agents that have found to be positive in one of these systems have subsequently been demonstrated to be promoters [191,192]. However, it has not been possible to extensively validate this approach yet, because of the paucity of proven neoplasm promoters outside of the class of phorbol compounds. Therefore, at present, it cannot be concluded that a positive chemical will definitely be a promoter. Consequently, activity in this system indicates the advisability of an in vivo test for neoplasm promotion. A negative result in one of these tests can only be interpreted as limited evidence against a promoting action because of the many possible mechanisms for promotion apart from membrane effects.

A nongenotoxic chemical that is active in an in vivo test for neoplasm promotion must be regarded as being likely to produce an increase in neoplasia upon chronic adminstration of high levels. Therefore, the finding of a positive effect indicates the need for further safety testing. This should include a multi-dose administration following a short course of the appropriate DNA-reactive carcinogen, which may permit the delineation of a no-effect level.

Neoplasm promoters are almost certain to be negative in the limited in vivo bioassays for carcinogenicity and, therefore, no point is served by submitting the chemical to testing of this type; the only resort is chronic bioassay. Since the finding of promoting activity indicates potential carcinogenicity, a chronic bioassay should be undertaken with a view to establishing a possible no-effect or threshold level by means of testing over a broad dose range. The finding of a no-effect level in chronic bioassay coupled with evidence for a lack of DNA-reactivity and a promoting action would lead to a health hazard assessment distinctly different of that for DNA-reactive carcinogens (sect. 9).

7.6 Limited in vivo bioassays

This stage of evaluation employs tests that will provide further evidence of the potential hazard of chemicals having limited evidence for genotoxicity without the necessity of undertaking a full-scale chronic bioassay. Also, certain of the tests will provide relative potency ratings when the design includes positive controls.

At this stage, the in vivo tests recommended are those that will provide definitive evidence of carcinogenicity, in a relatively short period (i.e. 30 weeks or less). Unlike the in vitro tests, these are not applied as a battery but rather used selectively according to the information available on the chemical. Of particular value in test selection would be pharmacodynamic data on tissue localization of the drug. This might include autoradiographic study of tissue distribution [193].

As in all in vivo carcinogenicity bioassays, such tests must be designed with an adequate number of animals in each group, including suitable positive, untreated, and where indicated, vehicle controls. At the end of the study, appropriate statistical procedures are used to establish whether or not a given treatment group had evidence of a positive response. To this end, consideration is given to percent incidence, multiplicity of lesions of a given type, and importantly also, the latent period for neoplasm development in the treated groups versus appropriate controls. As in any in vivo bioassay, with the rapid tests described, it is important to plan for termination of the test before the control group exhibits an increase in spontaneous lesions that would decrease the effective detection of any induced neoplasms.

An important point that must be kept in mind in the use of these rapid bioassays is that while positive results are a strong indication of carcinogenicity, negative results do not denote noncarcinogenicity, since the chemical may have another target site.

7.61 Altered foci induction in rodent liver

Research in a number of laboratories has established that during liver carcinogenesis several distinct hepatocellular lesions precede the development of carcinomas [193a]. The earliest of these, the altered focus, when sufficiently developed, can be demonstrated in routine histologic tissue sections. Altered foci are abnormal in a number of properties that permit their reliable and objective identification at early stages by more sensitive techniques. Altered foci in rat liver, and somewhat less regularly in mouse liver, display abnormalities in the enzymes gamma-glutamyl transpeptidase, glucose-6-phosphatase and adenosine triphosphatase that can be used for histochemical detection [194,195]. Another important marker for foci that permits histochemical identification is their exclusion of cellular iron [196]. This latter property is more sensitive in some situations than the enzyme abnormalities and also, unlike the enzyme abnormalities, characterizes mouse and hamster liver lesions in addition to rat.

Induction of altered foci (or nodules) in rodent liver has been used by several groups as a rapid means for detecting carcinogens [195–199]. With known liver carcinogens, foci have been detected within three weeks of carcinogen exposure and in high numbers by 12 to 16 weeks of exposure. Therefore, one approach is that of 12 to 16 weeks exposure to the test chemical (dietary or gavage), with subcutaneous injection of iron dextran (imferon) during the last two weeks to produce the hepatic siderosis that permits delineation of iron-excluding foci. Sections can be stained for both gamma-glutamyl transpeptidase and iron and the number of foci in standard sections from all lobes determined.

Another means of detecting precancerous liver lesions, which was developed by Farber and coworkers [197], is based upon the resistance of altered cells to the cytotoxic effect of chemicals. In this approach, administration of the test chemical is followed by exposure to a cytostatic agent and partial hepatectomy. The cytostatic agent is preferentially metabolized by normal liver cells and affects them so that they cannot proliferate in response to the partial hepatectomy. In contrast, the cells in the liver altered by the test chemical proliferate and become extremely conspicuous [197–199].

Another finding that may have application in carcinogen screening involves the observation by Remandet et al. [182] that induction of gamma-glutamyl transpeptidase-positive foci is accompanied by an

elevation of serum levels of the enzyme. This offers the possiblity of monitoring animals for hepatocarcinogenic effects and determining the most appropriate time of termination.
Since the liver is involved in the biotransformation of so many drugs, these tests for induction of the early stages of hepatocarcinogenesis would be the first choice for a limited in vivo bioassay for drugs. In such systems (table 4), several drugs have been positive [199a] including the antihistamine methapyrilene [200], while a series of benzodiazepine tranquilizers have been negative [182]. A candidate class for testing in this system is the beta-blocking anti-hypertensive agents, one of which has been reported to rapidly induce liver neoplasms in female rats [200a]. The positive results in this assay correlate well with carcinogenicity, but negative results do not, since both clofibrate and phenobarbital produce liver neoplasms upon chronic administration yet have been negative. Nevertheless, the negative findings in this assay provide evidence that these agents are not DNA-reactive.

Table 4
Pharmaceuticals tested in limited in vivo bioassays.

(1)	*Altered foci in rodent liver*	
	Positive:	Negative:
	azaserine	clofibrate
	hycanthone methylsulfo-nate	diazepam
	methapyrilene	oxazepam
		phenobarbital
		piperidine
		acetaminophen
(2)	*Pulmonary neoplasms in the mouse*	
	Positive:	Negative:
	cyclophosphamide	adriamycin
	dapsone	aminopterin
	hydralazine	cortisone
	isoniazid	emetine
	melphalan	myleran
		phenformin
		phenazopyridine
		tolbutamide
(3)	*Breast cancer in the female Sprague-Dawley rat*	
	Positive:	Negative:
	procarbazine	dapsone
		acetaminophen
(4)	*Skin neoplasms in the mouse*	
	Positive:	Negative:
	Nitrogen mustard	proflavin

7.62 Pulmonary neoplasm induction in mice

Andervont and Shimkin pioneered with the model involving the development of lung neoplasms in specific sensitive strains of mice, especially the A/Heston strain and related strains like the A/J. In this system, agents are administered orally or by injection and the incidence and multiplicity of pulmonary neoplasms is determined grossly, usually within 20–40 weeks [201]. A singular advantage of this assay system is that, in addition to an end-point measuring the percent of animals with neoplasms compared to controls, the multiplicity of neoplasms is a further parameter of carcinogenic potency. Another useful aspect of this assay is that significant results are obtained in as short a time as 30–35 weeks, and sometimes faster. Extension of the test for a longer period is not desirable since the incidence of pulmonary neoplasms in control animals increases rapidly after 35 weeks, and thus the test loses sensitivity. Most chemicals that are active in this system are also carcinogenic in other longer, chronic animal tests. The induction of lung neoplasms, however, is structure-dependent, and not all types of chemical carcinogens yield a positive response. Thus, as has been emphasized, a negative response in this rapid in vivo test, does not necessarily mean a compound can be considered safe.
Several drugs have produced pulmonary neoplasms in mice (table 4). Other structurally-related pharmaceuticals might, therefore, be positive in this system.

7.63 Breast cancer induction in female Sprague-Dawley rats

The rapid induction of neoplasia in the mammary gland of young female, random-bred Wistar and, to a greater degree, Sprague-Dawley rats was discovered by Shay, and elegantly extended by Huggins. With many carcinogens, the optimal age for effective breast cancer induction is about puberty or days 45–60 of age, when there appears to be maximal cell proliferation [202]. With some agents, the age factor is not so crucial and induction of neoplasms is effective using the typical protocol for a chronic bioassay [203]. A positive response is indicated by an increase in the incidence and multiplicity of rats with mammary neoplasms. With powerful carcinogens, especially select polycyclic hydrocarbons, arylamines, or nitrosoureas, mammary gland neoplasms are induced rapidly, in less than nine month. The multiplicity

of mammary neoplasms provides an additional quantitative criterion to denote relative strength of the carcinogenic stimulus. An advantage of this system is that the lesions can be visualized or palpated without killing the test animals.
Relatively few drugs have been tested in this system (table 4).

7.64 Skin neoplasm induction in mice

The induction of skin neoplasms in mice has been extensively used as a bioassay [204]. The carcinogenicity of a limited number of chemicals and crude products can readily be revealed when continuous application of the material to the skin of mice produces papillomas or carcinomas, or yields sarcomas upon subcutaneous injection. Any activity in producing neoplastic conversion also can be rapidly determined by the concurrent or sequential application of a promoter, such as one of the phorbol esters. Tars from coal, petroleum, or tobacco [205] are active in such systems, as are the pure polycyclic aromatic hydrocarbons and congeners contained in such products. Mouse skin responds positively because it appears to have the enzymes necessary for metabolic activation. On the other hand, many chemicals that are positive in this system rarely yield visceral neoplasms, such as in the liver, principally because the chemicals are detoxified systemically. However, lung and lymphoid tissues in sensitive mouse strains can be secondary sites of neoplasms. By selective breeding, strains of mice (SENCAR) were developed which demonstrate exquisite sensitivity to carcinogens and promoters for mouse skin [206].
A positive response in this system, as in others, is indicated by an increase in the number and multiplicity of papillomas and carcinomas in a statistically significant manner over controls treated and maintained identically, except for the presence of the test compound. Such comparisons can involve also the latent period of appearance and permanence of neoplasms. As with the induction of breast cancer in female rats, the lesions can be visualized or palpated without killing the test animals.
At least several hundred chemicals or materials have been tested on mouse skin. This system is useful primarily for chemicals such as polycyclic hydrocarbons and direct-acting chemical carcinogens such as sulfur or nitrogen mustard, bis (chloromethyl)ether, propiolactone, and alkylnitrosoureas. Arylamines and related carcinogens by them-

selves usually do not provide a positive response on mouse skin although there are some exceptions. This approach has not been widely used for pharmaceuticals (table 4) and no specific applications are presently evident.

7.7 *Decision point* 3

Proven activity in more than one of the limited in vivo bioassays may be considered unequivocal qualitative evidence of carcinogenicity.
A definite positive result in one of the limited in vivo bioassays together with two positive results in the battery of genotoxicity tests, also indicates potential carcinogenicity. This is true especially if the results were obtained with moderate dosages and more so, if there was evidence of a good dose response, particularly as regards the multiplicity of the in vivo lesions.
Positive results in one in vivo bioassay and in only one in vitro test makes the agent highly suspect, but further testing is indicated.
A negative result in any of these limited bioassays does not constitute proof of noncarcinogenicity since they all have definite limitations stemming from the organ selectivity of carcinogens.

7.8 Chronic bioassay systems

In the decision point approach, chronic bioassay is generally used as a last resort for resolving questionable results in the more limited testing or in the case of compounds that are negative in the preceding stages of testing, for the final determination of carcinogenicity. In the application to pharmaceuticals, however, it will usually be necessary in order to obtain marketing approval, to perform a chronic bioassay on drugs that were not eliminated by their adverse effects in the preceding stages of testing.
The ultimate goal of a carcinogenicity bioassay is to provide data that permit evaluation of the carcinogenic hazard for humans. Because of the fact that virtually all chemicals that have led to cancer in humans have also been carcinogenic in animal models, it has generally been assumed that chemicals that are carcinogenic to rodents represent a cancer hazard to humans. The accumulation of knowledge on the responses of rodents to high level exposure to chemicals (sect. 7.92) and current understanding of the varied mechanisms of carcinogenesis, as

discussed in sections 4 and 5, suggest that this may not always be the case. Therefore, the chronic bioassay must be designed to yield information that can be properly applied to hazard assessment.
The detailed methodology of chronic bioassay is described in several monographs [15, 16, 19, 23] and generally, regulatory agencies have established criteria for an acceptable test [14]. Some critical elements that bear upon eventual interpretation will be reviewed here. A point that should be borne in mind in the design of chronic bioassay is that the test should be constructed to be appropriate to the drug.

7.81 Purity of drug to be tested

Impurities in a chemical may modify, i. e., enhance or diminish, toxicity and carcinogenicity. Moreover, minor constituents can actually be responsible for carcinogenicity at high level testing. Therefore, it is essential to have detailed information on the composition of the test material. Where the drug is a formulation with other constituents, it may be useful to test both the formulation and the pure main component. If only the formulation is tested, positive results may be difficult to assess. Also, data must be available on the stability of the drug under the conditions of administration to animals.

7.82 Selection of animals

A comprehensive assessment of animal carcinogenicity is best performed on more than one species. Indeed, even certain powerful carcinogens that affect humans are not active in all strains or all species [37]. For example, 2-naphthylamine, for reasons that have not been established, is a weak carcinogen in most strains of rats, but is rather active in mice and hamsters. On the other hand, the mold product aflatoxin B_1, which is suspected as a cause of liver cancer in humans, does not elicit a neoplastic response in mouse liver under the customary bioassay conditions, although administration to newborn mice has produced a carcinogenic effect to the liver and in older mice to the lung.
The selection of the test species should be made with specific regard to the properties of the pharmaceutical to be tested. For example, in the case of a medication that has estrogenic properties, current views are that such an agent cannot be evaluated properly for carcinogenic potential in a rodent system that does not have an estrus cycle comparable

to that of the human female. Nevertheless, drugs of this type can be tested in conventional species for possible carcinogenic effects in non-endocrine target organs.

Another consideration of particular importance in selection of the species for testing of cyclic or heterocyclic drugs with extracyclic nitrogens is the acetylation activity of animals. Humans display a genetic polymorphism in activity of N-acetyltransferase [206a]. This is mimicked in rabbits [207], but most other species can be classified as either rapid or slow acetylators [208]. Thus, for drugs that are substrates for acetylation such as isoniazid, hydralazine and procainamide, a particular rodent species will provide a model only for humans with the corresponding acetylation activity. For this reason, it may be desirable to test the drug in both slow and rapid acetylating species.

Early carcinogenicity studies used larger animals such as rabbits and dogs. Because it is now known that the duration of the overall carcinogenic process is to some extent proportional to the lifespan of a given species, model bioassay systems have concentrated on small rodents such as mice, rats, or hamsters of various strains, which have an average lifespan, under good conditions, of two to three years and develop neoplasms within a shorter time frame. The use of large species, including primates, is dictated only by special considerations, such as obtaining data that might be particularly relevant to human effects.

There are numerous strains of inbred and non-inbred rats, mice and hamsters. Although some have been routinely used in various national testing programs, this does not mean that which they are necessarily the most appropriate.

Among the rats strains used for carcinogen bioassay, the Fischer strain is more sensitive than the Sprague-Dawley to aromatic amines. The non-inbred Osborne-Mendel strain displays a particular sensitivity to certain chlorinated hydrocarbons [208a].

With mouse strains, a particular problem is the high susceptibility of some to spontaneous neoplasms, which can complicate the interpretation of results. In particular, several strains of mice readily develop liver neoplasms [208b]. This response has been criticized as not always reflecting a true carcinogenic risk to humans, even though a number of human carcinogens yield mainly liver neoplasms in mice [208c]. As discussed, a positive response in such mouse strains may be seen also with a nongenotoxic promoter or estrogenic hormones.

Non-inbred Syrian hamsters offer several advantages in bioassay (see

[209]). With aromatic amines and azo dyes, hamsters develop urinary bladder neoplasms when high levels of chemical are given. Thus, the hamster can serve as a model for the mechanism of the induction of bladder cancer by such chemicals in humans. Hamsters are quite sensitive to liver carcinogenicity by dimethylnitrosamine [210] because of their deficiency in repair of alkylation damage to DNA [211]. Hamsters display a profile of bile acids that is more similar to the human pattern than that of rats or mice [212].

An important factor in species selection is to obtain an animal model whose metabolism and disposition of the drug, as well as the pharmacologic effects produced by it, are as similar as possible to those that would occur in humans. Where metabolism and pharmacokinetic data show that the customary rodents are not appropriate, they should not be used.

Newborn animals often are more susceptible to chemical carcinogens than older animals [213]. Also, a number of carcinogens have been active with transplacental exposure [214, 215]. The use of perinatal exposures in carcinogen testing is receiving increasing attention [216], especially in the light of human findings with diethylstilbestrol. The mechanisms, however, may be complex, as in the tests of saccharin. Here, 2-generation exposure to 5% saccharin led to a modest yield of bladder cancer in males in both generations, but not females [217]. Since it is now known that saccharin is not DNA-reactive [218] and acts by a promoting mechanism [187–189], the results of the 2-generation test may stem from an artifactual metabolic imbalance [219, 220] that seems to have no counterpart in humans consuming foods containing commonly used levels of saccharin. Thus, high level exposure tests over 2-generations need special justification in respect to potential human exposure modalities. In the case of drugs intended for use during pregnancy, there may be good reasons for a 2-generation study. In such a study, the litter may be the proper experimental unit, rather than the individual animal as in conventional chronic bioassays [220a].

7.83 Conduct of bioassay

Detailed recommendations on animal husbandry have been made [16] and only a few specifics will be mentioned here.

Adventitious, carcinogens can enter the environment of test animals from a variety of sources [221], especially the food [222,223]. These sources must be closely monitored.

In recent years, consideration has been given to formulating a standard diet for carcinogen bioassay that is uniform not only in percent composition of nutrient elements, as are commercial diets, but also in specific ingredients. A few laboratories have used a semi-purified diet formulation. However, it has been found in a number of instances with known carcinogens that such chemicals are often more toxic when mixed with a semi-purified diet than when fed in a commercial good-quality chow. The exact reason for this difference is not known and might stem from multiple effects such as the presence in laboratory chows of natural fibers and certain enzyme modifiers leading to an alteration of toxicity.

An important consideration regarding the diet is whether it will be nutritionally adequate under conditions of high level exposure to test chemicals where the agent may reduce food consumption or interfer with utilization of essential nutrients. An example of the latter problem occurs with feeding of high levels of butylated hydroxytoluene where apparent depletion of vitamin K leads to a hemorrhagic disorder [223a]. Similarly, administration of drugs has been reported to decrease liver vitamin A stores [233b].

7.84 Administration of test drug

Test chemicals can be administered in many possible ways including oral, cutaneous, subcutaneous injection, intraperitoneal injection, intravenous injection, bladder instillation or implantation in pellets, intrarectal instillation and inhalation or intratracheal instillation [12]. Drugs intended for oral use are best tested by this route, either in feed or by gavage. For those intended for other routes of administration, testing in the corresponding manner would seem logical. Positive findings using alternatives to oral administration constitute reliable evidence of carcinogenicity, but negative results are often not considered acceptable evidence of noncarcinogenicity. Consequently, nonparenteral routes of administration must be well justified.

For oral administration, test drugs can be mixed in food, added to drinking water, or given by stomach tube. Because chemicals can be mixed with the diet in almost any amount, artifacts may be obtained when a high percentage of the diet is an inert chemical, nutritionally speaking, because all other essential food components would be available in lower amounts unless their concentrations were increased or food consumption was increased by the animal to compensate. If the test agent affects the palatability of the diet, nutritional deficiencies may develop. This can be avoided by proper design involving dietary adjustment to ensure the proper balance of macro- and micronutrients. In testing a drug, it should not account for more than 5 % of the diet. In fact, a top level of 1 % probably would suffice to detect the carcinogenicity of even a weakly active DNA-reactive carcinogen.

The use of gavage or intragastric intubation offers the advantage of quantitative administration of a given dose since rodents do not regurgitate. Moreover, the dose regimen can be established to mimic intended human use. If the drug to be tested is lipophilic, it will require a suitable vehicle. One approach is to use a suspension of finely powdered product in a jelling agent such as agar, methylcellulose, or other such polysaccharide. This allows the preparation of reasonably stable suspensions so that aliquots can be drawn and inoculated into the stomach. A gel-based diet has been formulated that can be used to incorporate diverse substances satisfactorily and can be administered in a reasonably quantitative manner by placing the gel in the cages even without the use of food containers [224].

When a solution is deemed preferable for gavage, edible oils and fats have been utilized. This is an acceptable practice, but it must be realized that use of such oils or fats will contribute a sizable load of additional calories and fat [225]. These nutritional changes may have undesirable consequences in their own right, inasmuch as it is known that the effect of a number of carcinogens, specifically for targets such as mammary gland, large bowel, prostate, endometrium, and probably pancreas, is potentiated by dietary fat [226]. It is essential, therefore, to always utilize not only adequate numbers of vehicle control animals, but to employ a suitable positive control compound administered both in aqueous solution and in the oil vehicle in order to assess the potential modulating impact of the vehicle.

In most studies in the past, the test chemical has been given at a constant dose. A more precise form of dosimetry is to adjust the dose periodically to maintain a constant exposure per unit body weight.

7.85 Dose selection

Several considerations dictate that carcinogen bioassays must include at least one carefully selected high dose level. Firstly, carcinogens, like other toxins, have dose-related effects [227–229] such that the higher the dose level, the more elevated is the response – in this instance, induction of neoplasms. In addition and importantly, the higher the dose, the shorter the time to appearance of neoplasms. Thus, a positive result is seen in a shorter time with a high dose of a carcinogen, with a consequent reduction in the time on test. With two different chemicals exhibiting identical yields of neoplasms, it can be stated that the one inducing neoplasms faster is the more powerful carcinogen.

Carcinogen bioassays were standardized with high-level dosing based on two major considerations. The first and most important one was the realization that high doses were essential to obtain evidence of carcinogenicity with certain known human carcinogens such as 2-naphthylamine. Likewise with phenacetin, tests at substantial doses failed to disclose a carcinogenic effect, although this drug has the structure of an aromatic amine and is mutagenic. It was only when a test was conducted at a carefully selected high dose that the carcinogenic effect was noted, and the human condition reproduced in an animal model. The second reason was that in a series of bioassays involving five or six dose levels, evidence of carcinogenicity was obtained over a range of dosages with powerful carcinogens, but with somewhat weaker carcinogens, the effect was clearly apparent only at the highest dose levels [230]. These considerations together led to the formulation in 1967 [15] of procedures in which preliminary dose ranging was used to delineate the so-called maximally tolerated dose (MTD). Subsequent bioassays were conducted at that dose level, and also at a lower dose level such as ½ or ⅓ the MTD. The lower dose was designed to provide insurance of good survival of at least one group over a chronic 2-year test in case the MTD was somewhat too high and might have led to premature mortality of animals at that level. The approach was designed as a qualitative screening test, albeit, where animals survived at both dose levels, some indication of a dose response might be obtained.

The considerations involved in dose selection have been reviewed by a working group of the International Life Sciences Institute [231]. Most published guidelines recommend that the high dose in a bioassay should produce some toxic effect, usually measured by depression of weight gain, optimally of 10 % compared to controls. This is essentially the concept of the MTD. Procedures for estimating the MTD have been described [12,16].

An issue of increasing importance is whether high doses that produce major pathological changes such as chronic hepatotoxicity or bladder stones should be considered to be beyond the maximally tolerated dose.

7.9 Chronic bioassay

The long-term test is the fifth step in the decision point approach (table 3). At this stage, considerable information should be available on the drug. Presumably, it is not DNA-reactive or it would have been eliminated at previous steps, unless its specific use warranted further development. Therefore, the species and strain of animals must be selected from the perspective of identifying epigenetic agents. Some relevant information may have been obtained at stage C, tests for promoters. The design of a chronic bioassay for non-genotoxic agents must include at least 3 and preferably 4 or 5 dose levels, including the MTD. This is so that the dose-response and threshold can be established for eventual risk assessment (sect. 9). In such a multidose test, the lower dose groups should contain larger numbers of animals to increase the statistical power.

7.91 Conduct of study

A chronic bioassay must be performed according to the United States Food and Drug Administration Good Laboratory Practices [232] in order to obtain acceptance by that agency. Likewise, the requirements of other governments must be met to assure the widest possible acceptance of the data.

The details of the conduct of the study must be formalized before beginning. This includes a predetermination of how certain contingencies will be dealt with, so that changes in the protocol or decisions regarding problems will not appear to be arbitrary or intended to manipulate the results. In this regard, it is particularly important to ensure that untreated control animals survive no longer than any of the experimental groups. Therefore, while the plan might be to conduct a lifetime or a 2-year test, it may be advisable to plan autopsy of all of the animals if after the 21-month period, for example, any one group has lost 75–80 % of the animals, especially if this loss appears to be due to causes unrelated to treatment, such as pulmonary disease.

If, during the study, the weight gain of the animals at the selected MTD, appears inappropriate, for example, as high as that of the controls (dosage too low) or below the target 10 % weight depression (dosage too high), the study should be discontinued as a chronic bioassay. The animals may be further studied with regard to metabolism and pharmacokinetics to determine the basis for the discrepancy between the preliminary studies and the intended chronic bioassay such that the repeat can be better designed. An example of a study in which the high dose was inappropriate is the report by Flaks and Flaks (232a) of induction of liver tumors in mice by acetaminophen at a dose which killed 10–50 % of the mice within the first 48 hs.

7.92 Evaluation of results

In the evaluation of a chronic bioassay, the main, but not the only, consideration is whether the test chemical produced a carcinogenic effect. The effect of treatment on survival is determined from life table methods [233].

The assessment of carcinogenicity is properly made only by experts. The interpretation of a long-term study in animals in which many problems may exist and the outcome may be exceedingly complex requires experienced professional judgment. Following the practices of most regulatory agencies, a carcinogenic effect may be defined as an increase in the development of neoplasms manifested by (1) the development of types of neoplasms not seen in controls, (2) an increased incidence of the types of neoplasms occurring in controls, (3) the occurrence of neoplasms earlier than in controls, and (4) an increased multiplicity of neoplasms in individual animals.

In the application of these criteria, we consider an increase in benign plus malignant neoplasms, as well as in malignancies only, to be evidence of carcinogenicity. This is based on the current understanding that benign neoplasms may be a step in progression to malignancy and, in any event, are a manifestation of the effect of the chemical. Moreover, the distinction between benign and malignant neoplasms often rests on histologic criteria which are not fully agreed upon and are applied according to the experience of the pathologist. A more-or-less theoretical question is whether an increase in only benign neoplasms should be accepted as evidence of carcinogenicity. In limited studies that have yielded an increase in only benign neoplasms, the agents have often been found to produce malignant neoplasms in more rigorous studies or in another species. In fact, there is no scientific explanation why a chemical should give rise to only benign neoplasms. Therefore, an increase in only benign neoplasms cannot be dismissed, although such an occurrence in properly designed bioassays is rare, if not non-existant.

In chronic bioassays with chemicals that are DNA-reactive, the results are generally clear cut. Typically, malignant neoplasms are produced in several organs in high incidence and within a relatively short latent period. With carcinogens that are not clearly genotoxic, a few have produced a strong carcinogenic effect in a short period, for example the antihistamine methapyrilene [234]. Nevertheless, most epigenetic carcinogens, which should be the predominant type revealed in chronic bioassay under the approach described here, often produce neoplasms in only one species, in low yield and at few sites, especially those with a significant spontaneous background of neoplasms. The first step in the evaluation of data on neoplasms is to determine if a true increase occurred.

Various types of statistical analyses have been applied to neoplasm incidence data [235–237], and adjustment for intermittent, treatment-unrelated mortality [238–240]. In a given experiment, when statistical significance is not achieved by comparing any one group, such as the group given the MTD, to the controls, the data for a number of doses can be examined for trend [241].

Bioassays sometimes present a low but statistically significant increase of a specific neoplasm, such as adrenal pheochromocytoma in relation to simultaneous controls. However, many such spontaneously-arising neoplasms exhibit widely varying incidences in control animals, both in the same and different laboratories [242, 243]. Consequently, the incidence in the experimental group may represent only normal fluctuation. To assess this, the incidence in the treated group should also be compared to historic controls in the same laboratory and in other laboratories. Principles for the use of historic controls have been proposed by the Task Force of Past Presidents of the Society of Toxicology [244] as follows: (1) if the incidence rate in the concurrent control group is lower than in the historical control groups, but the incidence rates in the treated groups are within the historical control range, the differences between treated and control groups are not biologically significant; (2) if the incidence rates in the treated groups are higher than the historical control range but not statistically significantly greater than the concurrent control incidence, the conclusion would be that there is no relation to treatment, but with the reservation that this result could be a false negative resulting from some flaw; (3) if the incidence rates in the treated groups are significantly greater than in the concurrent controls, and greater than the historical control range, a treatment effect may be present which is unlikely to be a false positive test.

There are other possible outcomes of bioassays that represent problems in evaluation. One is the situation in which no increase in any specific type of neoplasm occurred, but the overall incidence of neoplasms in treated animals was greater than in controls. Or there may be an increase in total neoplasms in only one sex, as occurred in the FDA study of amaranth (red dye No. 2), which lead to amaranth being judged unsafe and banned [245]. At present, there is no mechanistic explanation for such effects and they should not be accepted as definite evidence of carcinogenicity. A recently documented problem in bioassays is a negative correlation between proliferative hepatocellular lesions and lymphoma in rats and mice [246, 247]. An implication of this phenomenon is that a drug that inhibits lymphoma development may permit the occurrence of increased numbers of liver tumors.

Even in cases where a statistically significant increase in neoplasms is found, it is very important to be aware of possible pitfalls in the conduct of the test and to carefully assess the biological significance of the results. A rather obvious defect in test conduct is to allow test animals to outlive controls such that the incidence of spontaneous neoplasms is greater in the treated animals due to their more advanced age (sect. 7.91). Reference to historical controls can place such findings in perspective.

Other problems include effects of adventitious carcinogens to which the test animals might be exposed. This could involve formation of a carcinogen by reaction of the test material with dietary constituents during improper storage. Another problem that should be kept in mind is the fact that rodents are coprophagic. There ist no practical way to overcome this in chronic bioassay. Consequently, the animals may ingest primary metabolites of the test material or secondary metabolites generated by bacteria. Also, since at least one chemical at high dose, saccharin, has been shown to alter intestinal function [220], it is possible that test animals may be exposed to unusual bacterial products.

A situation that has not received much attention is the effect on nutrition of high level exposures to chemicals. As discussed in section 7.83, availability of specific nutrients can be compromised by test chemicals and nutrition generally can be affected by reduced food consumption. The role of such alterations on the development of spontaneously-occurring neoplasms needs further consideration.

In the case of a positive bioassay with an impure material (sect. 7.81), consideration must, of course, be given to the possibility that the effect was not due to the substance of prime concern.

An increasingly important aspect in the evaluation of a positive bioassay is the mechanism(s) by which neoplasms are produced. This issue is illustrated by the situation in which treatment of animals with high dose levels of a chemical yields an increase in neoplasms and, at the same time, other abnormalities which could actually be responsible for the neoplastic response (sect. 7.85). One well recognized example is where the test chemical leads to crystalluria and the deposition of stones in the excretory organs as a primary effect. It is established that the presence of foreign bodies in the kidney, renal pelvis, or urinary bladder can elicit a neoplastic responce [247] and therefore, the possibility must be considered that where crystalluria and/or urolithiasis are present that the induction of neoplasms is secondary to this phenomenon and not necessarily indicative of a true carcinogenic effect of the chemical. A number of other examples have been described [248]. The question is whether such chemicals should be called carcinogens at all. Correctly or not, current practice by many regulatory groups is to do so. Consequently, it is imperative to distinguish such effects, which have been categorized as epigenetic, from the effects of DNA-reactine carcinogens in the final evaluation of all data.

8 Final evaluation

If the decision point approach has led to properly conducted long-term bioassays interpreted according to the criteria in the preceding section, then fairly definitive data on carcinogenicity in the animal models would be obtained. In the final evaluation of a carcinogenicity bioassay, all relevant information on biochemical and biological effects of the drug must be considered for a proper mechanistic interpretation of findings.

Where the drug was noncarcinogenic and showed no evidence of

DNA-reactivity, then the final evaluation is obvious. A finding of noncarcinogenicity for a truly DNA-reactive agent has not occurred in our experience and, therefore, should be viewed cautiously. It could be due to the use of doses that were too low. Unfortunately, a not unusual finding in the commonly used two dose studies is an equivocal carcinogenic effect [249]. If the structure of the drug is consistent with a DNA-reactive agent and the genotoxicity data are sound, it should be accepted as a probable carcinogen. If the drug is nongenotoxic and the study was adequate, with more than two doses, and one species was negative, the equivocal results may be discounted. This is especially justified if the equivocal results involved a spontaneously-occurring tumor type.

On the other hand, if the drug is found to be carcinogenic according to the criteria in section 7.92, the basis for this must be carefully evaluated using the results of the assessment of structure and biochemical properties and those from testing for genotoxicity. Convincing positive results in the in vitro tests coupled with documented carcinogenicity permits classification of the drug as a DNA-reactive carcinogen. In chronic bioassays, DNA-reactine carcinogens often produce predominantly malignant neoplasms in more than one organ and in high yield with a relatively short latent period. Such results, therefore, support the interpretation that interaction with DNA is the mode of action.

With genotoxic agents, however, it is still possible that carcinogenicity may result largely from nongenotoxic effects, such as cytotoxicity. This may be the case with phenacetin, where cancer in the urinary tract occurs only under nephrotoxic conditions. Likewise, with acetaminophen, recently shown to increase liver neoplasms in mice under conditioner of extreme toxicity (sect. 7.91) but not otherwise [250, 250a], and apparently genotoxic only under conditions of toxicity [251], the chronic hepatotoxicity of the drug at high dose may be the prime factor.

Carcinogenicity in the absence of evidence of DNA-reactivity can occur through a number of mechanisms as discussed in section 4. A variety of carcinogenic effects are consistent with a nongenotoxic effect: a small increase in neoplasms that occur spontaneously, an increase in spontaneously-occurring neoplasms only late in life, production of neoplasms only in hormonally-responsive tissues, occurrence of neoplasms only in severely damaged tissues. With drugs, such effects have involved increases in adrenal pheochromocytomas, pituitary adenomas thyroid neoplasms in rats and liver neoplasms, particularly in mice, as discussed in sect. 7.3. Specific mechanistic studies are required to provide the supporting evidence for classification of a drug with such a profile as an epigenetic carcinogen. Investigation on this should focus on study of genotoxic and other effects in the target organ(s) revealed in the bioassay [251a].

9 Health hazard assessment

The decision point approach described here, involves a series of steps which may be followed in the testing of a pharmaceutical. Based on such systematic in vitro and in vivo tests, a sequence of decisions can be made regarding animal carcinogenicity and human health hazards (table 5).

In the case of a drug that proves to be an experimental carcinogen, appropriate data bearing on the question of whether it is operating by a genotoxic or epigenetic mechanism is essential to human hazard assessment. In the past, and to a lesser degree at present, animal carcinogens were judged to be hazards regardless of their properties, modes of action or conditions of exposure. Better understanding of the proccsses of neoplasm induction in rodents, however, now permits a case by case analysis of the potential hazard of a carcinogen.

An important element in forming a picture of potential hazard is the degree of experimental evidence that the agent is carcinogenic. This has been discussed by Griesemer and Cueto [252]. A useful means of quantitative evaluation of long-term bioassay data has also been proposed by Squire [253]. Consideration is given to the numer of species and tissues affected, pathologic nature and serverity of the neoplasms, latent periods, dose-response relationships, and considerable weight is assigned to genotoxicity. As discussed herein, these are all important parameters in evaluating the overall carcinogenicity of a drug.

The effects of DNA-reactive carcinogens are not easily predictable and vary with the developmental and hormonal status of the animal. Also, they are heavily influenced by other factors such as cocarcinogens and promoters. Moreover, humans are highly heterogenous, live under different environmental conditions, and under varied dietary and other exposures, including occupational. Thus, they necessarily have widely varying response patterns to any given carcinogenic chal-

Table 5
Application of decision point approach results.

Stage B: Genotoxicity	Stage C: Promoting activity	Stage D: Activity in limited in vivo bioassay	Stage E: Carcinogenicity	Evaluation and recommendation	Hazard assessment
+				Likely carcinogen; can confirm at stage D	Caution for human exposure
+		+		DNA-reactive carcinogen	Control human exposure
–				Possible epigenetic agent; perform tests for promotion if exposure warrants	Not a qualitative carcinogenic hazard
–	+			Epigenetic agent; delineate dose-response and perform chronic bioassay over dose range	May have safe level of exposure
–	+		+	Epigenetic carcinogenic agent	If no effect level found, that is a guide to a safe level of exposure

lenge. For example, although it is clear that heavy cigarette smoking leads to lung cancer, it is also enident that even with equal numbers of cigarettes smoked per day, the response of different individuals varies a great deal. The smoking method, the inhalation depth, the nutritional status of the individual, the biochemical activation and detoxification systems, the ciliary and mucous clearance systems, and defense mechanisms all play a role in determining the overall outcome in this carcinogenic situation. Additional examples with other carcinogens or carcinogenic events could be cited. The specific cellular and molecular systems in humans, nevertheless, are essentially the same as equivalent cellular and molecular systems in experimental laboratory animals. Since these systems determine how a given exogenous chemical interacts, it is to be expected that humans exposed to genotoxic carcinogens would not react in a qualitatively differently manner from the animal models, although quantitative aspects and tissues affected may differ because of metabolic pathways and related elements. Therefore, it must be accepted that a DNA-reactive carcinogen active in one or more animal models represents a hazard to at least some individuals of *Homo sapiens*. It is probably for these reasons that most human carcinogens are genotoxic [254]. Accordingly, well-defined DNA-reactive carcinogens, in general, should be presumed to be qualitative human hazards.

Nevertheless, even with DNA-reactive carcinogens, their effects show major quantitative differences; for example, in liver carcinogenesis, four different carcinogens are recognized to vary greatly in the exposures required to induce a significant incidence of liver neoplasms; acetamide requires 12,500 ppm for 15 months, safrole 5,000 ppm for 12 months, diethylnitrosamine 40 ppm for 6 months, and aflatoxin B_1 0.015–0.1 ppm for 6–12 months. Thus, the last two, and especially aflatoxin B_1, represent a considerably higher risk for human liver cancer development than do the first two chemicals. It is probably for this reason that aflatoxin B_1 is associated with cancer in humans whereas exposure to safrole in sassafras is not.

All DNA-reactive carcinogens have displayed thresholds or no-observed-effect levels in animal studies [255]. With several potent carcinogens, it has been quite low whereas with weaker carcinogens, it has been rather high. This suggests that even with DNA-reactive carcinogens, there may be safe levels of human exposure. For a pharmaceutical with important benefits, therefore, a quantitative risk assessment may be indicated.

Different models have been developed for quantitative extrapolation of animal data to prediction of human risk [256–262]. These can yield quite different estimates and at present, it is not possible to say which of these models is most appropriate. Based on available data showing that humans are not more sensitive than experimental animals to chemical carcinogens [263], a simple linear extrapolation can be used to obtain a crude upper limit to the true risk. Current understanding of the complexities of chemical carcinogenesis (figure 1) and the kinetics of key steps, however, suggests that departure from linearity and a threshold should exist at low doses. Indeed, for the more simple process of chemical mutagenesis in vivo, a drop below linearity at low doses has been demonstrated [264]. Such facts have recently been taken into account in formulating a realistic model [265] that is appropriate for low-dose extrapolation. In the case of a DNA-reactive and carcinogenic drug that was unique and extremely important, such a risk assessment should be considered.

Epigenetic carcinogens are diverse in their modes of action and, therefore, generalizations must be cautious. Nevertheless, their hazards overall appear to differ from those of DNA-reactive carcinogens. They are often more limited in their carcinogenic effects than genotoxic carcinogens and often weaker. Where testing has been adequate, rather high thresholds have been found [255]. These caracteristics probably are the basis for the fact that epigenetic carcinogens such as phenobarbital have been used extensively as medications without causing cancer [266].

In the decision point approach testing, an objective for epigenetic carcinogens is to delineate a threshold (sect. 7.9). This is important for quantitative risk assessment because there are a few epigenetic agents that have shown low level effects. For a useful candidate drug, if a threshold above the intended human dose has been found, a hazard assessment may be undertaken.

To formulate a perspective on human hazard, a number of data points are useful, as follows: species/strain/organ affected; dose-response characteristics; biological effects at carcinogenic doses and the likelihood of similar effects in humans; pharmacokinetics in susceptible species compared to humans; nature of human exposure, intermittent versus continuous; probability of prior genotoxin exposures or presence of initiated cells in the potential human target organ.

If the evaluation indicates the absence of an obvious hazard to humans, a quantitative risk assessment may be performed. The various mathematical treatments discussed above for risk assessment of carcinogens were all derived from data involving bioassays of DNA-reactive carcinogens, and therefore, these are not applicable to epigenetic agents. Efforts are beginning to be made to address this deficiency. Park and Snee [267] have developed a comprehensive model for risk assessment which combines statistical and toxicological data into a quantitative framework. Our own view is that since the carcinogenicity of epigenetic agents often involves some form of chronic toxicity, a reasonable alternative for risk assessment would follow the procedures used to establish an acceptable daily intake (ADI), as for chemicals displaying other forms of chronic toxicity. In general, the ADI is obtained by dividing the no-observed-effect level by a safety factor. Often 100 is used on the basis of a factor of 10 to allow for variability of individual responses within the test species and a second factor of 10 for differences in response between the test species and humans [268].

This latter factor is now known to be adequate in light of interspecies comparisons of carcinogenicity data [206 a]. Where the toxicity data are not complete, a larger safety factor of 1,000 may be warranted.

10 Concluding Remarks

Sizeable progress has been made in the basic sciences underlying carcinogenesis. It is now recognized that some of the diverse types of chemicals capable of causing cancer do so because their specific chemical structure is that of an electrophilic reactant or can give rise to such after metabolic activation. Also, there has been further insight into the molecular target of such reactive carcinogens, namely, the genetic material in the cell, DNA.The interaction with DNA has provided the necessary connection to relate genotoxicity to carcinogenicity. This relationship forms the basis for utilizing the property of genotoxicity in the assessment of potential carcinogenicity. Relatively few pharmaceuticals, mostly cancer chemotherapeutic agents, are DNA-reactive.

In addition to genotoxic carcinogens, there are other agents capable of increasing cancer through mechanisms other than DNA-reactivity. These have been referred to as epigenetic carcinogens because their mode action, at least initially, involves effects on cellular constituents other than DNA.

Current evidence shows that most classes of agents operating through epigenetic modes of action, in particular promoters, do so in a reversible, highly dose-dependent fashion. Thus, there are sound theoretical and empirical reasons for treating such agents in an entirely distinct manner as regards health risk analysis. It is expected that further research will better substantiate the distinctions between different types of carcinogens, permitting control of potentially harmful substances without a categorical ban of all agents that increase cancer in experimental animals by whatever mechanism.

This is not to say that epigenetic agents are automatically of no concern. Indeed, there is substantial evidence relative to the causes of the currently prevailing main human cancers showing that these cancers are due as much to the presence of agents operating through epigenetic mechanisms as to DNA-reactive carcinogens. Examples include cancer of the lung due to smoking of cigarettes; and cancer of the colon, breast, prostate, and perhaps pancreas due to certain dietary habits, especially as regards the level of fat consumed. Thus, there is hope that

these major types of cancer can be controlled by modifying the environment not only with respect to genotoxic carcinogens, but also in its contribution of epigenetic carcinogens, thereby, lowering the complex multiple risk factors for these diverse human cancers.
The situation with nongenotoxic drugs, of course, is somewhat different from the just described, lifestyle-related factors. Many such drugs although carcinogenic in animal models, have been used safely for treating human diseases. A principal example is phenobarbital. Drugs such as clofibrate, fenofibrate, and gemfibrozil likewise are not DNA-reactive but seem to induce cancer in experimental animals through mechanisms involving peroxisomes and effects on cytoplasmic enzymes. More information is needed on dose-response reationships with such drugs, especially in relation to the possible occurrence of nonlinearity. This is important because despite extensive use by humans, there appears to be no obvious relationship between drug use and cancer. One possible explanation might be the fact that drug use, while in some instances quite chronic, started late in life to control conditions such as hypertension, coronary disease risk, and arthritis, all of which arise later in the second half of the lifespan. With genotoxic drugs, risk for disease would most likely still obtain under conditions of intake by older individuals, and this has been proven in patients on cancer chemotherapy where the clinician must be ever alert to the possibility of iatrogenic induction of second cancers. However, for nongenotoxic drugs, at this point no data from animal bioassays or from human observations demonstrate that cancer is seen when older individuals use such products, even continuously. More data are needed in this area through clinical observations, animal bioassay, and above all, an understanding of mechanisms which would be helpful in rationally assessing possible human hazard.

11 Acknowledgments

We wish to thank Ms. Laura Karabaic for her help in preparing and editing this paper.
This research has been supported by PHS grant number CA-17613 awarded by the National Cancer Institute, DHHS.

References

1 R. Truhaut (ed.): Potential Carcinogenic Hazards from Drugs (Evaluation of Risk). UICC Monogr. Series, vol. 7. Springer Verlag, Berlin, New York 1967.

2D. Schmähl, C. Thomas and R. Auer: Iatrogenic Carcinogenesis. Springer Verlag, Berlin 1977.

3 R. H. Adamson and S. M. Sieber: Environ. Health Perspect. *39,* 93 (1981).

4 R. Hoover and J. F. Fraumeni: Cancer *47,* 1071 (1981).

5 J. H. Weisburger and G. M. Williams, in: Cancer Medicine, 2nd ed., p. 42. Eds. J. F. Holland and E. III Frei. Lea and Febiger, Philadelphia 1982.

6 International Agency for Research on Cancer. IARC Monographs on the Evaluation of the Carcinogenic Risk of Chemicals to Humans. IARC Monographs Supplement 4. IARC, Lyon 1982.

7 Th. Thiede and B. C. Christensen: Acta Med. Scand. *185,* 133 (1969).

7a K. Rieche: Carcer Treat. Rev. II, 39 (1984).

8 A. L. Herbst, H. Ulfelder and D. C. Poskanzer: N. Engl. J. Med. *284,* 878 (1971).

9 N. Hultengren, C. Lagergren and A. Ljungquist: Acta Chir. Scand. *130,* 314 (1965).

10 G. M. Williams, in: Verterinary Pharmacology and Toxicology, p. 82–92. Eds. Y. Rucherbusch, P. L. Toutain and G. D. Koritz. MTP Press Limited, Boston 1983.

11 J. H. Weisburger and G. M. Williams: Science *214,* 401 (1981).

12 J. H. Weisburger and G. M. Williams, in: Chemical Carcinogens. 2nd Ed. p. 1323. Ed. C. E. Searle. American Chemical Society, Washington 1984.

13 A. J. Lehman (ed.): Appraisal of the Safety of Chemicals in Foods, Drugs, and Cosmetics. Assoc. Food and Drug Offic. of the United States, Baltimore, MD, 1959.

14 US Food and Drug Administration. Toxicological Principles For the Safety Assessment of Direct Food Additives and Color Additives Used in Food. US Food and Drug Administration, Washington, DC, 1982.

14a E. K. Weisburger: Prog. exp. Tumor Res. 26, 187 (1983).

15 J. H. Weisburger and E. K. Weisburger: Meth. Cancer Res. *1,* 307 (1967).

16 J. M. Sontag, N. P. Page and U. Saffiotti: Guidelines for Carcinogen Bioassay in Small Rodents. National Cancer Institute, Bethesda, MD, 1975.

17 National Toxicology Program Board of Scientific Counselors. Report of the Ad Hoc Panel on Chemical Carcinogenesis Testing and Evaluation, 1984.

18 Office of Science and Technology Policy. Chemical Carcinogens; Notice of Review of the Science and its Associated Principles, part. II. Federal Register, 49, 1984.

19 Health and Welfare Canada. The Testing of Chemicals for Carcinogenicity, Mutagenicity, Teratogenicity. Health and Welfare, Canada, 1973.

20 Committee of Health Council of the Netherlands. Report of the Evaluation of the Carcinogenicity of Chemical Substances. Government Publishing Office, The Hague 1980.

21 I. Berenblum (ed.): Carcinogenicity Testing. UICC Tech. Rept. Series, vol. 2. Union Internationale Contre le Cancer, Geneva, 1969.

22 P. Shubik, D. B. Clayson and B. Terracini (eds.): The Quantification of Environmental Carcinogens. UICC Tech. Rept. Series, vol. 4. Union Internationale Contre le Cancer, Geneva, 1970.

23 International Agency for Research on Cancer. Long-term and Short-term Screening Assays for Carcinogenesis: A Critical Appraisal, Suppl. 2. IARC, Lyon, France, 1980.

24 World Health Organization. Principles for the Testing and Evaluation of Drugs for Carcinogenicity, Report of a WHO Scientific Group. World Health Organization Technical Report Series No. 426, 1969.
25 F. A. de la Iglesia, R. S. Lake and J. E. Fitzgerald: Drug Met. Rev. *11,* 103–146.
26 D. Clayson: Chemical Carcinogenesis. Little, Brown and Co., Boston 1962.
27 J. C. Arcos and M. F. Argus: Chemical Induction of Cancer, vols. I, II A, II B, III. Academic Press, New York 1968–83.
28 International Agency for Research on Cancer. IARC Monographs on the Evaluation of the Carcinogenic Risk of Chemicals to Humans, vols. 1–33. IARC, Lyon.
29 C. E. Searle (ed.): Chemical Carcinogens, 2nd ed., American Chemical Society, Washington 1984.
30 G. M. Williams: J. Ass. Off. Anal. Chem. *62,* 857 (1979).
31 R. Kroes, in: Environmental Carcinogenesis: Occurrence, Risk Evaluation and Mechanisms, p. 287. Eds. P. Emmelot and E. Kriek, Elsevier/North-Holland Biomedical Press, Amsterdam, 1979.
32 J. H. Weisburger and G. M. Williams, in: Cassarett and Doull's Toxicology: The Basic Science of Poisons, 2nd ed., p. 84. Eds. J. Doull, C. Klaassen, and M. Amdur. Macmillan, New York 1980.
33 A. C. Kolbye, in: The Predictive Value of Short-Term Screening Tests in Carcinogenicity Evaluation. Eds. G. M. Williams, R. Kroes, H. W. Waaijers and K. W. van de Poll. Elsevier/North-Holland, Amsterdam 1978.
34 W. T. Stott and P. G. Watanabe: Drug Metab. Rev. *13,* 853 (1982).
35 G. M. Williams: Fund. Appl. Toxicol. *4,* 325 (1984).
35a E. K. Weisburger: Prog. Drug. Res 26, 143 (1982).
36 A. H. Conney: Cancer Res. *42,* 4875 (1982).
37 J. H. Weisburger and G. M. Williams, in: Cancer: A Comprehensive Treatise, 2nd ed., p. 241. Ed. F. F. Becker. Plenum Press, New York, 1982.
38 P. L. Grover (ed.): Chemical Carcinogens and DNA, vols. 1–2. CRC Press, Boca Raton, FL, 1979.
39 P. Brookes: Br. Med. Bull. *36,* 1 (1980).
40 R. A. Weinberg: Adv. Cancer Res. *36,* 149 (1982).
41 J. M. Bishop: Adv. Cancer Res. *37,* 1 (1982).
42 I. Berenblum: Carcinogenesis as a Biological Problem. In: Frontiers of Biology, vol. 34. North-Holland Publishing Company Amsterdam 1974.
43 L. Foulds: Neoplastic Development 1. Academic, New York 1969.
44 C. F. Springgate and L. A. Loeb: Proc. natn. Acad. Sci. USA *70,* 245 (1973).
45 E. C. Miller and J. A. Miller: Cancer *47,* 2327 (1981).
46 H. V. Malling: Mutat. Res. *3,* 537 (1966).
47 B. N. Ames, C. Yanofsky, in: Chemical Mutagens, Principles and Methods for their Detection, vol. 1. Ed. A. Hollaender. Plenum Press, New York 1971.
48 A. Hollaender and F. J. de Serres (eds.): Chemical Mutagens, Principles and Methods for Their Detection, vols. 1–7. Plenum Press, New York, 1971–82.
49 T. Sugimura, S. Sato, M. Nagao, T. Yahagi, T. Matsushima, Y. Seino, M. Taeuchi and T. Kawachi, in: Fundamentals in Cancer Prevention, p. 191–215. Eds. P. N. Magee, S. Takayama, T. Sugimura, and T. Matsushima. Japan Scientific Societies Press, Tokyo 1976.
50 A. C. Upton, D. G. Clayson, J. D. Jansen, H. Rosenkranz and G. M. Williams: Mutat. Res. *133,* 1 (1984).
51 R. P. Mason and C. F. Chingnell: Pharmacol. Rev. *33,* 189–193 (1982).
52 C. S. Moody and H. M. Hassan: Proc. Natn. Acad. Sci. *79,* 2855 (1982).
53 J. Yavelow, M. Gidlund and W. Troll: Carcinogenesis *3,* 135 (1982).
54 M. K. Logani, V. Solanki and T. J. Slaga: Carcinogenesis *3,* 1303 (1982).

55 L. R. Barrows and R. C. Shank: Toxicol. Appl. Pharmacol. *60,* 334 (1981).
56 L. R. Barrows and P. N. Magee: Carcinogenesis *3,* 349 (1982).
57 T. Sugiyama, in: Recent Topics in Chemical Carcinogenesis, p. 393. Eds. S. Odashima, S. Takayama and H. Sato. University Park Press, Tokyo 1975.
58 J. A. DiPaolo and M. C. Popescu: Am. J. Pathol. *85,* 709 (1976).
59 M. Ishidate, Jr., T. Sofuni and K. Yoshikawa, in: Mutation, Promotion, and Transformation In Vitro, p. 95. Eds. N. Inui, T. Kurski, M. A. Yamada and C. Heidelberger. Japan Scientific Societies Press, Tokyo 1981.
60 International Agency for Research on Cancer. Approaches to Classifying Chemical Carcinogens According to Mechanism of Action. IARC Technical Report No. 83/001. IARC, Lyon 1983.
61 F. W. Sunderman and K. S. McKully: Cancer Invest. *1,* 469 (1983).
62 R. A. Zakour, T. A. Kunkel and L. A. Loeb: Environ. Health Perspect. *40,* 197 (1981).
63 J. K. Reddy and N. D. Lalwani: CRC Crit. Rev. Toxicol. *12,* 1 (1983).
64 M. R. Ciriolo, I. Manelli, G. Ratilio, V. Borzatta, M. Cristofari and L. Stanzani: FEBS Letta. 144, 264 (1982).
64a K. Furukawa, S. Numoto, K. Furuya, N. T. Furukawa and G. M. Williams: Cancer Res., in press.
65 Interagency Regulatory Liason Group. J. natn. Cancer Inst. *65,* 241 (1979).
66 R. J. Pienta, L. M. Kushner and L. S. Russell: Regul. Toxicol. Pharmacol. *4,* 249 (1984).
67 J. H. Weisburger and G. M. Williams, in: Structural Correlates of Carcinogenesis and Mutagenesis, p. 45. Eds. I. M. Asher and C. Zervos. FDA Office of Health Affairs, Rockville, MD, 1978.
68 J. H. Weisburger and G. M. Williams: Science *214,* 401 (1981).
69 G. M. Williams and J. H. Weisburger: Ann. Rev. Pharmac. Toxic. *21,* 393 (1981).
70 J. Ashby: Br. J. Cancer *37,* 904 (1978).
71 E. K. Weisburger: NCI Monogr. *58,* 1 (1981).
72 B. L. van Duuren: J. Environ. Pathol. Toxicol. *3,* 11 (1983).
72a J. D. Prejean and J. A. Montgomery: Drug. Met. Rev. 15, 619 (1984).
73 S. D. Nelson, W. L. Nelson and W. F. Trager: J. Med. Chem. *21,* 721 (1978).
74 G. Eisenbrand, in: Das Nitrosamin-Problem, p. 213. Ed. R. Preussman. Verlag Chemie GmbH, Weinheim 1983.
74a C. S. Street, R. E. Cimprich and J. L. Robertson Scand. J. Gastroenterol *19* (Suppl. 101), 109, 1984.
75 W. C. Lutz: Mutat. Res. *65,* 289 (1979).
76 M. Hollstein and J. McCann: Mutat. Res. *65,* 133 (1979).
77 G. M. Williams, R. Kroes, H. W. Waaijers and K. W. van de Poll (eds.): The Predictive Value of Short-term Screening Tests in Carcinogenicity Evaluation. Elsevier/North-Holland Biomedical Press, Amsterdam 1980.
78 H. F. Stich and R. H. C. San (eds.): Short-Term Tests for Chemical Carcinogens. Springer Verlag, New York 1981.
79 N. Inui, T. Kuroki, M. A. Yamada, C. Heidelberger (eds.): Mutation, Promotion, and Transformation In Vitro. Gann Monogr. Cancer Res. No. 17. Japan Scientific Societies Press, Tokyo 1981.
80 V. Dunkel, V. Ray and G. M. Williams (eds.): Cellular Systems for Toxicity Testing. Ann. N. Y. Acad. Sci. *407,* New York Acad. Sci., New York 1983.
81 National Research Council. Identifying and Estimating the Genetic Impact of Chemical Mutagens. National Academy Press, Washington 1983.
82 H. S. Rosenkranz and W. T. Speck: Biochem. Biophys. Res. Commun. *66,* 520 (1975).
83 G. M. Williams, G. Mazue, C. A. McQueen and T. Shimada: Science *210,* 329 (1980).

84 G. Brambilla, M. Cavanna and S. Deflora, in: Chemical Carcinogenesis, p. 193. Ed. C. Nicolini. Plenum Press, New York 1982.
85 A. Martelli, E. Fugassa, A. Voci and G. Brambilla: Mutat. Res. *122,* 373 (1983).
86 International Life Sciences Institute. Current Issues in Toxicology, p. 159. Ed. H. C. Grice. Springer Verlag, New York 1984.
87 P. C. Hanawalt, E. C. Friedberg and C. F. Fox (eds.): DNA Repair Mechanisms. Academic Press, New York 1978.
88 H. F. Stich, R. H. C. San and J. H. Freemann, in: Short-Term Tests for chemical Carcinogens, p. 65. Eds. H. F. Stich and R. H. C. San. Springer-Verlag, New York, Heidelberg 1981.
89 G. M. Williams: Cancer Res. *37,* 1845 (1977).
90 A. D. Mitchell, D. A. Casciano, M. L. Meltz, D. E. Robinson, R. H. C. San, G. M. Williams and E. S. Vonhalle: Mutat. Res. *123,* 363 (1983).
91 G. M. Williams, in: Chemical Mutagens: Principles and Methods for their Detection, vol. 6, p. 61. Eds. A. Hollaender and A. DeSerres. Plenum Press, New York 1980.
92 G. S. Probst, R. E. McMahon, L. E. Hill, C. Z. Thompson, J. K. Epp and S. B. Neal: Environ. Mut. *3,* 11 (1981).
93 G. M. Williams, M. F. Laspia and V. C. Dunkel: Mutat. Res. *97,* 359 (1982).
93a M. O. Bradley, G. Dysant, K. Fitzsimmons, P. Horback, J. Lewin and G. Wolf, Cancer Res. *42,* 2592 (1982).
94 R. Devoret: Prog. Nucleic Acid Res. *29,* 252 (1981).
95 P. Quillardet, O. Huisman, R. D'Ari and M. Hofnung: Proc. natn. Acad. Sci. *79,* 5971 (1982).
96 R. M. Schaaper, B. W. Glickman and L. A. Loeb: Cancer Res. *42,* 3480 (1983).
97 B. M. Ames and L. Haroun, in: Short-Term Tests for Chemical Carcinogens. Eds. H. F. Stich, R. H. C. San. Springer Verlag, New York, Heidelberg 1981.
98 E. D. Levin, E. Yamasaki and B. N. Ames: Mutat. Res. *94,* 315 (1982).
99 J. Hyman, Z. Leifer and H. S. Rosenkranz: Mutat. Res. *74,* 107 (1980).
100 H. S. Rosenkranz and L. A. Poirier: J. Natn. Cancer Inst. *62,* 873 (1979).
101 D. E. Levin, H. Hollstein, M. F. Christman, E. A. Schwiers and B. N. Ames: Proc. Natn. Acad. Sci. USA *79,* 74445 (1982).
102 T. Matsushima, T. Sugimura, M. Nagao, T. Yahagi and A. Shirai, in: Short-Term Test and Systems for Detecting Carcinogens, p. 273. Eds. K. H. Norpoth R. C. Garner. Springer Verlag, New York 1980.
103 H. Bartsch, C. Malaveille, A. M. Camus, G. Martel-Planche, G. Brun, A. Hautefeuille, N. Sabadie, A. Barbin, T. Kuroki, C. Drevon, C. Piccoli and R. Montesano: Mutat. Res. *76,* 1 (1980).
104 L. Siminovitch: Cell *7,* 1 (1976).
105 P. Howard-Flanders: Mutat. Res. *86,* 307 (1981).
106 E. H. Y. Chu, in: Cellular Systems for Toxicity Testing. Eds. V. Dunkel, V. Ray and G. M. Williams. Ann. N. Y. Acad. Sci. *407,* 221 (1983).
107 E. Huberman, in: Screening Tests in Chemical Carcinogenesis, p. 521, Eds. R. Montesano, H. Bartsch and L. Tomatis. IARC, Lyon 1976.
108 D. F. Krahn and C. Heidelberger: Mutat. Res. *46,* 27 (1977).
109 A. W. Hsie, in: The Predictive Value of Short-Term Screening Tests in Carcinogenicity Evaluation, p. 89. Eds. G. M. Williams, R. Kroes, H. W. Waaijers and K. W. van de Poll. Elsevier/North-Holland, Amsterdam 1980.
110 D. Clive, K. O. Johnson, J. F. S. Spector, A. G. Batson and M. M. Brown: Mutat. Res. *59,* 61 (1979).
111 P. J. Davies and J. Perry: Genet. Res. *24,* 311 (1979).
112 E. L. Hubermann, L. Sachs, S. K. Yang and H. V. Gelboin: Proc. Natn.

Acad. Sci. USA *73,* 607 (1976).
113 D. Clive, A. G. Batson and N. T. Turner, in: The Predictive Value of Short-Term Screening Tests in Carcinogenicity Evaluation, p. 103. Eds. G. M. Williams, R. Kroes, H. W. Waaijers and K. W. van de Poll. Elsevier/North-Holland, Amsterdam 1980.
114 R. H. C. San and G. M. Williams: Proc. Soc. Exp. Biol. Med. *156,* 534 (1977).
115 C. C. Harris, H. Autrup, A. Haugen, J. Lechner, B. F. Trump and I. C. Hsu, in: Host Factors in Human Carcinogenesis, p. 497. Eds. H. Bartsch, B. Armstrong and W. Davis. IARC Sci. Publ. No. 39, International Agency for Research on Cancer, Lyon 1982.
116 R. S. Langenbach and L. Oglesby, in: Chemical Mutagens, vol. 8, p. 55. Ed. F. DeSerres. Plenum Press, New York 1983.
117 I. Schmeltz, J. Tosk and G. M. Williams: Cancer Lett. *5,* 81 (1978).
118 C. A. H. Bigger, J. E. Tomaszewski and A. Dipple: Science *209,* 503 (1980).
119 R. E. McMahon, in: Differentiation and Carcinogenesis in Liver Cell Cultures. Eds. C. Borek and G. M. Williams. Annals New York Academy of Sciences, vol. 349, p. 46. N. Y. Acad. Sci., New York, 1980.
119a C. Tong, M. Fazio and G. M. Williams: Proc. Soc. Exp. Biol. Med. *167,* 572 (1981).
120 C. Tong and G. M. Williams: Mutat. Res. *74,* 1 (1980).
120a J. J. Yunis: Science 221, 227 (1983).
120b S. Wolff: Ann. Rev. Genet. *11,* 183 (1977).
121 P. E. Perry, in: Chemical Mutagens: Principles and Methods for their Detection, vol. 6, p. 1–39. Eds. A. Hollaender and F. J. DeSerres. Plenum Press, New York 1980.
122 A. A. Sandberg (ed.): Sister Chromatid Exchange. Alan R. Liss, New York 1982.
122a B. Lambert, U. Ringborg and A. Lindblad: Mutat. Res. 59, 295 (1979).
123 S. A. Latt, J. Allen, S. E. Bloom, A. Carrano, E. Falke, D. Kram, E. Schneider, R. Schreck, R. Tice, B. Whitfield and S. Wolff: Mutat. Res. *87,* 17 (1981).
124 R. Thust, R. Warzok, E. Grund and J. Mendel: Mutat. Res. *51,* 397 (1978).
125 A. D. Klingerman, C. S. Stephen and G. Michalopolos: Environ. Mut. *2,* 157 (1980).
126 S. Ved Brat and G. M. Williams: Cancer Lett. *17,* 213 (1982).
127 R. Dean, G. Bynum, D. Kram and E. L. Schneider: Mutat. Res. *74,* 477 (1980).
128 A. L. Meyer and B. J. Dean: Mutat. Res. *91,* 47 (1981).
129 C. Tong, S. Ved Brat and G. M. Williams: Mutat. Res. *91,* 46 (1981).
130 H. J. Evans: Ann. N. Y. Acad. Sci. *407,* 131 (1983).
131 Y. Berwald and L. Sachs: Nature *200,* 1182 (1963).
132 J. A. DiPaolo, in: Environmental Carcinogenesis, p. 365 Eds. P. Emmelotand E. Kriek. Elsevier/North-Holland, Amsterdam 1979.
133 R. Pienta, in: Chemical Mutagens: Principles and Methods for their Detection, vol. 6, p. 175–202. Eds. A. Hollaender and F. J. DeSerres. Plenum Press, New York 1980.
134 J. C. Barrett, D. G. Thomassen and T. W. Hesterberg: Ann. N. Y. Acad. Sci. *407,* 291 (1983).
135 N. Mishra, V. Dunkel and M. Mehlman (eds.): Mammalian Cell Transformation By Chemical Carcinogens, in: Advances in Modern Environmental Toxicology, vol. 1. Senate Press, Princeton, NJ, 1980.
136 P. B. Fisher and I. B. Weinstein, in: Carcinogens in Industry and the Environment, p. 113. Ed. J. M. Sontag, Marcel Dekker, New York 1981.
137 C. Borek: Ann. N. Y. Acad. Sci. *407,* 284 (1983).
138 M. P. Meyer and J. B. Aust: Science *224,* 1445 (1984).

139 P. Nettesheim and J. C. Barrett, in: CRC Critical Reviews in Toxicology, vol. 12, p. 215. Ed. L. Golberg. CRC Press, Florida, 1984.
140 T. Shimada, K. Furukawa, D. M. Kreiser, A. Cawein and G. M. Williams: Cancer Res. *43,* 5087 (1983).
141 C. Drevon, C. Piccoli and R. Montesano: Mutat. Res. *89,* 83 (1981).
142 R. Pienta, in: The Predictive Value of Short-Term Screening Tests in Carcinogenicity Evaluation, p. 149. Eds. G. M. Williams, R. Kroes, H. W. Waaijers and H. W. van de Poll. Elsevier/North-Holland, Amsterdam 1980.
143 J. C. Barret, A. Wong and J. A. McLachlan: Science *212,* 1402 (1981).
144 M. M. Reidenberg and J. J. Martin: Drug Metab. Disposit. *2,* 71 (1974).
145 C. A. McQueen, C. J. Maslansky, I. B. Glowinsky, S. B. Rescenzi, W. W. Weber and G. M. Williams: Proc. Natn. Acad. Sci. USA *79,* 1269 (1982).
146 E. C. McCoy, E. J. Rosenkranz, L. A. Petrullo and H. S. Rosenkranz: Mutat. Res. *90,* 21 (1980).
147 H. S. Rosenkranz, G. Klopman, V. Chankong, J. Pet-Edwards and Y. Y. Haimes: Environ. Mut. *6,* 231 (1984).
148 B. A. Bridges, B. E. Butterworth and I. B. Weinstein (eds.): Indicators of Xenotoxic Exposure. Bambary Report 13. Cold Spring Harbor Laboratory, New York 1982.
149 G. M. Williams, M. F. Laspia, H. Mori and I. Hirono: Cancer Lett. *12,* 329 (1981).
150 J. C. Mirsalis, C. K. Tyson, and B. E. Butterworth: Environ. Mut. *4,* 553 (1982).
150a J. Dayan, M. C. Crojer, S. Bertozzi and S. Lefrancois: Mutat Res. 77, 301 (1980).
150b W. K. Lutz, W. Jaggi and Ch. Schlatter: Chem.-Biol. Interact. 42, 251 (1982).
151 V. A. Drill: Fd Cosmet. Toxicol. *19,* 607 (1981).
151a L. Ekman, E. Hansson, N. Havu, E. Carlsson and C. Lundberg. Scand. J. Gastroenterol. 20 (Suppl. 108), 53, 1985.
152 J. M. Ward, R. A. Griesemer and E. K. Weisburger: Toxicol. Appl. Pharmacol. *51,* 389 (1979).
153 G. M. Williams: Biochim. biophys. Acta Rev. Cancer *605,* 167 (1980).
154 C. Peraino, R. J. M. Fry and E. Staffeldt: Cancer Res. *31,* 1506 (1971).
155 H. Mori, T. Tanaka, A. Nishikawa, M. Takahashi and G. M. Williams: Gann *72,* 798 (1981).
156 H. Shinozuka, B. Lombardi and S. E. Abanobi: Carcinogenesis *3,* 1017 (1982).
157 H. S. Taper: Cancer *42,* 462 (1978).
158 I. R. Wanless and A. Medline: Lab. Invest. *46,* 313 (1982).
159 R. Cameron, K. Imaida, H. Tsuda and N. Ito: Cancer Res. *42,* 2426 (1982).
160 J. D. Yager and R. Yager: Cancer Res. *40,* 3680 (1980).
161 R. Schulte-Hermann: CRC Crit. Rev. Toxicol. *3,* 97 (1974).
162 G. M. Williams: Fd Cosmet. Toxicol. *19,* 577 (1979).
162a A. Sivak: Mutat. Res. *98,* 377 (1982).
163 I. B. Weinstein, A. D. Horowitz, R. A. Mufson, P. B. Fisher, V. Ivanovic and E. Greenbaum, in: Cocarcinogenesis and Biological Effects of Tumor Promoters, vol. 7, p. 599. Eds. E. Hecker, N. E. Fusenig, W. Kunz, F. Marks and H. W. Thielmann. Raven, New York 1982.
164 P. M. Blumberg, K. B. Delclos, J. A. Dunn, S. Jaken, K. L. Leach and E. Yeh: Ann. N. Y. Acad. Sci. *407,* 303 (1983).
165 S. Mondal and C. Heidelberger: Nature *260,* 710 (1976).
166 C. Lasne, A. Gentil and I. Chouroulinkov: Nature *247,* 490 (1974).
167 J. A. Poiley, R. Raineri and R. J. Pienta: Br. J. Cancer *39,* 8 (1979).
168 A. Sivak and A. S. Tu: Cancer Lett. *10,* 27 (1980).

169 A. R. Kinsella and M. Radman: Proc. Natn. Acad. Sci. USA *12,* 6149 (1978).
170 J. M. Parry, E. M. Parry and J. C. Barrett: Nature *294,* 263 (1981).
171 J. E. Trosko, L. P. Yotti, B. Damson and C.-C. Chang, in: Short-term Tests for Carcinogens, p. 420. Eds. H. F. Stich and R. H. C. San. Springer Verlag, New York 1981.
172 J. E. Trosko, L. P. Yotti, S. T. Warren, B. G. Tsushimoto and C. C. Chang, in: Cocarcinogenesis and Biological Effects of Tumor Promoters. Eds. E. Hecker, N. E. Fusenig, W. Kunz, F. Marks and H. W. Thielmann. Carcinogenesis – A Comprehensive Survey, vol. 7, p. 565. Raven Press, New York.
173 A. W. Murray, D. J. Fitzgerald and G. R. Guy, in: Cocarcinogenesis and Biological Effects of Tumor Promoters. Eds. E. Hecker, N. E. Fusenig, W. Kunz, F. Marks and H. W. Thielmann. Carcinogenesis – A Comprehensive Survey, vol. 7, p. 587. Raven Press, New York 1982.
174 G. M. Williams: Ann. N. Y. Acad. Sci. *349,* 273 (1980).
175 G. M. Williams, S. Telang and C. Tong: Cancer Lett. *11,* 339 (1981).
176 D. Schmähl: Arch. Toxicol., Suppl. *4,* 29 (1980).
177 G. M. Williams and K. Furuya: Carcinogenesis *5,* 171 (1984).
178 F. J. Stevens and C. Peraino, in: Cocarcinogenesis and Biological Effects of Tumor Promoters. Eds. E. Hecker, N. E. Fusenig, W. Kunz, F. Marks and H. W. Thielmann. Carcinogenesis – A Comprehensive Survey, vol. 7, p. 105. Raven Press, New York 1982.
179 H. C. Pitot and A. E. Sirica: Biochim. Biophys. Acta *605,* 191 (1980).
180 T. Kitagawa and H. Sugano: Gann *69,* 679 (1978).
181 K. Watanabe and G. M. Williams: J. Natn. Cancer Inst. *61,* 1311 (1978).
182 B. Remandet, D. Gouy, J. Berthe, G. Mazue and G. M. Williams: Fund. appl. Toxicol. *4,* 152 (1984).
182a H. Tsuda, T. Sakata, T. Masui, K. Imaida and N. Ito: Carcinogenesis 5, 525 (1984).
183 D. Hoffmann, I. Schmeltz, S. S. Hecht and E. L. Wynder, in: Polycyclic Hydrocarbons and Cancer, vol. 1, Eds. H. V. Gelboin and P. O. Ts'o. Academic Press, New York 1978.
184 L. Diamond, T. G. O'Brien and W. M. Baird: Adv. Cancer Res. *32,* 1 (1980).
185 M. Gottardis, K. M. Verdeal., E. Erturk and D. Rose: Eur. J. Cancer Clin. Oncol. *18,* 1395 (1982).
186 B. Highman, M. J. Norvell and T. E. Shellenberger: J. Environ. Path. Toxic. *1,* 1 (1977).
187 R. M. Hicks: Br. Med. Bull. *36,* 39 (1980).
188 K. Nakanishi, A. Hagiwara, M. Shibata, K. Imaida, M. Tatematsu and N. Ito: J. Natn. Cancer Inst. *65,* 1005 (1980).
189 G. Murasaki and S. M. Cohen: Cancer Res. *43,* 182 (1983).
190 B. S. Reddy, L. A. Cohen, G. D. McCoy, P. Hill, J. H. Weisburger and E. L. Wynder: Adv. Cancer Res. *32,* 237 (1980).
191 R. K. Jensen, S. D. Sleight, S. D. Aust, J. R. Goodman and J. E. Trosko: Toxicol. Appl. Pharmacol. *71,* 163 (1983).
192 G. M. Williams and S. Numoto: Carcinogenesis *5,* 1689 (1984).
193 S. Ullberg, L. Dencker, B. Danielsson, in: Developmental Toxicology, p. 125–163. Ed. K. Snell. Croom Helm, London 1982.
194 H. L. Stewart, G. M. Williams, C. H. Keysser, L. S. Lombard and R. J. Montali: J. Natn. Cancer Inst. *64,* 177 (1980).
195 M. A. Pereira: J. Am. Coll. Toxicol. *1,* 47 (1982).
196 G. M. Williams: Toxicol. Pathol. *10,* 3 (1982).
197 D. B. Solt, A. Medline and E. Farber: Am J. Path *88,* 595 (1977).
198 M. Tatematsu, T. Shirai, H. Tsuda, Y. Miyata, M. Hirose and N. Ito; J. Natn. Cancer Inst. *63,* 1411 (1979).
199 H. Tsuda, G. Lee and E. Farber: Cancer Res. *40,* 1157 (1980).

199a C. Peraino, R. J. M. Fry and D. D. Grulbe, in: Carcinogenesis, A Comprehensive Survey, vol. 2. Eds. T. J. Slaga, A. Sivak and R. K. Boutwell. Raven Press, New York 1978.
200 K. Furuya and G. M. Williams: Toxicol. Appl. Pharmacol. *74,* 63 (1984).
200a G. Ragnotti, M. Presta, L. Riboni and T. Avanella. J. Natl. Cancer Inst. *68,* 669, 1982.
201 M. B. Shimkin and G. D. Stoner: Adv. Cancer Res. *21,* 2 (1975).
202 J. Russo and I. H. Russo: Cancer Res. *40,* 2677 (1980).
203 C. J. Grubbs, J. C. Peckham and K. D. Cato; J. Natn. Cancer Inst. *70,* 209 (1983).
204 F. Homburger (ed.): Skin Painting Techniques and In Vivo Carcinogenesis Bioassays. In: Progress in Experimental Tumor Research, vol. 26. S. Karger, Basel 1983.
205 D. Hoffmann and E. L. Wynder: Cancer *27,* 848 (1971).
205a Y. C. Lin, J. M. Loring and C. A. Villee: Cancer Res. *42,* 1015 (1982).
206 R. K. Boutwell, A. K. Verma, C. L. Ashendel and E. Astrup, in: Cocarcinogenesis and Biological Effects of Tumor Promoters. Eds. E. Hecker, N. E. Fusenig, W. Kunz, F. Marks and H. W. Thielmann. Carcinogenesis – A Comprehensive Survey, vol. 7, p. 1. Raven Press, New York 1982.
206a R. A. Knight, M. J. Selin and H. W. Harris: Trans. Conf. Chemother. Tuberc. *18,* 52 (1959).
207 J. W. Frymoyer and R. R. Jacox: J. Lab. Clin. Med. *62,* 891 (1963).
208 W. W. Weber and I. B. Glowinski, in: Enzymatic Basis of Detoxication, vol. 11, p. 169. Academic Press, 1980.
208a M. D. Reuber and E. C. Glover: J. Natn. Cancer Inst. *44,* 419 (1970).
208b International Expert Advisory Committee to the Nutrition Foundation. The Relevance of Mouse Liver Hepatoma to Human Carcinogenic Risk. Nutrition Foundation, Washington 1983.
208c L. Tomatis: Ann. Rev. Pharmacol. Toxicol. *19,* 511 (1979).
209 F. Homburger: The Syrian Hamster in Toxicology and Carcinogenesis Research. Progress in Experimental Tumor Research, vol. 24. S. Karger, Basel 1977.
210 L. Tomatis and F. Cefis, Tumori *53,* 447 (1967).
211 R. Stumpf, G. P. Margison, R. Montesano and A. E. Pegg: Cancer Res. *393,* 50 (1979).
212 S. Emerman and N. Javitt: J. Biol. Chem. *242,* 661 (1967).
213 B. Toth: Cancer Res. *28,* 727 (1968).
214 L. Tomatis and U. Mohr (eds.): Transplacental Carcinogenesis. IARC Sci. Publ., No. 4. International Agency Research on Cancer, Lyon 1973.
215 J. Rice (ed.): NCI Monogr. *51,* 1 (1979).
216 H. C. Grice, I. C. Munro and D. R. Krewski: Fd Cosmet Toxicol. *19,* 373 (1981).
217 D. L. Arnold, C. A. Moodie, H. C. Grice, S. M. Charbonneau, B. Stavrio, B. T. Collins, P. F. McGuire, Z. Z. Zawidzska and I. C. Munroe: Toxicol. Appl. Pharmacol. *52,* 113 (1980).
218 J. Ashby, J. A. Styles, D. Anderson and D. Paton: Fd Cosmet. Toxicol. *16,* 95 (1977).
219 R. L. Anderson: Fd Cosmet Toxicol. *17,* 195 (1979).
220 J. Sims and A. G. Renwick: Toxicol. Appl. Pharmacol *67,* 132 (1983).
220a D. W. Gaylor, J. J. Chen, D. L. Greenman and Ch. Thompson. J. Natl. Cancer Inst. *74,* 803, 1985.
221 G. M. Williams, in: Toxicology Laboratory Design and Management for the 80s and Beyond, p. 14–19, Ed. A. S. Tegeris, Karger, Basel 1984.
222 H. C. Grice, D. J. Clegg, D. E. Coffin, M. T. LO, E. J. Middelton, E. Sandi, P. M. Scott. N. P. Sen, B. L. Smith and J. R. Withey, in: Carcinogens in Industry and the Environment, p. 49. Ed. J. M. Sontag Marcel Dekker, New York 1981.

223 E. A. Walker, M. Castegnaro, L. Griciute, M. Borzonyl and W. Davis (eds.): N-Nitroso Compounds: Analysis, Formation and Occurence, IARC Scientific Publication No. 31. IARC, Lyon 1980.
223a O. Takahashi, S. Hayashida and K. Hiraga: Fd Cosmet. Toxicol. *18,* 229 (1980).
223b M. A. Lev, N. Lowe and C. S. Lieber: Am. J. Clin. Nutr. 40, 1131 (1984).
224 G. N. Wogan and P. M. Newberne: Cancer Res. *27,* part. 1, 2370 (1967).
225 Report of the Ad Hoc Working Group on Oil/Gavage in Toxicology. The Nutrition Foundation, 1983.
226 Committee on Diet, Nutrition, and Cancer, Assembly of Life Sciences, National Research Council.'Diet, Nutrition, and Cancer'. National Academy Press, Washington, D. C., 1982.
227 W. R. Bryan and M. B. Shimkin: J. Natn. Cancer Inst. *3,* 503 (1943).
228 H. Druckrey, in: Potential Carcinogenic Hazards from Drugs, p. 60. Ed. R. Truhaut. UICC Monogr. Series, vol. 7. Springer Verlag, Berlin 1967.
229 R. Preussmann: Oncology *37,* 243 (1980).
230 M. B. Shimkin, J. H. Weisburger, E. K. Weisburger, N. Gubareff and V. Suntzeff: J. Natn. Cancer Inst. *36,* 915 (1966).
231 H. C. Grice: Current Issues in Toxicology. Springer Verlag, New York 1984.
232 Food and Drug Administration, Federal Register, part. II. Good Laboratory Practice Regulations; Proposed Rule, Department of Health and Human Services, vol. 49, 1984.
232a A. Flaks and B. Flaks: Carcinogenesis *4,* 363 (1983).
233 P. Armitage: Statistical Methods in Medical Research. Wiley, New York 1971.
234 W. Lijinsky, M. D. Reuber and B. N. Blackwell: Science *209,* 817 (1980).
235 J. Cornfield, K. Rai and J. van Ryzin: Arch. Toxicol., Suppl. *3,* 295 (1980).
236 N. Mantel: Arch. Toxicol., Suppl. *3,* 305 (1980).
237 D. W. Gaylor and D. G. Hoel, in: Carcinogens in Industry and the Environment, p. 97, Ed. J. M. Sontag. Marcel Dekker, New York 1981.
238 D. Cox: J. Res. Statist. Soc. *B 34,* 187 (1974).
239 A. S. Whittemore: Adv. Cancer Res. *27,* 55 (1977).
240 P. Tarone: Biometrika *62,* 679 (1975).
241 J. J. Gart, K. C. Chu and R. E. Tarone: J. Natn. Cancer Inst. *62,* 957 (1979).
242 R. E. Tarone, K. C. Chu and J. M. Ward: J. Natn. Cancer Inst. *66,* 1175 (1981).
243 J. K. Haseman, J. Huff and G. A. Boorman: Toxicol. Pathol. *12,* 126 (1984).
244 Task Force of Past Presidents. Fund. Appl. Toxicol. *2,* 101 (1982).
245 P. M. Boffey: Science *191,* 450 (1976).
246 J. Wahrendorf: J. Natn. Cancer Inst. *70,* 915 (1983).
247 S. S. Young and C. L. Gries, Fund. Appl. Toxicol. *4,* 632 (1984).
247a D. B. Clayson: J. Natn. Cancer Inst. *52,* 1685 (1974).
248 P. Grasso: Clin. Toxicol. *9,* 745 (1976).
249 G. H. Hottendorf and I. H. Pachter: Toxicol. Pathol. *10,* 22 (1982).
250 K. Hiraga and T. Fujii, Jpn. J. Cancer Res. *76,* 79 (1985).
250a H. Amo. and M. Matsuyama: Jpn. J. Hygiene 40, 567 (1985).
251 E. Dybing, J. A. Holme, W. P. Gordon, E. J. Soderlund, D. C. Dahlin and S. D. Nelson: Mutat. Res. *138,* 21 (1984).
251a G. M. Williams, in: Metabolic Principles of Toxicity Testing. Ed. J. Caldwell. Academic Press, New York 1985, in press.
252 R. A. Griesemer and C. Cueto, Jr., in: Molecular and Cellular Aspects of Carcinogen Screening Tests, p. 259. Eds. R. Montesano, H. Bartsch and L. Tomatis. International Agency for Research on Cancer, Lyon 1980.
253 R. A. Squire: Science *214,* 877 (1981).

254 G. M. Williams, in: Proceedings of the Second International Conference on Safety Evaluation and Regulation of Chemicals. S. Karger, New York, in press.
255 G. M. Williams and J. H. Weisburger, in: Toxicology the Basic Science of Poisons. 3rd Ed. Eds. J. Doull, C. D. Klaasen and M. O. Amdur. Macmillan Publ. Co., New York (1985).
256 H. A. Guess and K. S. Crump: Environ. Health Perspect. *22,* 149 (1982).
257 J. Cornfield: Science *198,* 693 (1983).
258 N. E. Day and C. C. Brown: J. Natn. Cancer Inst. *64,* 977 (1980).
259 D. Krewski and C. Brown: Biometrics *37,* 353 (1981).
260 F. W. Carlsborg: Fd Cosmet. Toxic. *19,* 361 (1981).
261 E. Crouch and R. Wilson: J. Toxicol. Envir. Health *5,* 1095 (1979).
262 R. Peto, M. C. Pike, N. E. Day, R. G. Gray, P. N. Lee, S. Parish, J. Peto, S. Richards and J. Wahrendorf: IARC Monogr. Suppl. *2,* 311 (1980).
263 G. M. Williams, B. Reiss and J. H. Weisburger, in: Carcinogenesis and Mutagenesis. Eds. L. Andrews, R. Lorentzen and G. Flamm. Princeton Scientific Press, Princeton, N. J., in press, 1985.
264 W. L. Russell, P. R. Hunsicker, G. D. Raymer, M. H. Steele, K. F. Stelzner and H. M. Thompson: Proc. Natn. Acad. Sci. USA *79,* 3589 (1982).
265 D. G. Hoel, N. L. Kaplan, M. W. Anderson: Science. *219,* 1032 (1983).
266 J. Clemmesen and S. Hjalgrim-Jensen: Ecotoxicol. Environ. Safety *1,* 457 (1978).
267 C. N. Park and R. D. Snee: Fundam. Appl. Toxicol. *3,* 320 (1983).
268 Office of Technology Assessment. 'Environmental Contaminants in Food', p. 39. Congress of the United States, Washington 1979.

Recent advances in drugs against hypertension

Neelima, B. K. Bhat and A. P. Bhaduri*
Division of Medicinal Chemistry, Central Drug Research Institute, Lucknow-226001, India

CDRI communication No. 3628
* To whom all correspondence should be made

1 Introduction

The usefulness of antihypertensive drugs in reducing morbidity and mortality due to cardiovascular disease and in providing protection against ischaemic heart disease is evident from the reports of their trials in patients with mild hypertension [1–3]. The successful control and management of hypertension requires a multidisciplinary approach and an understanding of the recent advances in drugs against hypertension which, in turn, requires a comprehensive knowledge of the various facets of current research in this field. Of these, clinical experience with the available drugs, pharmacology of antihypertensive drugs and the relationship between novel chemical structures and antihypertensive activity are particularly important. Although an elegant review on antihypertensive drugs has been published recently, covering the period 1969–80, [4], the necessity of the present article was felt in order to acquaint the readers with the recent advances in this field because since then a large number of publications have appeared every year.

2 Clinical studies

2.1 Diuretics

Diuretic-associated hypomagnesemia and saluresis, specially calciuresis and the effect of loop diuretics on carbohydrate metabolism have received considerable attention in clinical studies. Induction of hypomagnesemia by diuretics is known and the sudden deaths reported of patients with hypomagnesemia or of persons living in areas with soft water supplies deficient in magnesium have led to clinical studies in patients [5–7]. Exposure of these patients to either short-term or moderate dosage long-term treatment with a diuretic such as furosemide (I) confirmed hypomagnesemia in patients.

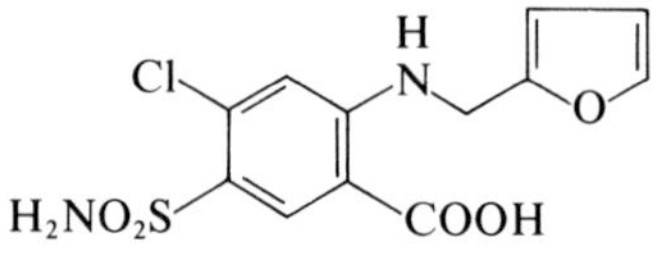

I

The magnesium depletion was compounded by the hospital diet and water supply low in magnesium. The common symptoms observed were depression, muscle weakness, refractory hypokalemia and arterial fibrillation refractory to digoxin treatment [7]. Fall in serum magnesium has been observed in patients with uncomplicated essential hypertension receiving piretanide (II), 12 mg/day [8]. Depletion of serum

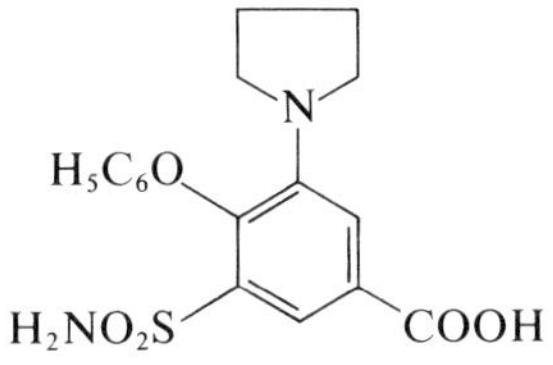

II

magnesium from 0.86 to 0.797 mmol/l in hypertensive patients after administration of a combination of hydrochlorothiazide (III), 100 mg/day and amiloride (IV), 10 mg/day, for 12 weeks has also been seen. It has been suggested that the fall is possibly due to a decrease in intracel-

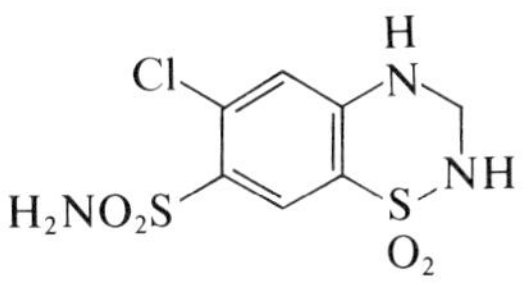

III

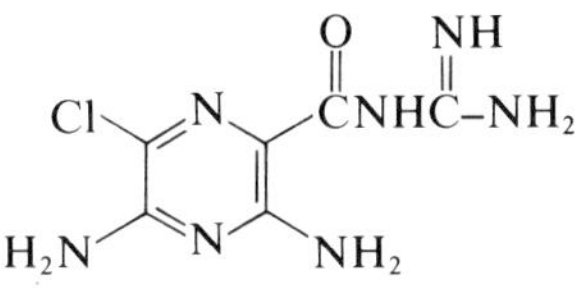

IV

lular magnesium which, in turn, may lead to diminution in intracellular potassium. These observations might have prompted the supplementing of antihypertensive formulation with magnesium stearate [9, 10].

The calciuretic property of furosemide has been observed at doses above 80 mg and simultaneous administration of furosemide, hypothiazide and triamterene (V) prevents excess of potassium excretion [11].

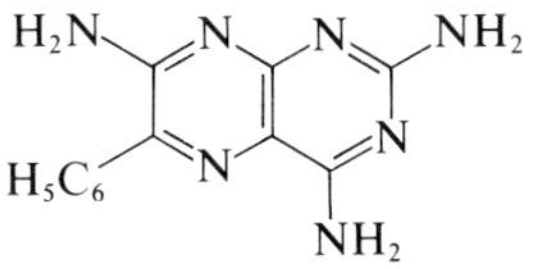

V

Hypercalcemic effects of diuretics such as brinaldix (VI) and hygroton (VII) appear after their chronic administration [12]. Single and multi-

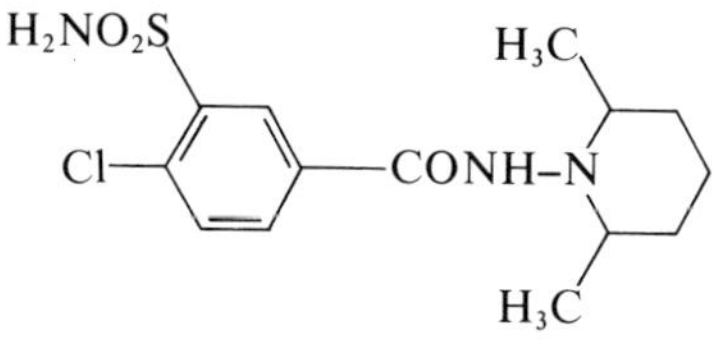

VI

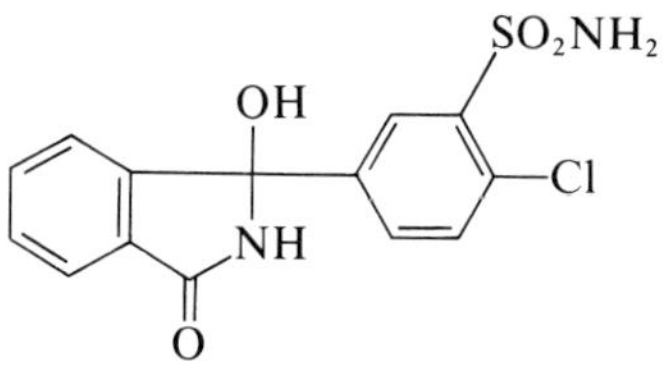

VII

ple oral doses of azosemide (VIII) in healthy volunteers increase urinary output and excretion of urinary electrolytes. The peak concentra-

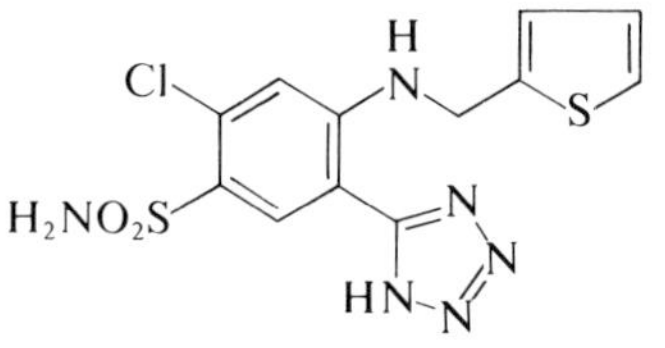

VIII

tion of the drug is observed after 3 to 4 hours of administration and its excretion in the urine is less than 10 % [13].

The effect of the loop diuretics, bumetanide (IX) and furosemide, in normal male subjects, has been monitored with reference to glucose,

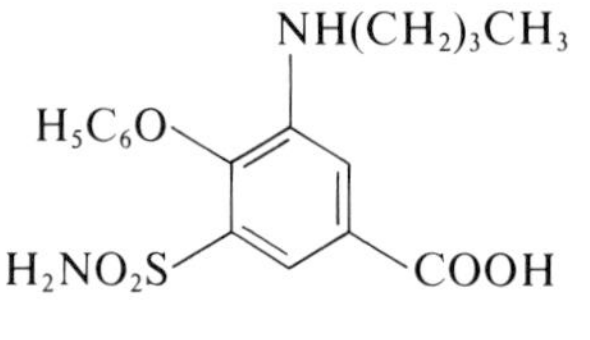

IX

insulin, glucagon and growth hormone responses to 5 hours glucose tolerance test [14]. Glucose tolerance significantly improves with bumetanide but not with furosemide. Plasma insulin, glucagon and growth hormone levels during the oral glucose tolerance test remain unaffected by either drug.

Indacrinone (X), a racemic mixture, is a potent diuretic which is effective in the treatment of mild to moderate hypertension and congestive heart failure [15, 16]. Studies with this drug in healthy men have been carried out to understand the pharmacological activity of the (+) and (–) enantiomers and of its (–) aromatic p-hydroxy metabolite (XI). It

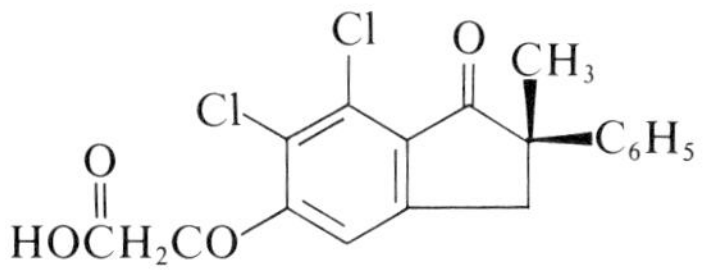

X

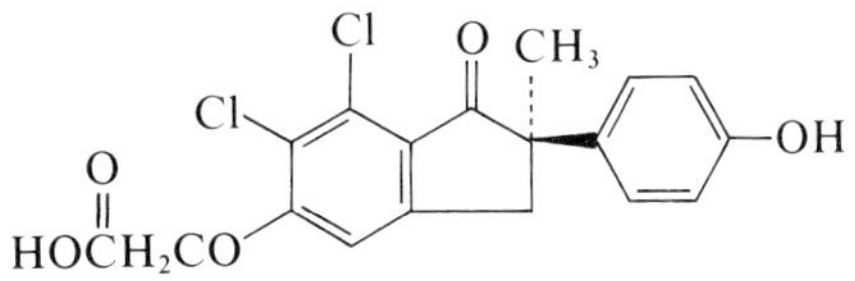

XI

has been found that natriuretic and diuretic activities are associated with (–) enantiomer and (–) p-hydroxylated metabolite, while the (+) enantiomer is responsible for potent hypouricemic effect [17].
The effect of furosemide on aged patients has received special attention [18]. Its diuretic effect in geriatric patients is reported to decrease and this decrease is both age-dependent and disease-dependent.

2.2 Renin-angiotensin system

The availability of potent angiotensin converting enzyme (ACE) inhibitors and the advent of angiotensin-II antagonists have made the direct renin blockade a distinct possibility. A decapeptide, Pro-His-Pro-Phe-His-Phe-Phe-Val-Tyr-Lys, has been found to inhibit human renin [19]. Specific inhibitors of human renin such as H-142, H-113 and H-77, obtained by modifying the N-terminal sequence of human angiotensinogen at the scissile Leu-Leu bond, are of interest [20, 21]. The inhibitor, H-142, after intravenous infusion at 1 mg/kg/hr in normal human subjects produces striking decrease in circulating angiotensin-I, angiotensin-II levels and a distinct fall in systolic and diastolic blood pressure [22].
Clinical mode of action of captopril (XII) is being still debated. Some

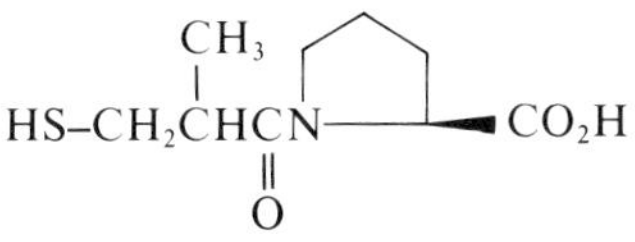

XII

believe that the activity is due to suppression of renin-angiotensin-aldosterone system [23] and the other hold the opinion that possibly augmented plasma kinin levels contribute towards the biological activity [24, 25]. Parasympathomimetic activity of captopril, 20 mg p. o. in normotensive subjects have been studied [26]. Bradycardia on facial immersion in water was attenuated by both captopril and endraphonium, a cholinesterase inhibitor. Acute effect of enalapril (XIII) in normal

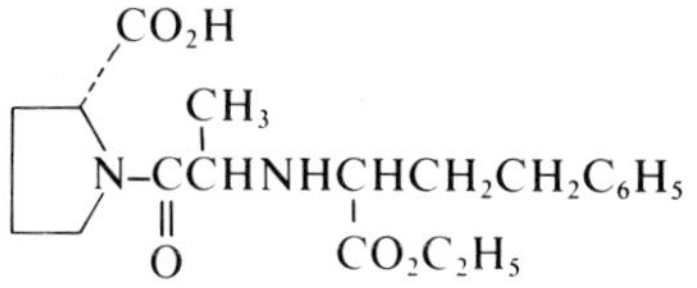

XIII

male volunteers after oral administration of 5 mg of the drug causes an increase in plasma renin from 7.3 ng/ml to 22.2 ng/ml and does not cause any change in plasma aldosterone system [27]. The immediate and long-term effect of enalapril and lysine analog MK-521 (XIV) at single daily dose in normal subjects have been studied.

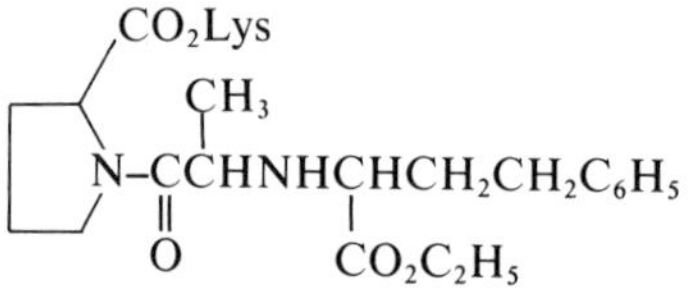

XIV

Both the drugs lower blood pressure equally throughout 24 hours without causing tachycardia and in both the cases the plasma renin concentration is significantly higher on day 8 than on day 1 [28]. Studies on the effect of enalapril in urinary excretion in healthy adults indicate no detectable disadvantageous urinary effects despite increased output of sodium ion and urinary volume [29]. In mildly sodium-depleted normotensive volunteers enalapril has been found to enhance baroreflex sensitivity to blood pressure increase [30]. Normal human subjects have been found to tolerate well 0.1–3.2 mg of HOE 498 (XV), a new orally active long-acting ACE inhibitor [31].

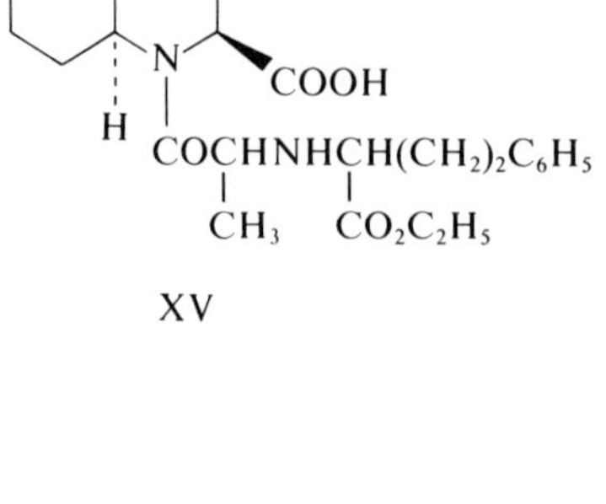

XV

2.3 β-Adrenoceptor blocking drugs

In the past decade β-blockers have attained prominence in the treatment of hypertension in European countries while their adoption in USA has been much slower [32]. The results of double-blind clinical trials with timolol (XVI), metoprolol (XVII) and propranolol (XVIII)

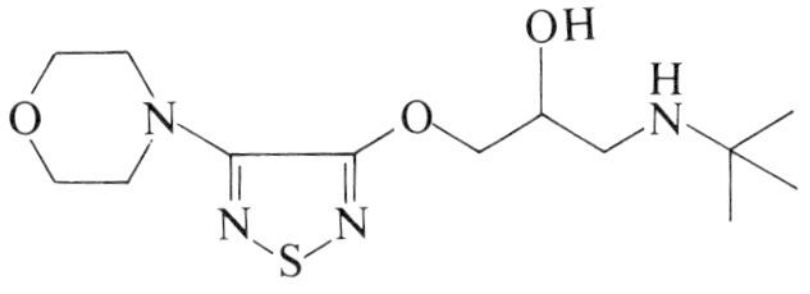

XVI

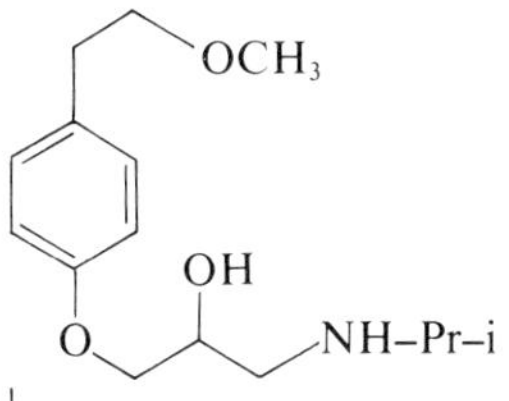

XVII

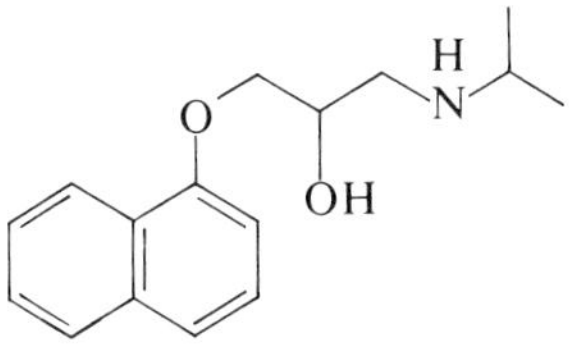

XVIII

indicate reduction in mortality rate by 44.6 %, 36 % and 26 % respectively [33–35]. Similarly, pindolol (XIX) has been found to be effective

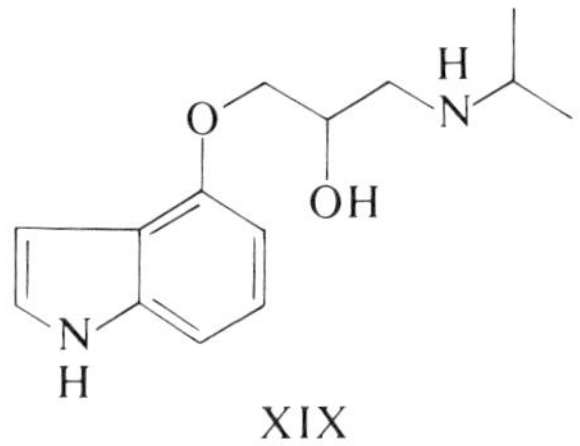

XIX

in reducing blood pressure both alone and in combination therapy [36]. Bopindolol (XX), structurally related to pindolol, at a dose of 1–4

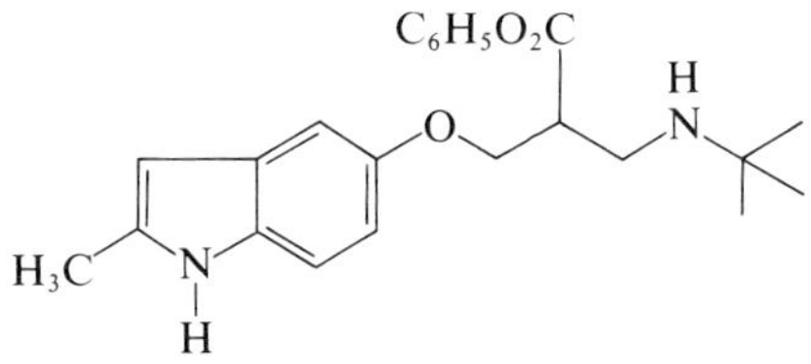

XX

mg in normal human volunteers causes significant reduction in the rise of systolic blood pressure. The reduction in the rise of heart rate observed with this drug is closely related [37]. Controlled trials with β-blockers in pregnancy hypertension reveal that these drugs are at least as effective as methyldopa and are safe for both mother and foetus [38]. In a double-blind study in patients with mild and moderate hypertension, with non-selective and cardio-selective β-blockers, it has been observed that some of the recommendations regarding initial dosing and frequency of administration of the drugs are not substantiated and should be reconsidered [39]. Effect of atenolol (XXI) in diabetic pa-

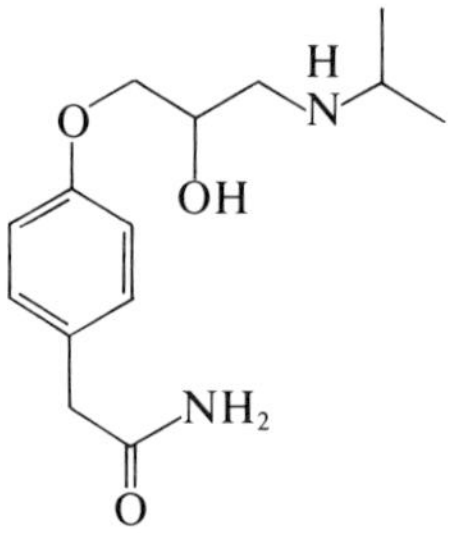

XXI

tients has been investigated [40]. Basal and reactive blood sugar values of non-insulin dependent diabetics, in either short- or long-term use of this drug, do not change significantly. Cardioselective β-blockers appear to have less tendency to reduce renal blood flow than non-selective ones. It has been concluded that the changes in renal blood flow in the presence of unaltered glomerullar filtration rate and normal biochemical indices of renal function remain to be elucidated [41]. The role of liver on the kinetics of metabolism of β-blockers such as propranolol, atenolol and sotalol (XXII) has been investigated in patients

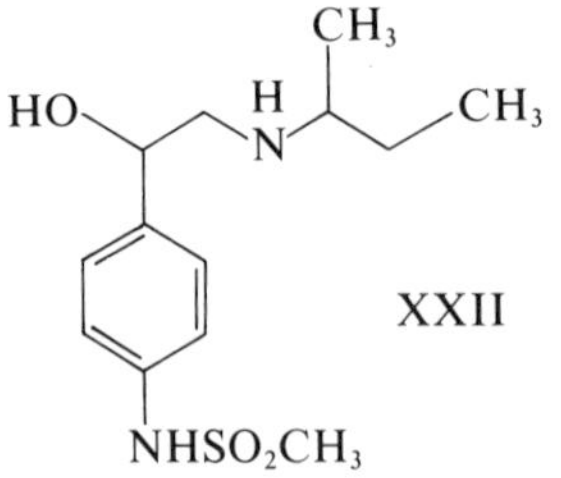

XXII

and it has been concluded that the kinetics of β-blockers which are eliminated unchanged, is independent of the liver and, therefore, may be recommended for patients with activated or impaired liver metabolism [42]. Studies on the effect of β-blockade on the pressure response to 'coffee plus smoking' in patients with mild hypertension suggest that the attenuation of the pressure effect of 'coffee plus smoking' by atenolol is a function of its β_1-adrenoceptor blocking activity [43]. The effect of β-blockers in asthma has been studied and its has been found that no β-blocker is completely safe for asthmatics and should be avoided if possible [44]. β-Adrenoceptor blocking drugs have also found use in psychiatry [45]. Studies on their effect on sleep in normal female subjects have revealed that metoprolol, propranolol and pindolol increase dreaming and early awakening while propranolol and pindolol increase awakenings in the night [46]. The frequency of the β-blocker withdrawal syndrome is uncertain and none of the hypothesis can be considered as explaining the cause of the syndrome [47]. The effect of cardioselective and non-cardioselective β-blockers on blood pressure, blood lipids and platelet aggregation have been studied. The rise in triglycerides in volunteers treated with mepindolol (XXIII) a non-car-

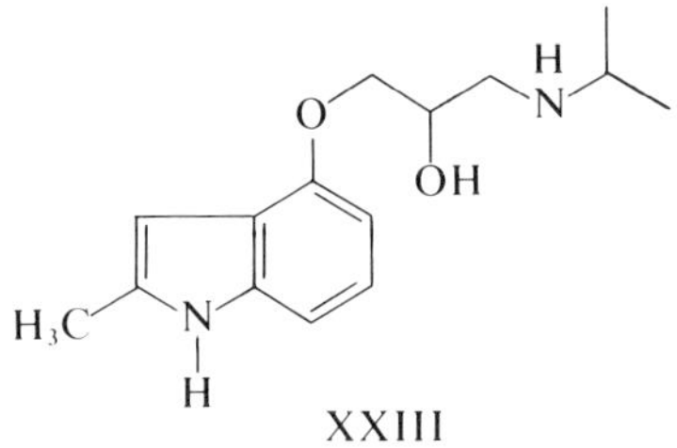

XXIII

dioselective β-blocker, is insignificant, and so also the changes observed in levels of cholesterol, free fatty acids and lipoproteins [48]. The antihypertensive effect of propranolol and oxprenolol (XXIV) in

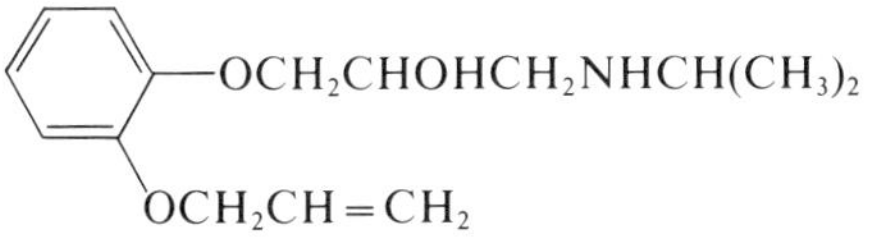

XXIV

man are attenuated by indomethacin [49] but indomethacin has not been found to influence the effect of a single dose of metoprolol (XVII) on exercise-induced tachycardia [50]. More than one mechanism have been proposed for the antihypertensive effect of β-blockers. One group of workers suggests that the effect is primarily attributable to the fall in cardiac output. However, the time course and the magnitude of the blood pressure fall indicate that it has little connection with the fall in arterial blood pressure [51]. The second suggestion, made by Bühler et al. [52], but debated by others [53–55], is that the blood pressure reduction by β-blockers is due to suppression of renin release. Despite this controversy, it is difficult to escape the conclusion that the reduction of renin leading to the reduction of circulating levels of angiotensin-II is at least partly responsible for the blood pressure fall. A third proposal relates to the fact that β-blockers in some manner interfere with central sympathetic outflow [53]. Yet a fourth possibility is that β-blockers act at presynaptic receptors to prevent neuro-transmitter release [56].

2.4 Vasodilators and calcium-entry blockers

Nifedipine (XXV) lowers blood pressure in man [57] and the clinical efficacy and tolerance of this drug can be enhanced by addition of a β-blocker [58], clonidine (XXVI) [59] or α-methyl dopa (XXVII) [60]. A

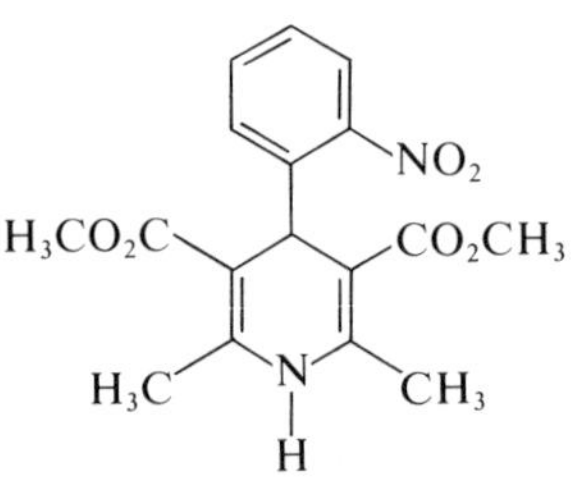

XXV

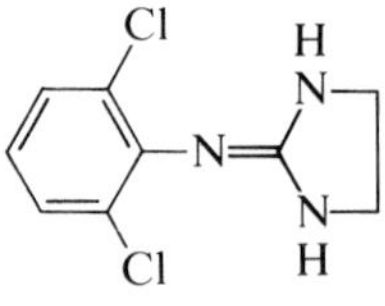

XXVI

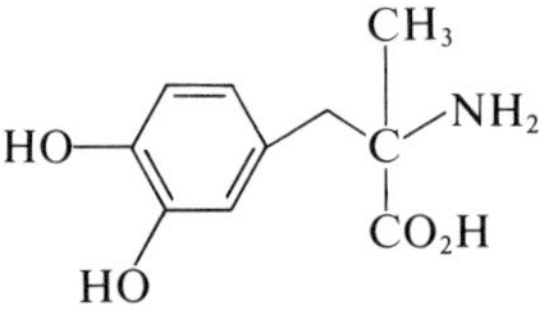

XXVII

couple of other vasodilators namely nisoldipine (XXVIII) and H-154/82 (XXIX), are claimed to be more potent than nifedipine and lower

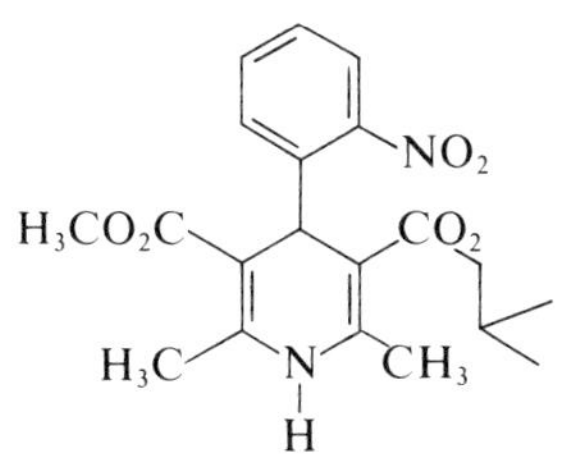

XXVIII

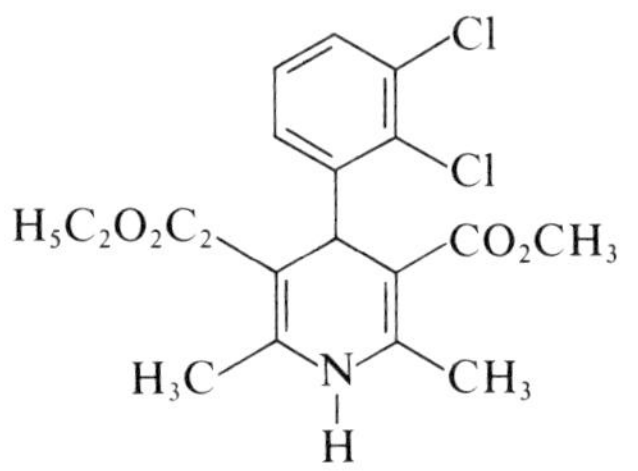

XXIX

blood pressure in man [61]. Doxazosin (UK-33274, XXX), an analog of prazosin (XXXI), lowers blood pressure in normal volunteers and

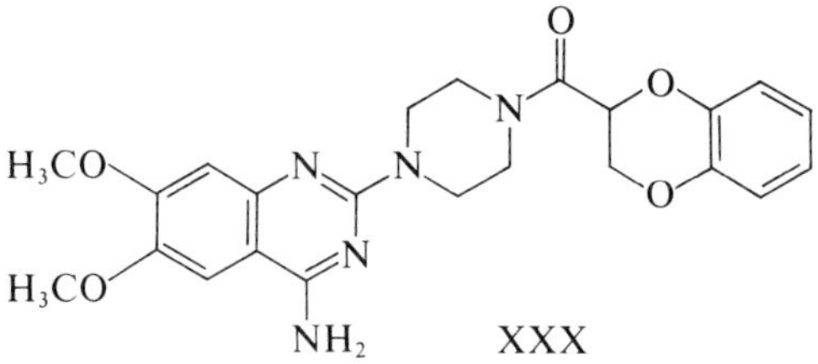

XXX

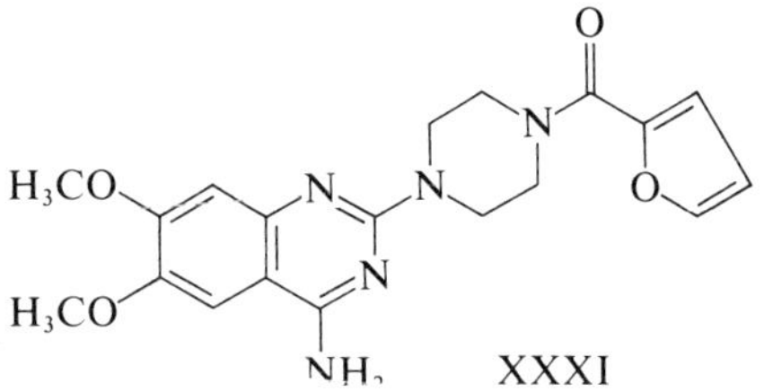

XXXI

increases heart rate [62]. Tolmesoxide (XXXII) has been reported to possess a short plasma half-life and a long duration of action [63].

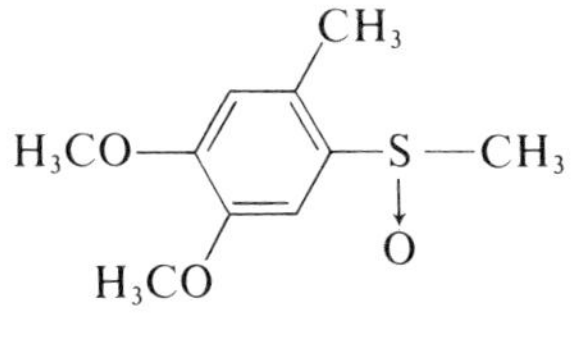

XXXII

The calcium entry blockers are particularly useful in therapy when hypertension and angina are present. Only three compounds, verapamil (XXXIII), nifedipine and diltiazem (XXXIV), have been approved by

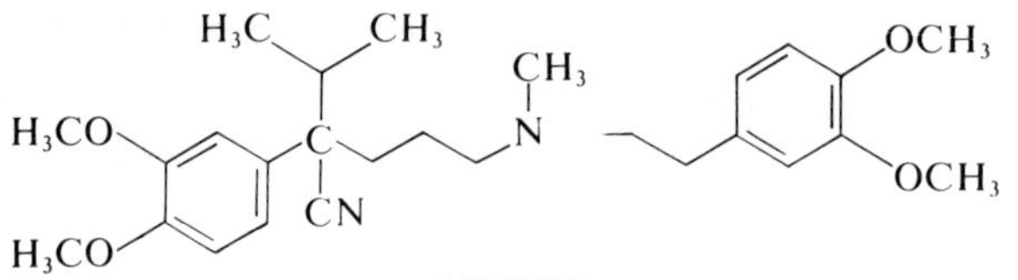

XXXIII

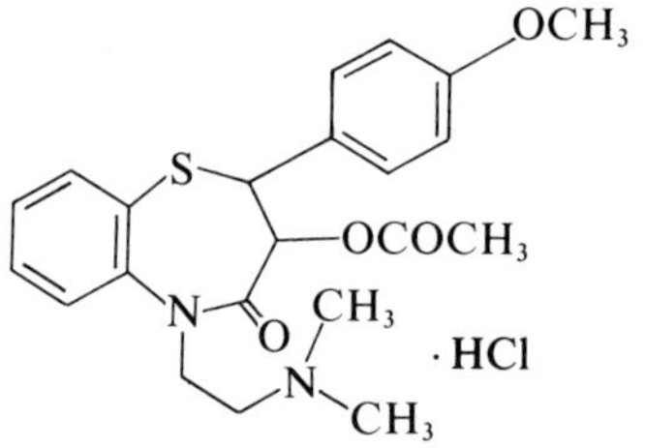

XXXIV

the FDA for the treatment of angina [64]. The results of the clinical studies with these drugs suggest that other disorders such as hypertension may be treated with these drugs since their mode of action may prevent the spasm of smooth muscles [65].

2.5 Centrally acting drugs

Combination therapy of clonidine, 0.15 mg three times a day for one week, and dihydralazine (XXXV), 25 mg orally three times a day for

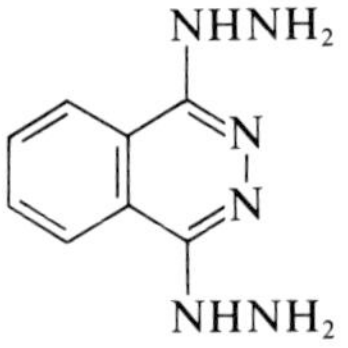

XXXV

one week, in patients with essential hypertension keeps the sympathetic tone within the normal range [66]. Clonidine reduces the mean arterial pressure in hypertensive patients and does not cause any adverse change in the blood lipid parameters [67]. At a dose of 75 mg, three times a day, for three days in healthy volunteers it diminishes the ergometroloading exercise (ELE)-induced rise in blood pressure while the combination therapy of this drug with indomethacin has no effect on blood pressure during ELE [68]. A combination of clonidine and prazosin, in a group of hypertensive patients with blood pressure over 140/90 mm Hg and who were using minoxidil (XXXVI), β-blockers

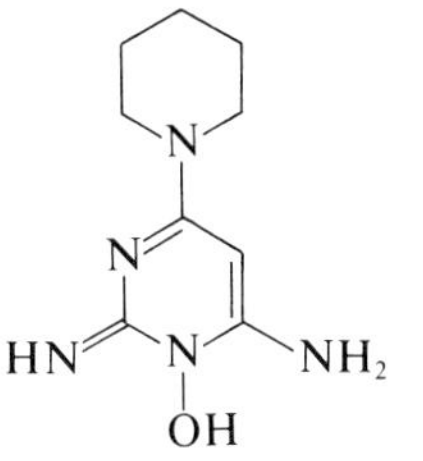

XXXVI

and diuretics, has been found to decrease the supine and standing systolic blood pressure at a dose of 0.2, 0.4, 0.6 and 0.8 mg/day. It also reduced the diastolic blood pressure in the standing position at a dose of 0.4–0.8 mg/day [69].

2.6 Miscellaneous

The serotonin antagonist ketansenin (RK 1468, XXXVII) exhibits antihypertensive effect in man [70]. Administration of 15-deoxy-16-hy-

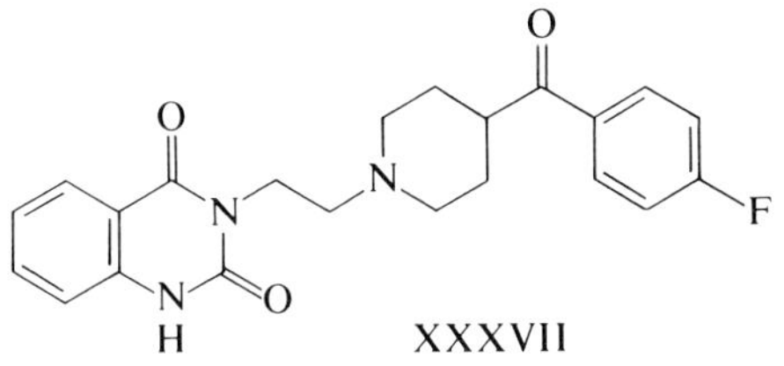

XXXVII

droxy-16-vinyl PGE_2 methyl ester in man causes hypotension and prolonged effects are observed with transdermal delivery [71]. Pharmacokinetic studies of the α_1-adrenergic antagonist, prazosin, in human reveal substantial binding of albumen, α_1- and glycoprotein. The variation in the plasma concentration between patients and between days in the same patients are very wide [72]. Treatment of hypertension with oral use of alkali metals, alkaline earth metals and ammonium salts of sulfites and bisulfites has been reported [73]. In humans 0.2 to 20 mg/kg of sodium bisulfite does not cause any serious side effects and the excretion rates of sodium, potassium, magnesium and calcium ions remain normal. A continuous dose of 5 % sodium bisulfite solution at a dropping rate of 15 drops/ml in an old man with generalized poor arterial circulation and afflicted by cardiac arrhythmia and hypertension, caused an improvement in the condition and no recurrence of his condition including hypertension has been observed even after approximately three years.

3 Advances in antihypertensive drug pharmacology

3.1 Diuretics

Furosemide appears to be the most extensively studied drug. It increases cyclic GMP in renal tissue, 10 minutes after its administration in rats, which returns to control level after 30 minutes [74]. The effect of furosemide on the gentamycin-induced nephrotoxicity in mice is interesting. Gentamycin sulfate at a dose of 50–100 mg/kg i. p. administered daily to healthy mice for 7–10 days causes cell damage of the proximal tubule, as is evident from electron microscopic examination and from urinary excretion of N-acetyl-β-D-glucasaminidase. Concomitant administration of furosemide at 5 mg/kg per day causes less tubular damage compared to that caused by gentamycin alone. However, similar effects have been observed with bumetanide and piretanide [75]. Furosemide alters the renin dependency of the blood pressure [76]. Studies on the role of renal prostaglandin in furosemide-induced diuresis indicate that renal PGE is necessary for furosemide to elicit optimal diuretic and natriuretic effects under condition of vasopressin infusion [77]. Effects of furosemide in combination with other drugs have also received attention. Furosemide in combination with indomethacin in salt-depleted dogs reduces the natriuretic effect possibly because of indomethacin-induced blockade of the renal vasodilation produced by furosemide [78]. The effect of diuretics such as furosemide, 2 mg/kg, ethacrynic acid, 2 mg/kg, and mannitol 1.5–2.0 g/kg, on renal glycine excretion in dogs have been studied. These drugs decrease urinary glycine excretion [79]. In regard to the excretion of sodium, calcium, magnesium and chloride ions no significant difference has been observed between these diuretics [80]. Furosemide and dichlorothiazide are able to increase the trans wall carotid artery potential in cats [81]. Hypokalemia in patients during furosemide and hydrochlorothiazide-induced diuresis has been mainly attributed to the elevated plasma concentrations of aldosterone and renin [82]. Increased urinary kallikrenin excretion has been observed with bumetanide 100 mg/kg i. v., in anaesthetized laparotomized rats [83].

3.2 Renin-angiotensin system

The effect of the inhibition of angiotensin-converting enzyme on thirst in rat has been studied [84]. Captopril, 0.5 mg/kg, has been found to block the synthesis of angiotensin-III only in the circulation; at higher doses (100 mg/kg) the blockade has been reported both in the blood and in the brain. The lower dose of captopril enhanced drinking in response to stimuli induced by isoprenaline, 0.1 mg/kg s. c., phenetol amine, 5 mg/kg s. c., and serotonin, 2 mg/kg s. c. The higher dose abolished drinking in response to stimuli. It has been suggested that the enhancement of drinking caused by lower doses of captopril, being a sensitive indicator, can be used as a parameter to ascertain whether renin-angiotensin system has participated in the regulatory response to a particular response to drink. In rats captopril, at about 1 mg/kg increases plasma renin activity from 1.6 to 4.5 ng/ml/hr but in combination with propranolol it does not change the activity. The chronic effect of oral administration of captopril, 63–83 mg/kg/day, on blood pressure, plasma renin activity, plasma renin concentration, plasma-aldosterone concentration and renal renin content has been studied in rats [85]. It has been found that this drug increases plasma renin activity and plasma and renal renin concentrations. Its long-term use lowers the blood pressure in normotensive rats with normal sodium intake. On account of the range of side effects associated with captopril and its similarity to penicillamine, it has been suggested that the thiol function may be responsible for these side effects [86]. This possibly prompted the synthesis of thiol free ACE inhibitors and of these enalapril has been subjected to clinical trials. This compound undergoes hydrolysis in vivo to enalaprilic acid. Intravenous administration of enalapril to rats, made hypertensive by continuous intravenous infusion of angiotensin-II for ten days, does not significantly lower blood pressure, and on the basis of this observation it has been suggested that there is no involvement of any important non-angiotensin mechanisms. Enalaprilic acid, when injected into the brain ventricles, has been found to reduce the blood pressure [87]. Investigations on the attenuation of vascular sympathetic responses in spontaneous hypertensive rats by captopril and enalapril and the involvement of α-adrenoceptor for the same reveal that both subtypes of adrenoceptors are involved in the converting enzyme inhibitor (CEI)-induced attenuation of sympathetic responses [88]. The mode of action of ACE inhibitors like captopril and enalapril remains unknown, though several mechanisms, such as the inhibition of the vasopressor action of renin-angiotensin system, involvement of sympathetic nervous system, increased prostaglandin activity, bradykinin accumulation and decreased vascular response to α-adrenergic agonists have been suggested [89–95].

The potential of direct renin blockade for designing antihypertensive compounds has been studied. Dog renin-specific antibody and fats fragment reduced blood pressure in salt-depleted normotensive and renally hypertensive dogs [96]. Renin secretion resulting from low frequency renal stimulation of nerve is not altered by α-adrenoceptor blockade in dogs [97]. Tanin, an enzyme capable of producing angiotensin-II directly from angiotensinogen, has been found to activate human amniotic fluid renin [98]. The mechanism of renin secretion continues to be a subject of current interest. Prostaglandins possibly mediate renin release secondary to their stimulation of the renal baroreceptors and activation of α-adrenergic receptors appears to involve a prostaglandin-mediated release of renin. Unlike α-adrenergic receptors, stimulation of β-adrenergic receptors involves a renin release independent of prostaglandin and it has been suggested that β-adrenergic receptor mediated release is functionally located distal to the angiotensin-II receptor controlling renin release [99]. Besides these, calcium influx has been found to play an inhibitory coupling role in the control of renin secretion. It has also been suggested that a kallikrenin-like enzyme may be responsible for the activation of prorenin [100–105].

Angiotensin itself has become a subject of interest recently. Studies in vitro and in vivo with 1-sarcosin-angiotensin-II analogs modified at position-4 for ascertaining their binding affinity to beef adrenocortical membranes have revealed that the hormone affinity is dependent on the electronegativity of the side-chain. Low electronegativity results in greater hormone affinity [106]. By modification of the N-terminal end of 5-leucine-angiotensin-II, it has been possible to identify in the isolated rabbit aorta the structural units responsible for inducing tachyphylaxis. Positively charged N-termines and the guanidine group of two arginine side-chains have been found essential for tachyphylatic

activity. Of the many derivatives studied, 1-sarcosin-5-leucine angiotensin-II has been found to cause marked tachyphylaxis [107].
The subcellular localization of renin and angiotensin-II in rat brain synaptosomes and nerve endings and the local generation of (Ileu[5]) angiotensin-I, -II and –III in the rat brain have been investigated. These studies suggest a greater accumulation of angiotensin-I and decrease in the level of angiotensin-II in stroke-prone spontaneously hypertensive rats. It is likely that higher synthesis and turnover rate of angiotensin-II occur in these animals [108].

3.3 Vasodilators and calcium channel inhibitors

Prazosin has a favourable effect on blood lipids [109, 110] but in combination with β-blockers it fails to exhibit the same effect. Investigations on the mode of action of hydralazine (XXXVIII) and propildazine (XXXIX) suggest their interaction with the smooth muscle recep-

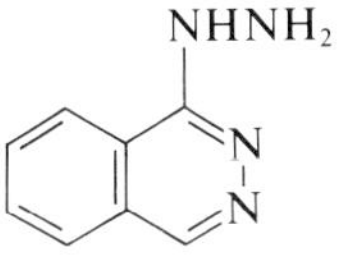

XXXVIII

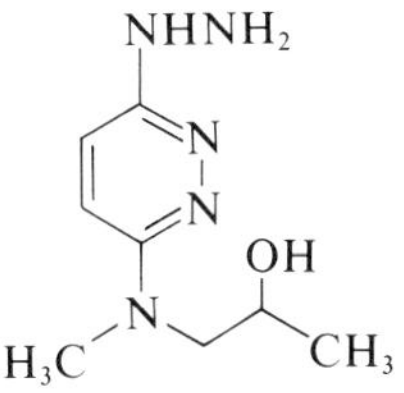

XXXIX

tor sensitive to endogenous ATP and adenosine [111]. The effect of combination therapy of a vasodilator and a β-adrenoceptor blocker on the left ventricular compliance and myocardial stiffness has been studied. In four week-old spontaneously hypertensive rats, hydralazine along with metoprolol reduces the mass-volume ratio and increases the compliance of the left ventricle. This combination, however, does not alter the myocardial stiffness and left ventricular collagel concentration [112]. The combination therapy of prazosin and isosorbide dinitrate (XL) in spontaneously hypertensive rats prevents cardiac hypertrophy [113]. Studies on pindolol (XIX), a β-adrenoceptor blocker, in healthy volunteers has revealed a direct vasodilatory effect at β-blocking doses of this drug, which has been explained on the basis of

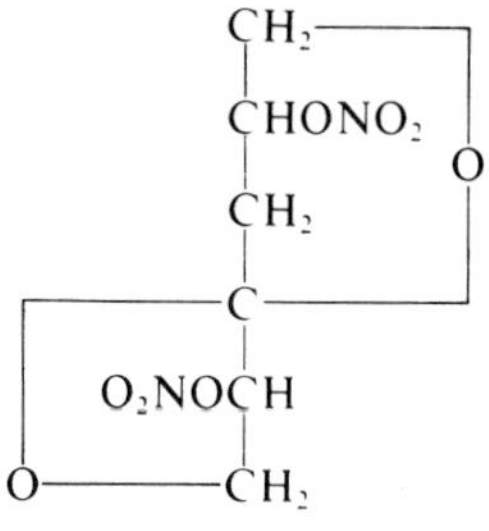

XL

stimulation of vascular β_2-adrenoceptors [114]. It is interesting that the vasodilatory effect of pindolol is significantly decreased by concomitant infusion of propranolol [115].

The effect of nifedipine (XXV) at 10 mg/kg i. v. on myocardial blood flow in coronary-occluded dogs has been investigated [116]. The results indicate that under controlled conditions, increasing exercise causes a progressive increase in myocardial blood flow in normally perfused areas and worsens subendocardial hypoperfusion in collateral dependent areas. However, in normally perfused areas nifedipine does not significantly alter myocardial blood flow at rest or during exercise. A large number of analogs of nifedipine have been synthesized and their profile of pharmacological activity has been studied. Niludipine (XLI), though similar to nifedipine in its pharmacological profile causes less cardiodepression in dogs [117]. Nicardipine (XLII) exhi-

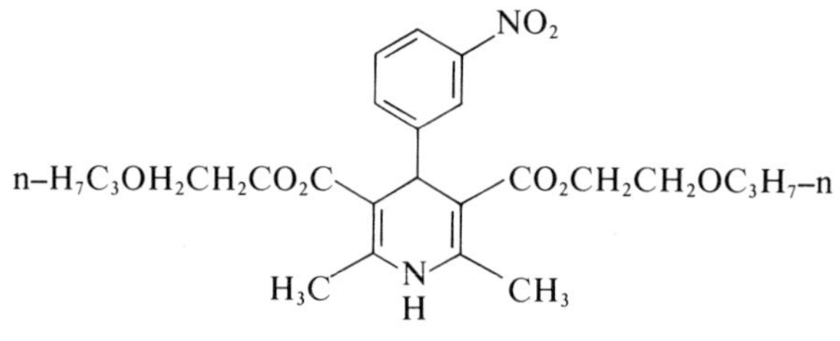

XLI

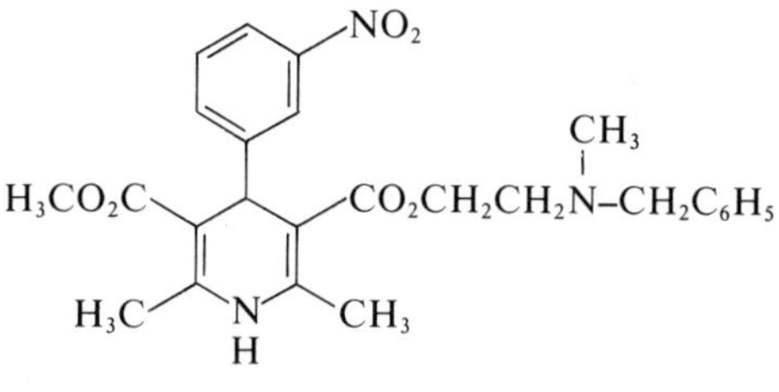

XLII

bits on potassium depolarized aortic stress, a high degree of calcium antagonistic property [118]. Nisoldipine (XXVIII), a more potent vasodilator than nifedipine, has marked effect on the blood flow in the ischemic myocardium [119–121]. Felodipine, an important peripheral vasodilator, significantly increases coronary sinus flow and reduces systemic and coronary vascular resistance at a dose of 10 mg p. o. in man [122]. The interaction of ^{14}C-felodipine with calmodulin, a calcium-binding protein capable of causing membrane activation, has been studied and a strong binding between them has been reported [123, 124]. Py 108-068 (XLIII) exhibits calcium antagonist properties

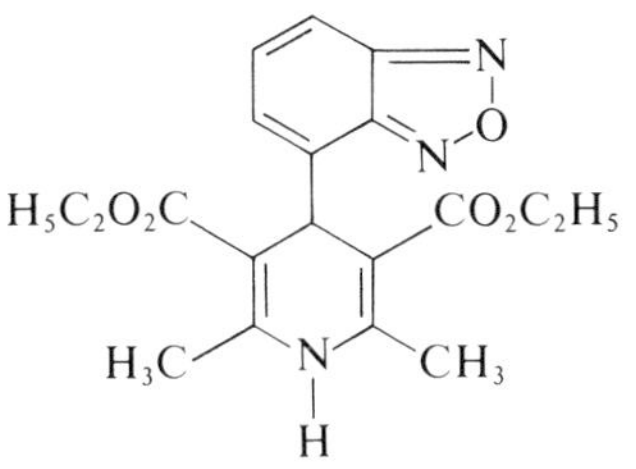

XLIII

on the coronary artery of the dog and in anaesthetized cats, reduces blood pressure, increases coronary blood flow and reduces heart rate [125].

3.4 β-Adrenergic system

The manifestation of the antihypertensive action of propranolol in spontaneously hypertensive rats possibly occurs through the release of an endogenous opioid [126]. The cardio-protective action of propranolol has been explained on the basis of the observed reduction in thromboxane-induced vasoconstriction and inhibition of platelet aggregation [127]. Chronic propranolol treatment increases the density of β-receptors on human lymphocytes and isoproterenol-induced cyclic AMP production after four weeks [128].
The β_1- and β_2-receptors have been further defined; innervated receptors which respond to neuronally released norepinephrine are classified as β_1 and hormonal receptors mediating responses to circulating epinephrine as β_2. A difference in the thermal sensitivity of solubilized β_1- and β_2-adrenoceptors has been reported [129, 130].

3.5 Centrally acting hypertensive agents

Clonidine, 100 µg/kg/day s. c. for nine days, can inhibit water intake in both controlled rats and in rats with hereditary diabetes insipidus [131]. Administration of clonidine along with the hypoglycemic drug, glibenclamide, in rabbit causes an early onset of hypoglycemia and produces a greater fall in blood sugar than that observed with glibenclamide alone [132].

The central regulation of blood pressure continues to be a subject of study. Intracerebroventricular administration of Leu-enkephalin produces a greater rise in blood pressure in spontaneously hypertensive rats than in normotensive animals [133]. The role of phenylethanolamine-N-methyl transferase in central blood pressure regulation is an emerging concept which requires further investigations [134]. It has been observed that tyrosine progenitor (XLIV) is orally more effective

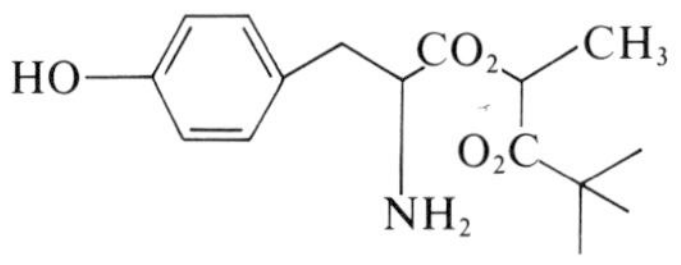

XLIV

than tyrosine in reducing blood pressure in spontaneously hypertensive rats and it is likely that the tyrosine may provide a new approach in elucidating the mechanism of central regulation of blood pressure [134]. The hypothesis that L-glutamic acid, serotonin and epinephrine may be involved in this regulation is attractive. Increase in serotonin level, induced by 1-(m-trifluoromethylphenyl) piperazine, has been reported to cause the lowering of pressure in spontaneously hypertensive rats [135].

3.6 Miscellaneous

The antihypertensive activity of E-643 (XLV) and tetrahydrobenzoxepine (XLVI) has been explained on the basis of their α_1-adrenoceptor blocking activity [136, 137]. In rats, the hypotensive effect of ketanserin (XXXVII), a serotonin antagonist, has been correlated with its peripheral α_1 rather than serotonin antagonism [138]. The dopamine agonist M-7 (XLVII) has been reported to reduce blood pressure and

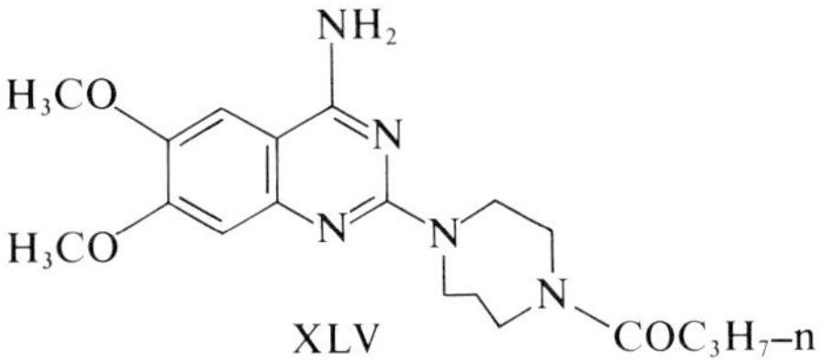

XLV

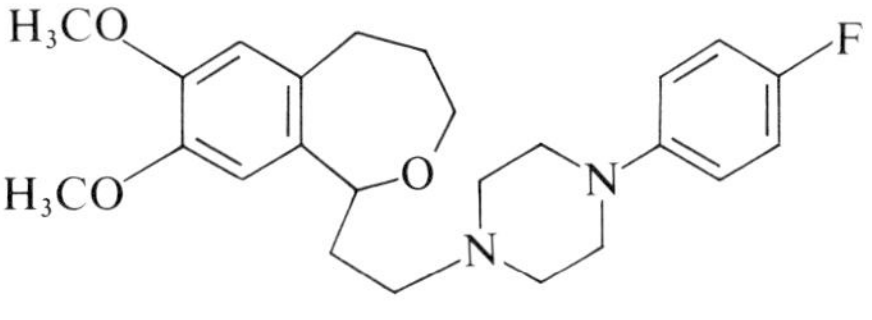

XLVI

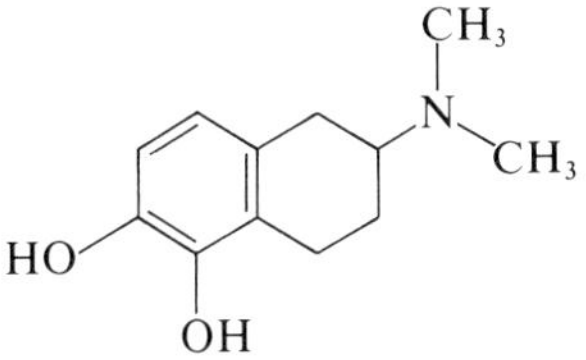

XLVII

heart rate in rats by stimulating presynaptic dopamine receptors on sympathetic neurons to the vasculature and presynaptic α_2-receptors on cardiac sympathetic nerve endings [139]. A comparative study of the effect of indoramin (XLVIII) and yohimbine (XLIX) at pre- and

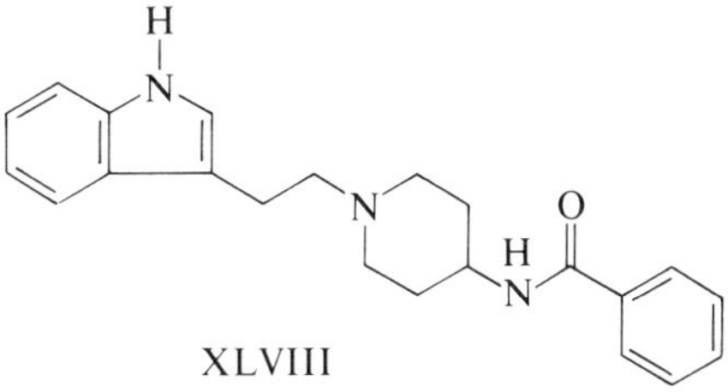

XLVIII

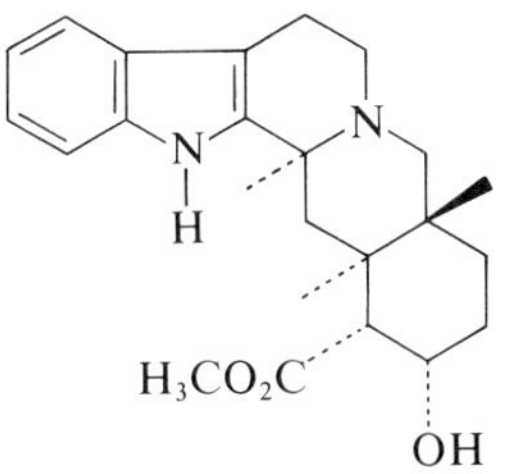

XLIX

post-synaptic α-adrenoceptors of rabbit pulmonary artery suggests a selective action of indoramin on post-synaptic α-adrenoceptors [140].

Trimazosin (L) in addition to its α_1-adrenoceptor blocking activity, has

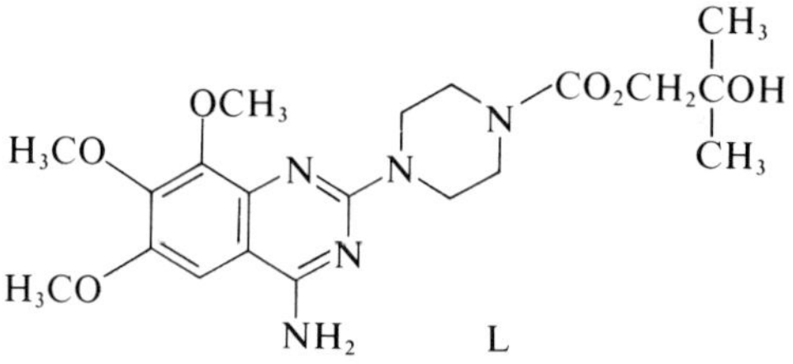

L

been found to elevate cyclic GMP levels in vascular smooth muscles and possibly this mechanism contributes toward its antihypertensive action [141]. The preferential α_2-adrenergic antagonist, yohimbine, at 4 mg/kg s. c., increases the serum renin activity and the rate of release of norepinephrine onto the granular juxtaglomerular cells of kidney, because of the blockade of prejunctional α_2-adrenergic receptors on the renal sympathetic nerves and/or centrally mediated activation of the peripheral sympathetic nervous system [142]. Studies in rats reveal that urapidil (LI) is an α-blocker with partial α_2-agonist activity [143].

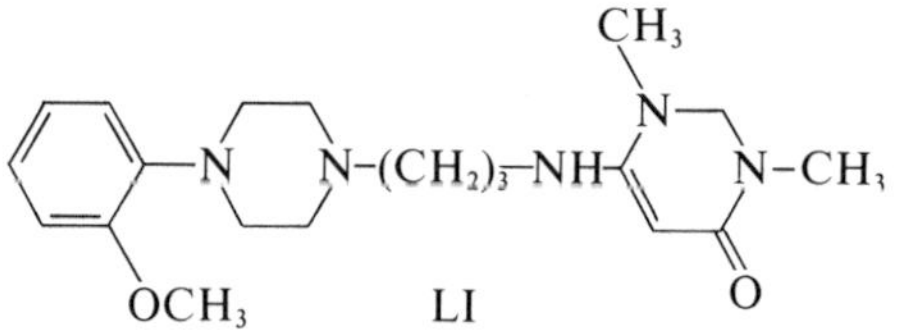

LI

Hypertension is associated with renal problems and renal failure in patients also produces other complications. The line of treatment in such cases requires an understanding of the interaction of drugs in these patients. Results of cimetidine (LII) administration twice daily for two

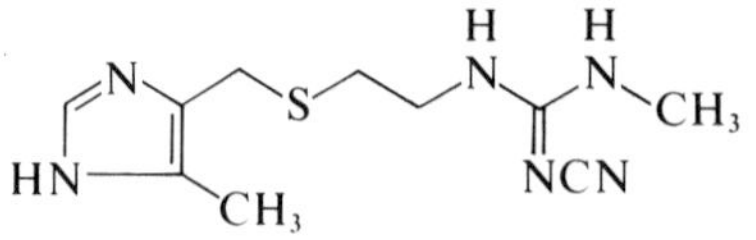

LII

weeks have indicated decreased immunoreactive parathyroid hormone and increased creatinine and urine concentration in patients with chronic renal failure [144]. In anaesthetized rats this compund at 20–200 μg/kg i. v. produces a long-lasting fall in the mean blood pressure and heart rate. In spontaneously hypertensive rats, bromocriptine (LIII) at a dose of 0.3 mg/kg causes a gradual long-lasting fall in mean

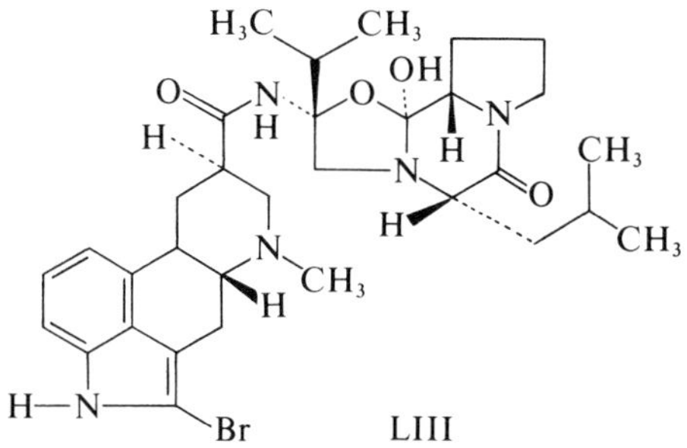

LIII

arterial pressure and the fall is related to the decrease in stroke volume index and cardiac index. No significant change in heart rate or total peripheral resistance index has been observed. The action of bromocriptin is due to a central dopaminergic mechanism rather than to its peripheral effects. In order to ascertain the role of antihypertensive polar renomedullary lipid in the pathogenesis of hypertension, a semi-synthetic compound namely 1-ox-hexadecyl-2-o-acetyl-sn-glycero-3-phosphorylcholine (HAGPC) has been investigated and it has been found to have a strong hypotensive action in rabbits [145]. The mechanism of the antihypertensive effect of D- and L-tryptophan has been studied, while the former does not cause any change in the blood pressure, the latter produces an antihypertensive effect in spontaneously hypertensive rats and it has been concluded that the antihypertensive effect is not mediated by brain serotonin [146]. The effect of clonidine on heart function in spontaneously hypertensive rats has been studied [147]. At a dose of 10, 30 and 50 μg/kg i. v. there is a dose-related reduction in the left ventricular systolic pressure and systolic wall stress.

4 Emerging antihypertensive agents

The patent literature about novel antihypertensive compounds regarding their biological activity is often very meagre. Despite this, an overview of novel compounds, claimed to have antihypertensive activity, would not be out of place for understanding the trends of research in medicinal chemistry to develop drugs against hypertension. The novel compounds of the last four years are, therefore, described.

4.1 Diuretics

The benzimidazole derivative (LIV) at 10 mg/kg in mice has been

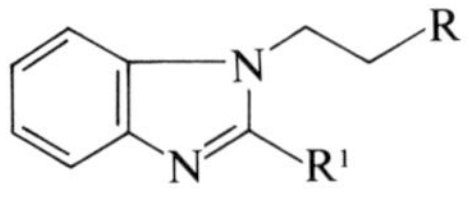

LIV

found to be slightly more effective than euphyllin [148]. The quinolone oxime derivative, M 12285 (LV) exhibits four times the potency of fu-

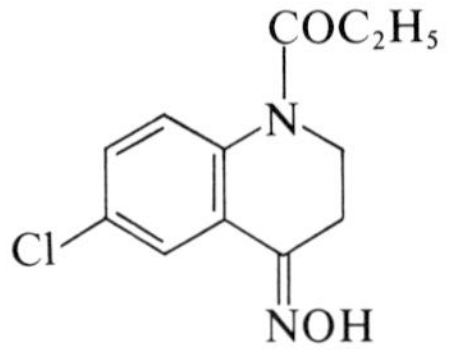

LV

rosemide and increases the sodium/potassium ratio in the urine. Since it does not increase urinary phosphate excretion, unlike furosemide, it does not act on proximal tubule. The LD_{50} of this compound is 3950 mg/kg p. o. in male mice [149].
Indapamide (LVI), an indoline derivative, at 0.1 to 30 mg/kg p. o.,

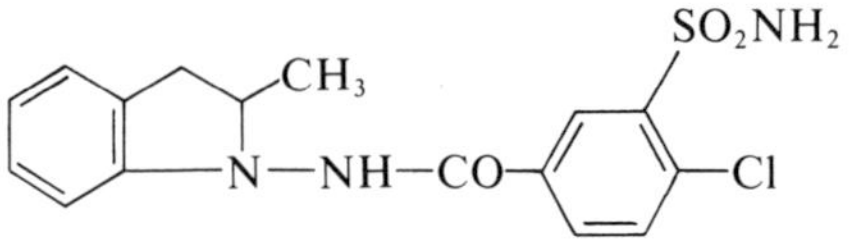

LVI

causes diuresis in spontaneously hypertensive rats and has been found to be superior to trichloromethiazide as an antihypertensive [150].
A riboflavin analog, 7,8-dimethyl-10-(3-chlorobenzyl)isoalloxazine (LVII) claimed to be a potassium sparing diuretic, is better than hydrochlorothiazide. In unanaesthetized spontaneously hypertensive rats, this compound at 0.5 mg/kg twice weekly for seven weeks, lowered the systolic blood pressure from 188 to 148 mm Hg [151]. A 1,2-dihydro-2-(3-pyridyl)-3H-pyrido[2,3-d]pyrimidin-4-one (LVIII) at

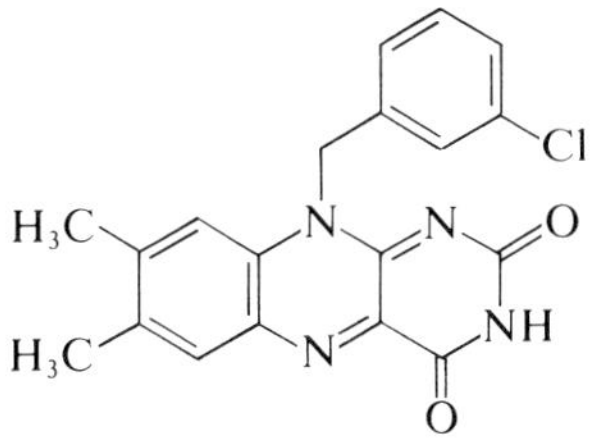

LVII

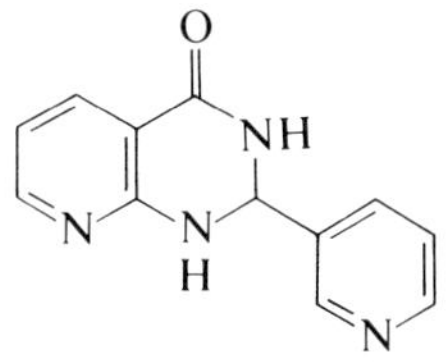

LVIII

27 mg/kg increased urine output by 250 % [152]. An increase in urine and sodium output by 211 % was induced by 0.1 mg/kg dose of a benzophenone derivative (LIX) in rats [153].

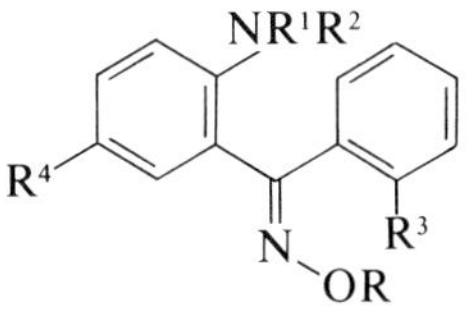

LIX

A dihydrophthalazin-1-ylacetic acid derivative (LX) exhibits strong

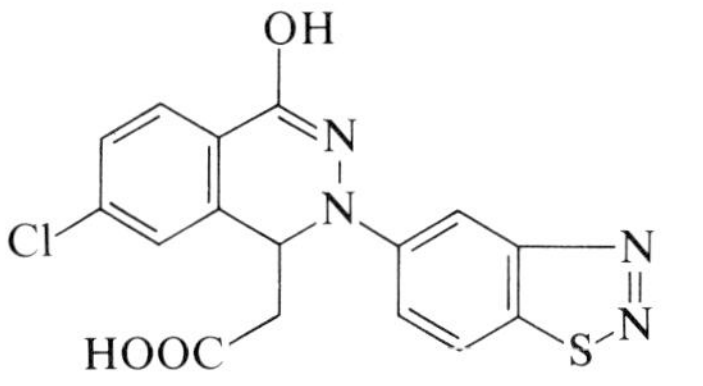

LX

diuretic activity in mice, rats, rabbits and dogs and is twice as active as furosemide in mice and dogs but is ten to twenty times more potent in rats and rabbits [154]. An enamine (LXI) has been claimed to be ten

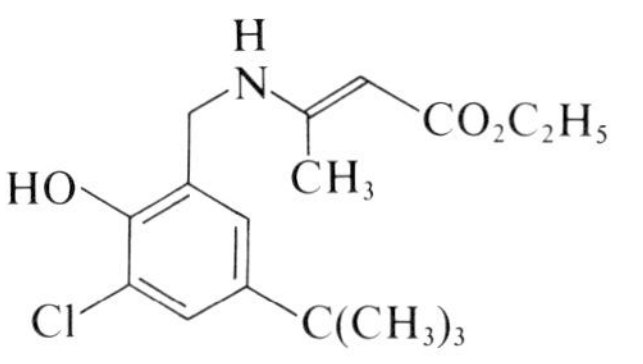

LXI

times as potent as furosemide [155]. The diuretic activity of the pentacyclic compound, indolo[2′,3′,3,4]pyrido[2,1-b]quinazolin-5-one (LXII) is superior to that of hypothiazide [156].

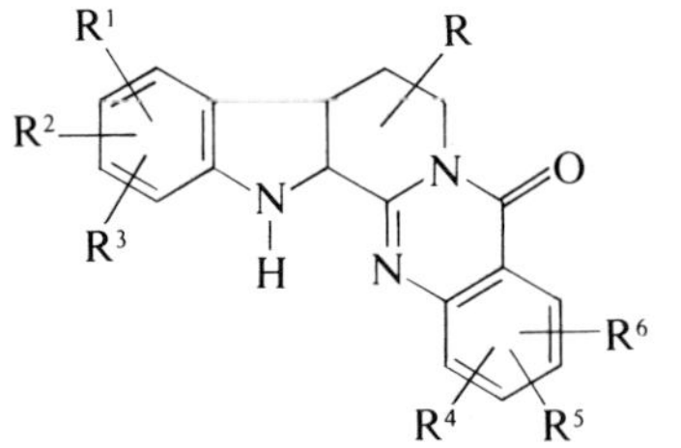

LXII

4.2 Angiotensin converting enzyme inhibitors (ACE)

High ACE inhibitory activity has been reported for N-phosphoryl-ala-Pro (LXIII) [157].

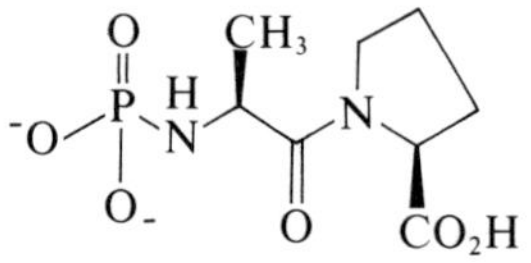

LXIII

Another proline derivative LXIV, exhibits better ACE inhibitory activ-

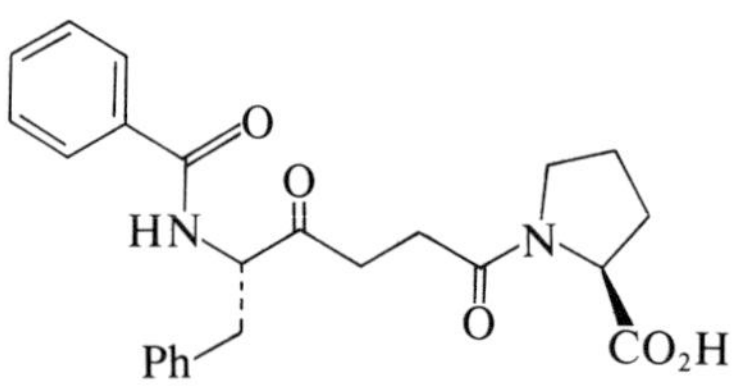

LXIV

ity than captopril [158]. Other novel inhibitors include compounds with and without sulfhydryl groups (LXV–LXVII) [159–161].
A peptide derivative of the bicyclic heterocyclic (LXVIII), reduces angiotensin-related blood pressure in rats with an ED_{50} of 0.2 mg/kg i. v. [162]. The imidazolidinone derivative (LXIX) at 1 mg/kg p. o. in rats, exhibits 60 ± 6 % angiotensin-converting enzyme inhibiting activity

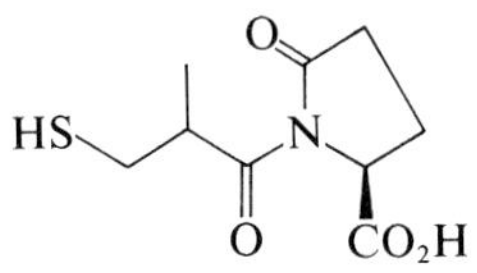

LXV

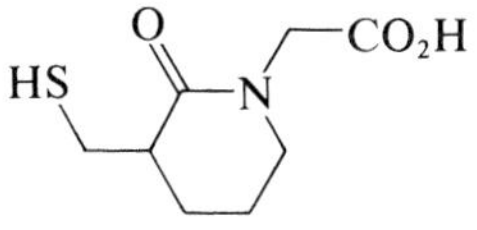

LXVI

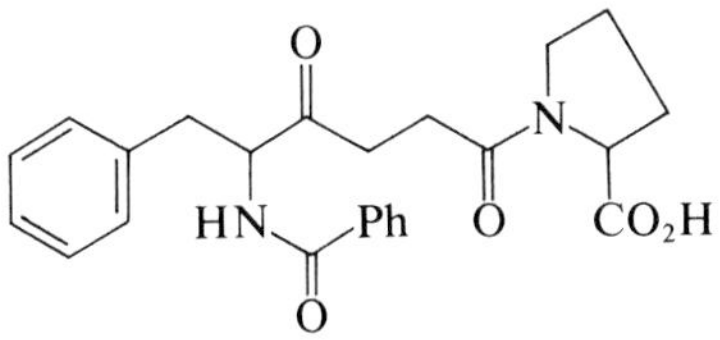

LXVII

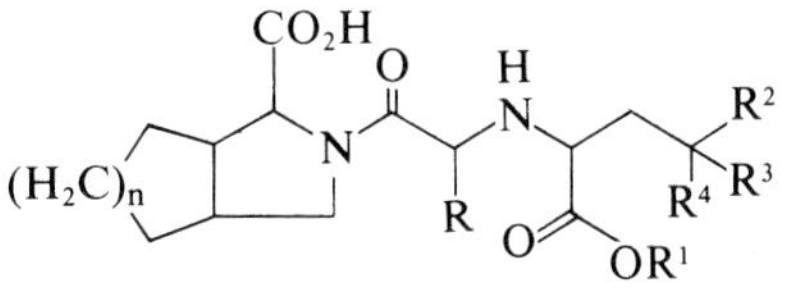

LXVIII

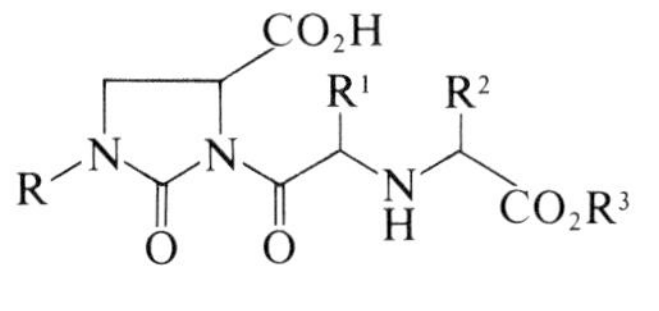

LXIX

[163]. The indoline derivative (LXX) at 1 mg/kg i. v. in rats causes 83 %

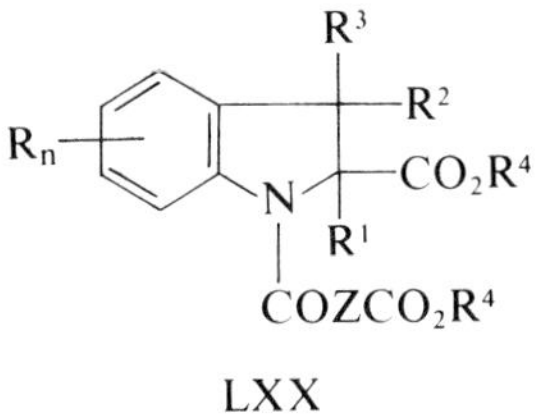

LXX

inhibition of angiotensin-induced increase of blood pressure [164]. A structurally related compound LXXI inhibits angiotensin converting

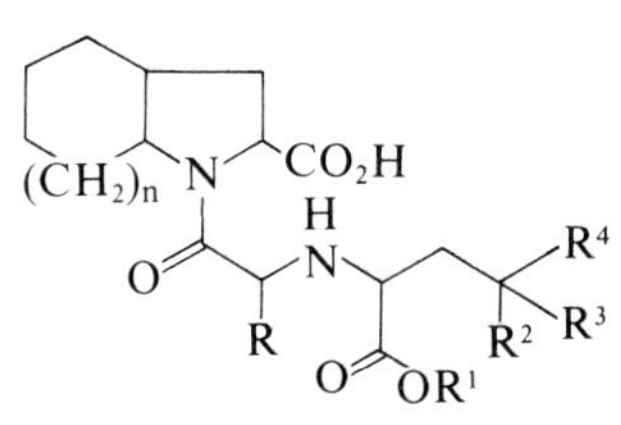

LXXI

enzyme in rats with an ED_{50} of 800 μg/kg [165]. The spiro derivative (LXXII) at 100 μg/kg in rats has been found to inhibit 95 % of the pressure response caused by 310 ng of angiotensin-I [166].

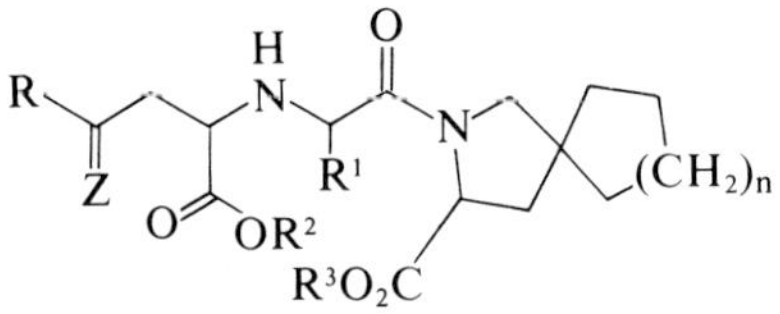

LXXII

4.3 Vasodilators

The newer vasodilators include RO-124713 (LXXIII), HL-725 (LXXIV) and the oxygen containing heterocycle (LXXV) [167–169]. The tricyclic compound GPA-1595 (LXXVI) possesses a similar profile of pharmacological activity as hydralazine [170].

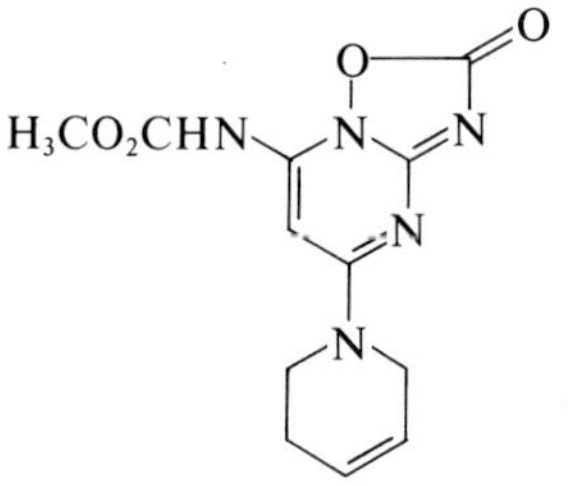

LXXIII

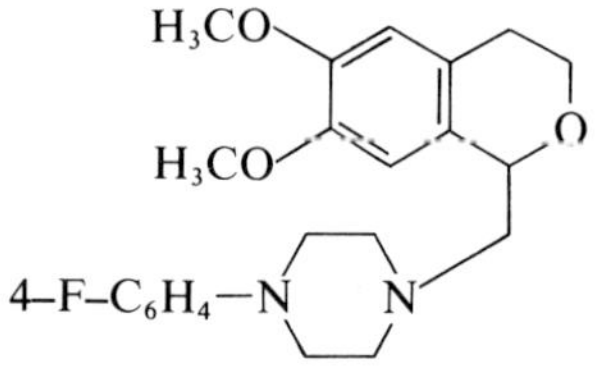

LXXV

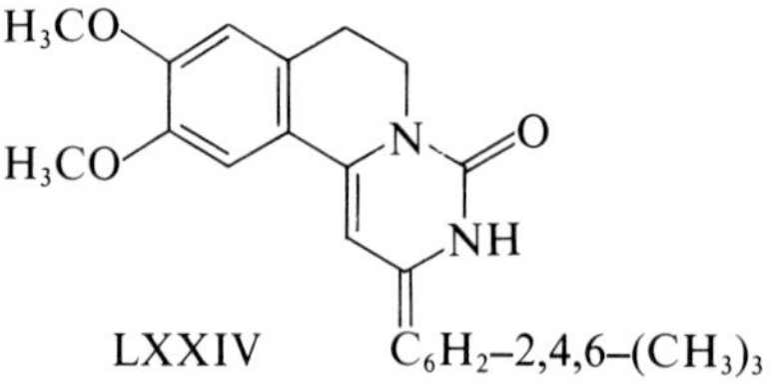

LXXIV

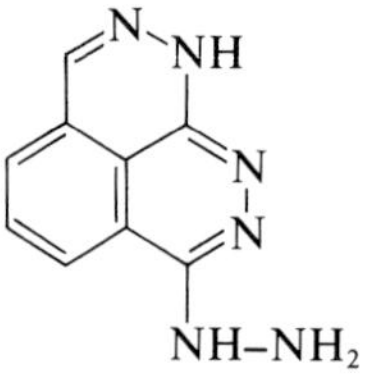

LXXVI

Besides this, the imidazolylpyridazine derivative (LXXVII), P-1134 (LXXVIII) and the triazolo[1,5-a]pyrimidine derivative (LXXIX) have shown encouraging results as vasodilators [171–173].

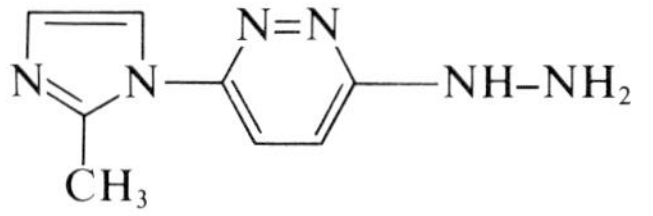

LXXVII

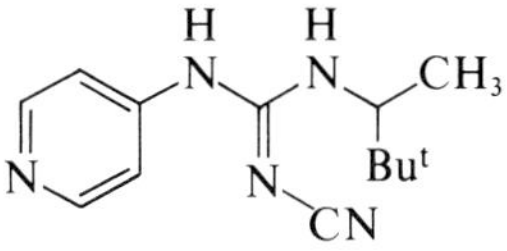

LXXVIII

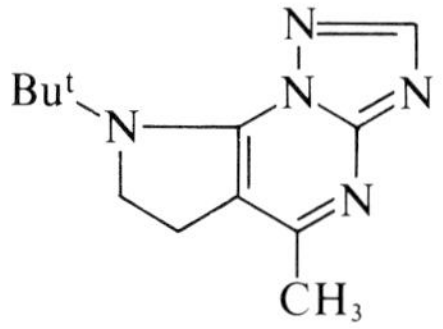

LXXIX

A central mode of action has been proposed for GYRI-11679 (LXXX) [174]. The xanthin derivative HWA-285 (LXXXI) has been found to cause preferential dilation of small arterioles in dogs [175].

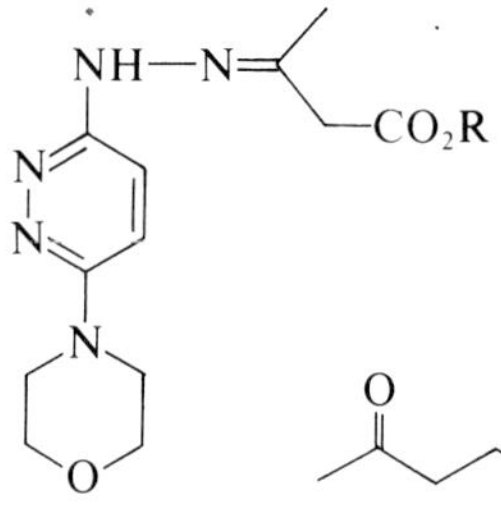

LXXX

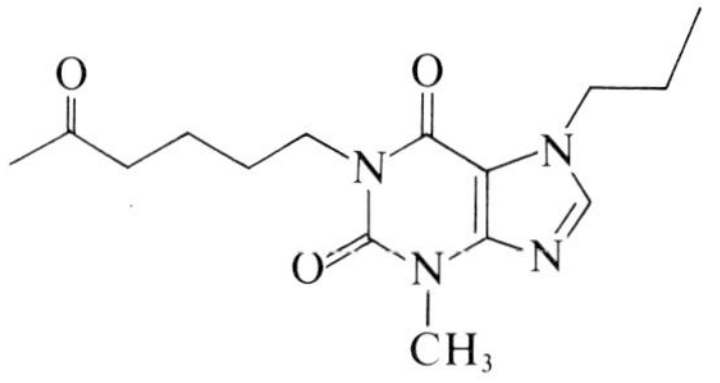

LXXXI

Quinazoline derivatives, ORH 3088 (LXXXII) and ORN 0676 (LXXXIII) cause vasodilation and in spontaneously hypertensive rats

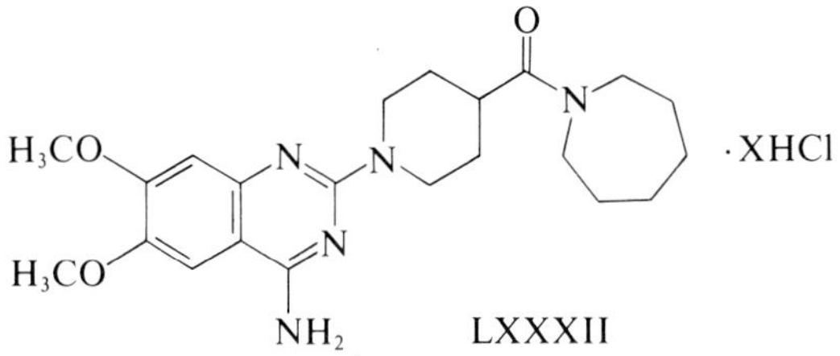

LXXXII

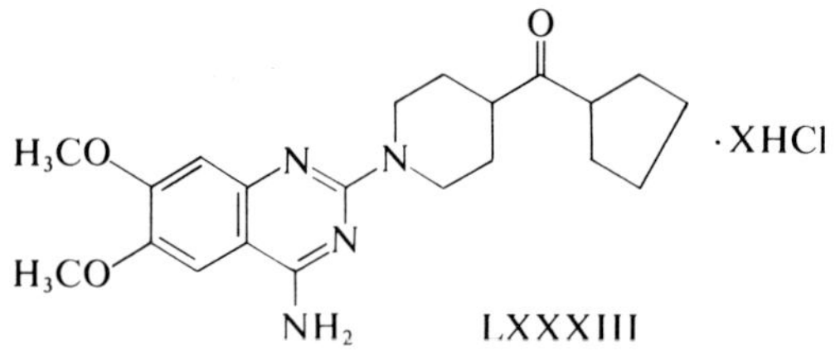

LXXXIII

at 100 μmol/kg p. o. exhibit marked antihypertensive effect [176]. The dihydropyridine derivative, LXXXIV in anaesthetized dog at 3–10 μg/

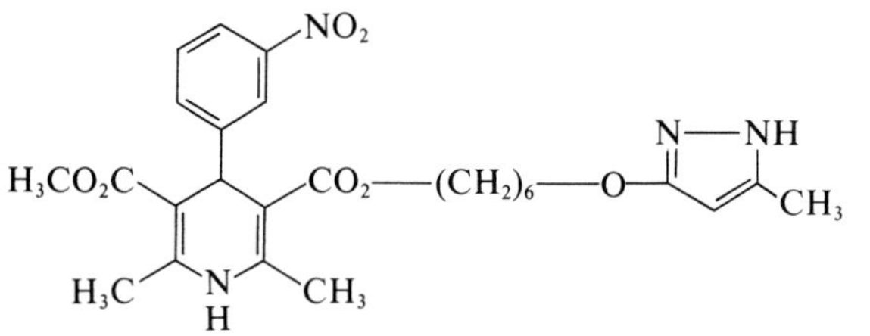

LXXXIV

kg exhibits antihypertensive activity equivalent to that of nifedipine (XXV) and nicardipine but has long duration of action and causes less increase in heart rate [177].

4.4 Centrally acting antihypertensive agents

ICI-106270 (LXXXV) is less potent than clonidine but also has less

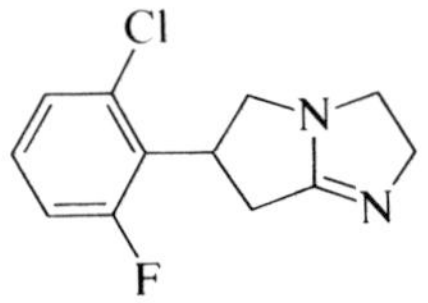

LXXXV

CNS-depressant action [178]. The mechanism of action of M-6434 (LXXXVI) appears to be due to direct α-adrenergic stimulation [179]. MPV-295 (LXXXVII) which represents a new class and possibly de-

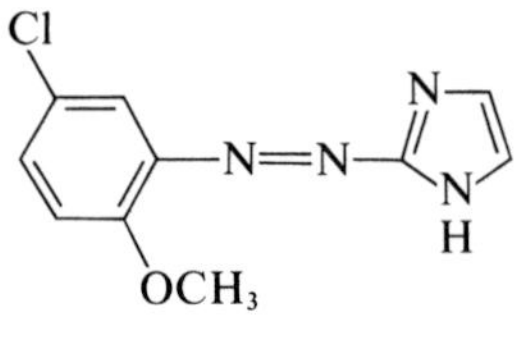

LXXXVI

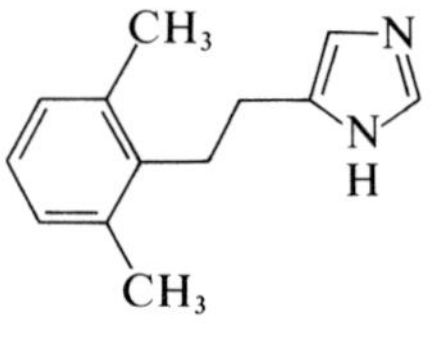

LXXXVII

creases the norepinephrine turnover [180, 181]. The thiourea LXXXVIII, is orally active in rats and dogs and causes less sedation

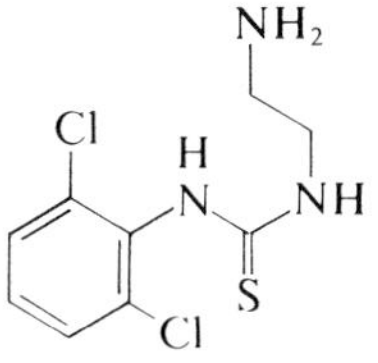

LXXXVIII

than clonidine [182]. B-HT-920 (LXXXIX), a more potent compound than B–HT-933 (XC) has been recently described [183]. LR-99853 (XCI), structurally similar to clonidine, possibly does not act via α-

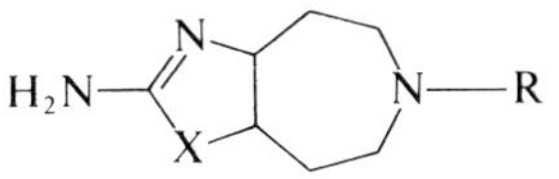

LXXXIX; R = Alkyl, X = S
XC; R = Et, X = O

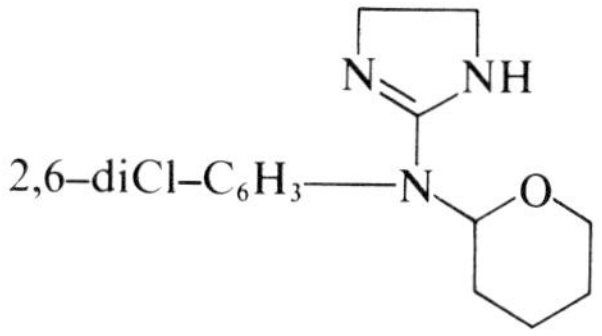

XCI

adrenoceptor stimulation [184]. BDF-6143 (XCII), has been characterized as the most selective and potent α_2-adrenoceptor blocking agent described todate [185].

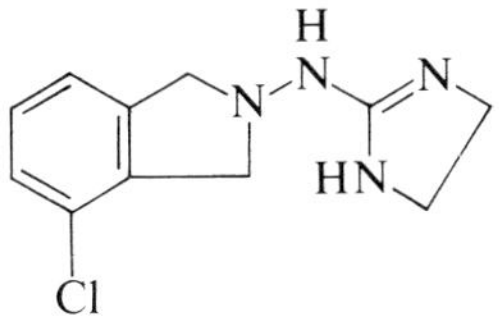

XCII

KF-4942 (XCIII), increased blood-flow in common carotid, vertebral and femoral arteries after intravenous administration in anaesthetized dogs, much less selectivity than that of flunarizine [186].

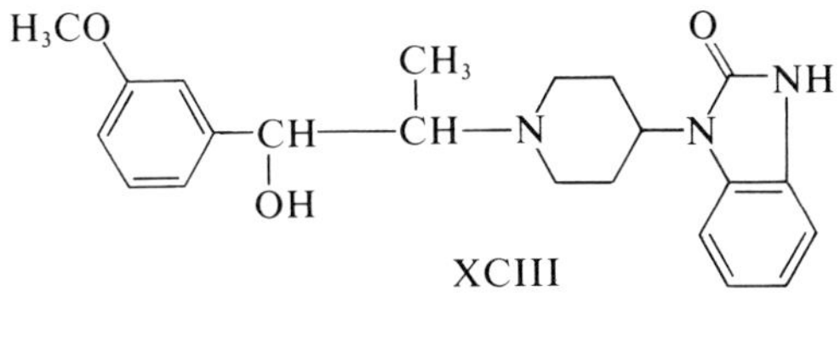

XCIII

4.5 β-Adrenoceptor blocking agents

Pacrinolol (XCIV) is a specific β_1-symapatholytic agent without intrin-

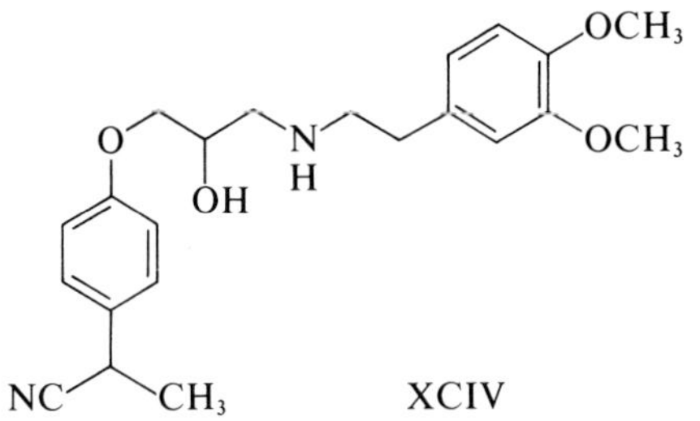

XCIV

sic activity but with marked antihypertensive activity. This preferential action on β_1-adrenoceptors have a potential advantage in patients with chronic lung disease or asthma, peripheral vascular disease and insulin requiring diabetes mellitus [187].

4.6 Miscellaneous

A new class of antihypertensive neutral lipids, 1-alkyl-2-acetyl-sn-glycerols, has been found to exhibit hypotensive activity in genetically hypertensive rats [188].
A novel antihypertensive agent, ORH-3088 (LXXXII) exhibited a dose-dependent decrease in blood pressure at 0.03–1.0 μmol/kg i. v. in goats. The profile of cardiovascular activity of this compound is similar to that of prazosin [189].
Cianergoline (XCV) is an effective antihypertensive in different models of experimental hypertension and in normotensive animals either

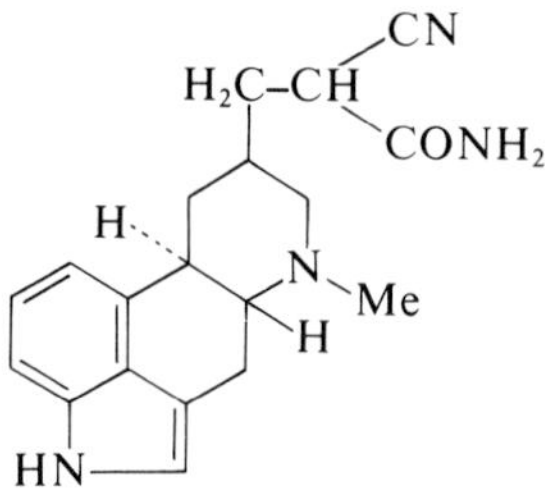

XCV

orally, intravenously or intraduodenally. Evidence was gained that the blood pressure lowering effect of cianergoline depends on the integrity of peripheral dopamine receptors [190].

U-54669-F (XCVI) is a new orally and parenterally active antihypertensive agent whose action has been suggested to be related to an effect on the peripheral sympathetic nerve terminals. Preclinical experience

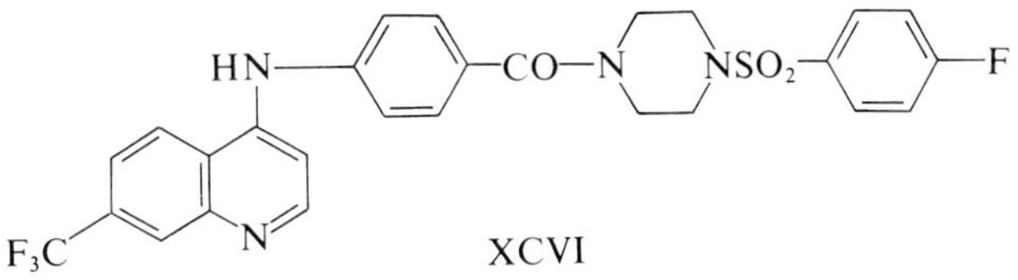

XCVI

with XCVIII indicated that it may be free of untoward side effects as cardiac activity, CNS and orthostatic hypotension [191].

Other compounds with a wide variety of molecular structures have been claimed as diuretics, ACE inhibitors, β-blockers, etc. However, these compounds are devoid of any meaningful pharmacological data. A chemical classification of these compounds along with their claimed pharmacological activities and literature references (in parentheses) are described.

DIURETIC

Alicyclic and aromatic compounds

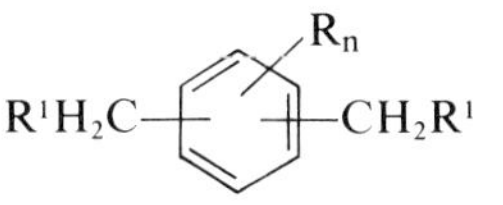

XCVII [192]

n = (0–4), R = CH_3, NO_2
R^1 = substituted cyclic amines

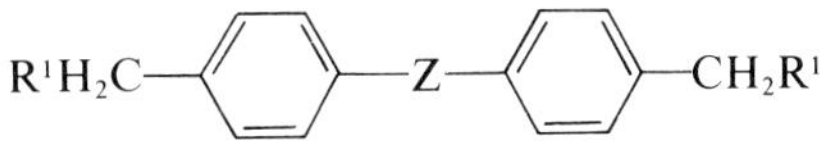

XCVIII [192]

Z = bond, CH_2, O or SO_2
R^1 = Substituted cyclic amines

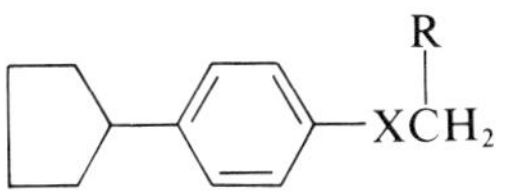

XCIX [193]

X = CO, CH_2
R = $NCH(CH_3)_2$
H

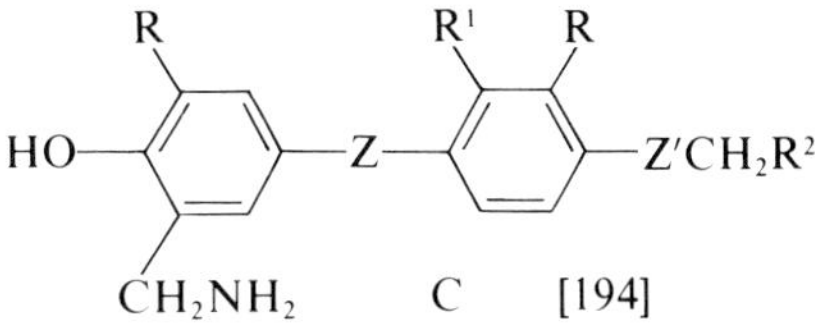

C [194]

R = H or Cl, R^1 = Cl, Z = Z′ = O,
R^2 = $CONH_2$, CO_2Et, CO_2H

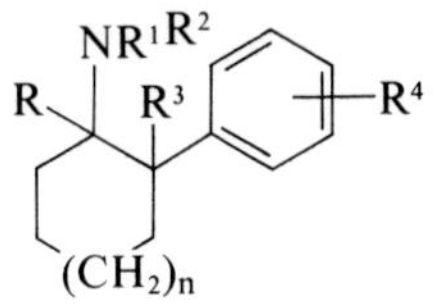

CI [195]

$R = R^1 = R^2 =$ alkyl,
$R^3 =$ H, alkyl, cycloalkyl,
phenyl or benzoylalkyl
$R^4 =$ H, OCH_3, OH; n = O–2

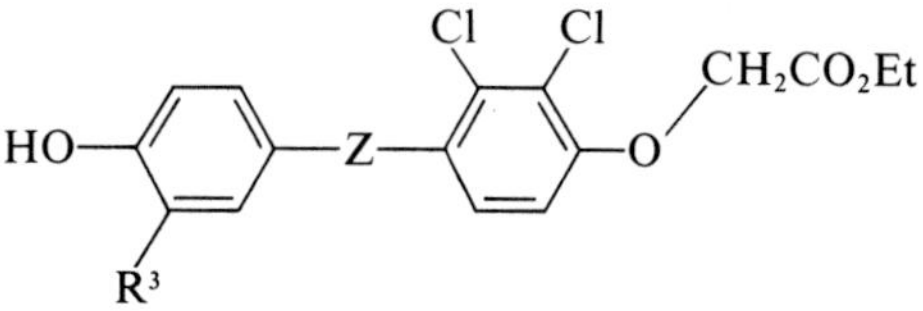

CII [194]

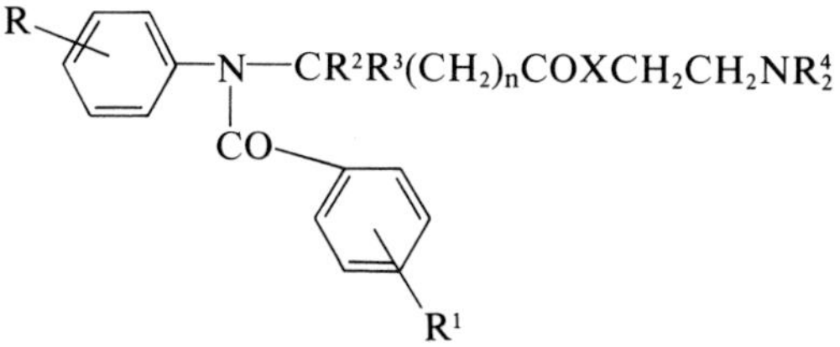

CIII [196]

$R = R^1 = R^2 =$ halo, alkoxy, alkyl
or H, $R^3 = R^4 =$ H or alkyl,
X = O, NH; n = 0, 1

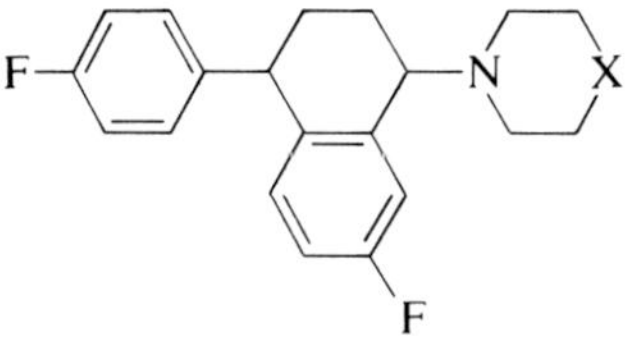

CIV [197]

$X = NCH_2CH_2OH$

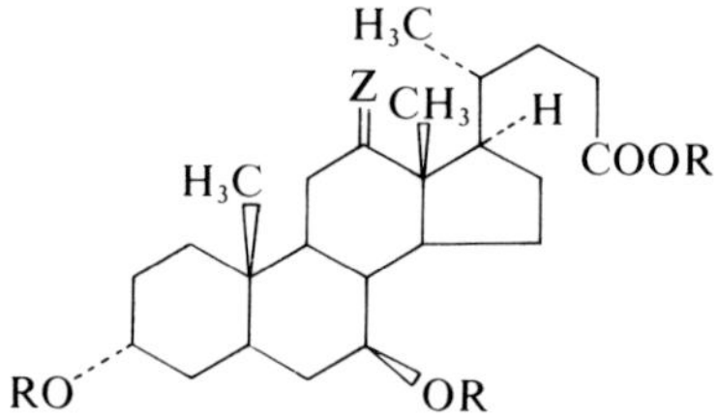

CV [198]

R = H, $(CH_3)_3Si$, Z = H_2, O

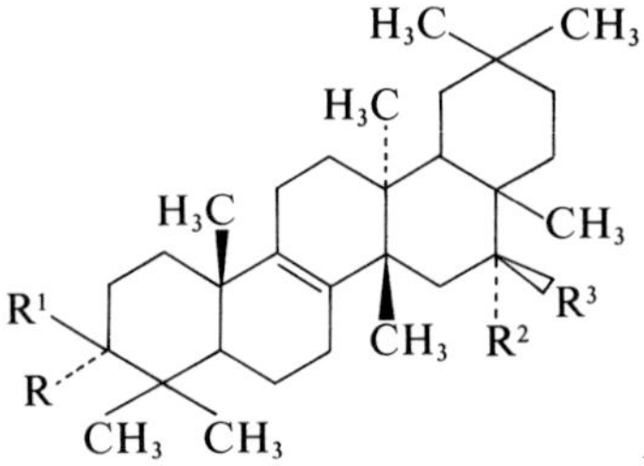

CVI [199]

$R = R^1 = O$, $R^2 = OH$, $R^3 = H$

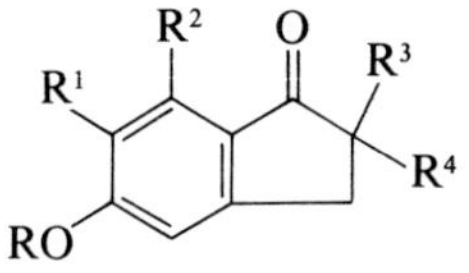

CVII [200]

$R = CH_2CH_2OH$, CH_2COCH_3, $CH_3CH(OH)$
$R^1 = R^2 =$ halo, CH_3, $R^3 =$ H, C_{1-6}
alkyl, $R^4 = C_{1-6}$ alkyl, C_{3-6}
cyloalkyl, aryl
CVIII, R = H, $R^1 = R^2 = Cl$, $R^3 = CH_3$
$R^4 = C_6H_5$ [201]
CIX, $R = CH_2CO_2H$, $R^1 = R^2 = Cl$,
$R^3 =$ alkyl, $R^4 = C_6H_5$ [201]

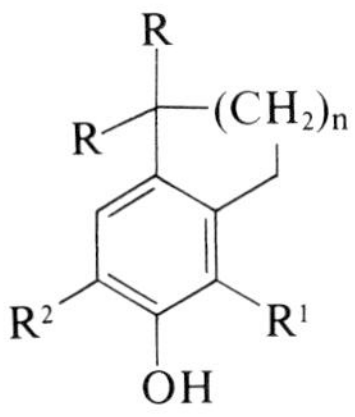

CX [202]

R = H, alkyl
n = 1, 2; $R^1 = R^2$ = Br, I etc.

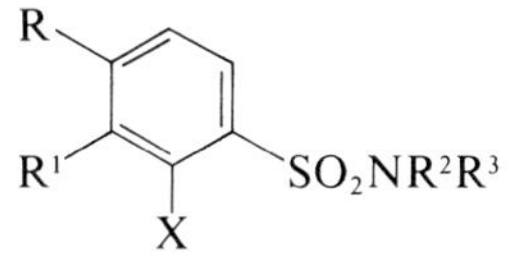

CXI [203]

R = halo, OH, alkyl, R^1 = halo
NO_2, NH_2 etc.
$R^2 = R^3$ = H, alkyl, X = halo

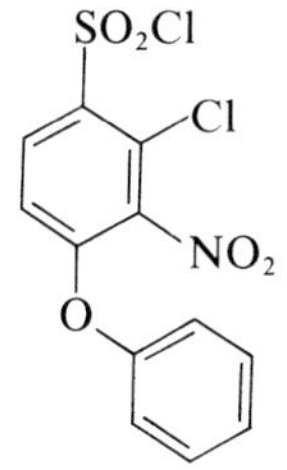

CXII [203]

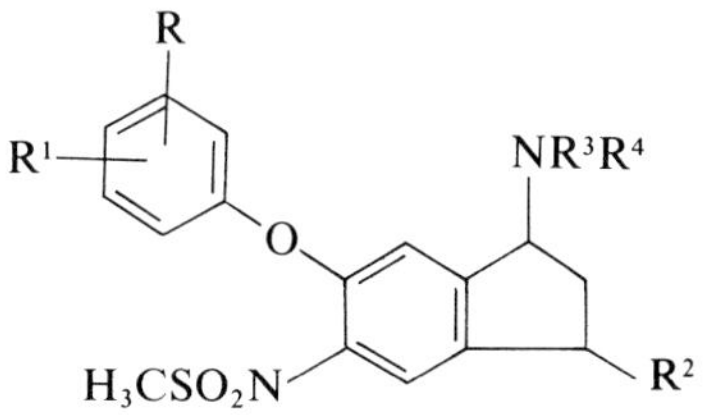

CXIII [204]

$R = R^1$ = H, Cl, F, R^2 = H, C_6H_5, R^3 = H,
alkyl, R^4 = H, alkyl, cyclohexyl,

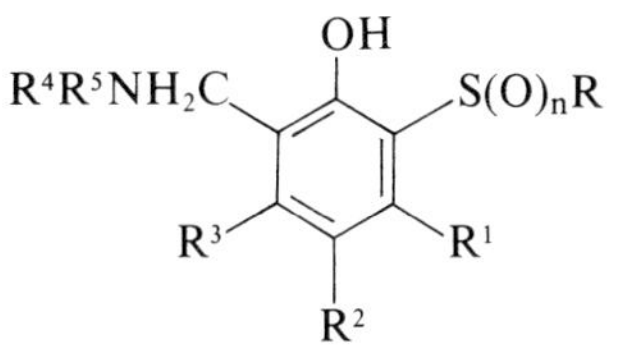

CXIV [205]

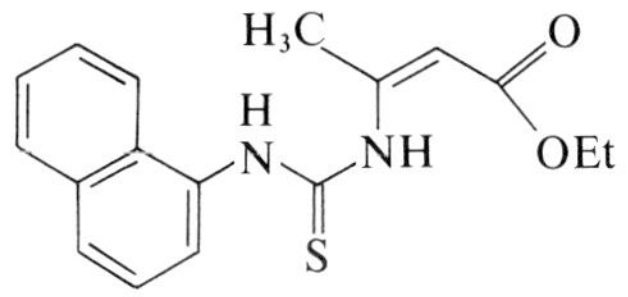

CXV [206]

R = alkyl, alkenyl, cycloalkyl, R^1, R^3 = H, alkoxy, alkyl, halo,
R^2 = H, alkyl, R^1R^2, R^2R^3 = alkylene, n = 1, 2
R^4R^5 = H, alkyl

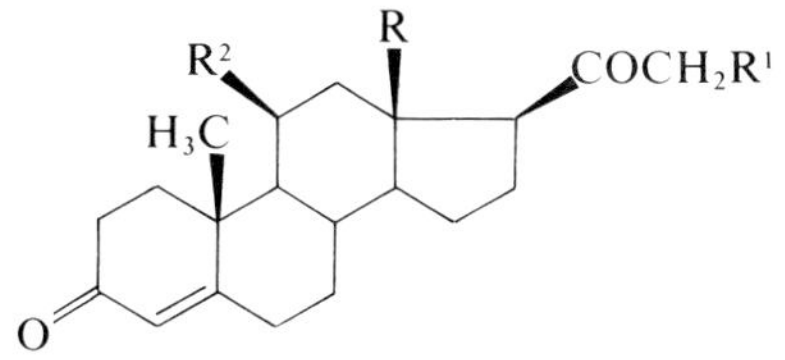

CXVI [207]

R = CH_2=C=CH, 2–alkynyl, CH≡
C–CH(OH), R^1, R^2 = H, OH

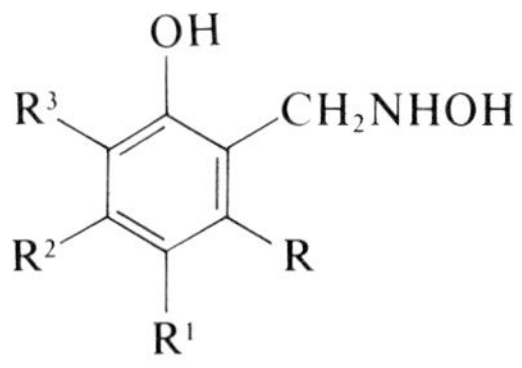

CXVII [208]

$R = R^2$ = H, alkyl, alkoxy, Cl
R^1 = Cl, alkyl, alkoxy, R^3 = H,
halo, alkyl, CF_3

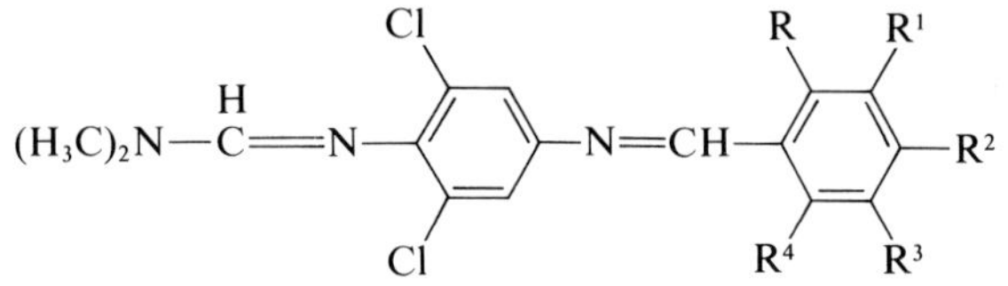

CXVIII [209]

R, R^1, R^3, $R^4 = H$, halo, OH, OMe, Me, CF_3, $R^2 = H$, halo, cyano, OH, OCH_3, CF_3, NHAc, Et_2N or R^1 $R^2 = OCH_2O$

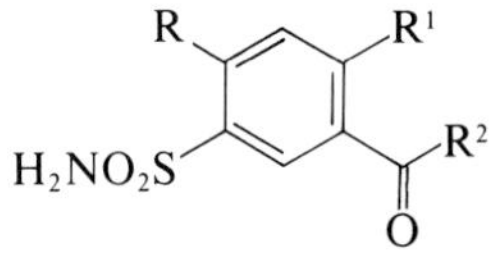

CXIX [210]

$R = Cl$, R^1 = cyclohexylamino, $R^2 = (CH_3)_2CHCH_2NH$

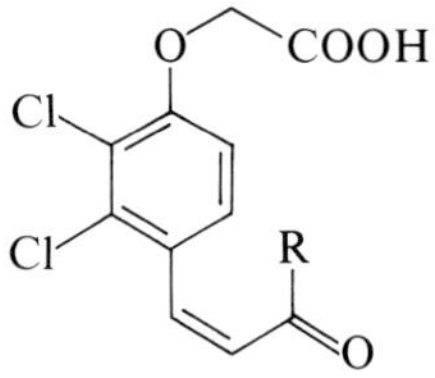

CXX [211]

Heterocyclic compounds with one hetero atom

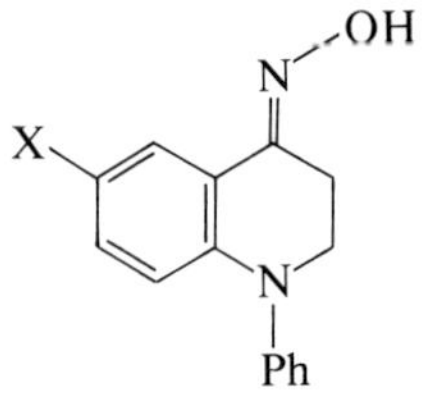

CXXI [212]

X = H or halogen

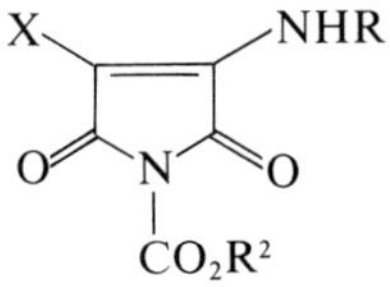

CXXII [213]

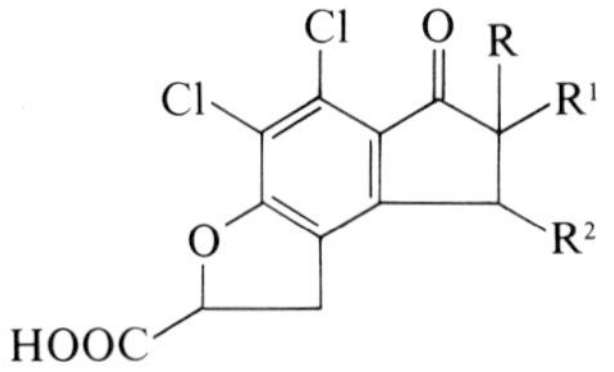

CXXIII [214]

$R = C_6H_5$, C_2H_5, $R^1 = H$ or CH_3, $R^2 = H$ or C_6H_5, $RR^1 = (CH_2)_4$

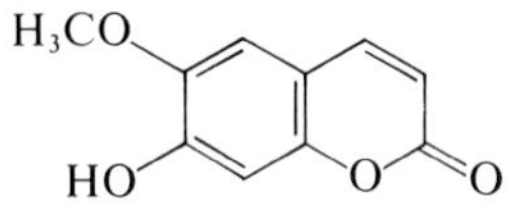

CXXIV [215]

Recent advances in drugs against hypertension   251

CXXV   [216]

CXXVI   [216]

CXXVII   [217]

R^1, R^2 = (un)substituted aryl or (un)substituted heteroaryl
R^3 = alkyl, cycloalkyl
R^4 = H, alkyl, aralkyl, amino

CXXVIII   [218]
R = CH_3, R^1 = C_2H_5

CXXIX   [219]

R, R^1, R^2 = (un)substituted alkyl or aryl

CXXX   [219]

R^3 = alkyl, aryl, CO_2H, CN, R^4, R^5 = H, cycloaliph·, aliph· Z = O, SO_x(X = 0–2), m, n = 0, 1

CXXXI   [220]
R = CH_3, CH_2OH, CO_2H

CXXXII   [221]

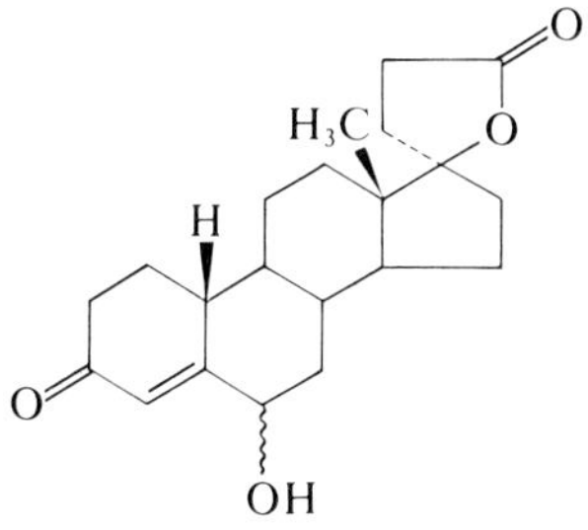

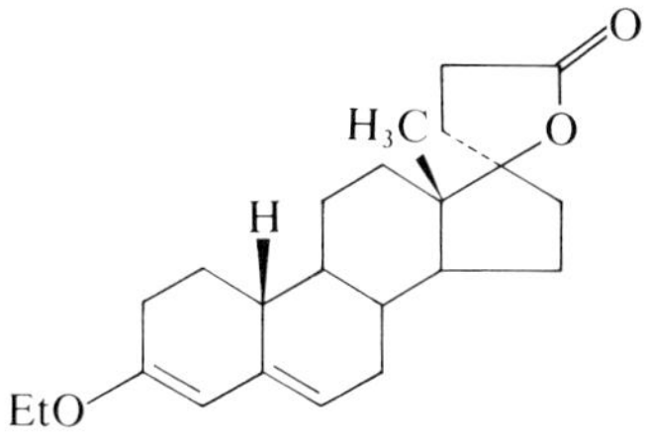

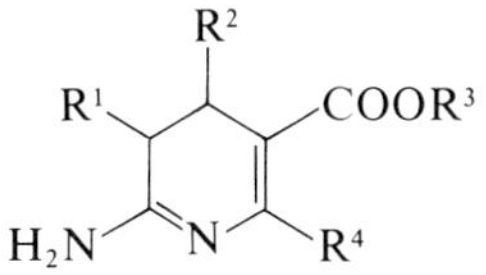

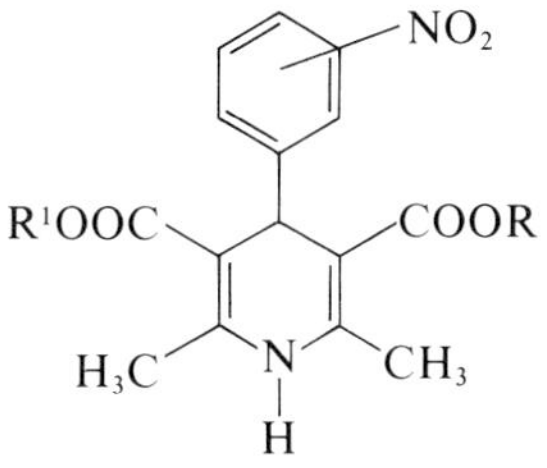

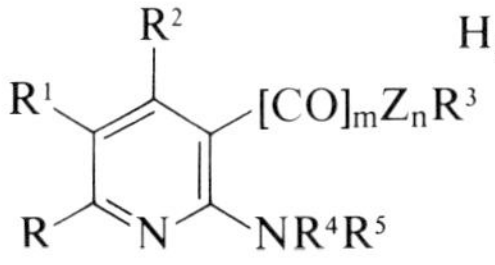

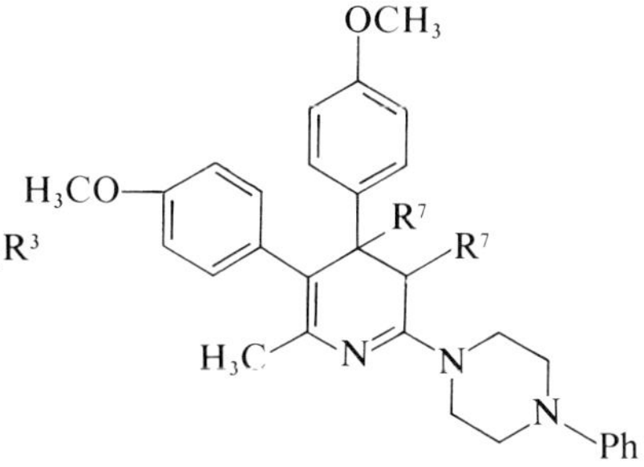

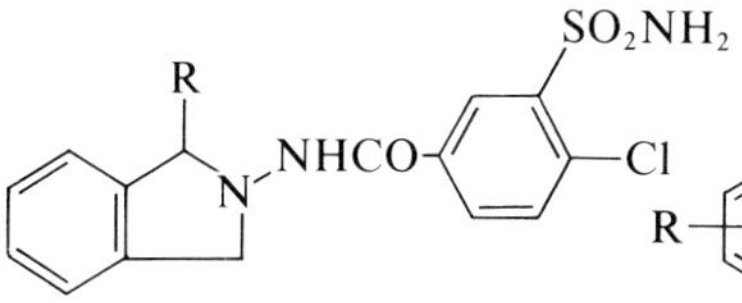

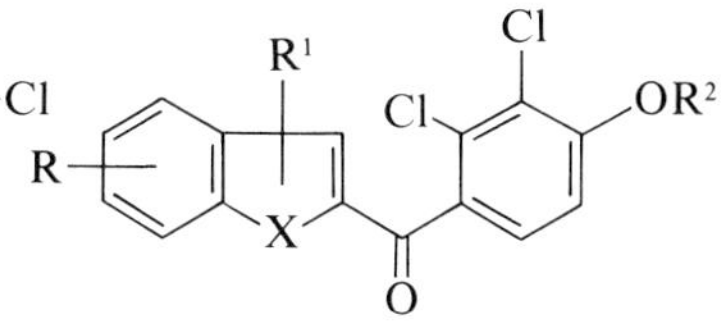

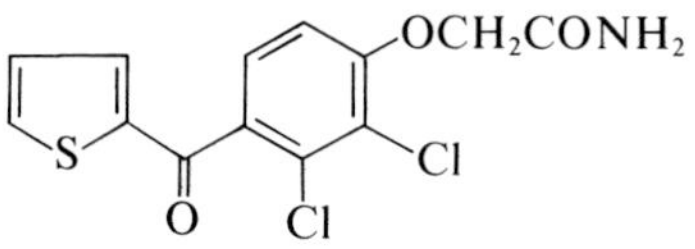

CXXXIII [221]

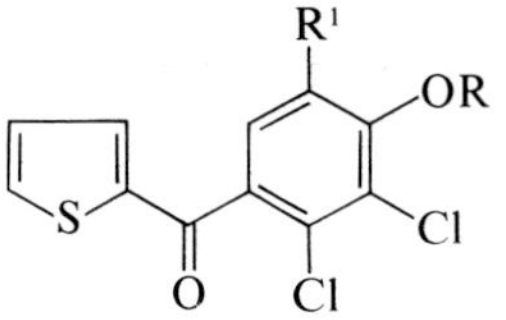

CXXXIV [222]

$R = CH_2CO_2CH_3$, $R^1 = H$ or COOH
or H

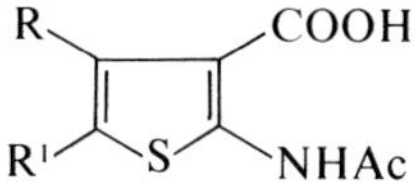

CXXXV [223]

R = aryl or arylalkyl
R^1 = alkyl

Heterocycles with two hetero atoms

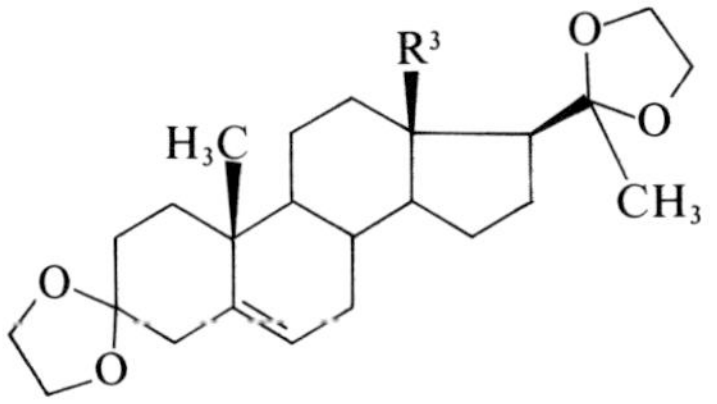

CXXXVI [207]

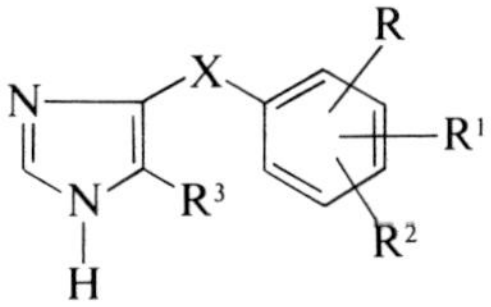

CXXXVII [224]

R, R^1, R^2 = H, Cl, Br, F, CH_3, Et, OCH_3, NH_2, R^3 = H, C_{1-7} alkyl, X = CO

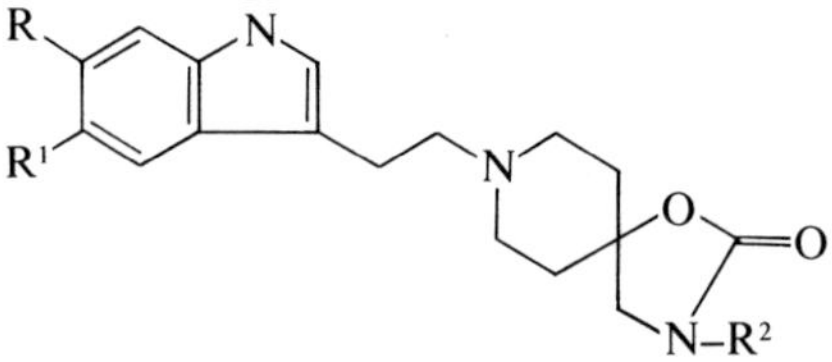

CXXXVIII [225]

R, R^1 = H, alkoxy, R^2 = alkyl

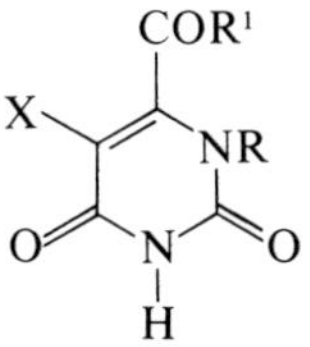

CXXXIX [213]

R = C_{1-4} alkyl,
X = H or halo
$R^1 = NH_2$, alkylamine, cyclicamines

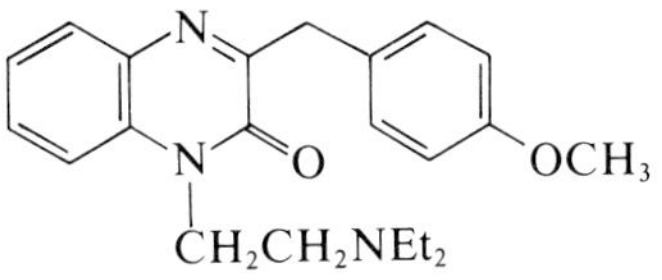

CXL [226]

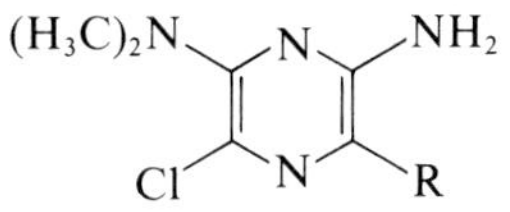

CXLI [227]

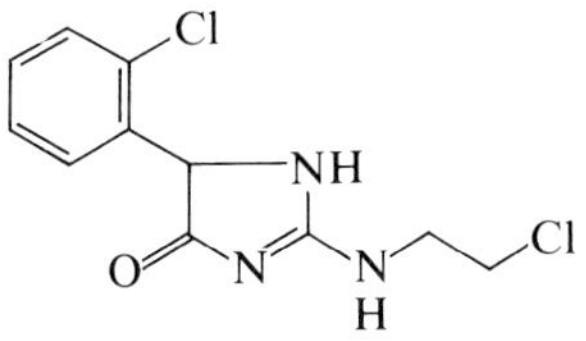

CXLII [228]

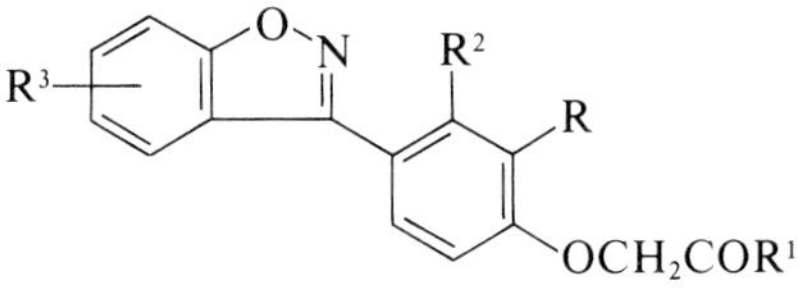

CXLIII [229]

R, R^2 = Cl, CH_3, Br, R^1 = H, C_{1-6} alkyl, R^3 = H, halo, C_{1-6} alkyl, NH_2, NO_2

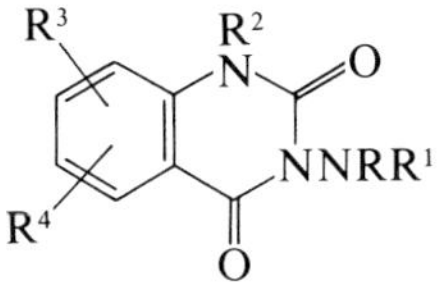

CXLIV [230, 231]

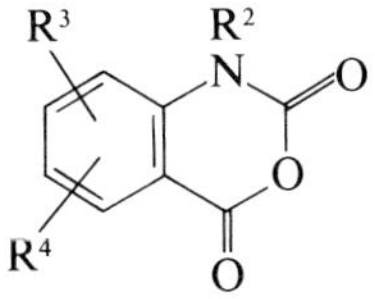

CXLV [231]

R = H, alkyl, acyl, R^1 = alkyl, acyl, aryl, NRR^1 = substituted cyclic amines, R^2 = H or alkyl, aryl, R^3 = R^4 – H, Cl, OCH_3

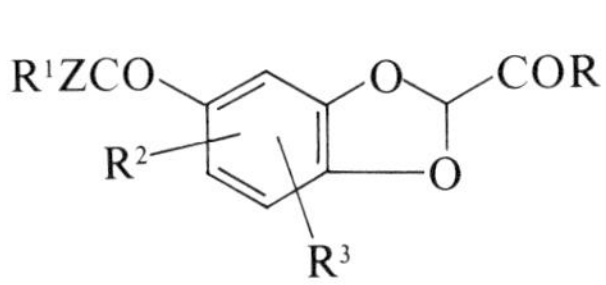

CXLVI [232]

R = R^1 = (un)substituted aromatic or heteroaromatic groups, R^2, R^3 = H, alkyl, alkoxy, Z = alkylene, alkylidene

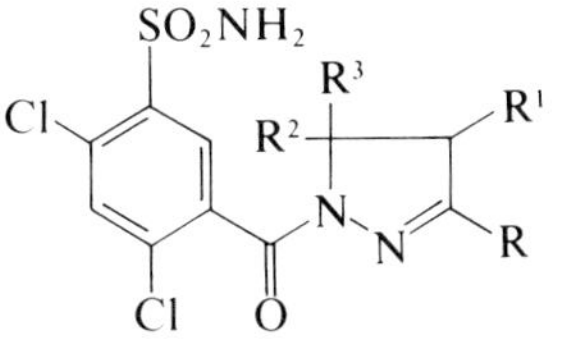

CLXVIII [233]

R = alkyl, R^1 = R^2 = R^3 = H

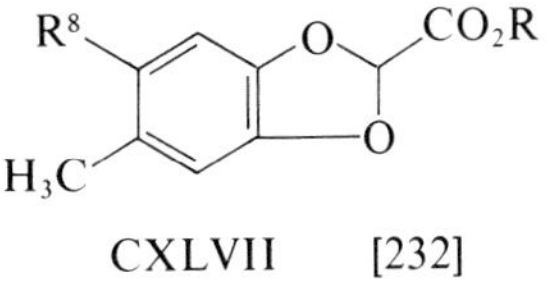

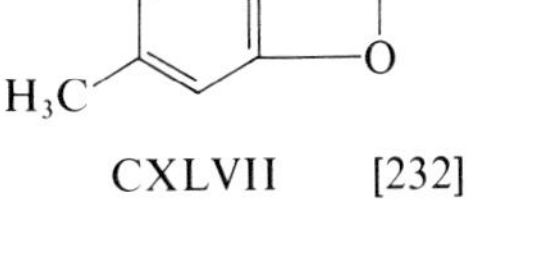

CXLVII [232]

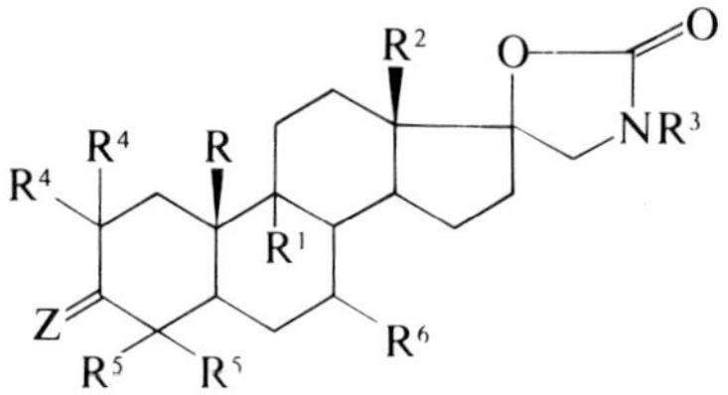

CXLIX [234]

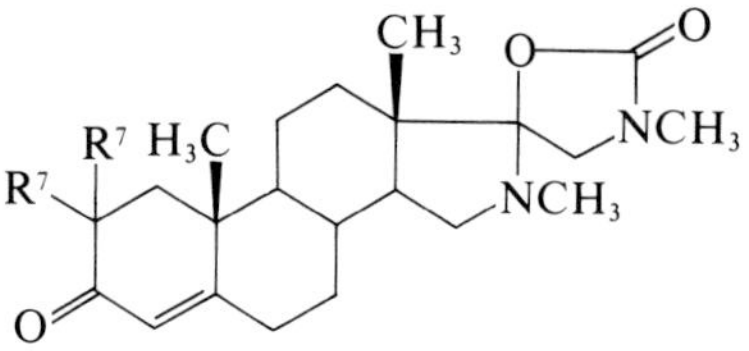

CL [234]

R = H, CH_3, R^1 = H, R, R^1 = bond, R^2 = alkyl, R^3 = alkyl, alkenyl, R^4, R^5 = H, CH_3, R^6 = H, alkylthio, arylthio, Z = O, NOH

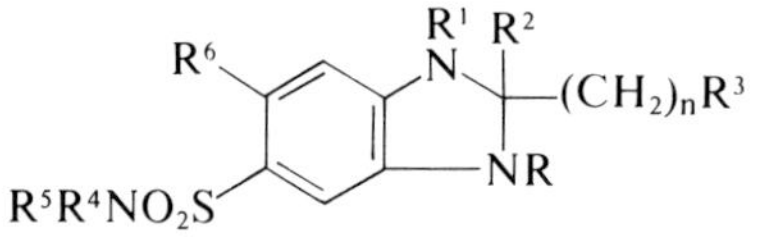

CLI [235]

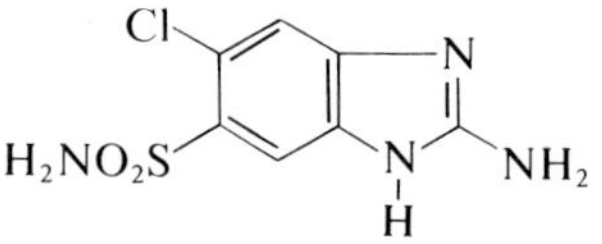

CLII [235]

R = R^1 = H, alkyl, substituted phenyl, R^2 = H, R^1,R^2 = bond,

R^3 = N, NH, OH, SH, R^4R^5 = H, alkyl, R^6 = halo, CF_3, n = O

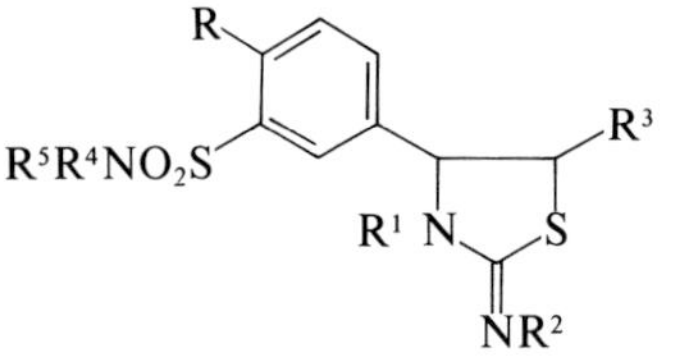

CLIII [236]

R = H, alkyl, halo, R^1 R^2 = alkylene, R^3 = H, alkyl, R^4, R^5 = H, alkyl, aryl

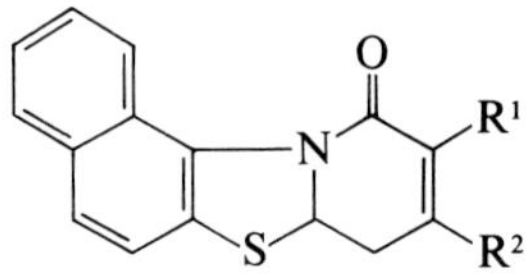

CLIV [206]

R^1 = H, R^2 = $CH_2CO_2C_2H_5$

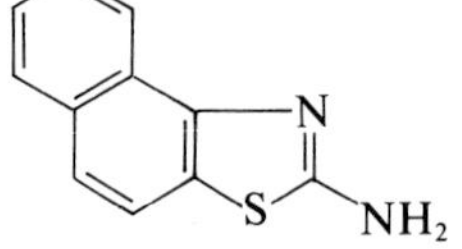

CLV [206]

Compounds containing more than two hetero atoms

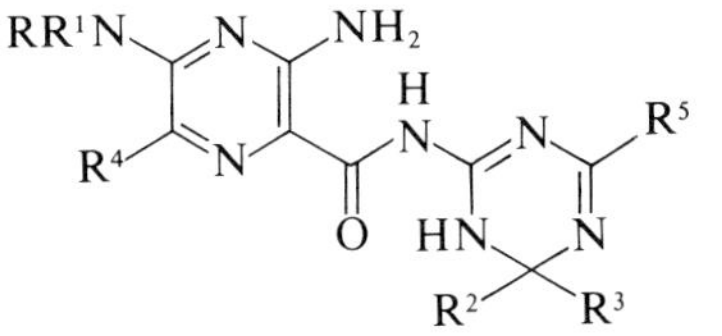

CLVI [237]

R = R¹ = H, alkyl, cycloalkyl, R², R³ = alkyl, R⁴ = halo, R⁵ = OH, SH, alkoxy

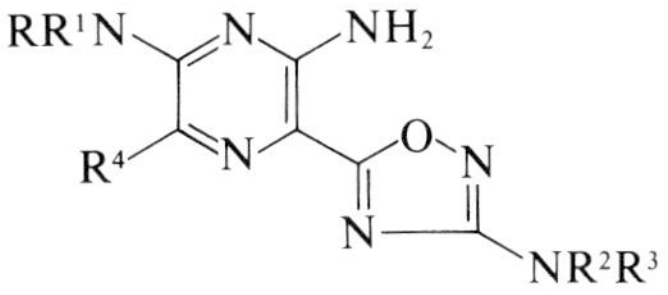

CLVII [227]

R²R³ = H, alkyl, RR¹ = alkylene, R⁴ = halo, cyano, phenyl

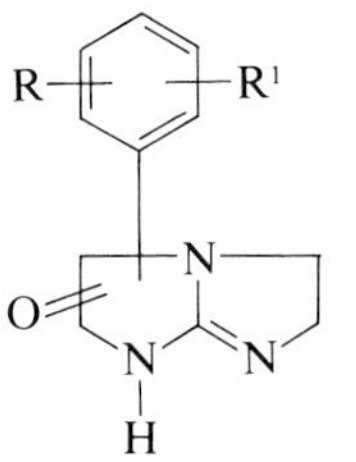

CLVIII [228]

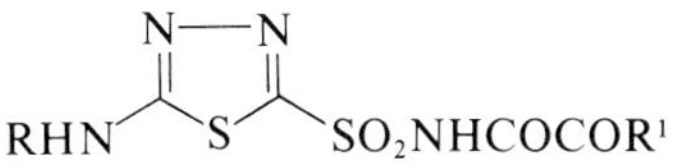

CLIX [238]

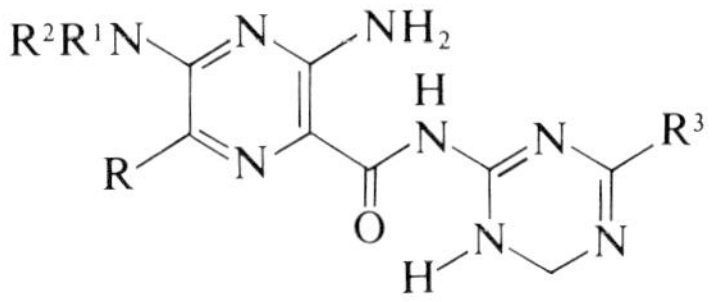

CLX [239]

R = halo, R¹ = R² = H, alkyl, alkenyl or cycloalkyl, R³ = OH

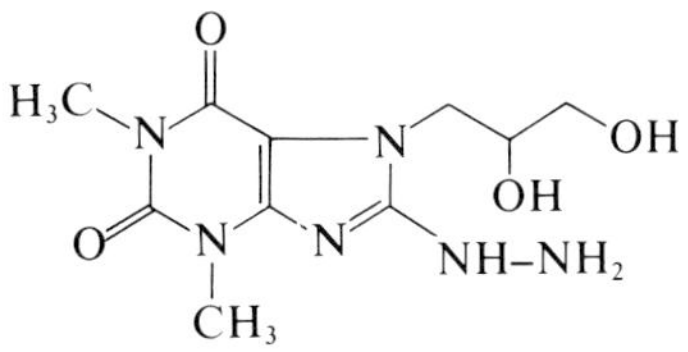

CLXI [240]

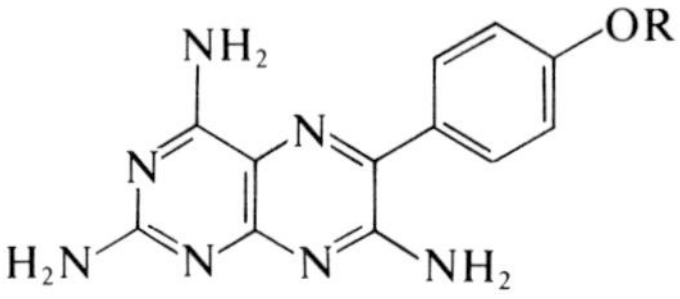

CLXII [241]

$R = H$, C_{1-6} alkyl or a cation

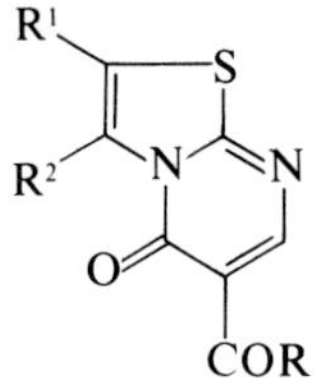

CLXIII [242]

$R = $ 5-tetrazolylamino, $R^1R^2 = (CH_2)_4$

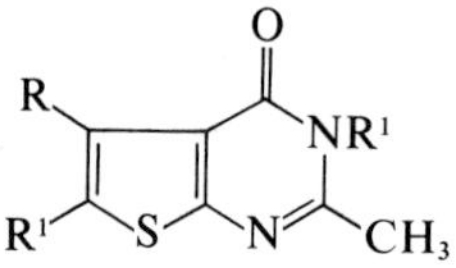

CLXIV [223]

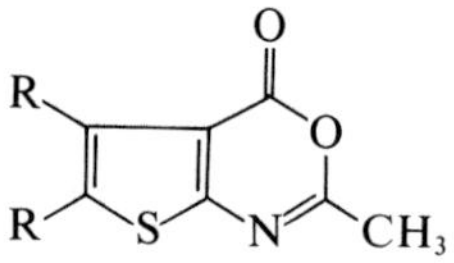

CLXV [223]

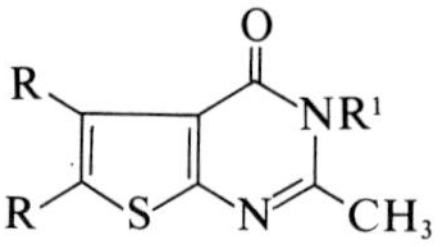

CLXVI [223]

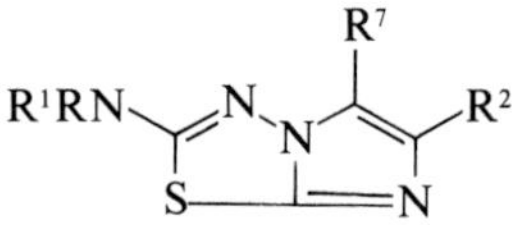

CLXVII

$R^{-} = CR^3 = CR^4CO{-}NR^5R^6$ [243]

CLXVIII: $R^7 = CHO$, $R^2 = Ph$, $R = CH_3$, $R^1 = Ac$ [243]

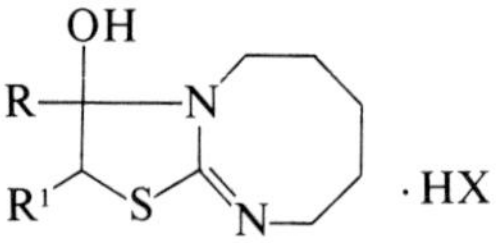

CLXIX [244]

$R = $ aryl or substituted aryl, $R^1 = H$, alkyl or aryl, $X = $ halo

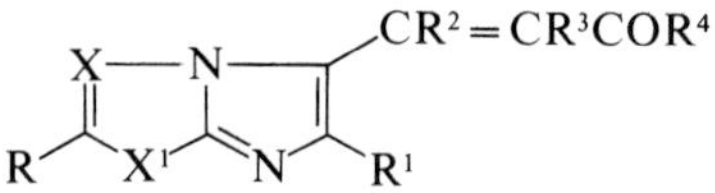

CLXX [245]

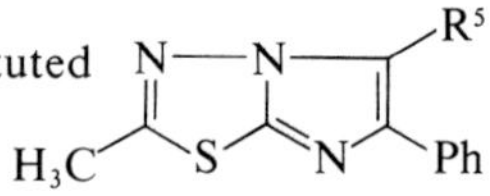

CLXXI [245]

$X = N$ or CH, $X^1 = S$, O, NH, alkylamino, R, $R^1 = $ (un)substituted alkyl, aryl, heterocyclic, alkylthio, amino, acyl etc.

RENIN-ANGIOTENSIN ENZYME INHIBITORS

Amino acid analogs

H-His-Ala-Asp-Gly-Val-Phe-Thr-Ser-Asp-Phe-Ser-Arg-Leu-Leu-Gly-Glu-Leu-Ser-Ala-Lys-Lys-Tyr-Leu-Glu-Ser-Leu-Ileu NH$_2$

CLXXII [246]

RP(O)(OR')(CH$_2$)$_n$CH[(CH$_2$)$_m$R^2]CO–X–OR^3

CLXXIII [247]

R^{3}P(O)(OR^2)CH$_2$NR^{1}CO–X–OR

CLXXIV [248]

CLXXV [249]

CLXXVI [250]

CLXXVII [251]

CLXXVIII [252]

CLXXIX [252]

R = R^1 = H or alkyl, R^2 = H, alkyl or aryl, R^3 = H, OH, R^4 = H, n = 0, 1

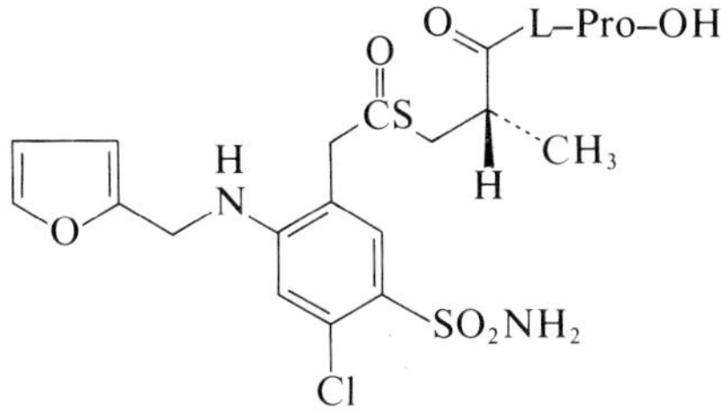

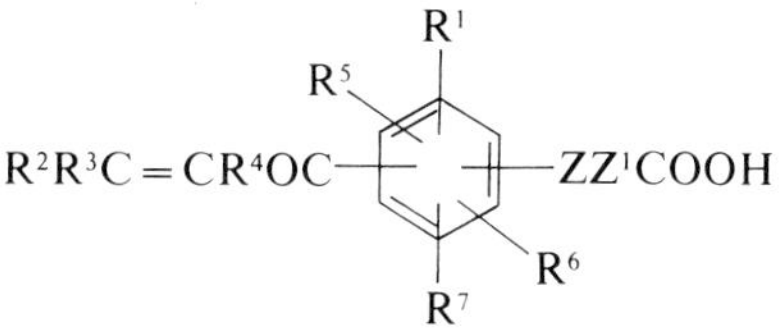

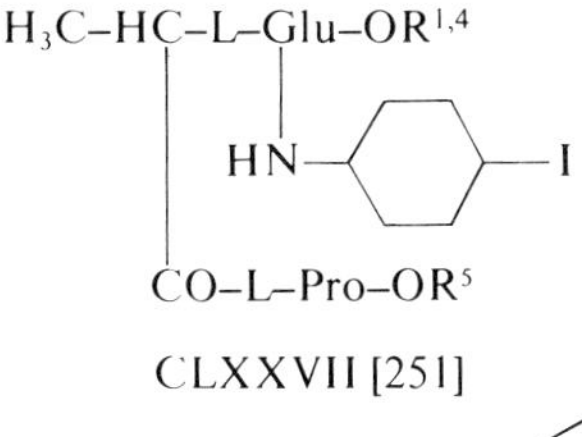

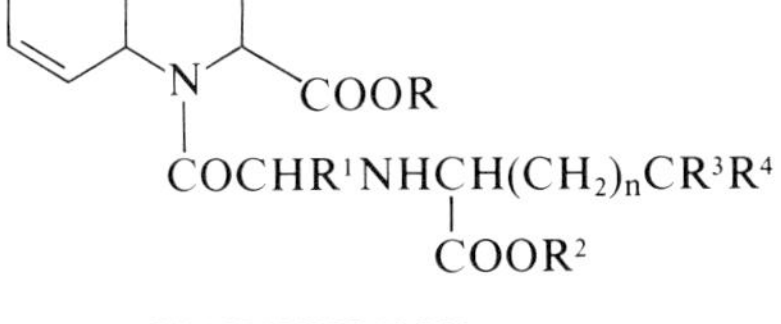

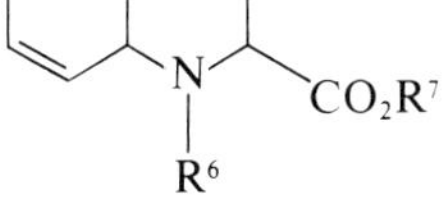

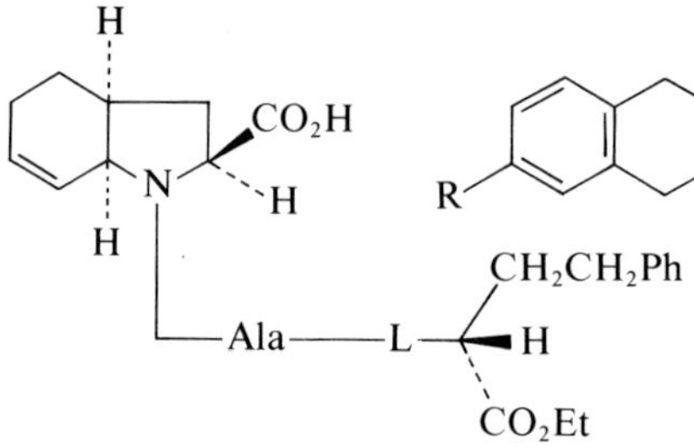

CLXXX [252]

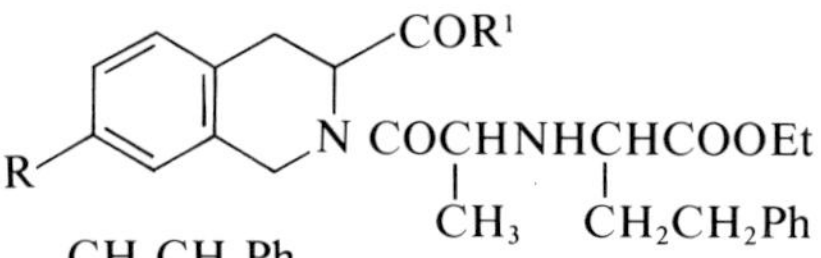

CLXXXI [253]

$R = R^1 = (CH_3)_3C$ etc.

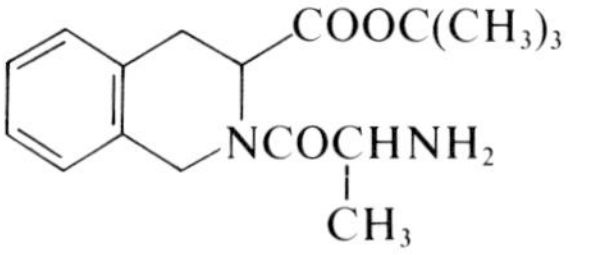

CLXXXII [253]

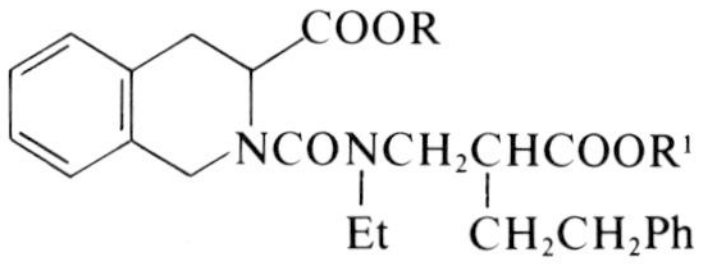

CLXXXIII [254]

$RR^1 = CH_3, CH_3(2R)$; H,H(2R), $PhCH_2$, Et(2R), H, Et (2R), $PhCH_2$, Et(2S), H, Et(2S)

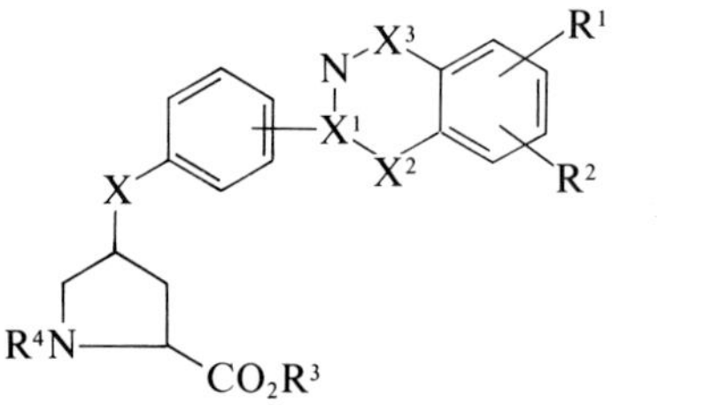

CLXXXIV [255]

X = O, S, X^1, X^2 = CHNH, CN, X^3 = CO, SO_2, R = H, alkyl, aryl, R^1, R^2 = H, halo, alkyl, alkoxy etc., R^3 = H, alkyl, aryl etc., $R^4 = R^5\,SCH_2CHR^6CO$ (R^5 = H, acyl, R^6 = H, alkyl or aryl etc.)

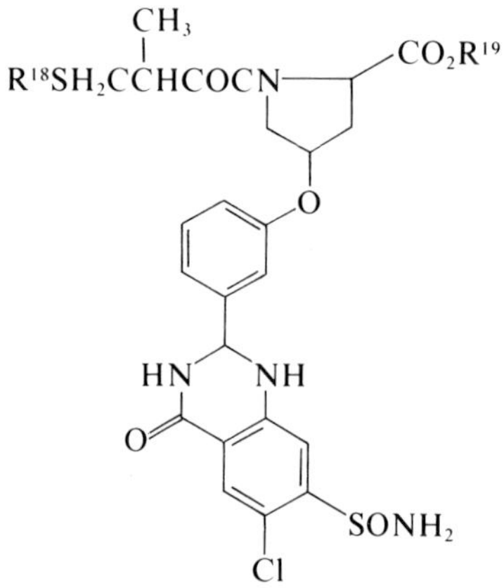

CLXXXV [255]

β-BLOCKERS

Alicyclic and aromatic compounds

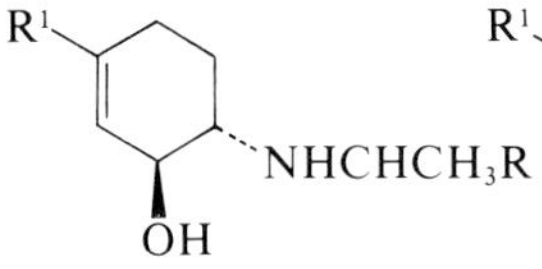

CLXXXVI [256]

R = alkyl or aryl
R^1 = aryl

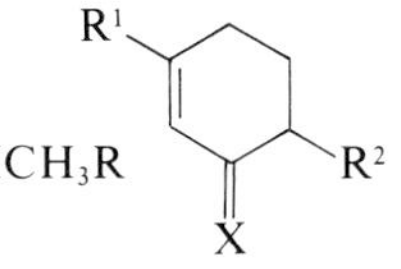

CLXXXVII [256]

$X = O$, NOH; R^1 = Ph
R^2 = H, NHAC

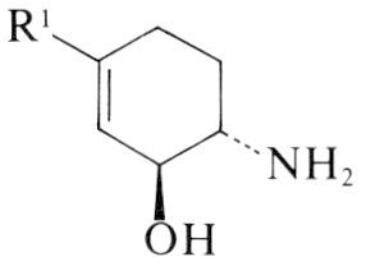

CLXXXVIII [256]

R^1 = Ph

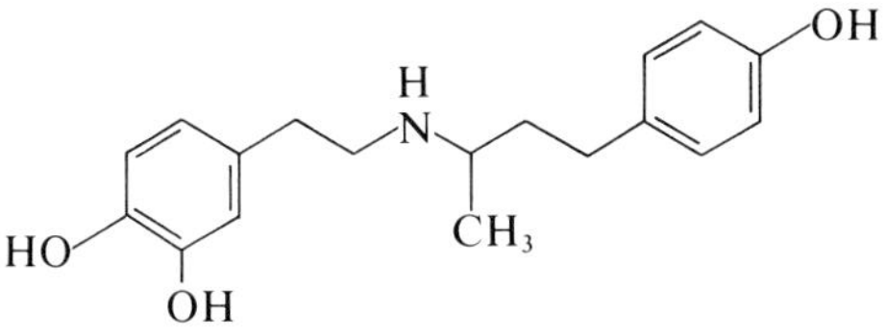

CLXXXIX [257]

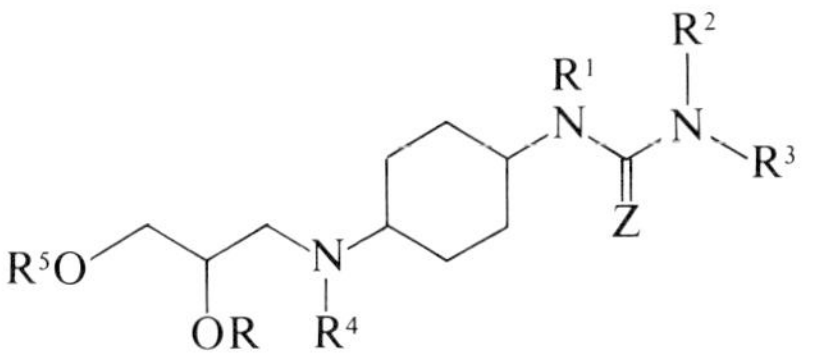

CXC [258]

R = H, alkanoyl or aroyl, R^1, R^2, R^3, R^4 = H or alkyl,
R^5 = carbocyclic, heterocyclic etc., $Z = O$, S

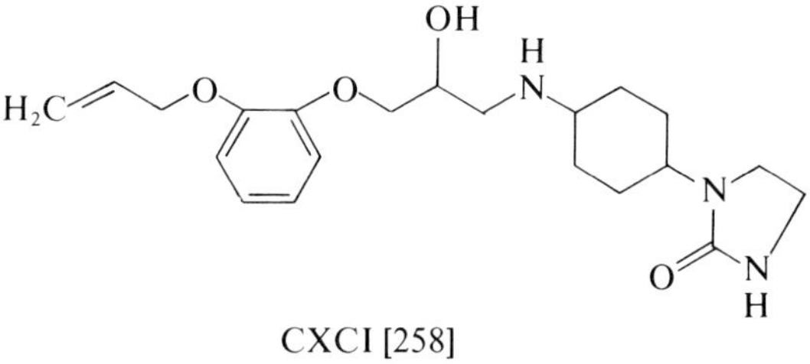

CXCI [258]

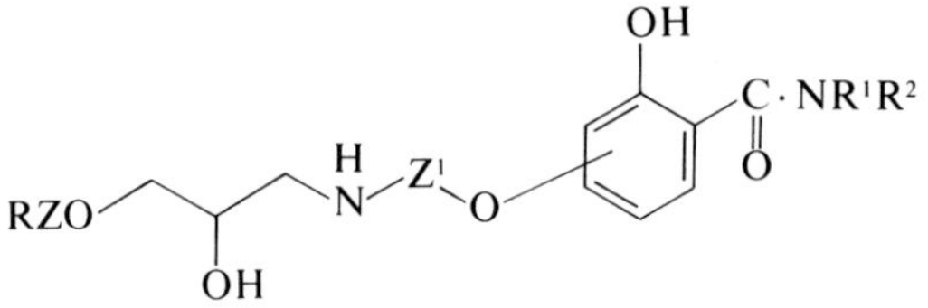

CXCII [259]

R = mono or bicyclic carbocyclic, R^1R^2 = H, alkyl, Z = C-bound azaphenylene, Z^1 = C_{2-5} alkylene

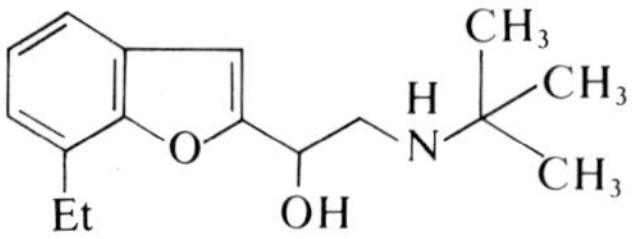

CXCIII [260]

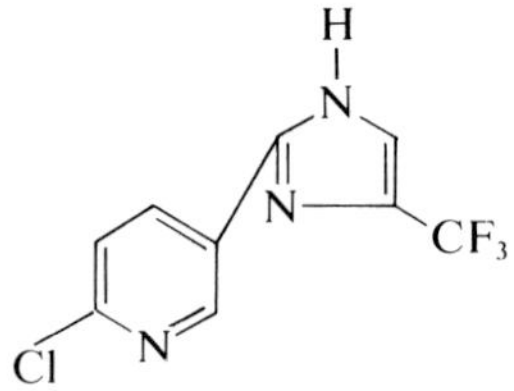

CXCIV [259]

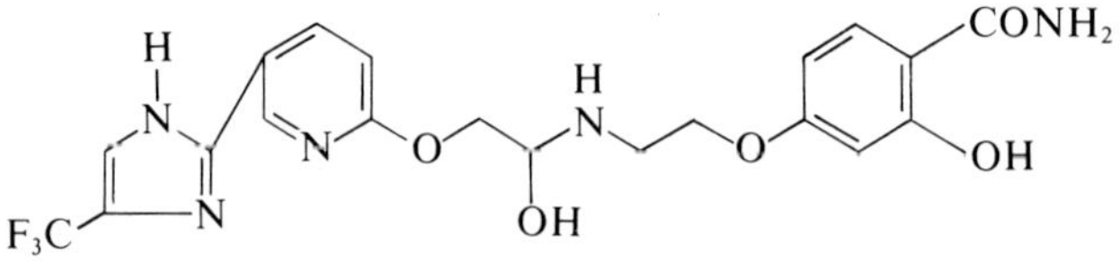

CXCV [259]

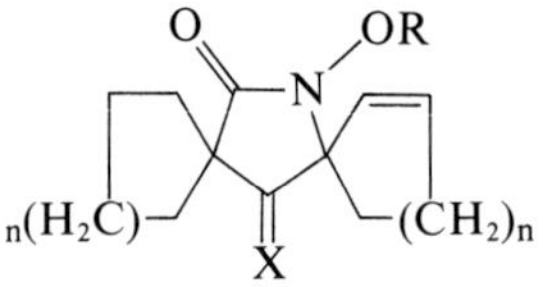

CXCVI [261]

R = $CH_2CH(OH)CH_2NR^1$ (R^1 = sec. or tertiary amine); n = 1, 2; X = O

VASODILATORS

Aromatic compounds

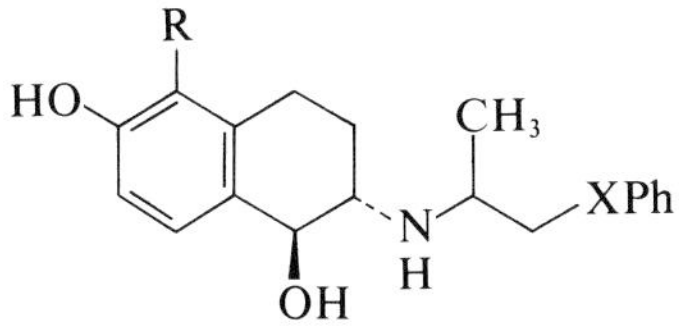

CXCVII [262]

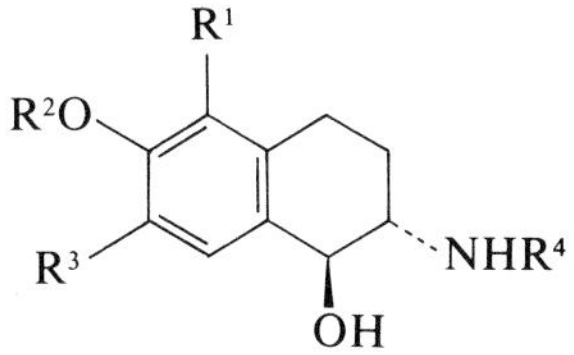

CXCVIII [262]

Heterocyclic compounds with one hetero atom

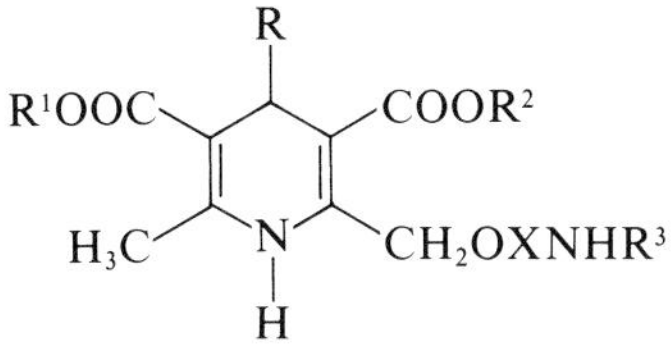

CXCIX [263]

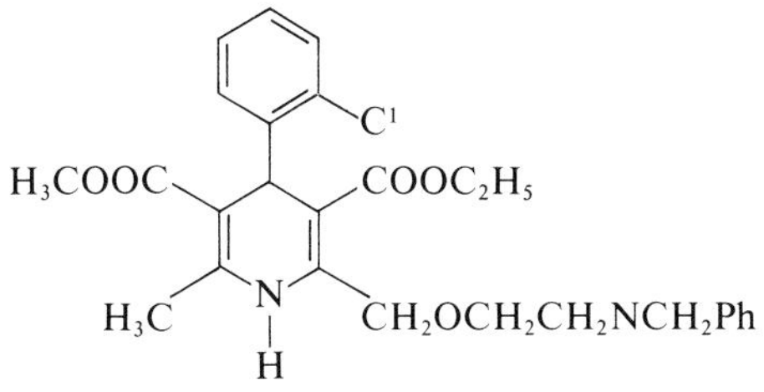

CC [263]

$X = (CH_2)_2$ or $(CH_2)_3$ etc., $R^1 = R^2 =$ alkyl or $CH_2CH_2OCH_3$ etc.
R = aryl, heteroaryl
R^3 = H, alkyl or aryl

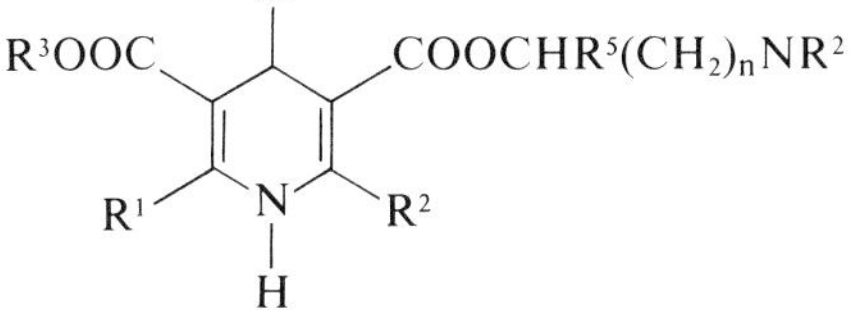

CCI [264]

n = 1, 2, 3, 4 or 5

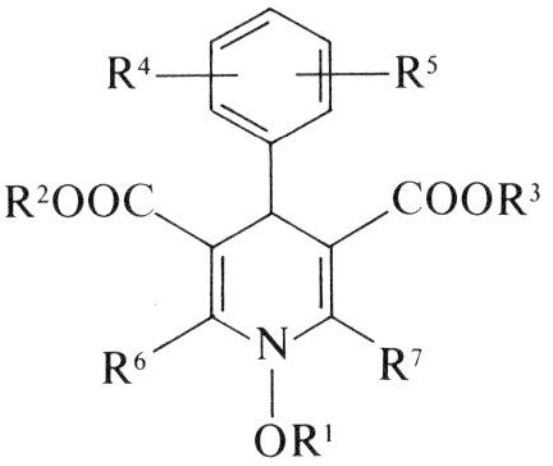

CCII [265]

R^1 = H, alkyl, alkenyl, arylalkyl, acyl,
R^2, R^3 = (un)substituted alkyl, alkenyl,
R^4R^5 = H, halo, NO_2
R^6, R^7 = alkyl

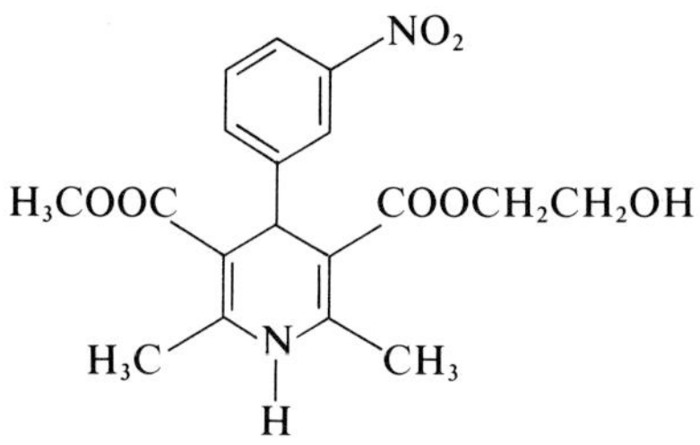

CCIII [266]

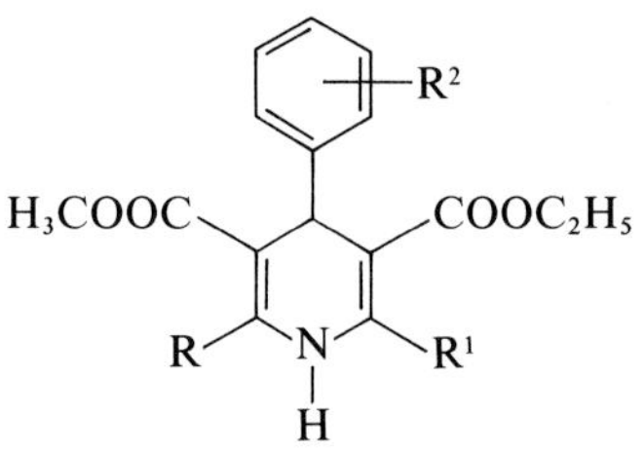

CCIV [267]

R, R^1 = alkyl, formyl etc.
$R^2 = NO_2$, halo, alkyl

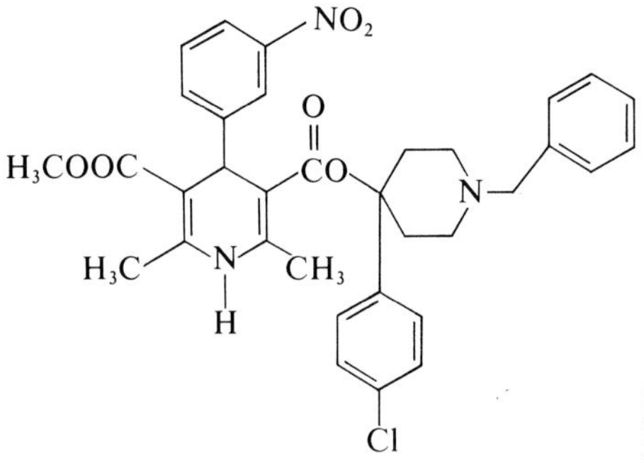

CCV [268]

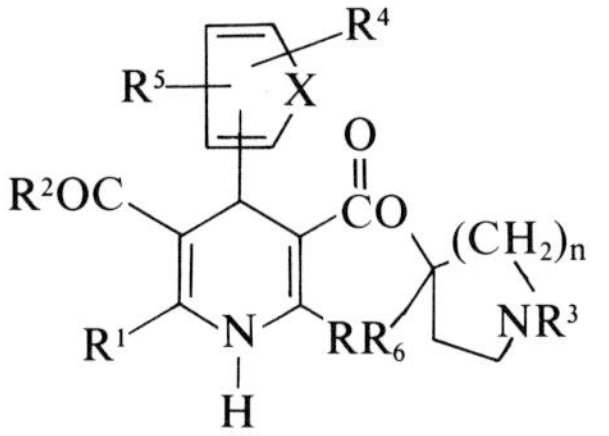

CCVI [268]

R, R^1, R^2 = alkyl, R^3 = H, alkyl, aralkyl, R^4, R^5 = H, halo, alkyl etc. X = O, S, NH, R^6 = aryl, heterocyclic, n = 1, 2

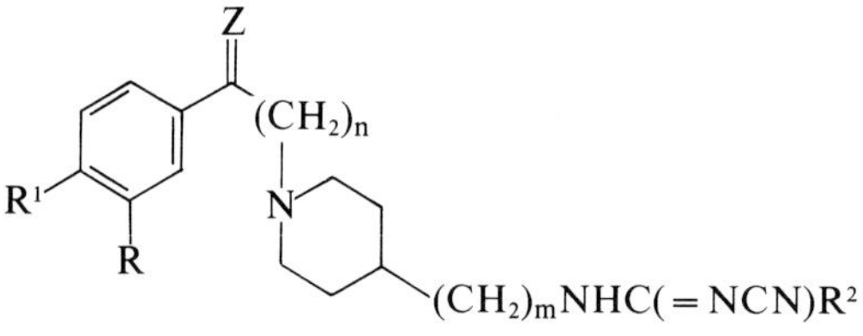

CCVII [269]

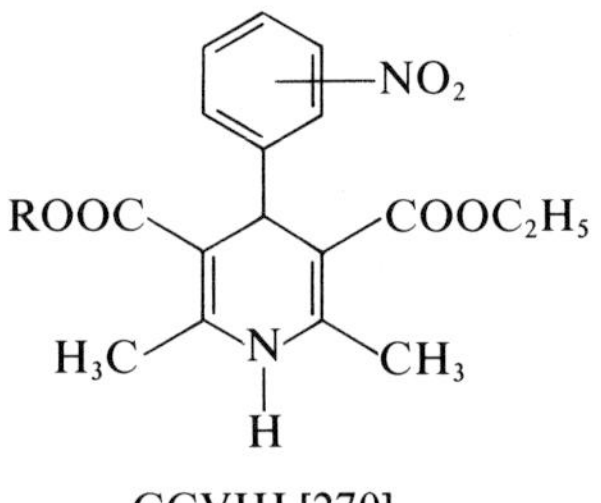

CCVIII [270]

R = alkyl or nitro alkyl

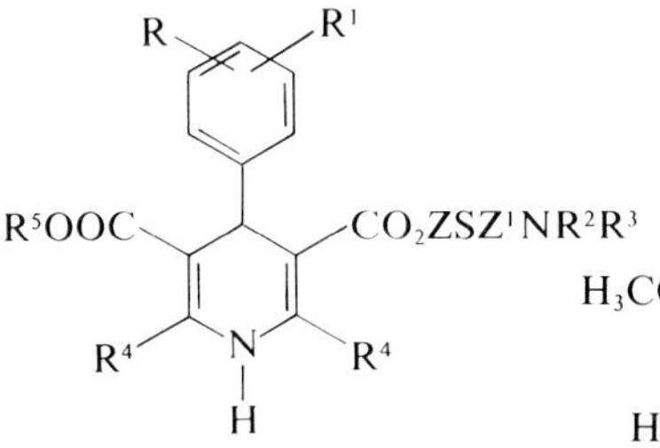

CCIX [271]

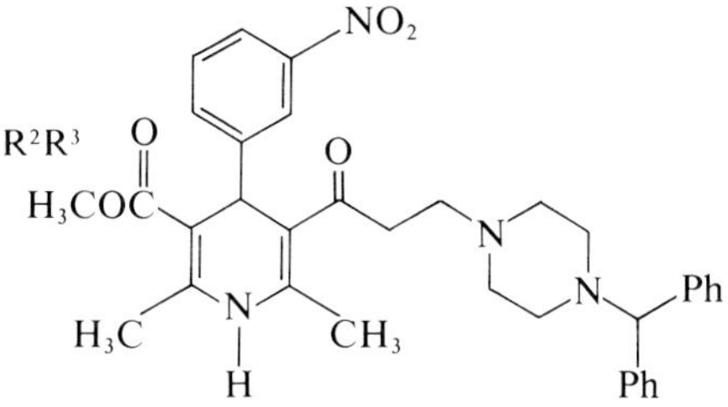

CCX [272]

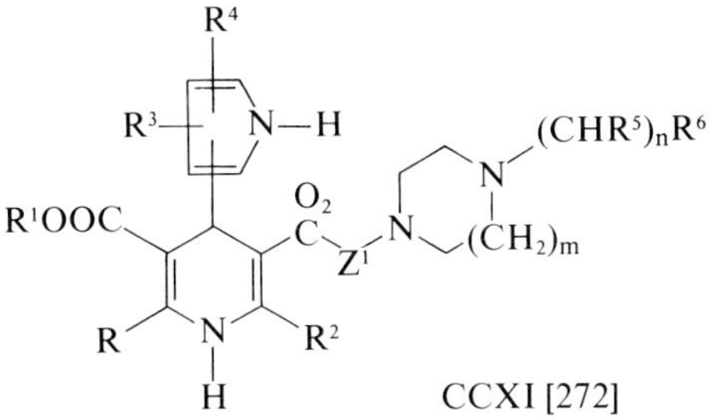

CCXI [272]

R, R[1], R[2] = alkyl, cycloalkyl, R[3], R[4] = H, halo, nitro etc.
R[5] = H, alkyl or aryl etc., R[6] = aryl, pyridyl, Z = O, S etc.
Z[1] = alkylene, m = 1–3, n = 0–2

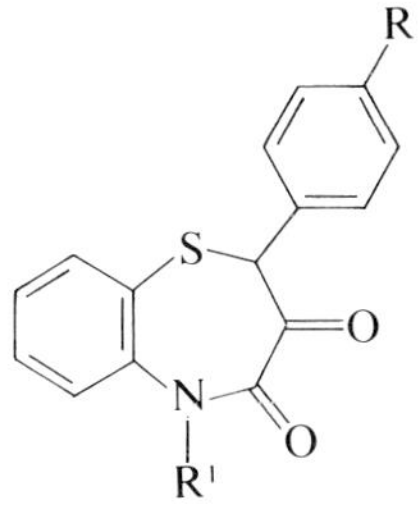

CCXII [273]

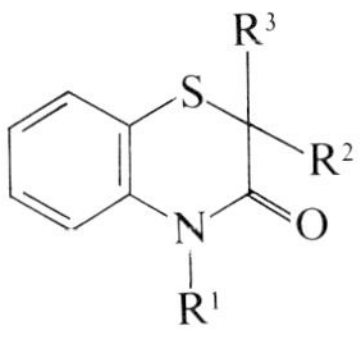

CCXIII [273]

R = H, halo, alkoxy,
R[1] = alkyl or dialkylamino

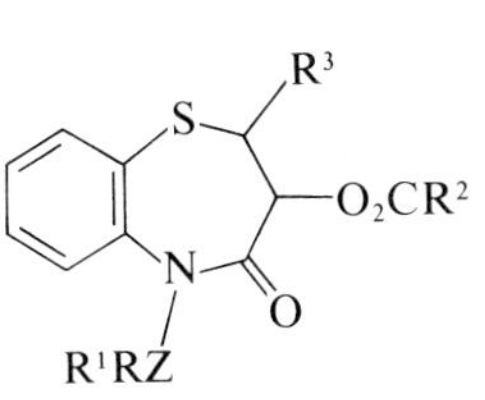

CCXIV [274]

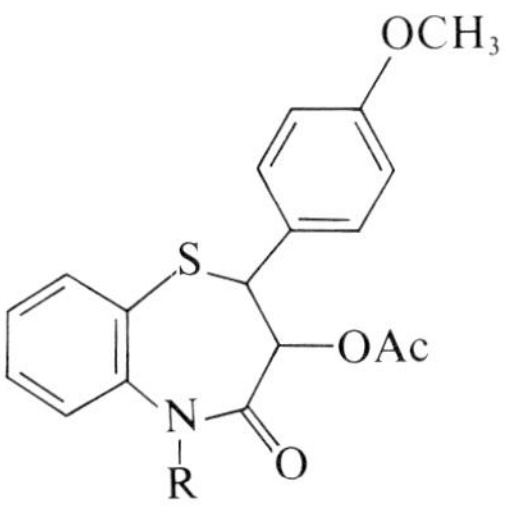

CCXV [274]

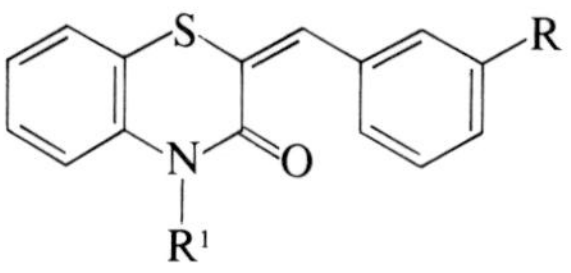

CCXVI [275]

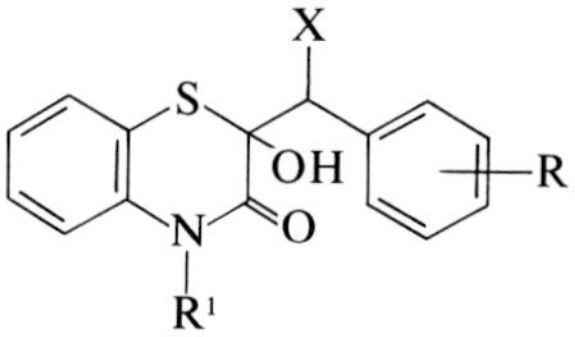

CCXVII [275]

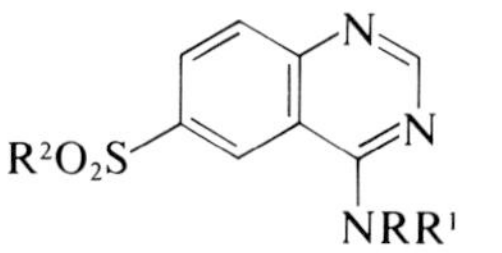

CCXVIII [276]

NRR[1] = piperidino, R[2] = NH$(CH_2)_4$NH

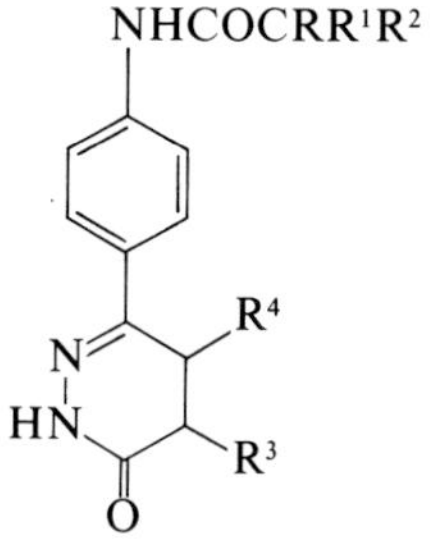

CCXIX [277]

R, R[1] = H, alkyl, R[2] = CO_2H, CN etc., R[3] = H, R[4] = H, alkyl

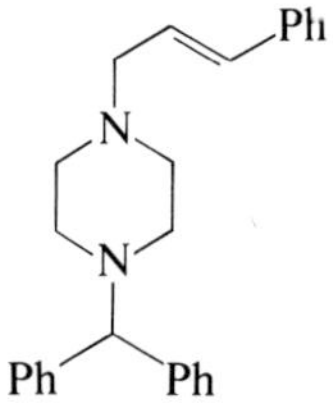

CCXX [278]

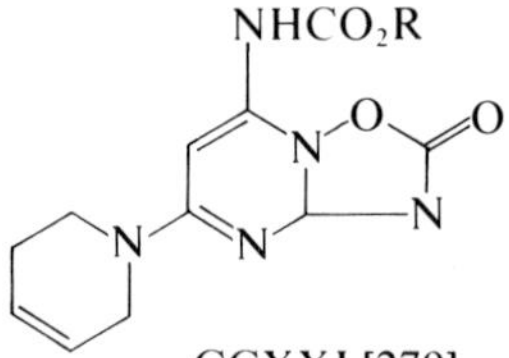

CCXXI [279]

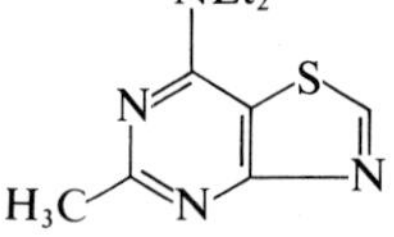

CCXXII [280]

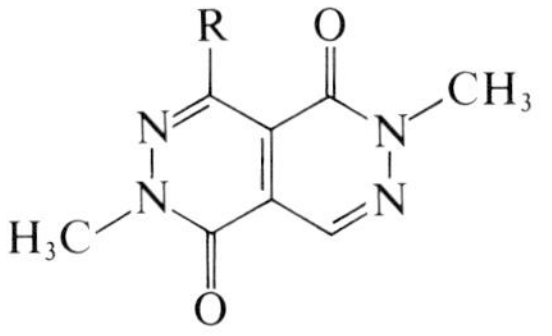

CCXXIII [281]

R = Cl, OCH_3 or $NHNH_2$ etc.

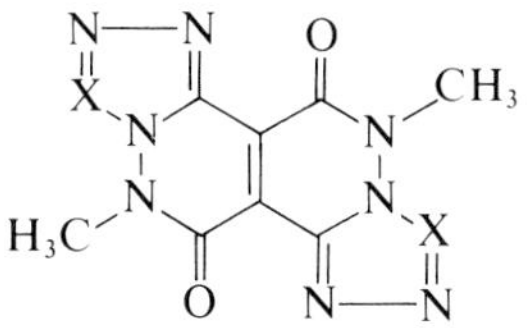

CCXXIV [281]

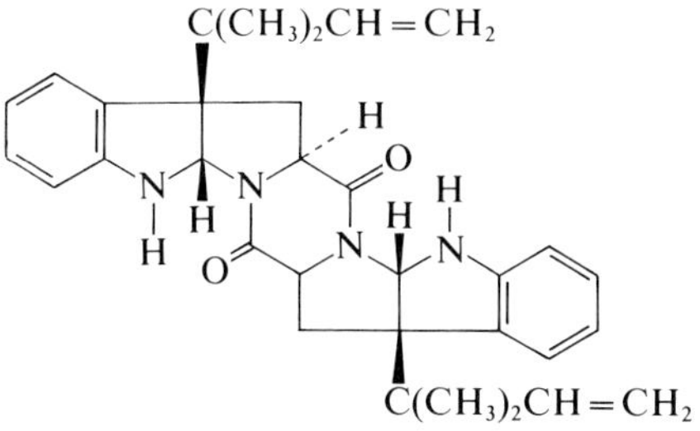

CCXXV [282]

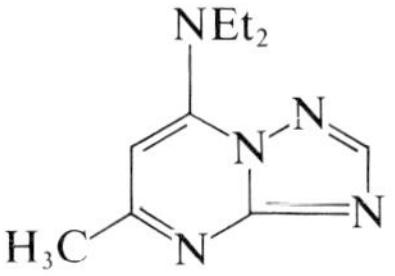

CCXXVI [283]

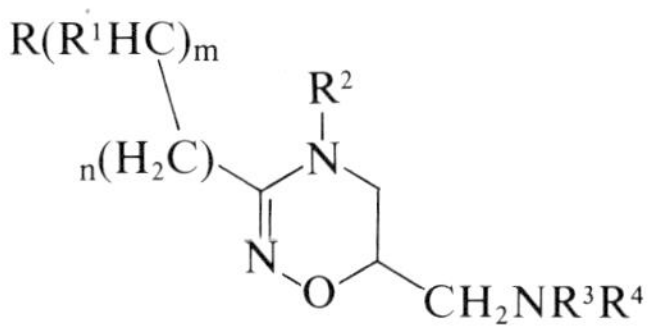

CCXXVII [284]

R = H, aryl, R^1 = alkyl, cycloalkyl,
R^2 = alkyl, acyl, H, R^3 = H, alkyl,
R^4 = C_{5-7} cycloalkyl, substituted aryl,
R^3R^4 = X, m, n = 0, 1, 2

α-ADRENOCEPTOR BLOCKING AGENTS

Aromatic compounds

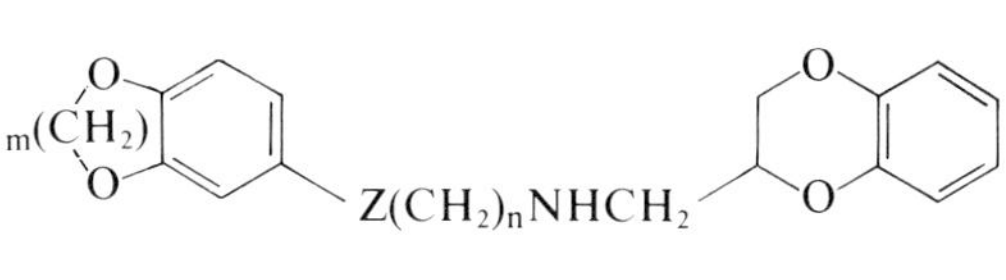

CCXXVIII [285]

m, n = 1–3, Z = $NRCH_2$ (R = H, alkyl etc.)

Heterocyclic compounds with one hetero atom

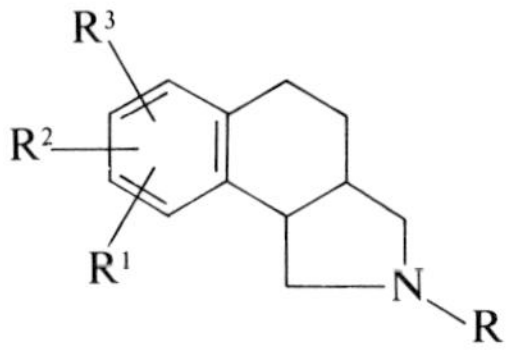

CCXXIX [286]

R = H, alkyl, $R^1 = R^3$ = H, OH, SCH_3 etc.
R^1R^2, R^2R^3 = OCH_2O, OCH_2CH_2O

Heterocyclic compounds with more than one hetero atom

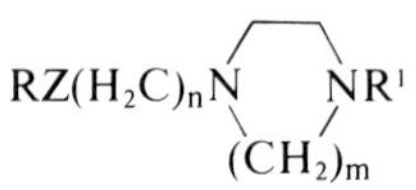

CCXXX [287]

R = aryl, R^1 = pyridopyrazinyl etc.
Z = CH_2, CO, O, n = 0–3, m = 1, 2

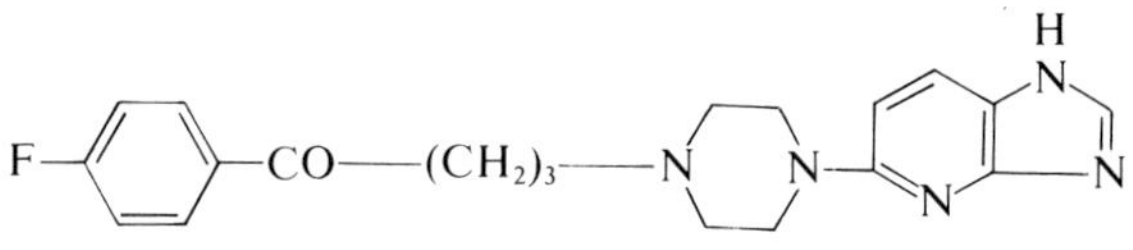

CCXXXI [287]

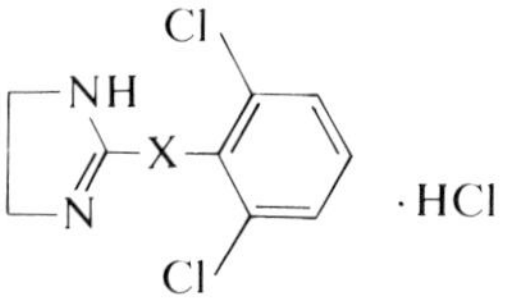

CCXXXII [288]

X = O, S or CH_2

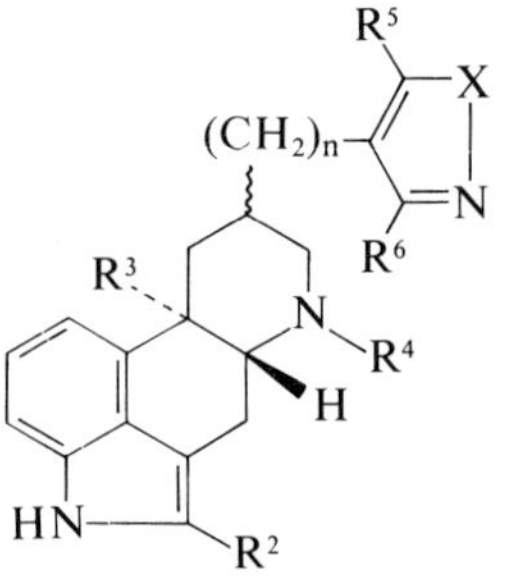

CCXXXIII [289]

R^1 = H or alkyl, R^2 = H, halogen or alkyl,
R^3 = H or alkoxy, R^4 = C_{1-4} hydrocarbyl,
R^5, R^6 = H, C_{1-3} alkyl, aryl etc., X = O etc.

5 Conclusions

The search for novel classes of antihypertensive agents is directed more towards generating diuretics and vasodilators and there is less interest in new angiotensin enzyme inhibitors, β-blockers and α-adrenergic blocking agents during the last four years. Studies on various clinical aspects of drugs and advances in the understanding of their pharmacology relate to diuretics, vasodilators, angiotensin converting enzyme inhibitors and β-blockers. The latter class of drugs, especially in combination with other antihypertensive agents, and its schedule of administration, have received considerable attention. The most important emerging area appears to be the development of calcium antagonists and their application in acute myocardial ischemia, myocardial infarction, cardioplegia, cardiovascular disorders, pulmonary hypertension, bronchial asthma and in the prevention of tissue damage is of major interest. Interesting results have been reported with various combinations of drugs and that more such results are expected with the growing understanding of disease conditions and of the pharmacological behaviour of drugs. It is disappointing that in recent publications on development of new antihypertensive agents, it is difficult to comprehend the scientific rationale employed in many cases for designing completely new classes of such agents. It is most likely that many of them are products of random screening. However, there have been improvement/advances in the methodology of clinical studies of disease conditions which may provide distinct clinical advantages in the treatment of hypertensive patients with drug(s). The trend of developments in this area suggests that by the end of this decade more knowledge regarding the application of calcium antagonists would be generated and this knowledge may bring a significant change in the schedule of treatment of hypertensive patients.

Acknowledgment

Thanks are due to Dr. S. Bhattacharya for rendering all possible help in the preparation of the manuscript.

References

1 The Management Committee: Lancet *1,* 1261 (1980).
2 A. Helgeland: Am. J. Med. *69,* 725 (1980).
3 Wld Hlth Org./I. S. H. Mild Hypertension Liaison Committee: Lancet *1,* 149 (1982).
4 O. Schier and A. Marxer: Progress in Drug Research, vol. 25, p. 9–132. Ed. E. Jucker. Birkhäuser Verlag, Basel 1981.
5 T. W. Anderson, W. H. LeRichc and J. S. Mackay: N. Engl. J. Med. *280,* 805 (1969).
6 M. D. Crawford, M. J. Gardener and J. N. Morris: Br. Med. Bull. *27,* 21 (1971).
7 S. John and W. Aideen: Br. Med. J. *285,* 1157 (1982).
8 W. P. Leary and A. J. Reyes: S. Afr. Med. J. *61,* 279 (1982).
9 Dainippon Pharmaceutical Co., Ltd. Jpn. JP 81, 145, 220 (1981) [C.A. *96,* 168717 (1981)].
10 C. D. Blume and P. H. Bonner: Eur. Pat. 100,061 (1984) [C.A. *100,*145005 (1984)].
11 V. B. Noskov and A. J. Grigor'ev: Fiziol. Chel. *9,* 515 (1983).
12 R. Janos, H. Anna, G. Gyorgy, J. Agnes, B. Karolyne, W. Gabor and G. Eva: Orv. Hétil. *123,* 159 (1982).
13 F. Kuzuya: Int. J. Clin. Pharm. Ther. Tox. *21,* 10 (1983).
14 D. S. Robinson, C. M. Nilsson, R. F. Leonard and E. S. Horton: J. Clin. Pharm. *21,* 637 (1981).
15 W. LaCorte, J. D. Irvin, A. K. Jain, P. B. Huber, J. R. Ryan, J. J. Schrogie, R. O. Davies and F. G. McMahon: Clin. Pharmacol. Ther. *25,* 233 (1979).
16 J. A. Tobert, G. Hitzenberger, I. James, J. Pryor, T. Cook, A. Bunlink, I. B. Holmes and P. M. Lutterbeck: Clin. Pharmacol. Ther. *29,* 344 (1981).
17 P. H. Vlasses, J. D. Irvin, P. B. Huber, R. B. Lee, R. K. Ferguson, J. J. Schrogil, A. G. Zacchei, R. O. Davies and W. B. Abrams: Clin. Pharmacol. Ther. *29,* 798 (1981).
18 A. L. M. Kerremans, T. Yuen, A. M. Cees and F. W. J. Gribnau: Clin. Pharmacol. Ther. *34,* 181 (1983).
19 J. Burton, R. J. Cody, J. A. Herd and E. Haber: Proc. natn. Acad. sci. *77,* 5476 (1980).
20 M. Szelke, B. J. Leckie, M. Tree, A. Brown, J. Grant, A. Hallett, M. Hughes, D. M. Jones and A. F. Lever: Hypertension *4,* 11 (1982).
21 M. Szelke, B. Leckie, A. Hallett, D. M. Jones, J. Sueiras, B. Atrash and A. F. Lever: Nature *299,* 555 (1982).
22 D. J. Webb, A. M. M. Cumming, B. J. Leckie, A. F. Lever, J. J. Morton, J. I. S. Robertson, M. Szelke and B. Donovan: Lancet *1,* 1486 (1983).
23 J. H. Laragh, D. B. Case, S. A. Atlas and J. E. Sealey: Hypertension *2,* 586 (1980).
24 A. Mimran, R. Targhetta and B. Laroche: Hypertension *2,* 732 (1980).
25 S. L. Swartz, G. H. Williams, N. K. Hollenberg, F. R. Crantz, T. J. Moore, L. Levine, A. A. Sasahara and R. G. Dluhy: Clin. Pharmacol. Ther. *28,* 499 (1980).
26 B. C. Campbell, A. Sturani and J. L. Reid: J. Hypertension *1,* 246 (1983).
27 Y. Kasai, K. Abe, M. Yasujima, J. Tajima, M. Seino, S. Chiba, K. Sato, T. Goto and K. Omata: Tohoku J. exp. Med. *141,* 417 (1983).
28 G. P. Hodsman, J. R. Zabludowski, C. Zoccali, R. Fraser, J. P. Morton, G. D. Murray and J. I. S. Robertson: Br. J. Clin. Pharmacol. *17,* 233 (1984).
29 W. P. Leary, A. J. Reyes and B. L. Vander: Curr. Ther. Res. *35,* 287 (1984).
30 H. Ibsen, B. Egan and S. Julius: J. Hypertension *1,* 222 (1983).
31 P. U. Witte, H. Metzger and R. Irmisch: IRCS Med. Sci., Libr. Compend. *11,* 1053 (1983).

32 J. I. S. Robertson: Drugs *25,* 9 (1983).
33 The Norwegian Multicenter Study Group: N. Engl. J. Med. *304,* 801 (1981).
34 A. Hjalmarson, J. Herlitz, I. Malek, L. Ryden, A. Vedin, A. Waldenstrom, H. Wedel, D. Elmfeldt, S. Holmberg, G. Hyberg, K. Sweberg, F. Waagstein, J. Waldenstrom, L. Wilhelmsen and C. Wilhelmson: Lancet *2,* 823 (1981).
35 β-Blocker Heart Attack Study Group: J. Am. Med. Ass. *246,* 2073 (1981).
36 L. M. Gonasun: Am. Heart J. *104,* 374 (1982).
37 D. R. Turner, F. J. Goodwin, A. R. Knight, D. W. Littlejohns, V. L. Sharman and D. W. Vere: Br. J. Clin. Pharmacol. *17,* 295 (1984).
38 P. C. Rubin: N. Engl. J. Med. *305,* 1323 (1982).
39 J. C. Petrie, T. A. Jeffers, A. K. Scott and J. Webster: Drugs *25,* 26 (1983).
40 L. Hansmann and K. M. Goebel: Drugs *25,* 73 (1983).
41 K. O'Malley, W. G. O'Callaghan, M. S. Laher, K. McGarry and E. O'Brien: Drugs *25,* 103 (1983).
42 E. A. Sotaniemi, R. O. Pelkonen, A. J. Arranto, S. Sako and M. Anttila: Drugs *25,* 113 (1983).
43 S. Freestone and L. E. Ramsay: Drugs *25,* 141 (1983).
44 D. S. Lawrence, J. N. Sahay, S. S. Chatterjee and J. M. Cruickshank: Drugs *25,* 232 (1983).
45 P. Turner: Drugs *25,* 262 (1983).
46 T. A. Betts and C. Alford: Drugs *25,* 268 (1983).
47 A. J. J. Wood: Drugs *25,* 318 (1983).
48 M. O. Luque, P. C. Fernandez, P. A. Escriba, V. M. Rodriguez, N. C. Martell and A. Fernandez-Cruz: Br. J. Clin. Pharmacol. *17,* 361 (1984).
49 A. Salvetti, F. Arzilli, R. Pedrinelli, P. Beggi and M. Motolese: Eur. J. Clin. Pharmacol. *22,* 197 (1982).
50 S. R. Smith, R. Gibson, D. Bradley and M. J. Kendall: Br. J. Clin. Pharmacol. *15,* 267 (1983).
51 R. Tarazi and H. Dustan: Am. J. Cardiol. *29,* 633 (1972).
52 F. R. Bühler, J. H. Laragh, L. Baer, E. D. Vaughan and H. R. Bruner: N. Engl. J. Med. *287,* 1209 (1972).
53 W. H. Birkenhager, P. W. DeLeeuw, A. Wester, T. L. Kho, R. Vandogen and H. E. Falke: Adv. intern. Med. Pediat. *39,* 117 (1977).
54 J. W. Hollifield and P. E. Slanton: R. Soc. Med. int. Congr. Symp. Ser. *44,* 17 (1981).
55 J. I. S. Robertson, F. R. Bühler, C. F. George, C. G. Geyskes, G. Leonetti, H. Lieban, A. Maggiove, T. O. Morgan, G. Muiesan, M. A. Waber und P. Weidmann: Clin. Sci. *48,* 109 (1975).
56 S. Z. Langer: Clin. Sci. *51,* 421 (1976).
57 O. L. Pederson, N. J. Christensen and K. D. Rämsch: J. Cardiovasc. Pharmacol. *2,* 357 (1980).
58 O. L. Pederson, C. K. Christensen, E. Mikkelsen and K. D. Rämsch: Eur. J. Clin. Pharmacol. *18,* 287 (1980).
59 Y. Imai, K. Abe, Y. Otsuka, N. Irokawa, M. Yasujima, K. Saito, Y. Sakurai, S. Chiba, T. Ito, M. Sato, T. Haruyama, Y. Miura and K. Yoshinaga: Arzneimittel-Forsch. *30,* 674 (1980).
60 M. D. Guazzi, C. Fiorentini, M. T. Olivari, A. Bartorelli, G. Necchi and A. Polese: Circulation *61,* 913 (1980).
61 B. Ek, M. Ahnoff, M. Hallbäck-Nordlander and B. Ljung: Naunyn-Schmiederberg's Arch. Pharmacol. *313,* R37 (1980).
62 H. L. Elliot, P. A. Meredith, D. J. Sumner, K. McLean and J. L. Reid: Br. J. Clin. Pharmacol. *13,* 699 (1982).
63 C. P. O'Boyle, M. Laher, E. T. O'Brien, K. O'Malley and J. G. Kelly: Eur. J. Clin. Pharmacol. *23,* 93 (1982).
64 R. G. Rahwan, D. T. Witiak and W. W. Muir: Annu. Rep. Med. Chem. *16,* 257 (1981).

65 B. Merz: J. Am. Med. Ass. *248,* 1285 (1982).
66 G. Planz: Cent. Blood Pressure Regul. Clonidine Workshop, p. 143–148. Eds. K. Hayduk and K. D. S. Bock. Darmstadt, Federal Republic of Germany (1983).
67 H. Moerl and C. Diehm: Cent. Blood Pressure Regul. Clonidine Workshop, p. 255–260. Eds. K. Hayduk and K. D. S. Bock. Darmstadt, Federal Republic of Germany (1983).
68 H. B. Steinhauser and R. Schollmeyer: Cent. Blood Pressure Regul. Clonidine Workshop, p. 261–269. Eds. K. Hayduk and K. D. S. Bock. Darmstadt, Federal Republic of Germany (1983).
69 H. C. Mitchell and W. A. Pettinger: Clin. Pharmacol. Ther. *30,* 297 (1981).
70 J. DeCree, J. Leempoels, W. DeCock, H. Geukens and H. Verhaegen: Angiology *32,* 137 (1981).
71 J. E. Birnbaum, P. Cervoni, P. S. Chan, S. M. L. Chen, M. B. Floyd, C. V. Grudzinskas and M. J. Weiss: J. Med. Chem. *25,* 492 (1982).
72 A. Grahnen, P. Seideman, B. Lindstrom, K. Haglund and C. von Bahr: Clin. Pharmacol. Ther. *30,* 439 (1981).
73 J. A. A. Alvarez: US Pat. 859,705 (1977) [C.A. *100,* 96711 (1984].
74 W. Y. Lee and K. C. Cho: K' at Ollik Taehak Uihakpu Nonmunjip *36,* 23 (1983) [C.A. *99,* 64014 (1983)].
75 P. H. Whiting, J. Petersen and J. G. Simpson: Br. J. exp. Path. *62,* 200 (1981).
76 F. H. H. Leenen: Can. J. Physiol. Pharmacol. *59,* 1002 (1981).
77 Y. Ayano, K. Yamasaki, H. Soejima and K. Ikegami: Urol. int. *39,* 25 (1984) [C.A. *100,* 114786 (1984)].
78 A. S. Nies, S. Fadul, J. Gal and J. G. Gerber: Adv. Prostaglandin Thromboxane, Leukotriene Res. *11,* 529 (1983).
79 V. V. Vorontsov and A. V. Aladyshev: Farmak. Toks. *45,* 52 (1982).
80 M. G. White, J. VanGelder and G. Eastes: J. Clin. Pharm. *21,* 610 (1981).
81 V. M. Erikov: Farmak. Toks. *44,* 597 (1981).
82 J. C. Melby: Int. Congr. Symp. Ser. R. Soc. Med. *44,* 55 (1981).
83 U. B. Olsen: Acta Pharmac. Tox. *49,* 321 (1981).
84 M. D. Evered and M. M. Robinson: J. Physiol. *348,* 573 (1984).
85 M. Ishii, A. Goto, K. Kimura, Y. Hirata, M. Yamakado, T. Takeda and S. Murao: Jap. Heart J. *24,* 623 (1983).
86 Editorial: Lancet *2,* 129 (1980).
87 C. S. Sweet, S. L. Gant, P. M. Reitz, E. H. Blaine and L. T. Ribeiro: J. Hypertension *1,* 53 (1983).
88 C. Richer, M. P. Doussau and J. F. Giudicelli: Eur. Heart J. *4,* 55 (1983).
89 A. Konrads, K. A. Meurer, W. Hummerick, G. Wambach, J. Kindler, H. Feltkamp and A. Helber: Am. J. Cardiol. *49,* 1558 (1982).
90 D. P. Clough, M. G. Collis, J. Conway, R. Hatton and J. R. Keddie: Am. J. Cardiol. *49,* 1410 (1982).
91 S. L. Swartz and G. H. Williams: Am. J. Cardiol. *49,* 1405 (1982).
92 W. Kiowski, P. van Brummelen, L. Hulthen, F. W. Amann and F. R. Bühler: Clin. Pharmacol. Ther. *31,* 677 (1982).
93 D. C. Kikta and M. J. Fregly: Hypertension *4,* 118 (1982).
94 C. S. Sweet, P. T. Arbegast, S. L. Gaul, E. H. Blaine and D. M. Gross: Eur. J. Pharmacol. *76,* 167 (1981).
95 T. Unger, B. Schull, W. Rascher, R. E. Lang and D. Ganten: Biochem. Pharmac. *31,* 3063 (1982).
96 E. Haber: Clin. Sci. *59,* 7 (1980).
97 J. L. Osborn, G. F. DiBona and M. D. Thames: Am. J. Physiol., F 620 (1982).
98 J. Gutkowska, P. Corvol, G. Thibault and J. Genest: Can. J. Biochem. *60,* 843 (1982).
99 B. Peter and W. Joergen: Am. J. Physiol. *245,* R906 (1983).

100 J. L. Osborn, G. F. DiBona and M. D. Thames: J. Pharmacol. exp. Ther. *216,* 265 (1981).
101 E. J. Johns: Br. J. Pharmacol. *73,* 749 (1981).
102 J. G. Gerber, R. D. Olson and A. S. Nies: Kidney int. *19,* 816 (1981).
103 R. D. Olson, A. S. Nies and J. G. Gerber: J. Pharmac. exp. Ther. *219,* 321 (1981).
104 P. C. Churchill, F. D. McDonald and M. C. Churchill: Life Sci. *29,* 383 (1981).
105 J. E. Sealey, A. Overlack, J. H. Laragh, K. O. Stumpe and S. A. Atlas: J. Clin. Endocr. Metab. *53,* 626 (1981).
106 G. Guillemette, M. Bernier, P. Parent, R. Leduc and E. Escher: J. Med. Chem. *27,* 315 (1984).
107 M. E. M. Oshiro, N. Miasiro, T. B. Paiva and A. C. M. Paira: Blood Vessels *21,* 72 (1984).
108 M. Paul, K. Hermann, M. Pritz, R. E. Lang, T. Unger and D. Ganten: J. Hypertension *1,* 9 (1983).
109 P. Leren, I. Eide, O. P. Foss, A. Helgeland, I. Hjermann, I. Holme, S. E. Kjeldsen and P. G. Lund-Larsen: J. Cardiovasc. Pharmacol. *4,* S 222 (1982).
110 Prazosin: Pharmacology, Hypertension and Congestive Heart Failure. Ed. M. D. Rawlins 1981.
111 C. Chevillard, B. Saiag and M. Worcel: Br. J. Pharmacol. *73,* 811 (1981).
112 W. Motz, R. Ippisch, G. Ringsgwandl and B. E. Strauer: Card. Adapt. Hemodyn. Overload, p. 364. Eds. J. Ruthard, R. W. Guelch and G. S. Kissling. Darmstadt, Federal Republic of Germany, 1983.
113 H. Ruskoaho: Acta Pharmac. Tox. *54,* 154 (1984).
114 R. C. Hill and P. Turner: Br. J. Pharmacol. *36,* 368 (1969).
115 P. C. Chang, G. J. Blauw and P. Van Brummelen: J. Hypertension *1,* 338 (1983).
116 R. J. Bache, X. Z. Dai and J. S. Schwartz: J. Am. Coll. Cardiol. *3,* 143 (1984).
117 K. Ogawa, Y. Wakamasu, T. Ito, T. Suzuki and N. Yamazaki: Arzneimittel-Forsch. *31,* 770 (1981).
118 M. Terai, T. Takenaka and H. Maeno: Biochem. Pharmac. *30,* 375 (1981).
119 S. Kazada, B. Garthoff, H. Meyer, K. Schlossmann, K. Stoepel, R. Towart, W. Vater and E. Wehinger: Arzneimittel-Forsch. *30,* 2144 (1980).
120 Drugs of Future *6,* 361 (1981).
121 D. C. Wartier, C. M. Meils, G. J. Gross and H. L. Brooks: J. Pharmacol. exp. Ther. *218,* 296 (1981).
122 T. H. Pringle, R. G. Murray, G. Johnsson, I. Hutton and T. D. V. Lawrie: Scot. Med. J. *26,* 88 (1981).
123 S.-L. Boström, B. Ljung, S. Mardh, S. Forsen and E. Thulin: Nature *292,* 777 (1981).
124 S. J. Thorens: Muscle Res. Cell. Motility *1,* 455 (1980).
125 R. P. Hof, E. Müller-Schweinitzer and P. Neumann: Br. J. Pharmacol. *73,* 196 (1981).
126 C. Farsang, M. D. Ramirez-Gonzalez, L. Tchakarov and G. Kunos: Acta Physiol. Hung. *62,* 167 (1983).
127 W. B. Campbell, K. S. Callahan, A. R. Johnson and R. M. Graham: Lancet *II,* 1382 (1981).
128 D. R. A. Lima, S. Kilfeather and P. Turner: Br. J. Clin. Pharmacol. *11,* 120 (1981).
129 L. J. Bryan, J. J. Cole, S. R. O'Donnell and J. C. Wanstall: J. Pharmac. exp. Ther. *216,* 395 (1981).
130 K. E. J. Dickinson and S. R. Nahorski: Life Sci. *29,* 2527 (1981).
131 A. Annie, A. Janie, D. M. Claude, R. Georges and M. J. Louis: J. Pharmacol. *12,* 277 (1981).
132 H. Mishra, B. Gupta, K. K. Saxena, V. K. Kulshrestha and D. N. Prasad: Indian J. Pharmacol. *14,* 301 (1982).

133 B. A. Schölkens, Th. Unger and D. Ganten: Naunyn-Schmiedeberg's Arch. Pharmacol. *313,* R 25 (1980).
134 J. J. Baldwin, G. H. Denny, G. S. Ponticellu, C. S. Sweet and C. A. Stone: Eur. J. Med. Chem. *17,* 297 (1982).
135 R. W. Fuller, T. T. Yen and N. B. Stamm: Clin. exp. Hypertens. *3,* (1981).
136 T. Kawasaki, K. Uezono, I. Abe, G. Nakamuta, M. Ueno, N. Kawazoe and T. Omae: Eur. J. Clin. Pharmacol. *20,* 399 (1981).
137 R. E. TenBrink, J. M. McCall, D. T. Pals, R. B. McCall, J. Orley, S. J. Humphrey and M. G. Wendling: J. Med. Chem. *24,* 64 (1981).
138 B. Persson, T. Hedner and M. Henning: J. Pharm. Pharmac. *34,* 442 (1982).
139 J. C. Clapham and T. C. Hamilton: J. Pharm. Pharmac. *34,* 644 (1982).
140 K. F. Rhodes, M. Stannard and J. F. Waterfall: Biochem. Pharmac. *32,* 3875 (1983).
141 J. W. Constantine, W. Lebel and R. Weeks: J. Cardiovasc. Pharmacol. *6,* 142 (1984).
142 S. L. Pfister and T. K. Keeton: Eur. J. Pharmacol. *97,* 247 (1984).
143 P. Bousquet, N. Decker, J. Feldman and J. Schwartz: J. Pharmacol. *14,* 465 (1983).
144 G. Garibotto, M. Giusti, S. Bagnasco and G. Deferrari: IRCS Med. Sci., Libr. Compend. *9,* 718 (1981).
145 F. Masugi, T. Ogihara, A. Otsuka, S. Saeki and Y. Kumahara: Jap. Circul. J. *48,* 196 (1984).
146 W. A. Wolf and D. M. Kuhn: Brain Res. *295,* 356 (1984).
147 W. Motz, R. Iplisch and B. E. Strauer: Cent. Blood Pressure Regul. Clonidine Workshop, p. 171–179. Eds. K. Hayduk and K. D. S. Bock. Darmstadt, Federal Republic of Germany, 1983.
148 B. A. Priimenko, B. A. Samura, S. N. Garmash and N. I. Romanenko: Khim.-Farm. Zh. *17,* 37 (1983).
149 H. Ohnishi, K. Yamaguchi, K. Takado, Y. Toyonaka, Y. Suzuki and Y. Orita: Drugs exp. Clin. Res. *7,* 823 (1981).
150 S. Morishita, K. Nishimura, E. Kato, H. Shirahase, S. Osumi, K. Kitao and T. Seki: Nippon Yakurigaku Zasshi *79,* 137 (1982).
151 D. Trachewsky and D. C. Kem: Clin. Sci. *63,* 129 (1982).
152 H. A. Parish, Jr., R. D. Gilliam, W. R. Purcell, R. K. Browne, R. F. Spirk and H. D. White: J. Med. Chem. *25,* 98 (1982).
153 Q. Branca, A. E. Fischli and A. Szente: Eur. Pat. 45,450 (1982) [C.A. *96, 199309* (1982)].
154 M. J. Cooling and M. F. Sim: Br. J. Pharmacol. *74,* 359 (1981).
155 L. Bruseghini, B. Ribalta and M. Jose: Span. Pat. 492,911 (1981) [C.A. *96,* 68615 (1982)].
156 C. Gyogyszer and V. T. Gyara Rt: Belg. Pat. 889,337 (1981) [C.A. *96,* 104584 (1982)].
157 R. E. Galardy: Biochem. biophys. Res. Commun. *97,* 94 (1980).
158 R. G. Almquist, W.-R. Chao, M. E. Ellis and H. L. Johnson: J. Med. Chem. *23,* 1392 (1980).
159 K. Imaki, S. Sakuyama, T. Okada, M. Toda, M. Hayashi, T. Miyamoto, A. Kawasaki and T. Okegawa: Chem. Pharm. Bull. *29,* 2210 (1981).
160 R. D. Smith, A. D. Essenburg, R. B. Parken, V. L. Nemeth, M. J. Ryan, D. H. Dugan and H. R. Kaplan: J. Med. Chem. *24,* 104 (1981).
161 R. F. Meyer, E. D. Nicolaides, F. J. Tinney, E. A. Lunney, A. Holmes, M. L. Hoefle, R. D. Smith, A. D. Essenburg, H. R. Kaplan and R. G. Almquist: J. Med. Chem. *24,* 964 (1981).
162 R. Henning, H. Urbach, R. Geiger, V. Teetz and R. Becker: German Pat. 3,211,676 (1983) [C.A. *100,* 121609 (1984)].
163 N. Yoneda, J. Kato, K. Hayashi, T. Ochiai and K. Kinashi: Eur. Pat. 95,163 (1983) [C.A. *100,* 174827 (1984)].

164 N. Gruenfeld: Eur. Pat. 93,084 (1983) [C.A. *100,* 138951 (1984)].

165 H. Urbach, R. Henning, V. Teetz, R. Geiger, R. Becker and H. Gaul: Eur. Pat. 84,164 (1983) [C.A. *100,* 139616 (1984)].

166 V. Teetz, H. Urbach and R. Becker: Eur. Pat. 90,341 (1983) [C.A. *100,* 85588 (1984)].

167 J.-C. Muller and H. Ramuz: Helv. Chim. Acta *65,* 1445 (1982).

168 V. P. Arya: Drugs of Future *7,* 390 (1982).

169 J. M. McCall, R. B. McCall, R. E. Ten Brink, B. Y. Kamdar, S. J. Humphrey, V. H. Sethy, D. W. Harris and C. Daenzer: J. Med. Chem. *25,* 75 (1982).

170 J. E. Francis, K. J. Doebel, P. M. Schutte, E. C. Savarese, S. E. Hopkins and E. F. Bachmann: Can. J. Chem. *57,* 3320 (1979).

171 G. Steiner, J. Gries and D. Lenke: J. Med. Chem. *24,* 59 (1981).

172 E. Arrigoni-Martelli, Chr. K. Nielsen, U. B. Olsen and H. J. Petersen: Experientia *36,* 445 (1980).

173 Y. Sato, Y. Shimoji, H. Fujita, H. Nishino, H. Mizuno, S. Kobayashi and S. Kumakura: J. Med. Chem. *23,* 927 (1980).

174 Z. Huszti, G. Szilagyi, P. Matyus and E. Kasztreiner: J. Neurochem. *37,* 1272 (1981).

175 O. Hudlicka, J. Komarck and A. J. A. Wright: Br. J. Pharmacol. *72,* 723 (1981).

176 H. Karppanen, I. Paakkari, P. Paakkari, L. Enkovaara, M. Svartstram-Fraser and E. Honkanen: Acta Med. Scand., suppl. *677,* 148 (1983).

177 M. Kamibayashi, S. Tsuchiya and K. Hiratsuka: US Pat. 4,419,518 (1983) [C.A. *100,* 103191 (1984)].

178 D. P. Clough, R. Hatton and S. J. Pettinger: Arzneimittel-Forsch. *31,* 1698 (1981).

179 H. Ohnishi, K. Yamaguchi, M. Satoh, M. Obato, A. Uemura, Y. Toyonaka and Y. Suzuki: Arzneimittel-Forsch. *31,* 1425 (1981).

180 A. J. Karjaininen and K. O. A. Kurkela: Abstr. int. Congr. Pharmacol. *8,* 10 (abstr.) (1981).

181 R. Lammintausta, E. McDonald and L. Nieminen: Abstr. int. Congr. Pharmacol. *8,* 12 (abstr.) (1981).

182 J. W. Tilley, P. Levitan, R. W. Kierstead and M. Cohen: J. Med. Chem. *23,* 1387 (1980).

183 R. Hammer, W. Kobinger and L. Pichler: Eur. J. Pharmacol. *62,* 277 (1980).

184 G. B. Fregnan and G. Ferni, IRCS Med. Sci.-Biochem. *8,* 548 (1980).

185 Drugs of Future *9,* 100 (1984).

186 Drugs of Future *9,* 116 (1984),

187 Drugs of Future *9,* 203 (1984).

188 M. L. Blank, E. A. Cress and F. Snyder: Biochem. biophys. Res. Commun. *118,* 344 (1984).

189 L. Eriksson, H. Karppanen, E. Honkanen and U. M. Kokkonen: Acta Pharmac. Tox. *54,* 158 (1984).

190 Drugs of Future *9,* 13 (1984).

191 Drugs of Future *9,* 36 (1984).

192 D. Dobrescu, M. Iovu, E. Cristea and A. W. Baloch: Rom. Pat. 66,758 (1979) [C.A. *95,* 132453 (1981)].

193 Z. Vejdelek and M. Protiva: Collect. Czech. Chem. commun. *48,* 642 (1983).

194 J. J. Plattner: US Pat. 4,389,416 (1983) [C.A. *99,* 139494 (1983)].

195 G. C. Helsley, W. Dornauer and L. Davis: US Pat. 287,212 (1981) [C.A. *96,* 34815 (1982)].

196 G. Lettieri, A. Larizza and G. Brancaccio: Boll. Chim. Farm. *120,* 208 (1981).

197 M. Rajsner and M. Protiva: Czech. Pat. 193,303 (1981) [C.A. *96,* 199731 (1982)].

198 A. Bonaldi and E. Molinari: Eur. Pat. 63,106 (1982) [C.A. *98,* 126470 (1983)].
199 S. H. Rizvi, A. Shoeb, R. S. Kapil and S. P. Popli: Phytochemistry *19,* 2409 (1980).
200 O. W. Woltersdorf, Jr., and E. J. Cragoe, Jr.: US Pat. 4,291,050 (1981) [C.A. *96,* 6467 (1982)].
201 E. J. Cragoe, Jr., H. W. R. Williams and O. W. Woltersdorf, Jr.: US Pat. 4,249,021 (1981) [C.A. *95,* 6906 (1981)].
202 H. Ott: Fr. Pat. 2,473,512 (1981) [C.A. *96,* 122458 (1982)].
203 Fujisawa Pharmaceutical Co., Ltd., Japan: Japan. Pat. 58,124,758 (1983) [C.A. *100,* 34278 (1984)].
204 C. Ruferfer and I. Boettcher: German Pat. 3,208,079 (1983) [C.A. *100,* 6113 (1984)].
205 H. Englert, H. J. Lang, M. Hropot and J. Kaiser: German Pat. 3,208,190 (1983) [C.A. *100,* 6079 (1984)].
206 K. C. Liu, L. C. Lee, B. J. Shih, C. F. Chen and T. M. Tao: Arch. Pharm. *315,* 872 (1982).
207 B. W. Metcalf and J. O'Neal Jonston: US Pat. 4,293,548 (1981) [C.A. *96,* 104603 (1982)].
208 F. Haviv: US Pat. 4,312,887 (1982) [C.A. *96,* 142446 (1982)].
209 W. E. Meyer, A. S. Tomeufcik and J. W. Marsico, Jr.: US Pat. 4,360,466 (1982) [C.A. *98,* 178975 (1983)].
210 Z. Brzowski, E. Pomarnacka-Jankowska and S. Angielski: Acta Pol. Pharm. *38,* 11 (1981).
211 A. Nuhrich, A. Carpy, B. Barbe, A. Marchand and G. Devaux: Eur. J. Med. Chem.-Chim. Ther. *16,* 556 (1981).
212 Mochida Pharmaceutical Co., Ltd. Japan: Japan. Pat. 8153,614 (1981) [C.A. *95,* 121144 (1981)].
213 J. Seres, I. Daroczi, G. Horvath and I. Szilagyi: Hung. Pat. 22,408 (1982) [C.A. *98,* 126144 (1983)].
214 O. W. Woltersdorf, Jr., S. J. DeSolms and E. J. Cragoe, Jr.: J. Med. Chem. *24,* 874 (1981).
215 Y. H. Park, Y. S. Kim and B. H. Cho: Taehan Yakrinhak Chapchi *19,* 17 (1983) [C.A. *100,* 167920 (1984)].
216 E. A. Brown: German Pat. 3,038,821 (1981) [C.A. *95,* 133253 (1981)].
217 H. Meyer, R. Sitt, G. Thomas and H. P. Krause: German Pat. 3,015,219 (1981) [C.A. *96,* 6604 (1982)].
218 B. Garthoff, S. Kazda, A. Knorr and G. Thomas: German Pat. 3,212,736 (1983) [C.A. *100,* 12657 (1984)].
219 U. Rosentreter, W. Puls and H. Bischoff: German Pat. 3,124,673 (1983) [C.A. *99,* 70765 (1983)].
220 G. Cignarella, P. Sanna, E. Miele, V. Anania and M. S. Desole: J. Med. Chem. *24,* 1003 (1981).
221 W. Liebenov and K. Mannhardt: German Pat. 2,923,345 (1980) [C.A. *95,* 7044 (1981)].
222 W. Liebenow and K. Mannhardt: German Pat. 2,942,643 (1981) [C.A. *95,* 97574 (1981)].
223 M. J. Kulshreshtha, S. Bhatt, M. Pardasani and N. M. Khanna: J. Indian Chem. Soc. *58,* 982 (1981).
224 A. J. Karjalaynen and K. O. A. K. Kurkela: Eur. Pat. 24,829 (1981) [C.A. *95,* 115545 (1981)].
225 A. F. Kluge, A. M. Strosberg and S. H. Unger: US Pat. 4,255,432 (1981) [C.A. *95,* 97773 (1981)].
226 A. G. Meetichemie: Belg. Pat. 886,874 (1981) [C.A. *95,* 162320 (1981)].
227 M. G. Bock, E. J. Cragoe, Jr., and R. L. Smith: Eur. Pat. 40,422 (1981) [C.A. *96,* 85592 (1982)].

228 Daüchi Seiyaku Co., Ltd.: Japan. Pat. 5883,693 (1983) [C.A. *99,* 139944 (1983)].
229 G. M. Shutske, L. L. Setescak and R. C. Allen: Eur. Pat. 45,078 (1982) [C.A. *96,* 217825 (1982)].
230 R. Baronnet, R. Callendret, L. Blanchard, O. Foussard-Blanpin and J. Bretaudeau: Eur. J. Med. Chem.-Chim. Ther. *18,* 241 (1983).
231 B. Balzac: Japan. Pat. 81,128,771 (1981) [C.A. *96,* 104275 (1982)].
232 E. Habicht and P. Zbinden: Eur. Pat. 64,027 (1982) [C.A. *98,* 179352 (1983)].
233 Z. Brzozowski and E. Pomarnacka: Acta Pol. Pharm. *37,* 373 (1980).
234 F. K. Szilagyi, S. Solyom, L. Toldy, I. Schafer, E. Szondy, J. Borvendeg and S. I. Hermann: Eur. Pat. 31,591 (1981) [C.A. *96,* 6946 (1982)].
235 B. V. Shetty and A. McFadden: US Pat. 4,420,487 (1983) [C.A. *100,* 174831 (1984)].
236 H. J. Lang and R. Muschaweck: Rom. Pat. 514,406 (1981) [C.A. *96,* 142843 (1982)].
237 O. W. Woltersdorf, Jr., S. J. DeSolms and R. L. Smith: US Pat. 4,272,537 (1981) [C.A. *95,* 132960 (1981)].
238 V. P. Chernykh, Z. P. Bulada, P. O. Benzuglyi, V. I. Makurina, L. M. Voronina and A. I. Bereznyakova: Farm. Zh. *5,* 33 (1981).
239 O. W. Woltersdorf and E. J. Cragoe, Jr.: US Pat. 4,277,602 (1981) [C.A. *95,* 150707 (1981)].
240 B. A. Priimenko, Yu. V. Strokin, B. A. Samura and A. K. Sheinkman: USSR Pat. 819,106 (1981) [C.A. *95,* 91201 (1981)].
241 H. Knauf, E. Mutschler and K. D. Volger: Can. Pat. 1,110,543 (1981) [C.A. *96,* 91651 (1982)].
242 S. B. Kadin: Fr. Pat. 2,470,132 (1981) [C.A. *96,* 35286 (1982)].
243 A. Ingendoh, H. Meyer and B. Garthoff: German Pat. 3,208,437 (1983) [C.A. *100,* 156587 (1984)].
244 W. B. Wright, Jr., J. Marsico, Jr., and W. Joseph: US Pat. 4,283,334 (1981) [C.A. *95,* 204013 (1981)].
245 H. Meyer, H. Horstmann, E. Moller and B. Garthoff: Eur. Pat. 41,215 (1981) [C.A. *96,* 122803 (1982)].
246 N. Fuji, W. Lee, H. Yajima, M. Haruaki, M. Moriga and K. Mizuta: Chem. Pharm. Bull. *31,* 35014 (1983).
247 D. E. Ryono and E. W. Petrillo, Jr.: German Pat. 3,309,014 (1983) [C.A. *100,* 86121 (1984)].
248 D. S. Karanewsky: U. K. Pat. 2,118, 944 (1983) [C.A. *100,* 103894 (1984)].
249 M. A. Ondetti and D. E. Ryono: Fr. Pat. 2,479,827 (1981) [C.A. *96,* 143324 (1982)].
250 R. D. Haugwitz: U. K. Pat. 2,061,931 (1981) [C.A. *96,* 20460 (1982)].
251 J. W. Ryan and A. Chung: Eur. Pat. 73,143 (1983) [C.A. *100,* 103897 (1984)].
252 H. Urbach, R. Henning, V. Teetz, H. Wissmann and R. Becker: German Pat. 3,210,496 (1983) [C.A. *100,* 139617 (1984)].
253 Takeda Chemical Industries, Ltd.: Japan. Pat. 58,188,857 (1983) [C.A. *100,* 138974 (1984)].
254 Tanabe Seiyaku Co., Ltd.: Japan. Pat. 58,185,565 (1983) [C.A. *100,* 138975 (1984)].
255 R. D. Haugwitz and P. W. Sprague: Eur. Pat. 95,584 (1983) [C.A. *100,* 175285 (1984)].
256 K. Itoh and Y. Oka: Chem. Pharm. Bull. *37,* 2106 (1983).
257 C. Maccarrone, E. Malta and C. Raper: J. Cardiovasc. Pharmacol. *6,* 132 (1984).
258 L. H. Werner and N. Ford: Eur. Pat. 96,006 (1983) [C.A. *100,* 156602 (1984)].
259 W. Fuhrer, F. Ostermayer and M. Zimmermann: German Pat. 3,314,196 [C.A. *100,* 68320 (1984)].
260 L. C. Blaber, D. T. Burden, R. Eigenmann and M. Gerold: J. Cardiovasc. Pharmacol. *6,* 165 (1984).

261 Neelima, M. Seth and A. P. Bhaduri: Indian J. Chem. (B) *22B,* 653 (1983).
262 A. Miyake, K. Itoh, N. Tada, M. Tanabe, M. Hirata and Y. Oka: Chem. Pharm. Bull. *31,* 2329 (1983).
263 S. F. Campbell, P. E. Cross and J. K. Stubbs: Eur. Pat. 89,167 (1983) [C.A. *100,* 6351 (1984)].
264 K. Araki, H. Ao, J. Inui and K. Aihara: Eur. Pat. 88,903 (1983) [C.A. *100,* 85591 (1984)].
265 M. Watanabe, K. Moto, Y. Takemoto, T. Hatta, T. Hashimoto and K. Yamada: Eur. Pat. 93,945 (1983) [C.A. *100,* 120891 (1984)].
266 Maruko Pharmaceutical Co., Ltd.: Japan. Pat. 58,201,764 (1983) [C.A. *100,* 138968 (1984)].
267 P. B. Berntsson and S. A. I. Carlsson: U. K. Pat. 2,120,251 (1983) [C.A. *100,* 138964 (1984)].
268 Yoshitomi Pharmaceutical Industries Ltd.: Japan. Pat. 58,216,158 (1983) [C.A. *100,* 174672 (1984)].
269 P. J. Cornu, C. Perrin, B. Dumaitre and G. Streichenberger: Fr. Pat. 82/8,648 (1982) [C.A. *100,* 156505 (1984)].
270 K. Hatayama, A. Nakazato, T. Ogawa, S. Ito and J. Sawada: Eur. Pat. 92,936 [C.A. *100,* 68180 (1984)].
271 P. B. Berntsson, S. A. I. Carlsson and B. R. Ljung: Eur. Pat. 95,451 (1983) [C.A. *100,* 103187 (1984)].
272 K. Meguro and A. Nagaska: PCT Int. Pat. 83,04,023 (1983) [C.A. *100,* 120900 (1984)].
273 Hamari Yokuhin Kogyo Co., Ltd.: Japan. Pat. 57,180,476 (1983) [C.A. *100,* 85732 (1984)].
274 S. Nagao, K. Kurabayashi, N. Futamura, H. Kinoshita and T. Takahashi: US Pat. 4,416,819 (1983) [C.A. *100,* 85733 (1984)].
275 Hamari Yakuhin Kogyo Co., Ltd.: Japan. Pat. 58,206,577 (1983) [C.A. *100,* 139162 (1984)].
276 H. Hidako, T. Sone, Y. Sasaki, T. Sugihara, S. Takagi and K. Sako: Eur. Pat. 46,572 (1982) [C.A. *96,* 217873 (1982)].
277 P. A. Rossy, M. Thyes, A. Franke, H. Koening, J. Gries, H. Lehmann and L. D. Dieter: German Pat. 3,209,159 (1983) [C.A. *100,* 6536 (1984)].
278 Eisai Co., Ltd.: Japan. Pat. 58,164,582 (1983) [C.A. *100,* 121107 (1984)].
279 M. Imfeld, J. C. Muller, H. Ramuz and P. Vogt: German Pat. 3,308,037 (1983) [C.A. *100,* 6545 (1984)].
280 H. Ohnishi, H. Kosuzume, Y. Suzuki and Ei. Mochida: PCT Int. Pat. 83,02,945 (1983) [C.A. *100,* 85713 (1984)].
281 G. Auzzi, F. Bruni, L. Cecchi, A. Lostanzo, V. L. Pecori and F. De Sio: Farmaco, Ed. Sci. *38,* 842 (1983).
282 S. Takase, Y. Kawai, Y. Ito, I. Uchida, H. Horiai, M. Iwam, M. Okamoto, K. Yoshida and H. Tanaka: Tennen Yuki Kagobutsu Toronkai Koen Yoshishu *26,* 94 (1983) [C.A. *100,* 135482 (1984)].
283 Mitsui Petrochemical Industries, Ltd.: Japan. Pat. 58,174,384 (1983) [C.A. *100,* 121100 (1984)].
284 G. Chinoin and T. G. R. Vegyeszeti: Belg. Pat. 885,634 (1981) [C.A. *95,* 43181 (1981)].
285 Mitsubishi Chemical Industries Co., Ltd.: Japan. Pat. 58,154,574 (1983) [C.A. *100,* 103366 (1984)].
286 J. F. K. Derbernardis, J. Daniel and W. J. McClellan: Eur. Pat. 95,666 (1983) [C.A. *100,* 103176 (1984)].
287 P. Neumann: German Pat. 3,318,617 (1983) [C.A. *100,* 10339 (1984)].
288 P. B. M. W. M. Timmermans, J. C. J. Mackaay, P. H. M. Fluitman and P. A. Van Zweiten: Pharmacology *19,* 294 (1979).
289 L. Bernardi, A. Temperilli, S. Mantegani, G. Traquandi and P. Salvati: Belg. Pat. 896,609 (1983) [C.A. *100,* 96714 (1984)]

Drug inhibition of mast cell secretion

By R. Ludowyke and D. Lagunoff

Department of Pathology, St. Louis University
School of Medicine, 1402, S. Grand Blvd. St. Louis, Mo 63104

1 Introduction

Interest in inhibitors of mast cell secretion derives from two purposes: one is the explication of the mechanism of secretion; the other is the development of clinically applicable drugs to prevent the unwanted consequences of mast cell secretion. Inhibition studies directed at an understanding of mast cell secretion are focused on postulated components of the process and thus typically utilize drugs whose actions are known. Studies in the second category are more likely to use new, potential drugs in an empirical mode. However, the distinction between the two types of studies should not be taken too seriously, since on the one hand the mechanism of action of drugs is not always as purported, and, on the other, empirically selected drugs may turn out to function through explicable actions on critical components of the secretory mechanism.

In approaching the voluminous literature on inhibitors of mast cell secretion, we have chosen to critically review the evidence for the action of selected groups of drugs on specific steps in the secretory mechanism. We will emphasize the importance of recognizing that inhibitors have different effects with different secretagogues. We in general take a dim view of premature efforts at laying out detailed sequences of events in the secretory process. So long as critical pieces of the mechanism remain either poorly defined or undefined, metabolic maps are likely to be misleading rather than heuristic.

The study of mast cell secretion and thus of its inhibitors takes place in a context created by parallel studies of secretion in a variety of cells. The debt of those who study mast cell secretion to the observations made in other secretory systems is great, and any thorough consideration of inhibitors of mast cell secretion would have to acknowledge the results from studies of other secretory systems. The dual constraints of time, ours, and space, the volume's, have conspired against us, so we will consider exclusively studies of inhibitors acting on tissue mast cells and their near relations, basophils and mucosal mast cells. In the interest of conciseness of expression we will at times refer to these three cell types collectively as mast cells.

2 Drugs that modify cell cAmp

The history of the development of evidence for a critical role of intracellular cAMP in mast cell secretion is characterized by complexity and controversy. After 15 years of intense study we are doubtless nearer a satisfactory description of the part cAMP plays, but the answers are not yet at hand. The first suggestion that cAMP was involved in mast cell and basophil secretion came from experiments showing that catecholamines known to stimulate adenylate cyclase and methylxanthines known to inhibit phosphodiesterases both inhibited histamine release [1]. These observations have been extensively confirmed [2–6]. The common effect of these two distinct classes of drugs in increasing intracellular cAMP was reasonably proposed as the basis for their action. Other agents capable of elevating intracellular cAMP have since been found to inhibit histamine release, these included, prostaglandin E_1 [3, 6–10] dibutyryl cAMP [3, 4], histamine [6, 11], forskolin [12] and cholera toxin [13]. Complicating an understanding of the function of cAMP in mast cell secretion are substantial discrepancies between the effects of different agents on cAMP levels and on histamine release [14–21]. The corollary of the proposal that elevated cAMP inhibited histamine release was that a decrease in cAMP could be a trigger for mast cell secretion, and with some secretagogues the predicted requisite fall in mast cell cAMP was found [22–26]. However with other secretagogues a rapid transient increase in cAMP is observed prior to the onset of secretion [27–29].

Yet additional complexities were added when it became apparent that methylxanthines in addition to inhibiting phosphodiesterases were also able to block stimulatory effects of adenosine on adenylate cyclase [30]. Marquardt et al. [31] have shown that adenosine stimulates histamine release from rat mast cells; yet adenosine inhibits basophil secretion [32]. At low concentrations, adenosine and a range of adenosine derivatives also inhibit mast cell secretion [33]. The most substantial support for an increase in cAMP as an intracellular mediator of mast cell secretory events comes from studies of IgE-dependent secretion in which an adenosine analogue 2′,5′-dideoxyadenosine, has been used to inhibit adenylate cyclase and thus prevent the rise in cAMP in stimulated mast cells and, in parallel, inhibit secretion [31–33]. Detracting from the cogency of these results is the observation that 2′,5′-dideoxyadenosine at a somewhat higher concentration inhibits secretion induced by the ionophore A23187 even though such secretion is not associated with any increase in cAMP [36]. Furthermore 9-beta-D-arabinofuranosyladenine, a generally more potent inhibitor of adenylate cyclase than 2′,5′-dideoxyadenosine, did not affect histamine secretion induced by anti-IgE or A23187, suggesting to Burt and Stanworth [36] that the inhibitory effect observed for 2′,5′-dideoxyadenosine might not be related to its effect on adenylate cyclase.

Table 1
Drugs that Modify Cell cAMP

Inhibitor	Secretagogue	Type of Cell	IC_{50} (mM)	Reference
Theophylline	Antigen	Rat Mast Cells	2.5	17
	Antigen	Rat Mast Cells	1*	19
	Antigen	Rat Mast Cells	10*	15
	Antigen	Rat Mast Cells	2	138
	Con A (10 ug/ml)	Rat Mast Cells	1.5*	140
	Chlortetracycline (50 uM)	Rat Mast Cells	1	109
	48/80 (1 ug/ml)	Rat Mast Cells	1*	10
	Antigen	Human Basophils	0.25*	1
	Antigen	Monkey Lung	0.07	4
	Antigen	Human Basophils	0.8*	A
Dibutyryl cyclic AMP	Antigen	Monkey Lung	0.1	4
	Con A (10 ug/ml)	Rat Mast Cells	0.4*	140
	Chlortetracycline (50 uM)	Rat Mast Cells	0.5	109
Isobutyl Methylxanthine	Antigen	Rat Mast Cells	0.2	17
	Chlortetracycline (50 uM)	Rat Mast Cells	0.2	109
	Con A (10 ug/ml)	Rat Mast Cells	0.09*	140
Prostaglandin E_1	Chlortetracycline (50 uM)	Rat Mast Cells	0.025	109
	Antigen	Human Lung	0.05	7
	Antigen	Human Basophils	0.007*	A
Prostaglandin E_2	Chlortetracycline (50 uM)	Rat Mast Cells	0.01	109
Prostaglandin $F_{2\alpha}$	Antigen	Human Lung	0.5	7
2’5’-dideoxyadenosine	Antigen	Rat Mast Cells	0.5*	35
	Antigen	Rat Mast Cells	0.25	34

* Indicates estimated value for IC_{50} (mM)

(A) J. A. Grant, L. Settle, E. B. Whorton and E. Dupree: J. Immunol. *117,* 450 (1976)

3 Drugs that alter phospholipid metabolism

3.1 Phospholipase A_2 and lysophospholipid production

Four aspects of phospholipid metabolism have been examined for their contributions to the secretory mechanism of mast cells. The first historically was the formation of lysophospholipids by the action of phospholipase A_2 on intrinsic membrane phospholipids. The initial observation was that exogenous phospholipase A_2 was capable of inducing histamine release [37]. Although doubt has been cast on the original experiment because of possible contamination of the enzyme preparation with another mast cell secretagogue, recent repetition with a homogeneous preparation of enzyme has confirmed the original finding [38]. The absence of reliable inhibitors of phospholipase A_2 has been a persistent problem in exploring the role of this enzyme. Neither bromophenacylbromide nor mepacrine can be considered a priori to be specific inhibitors of phospholipase A_2. Bromophenacyl bromide does not inhibit all phospholipase A_2' s [39, 40] and the rate of inhibition may be moderately slow [41]. Inhibition of histamine release [42, 44] is more likely to result from the reaction of the bromophenacyl bromide with sulfhydryl groups than a specific inhibition of phospholipase A_2. Quinacrine (mepacrine) is an even less likely candidate for a specific inhibitor of phospholipase A_2 [45]. Both compounds have been tested for their inhibitory action against phospholipase A_2 activity obtained from disrupted rat mast cells and found to produce no (bromophenacyl bromide) or limited (quinacrine) inhibition under circumstances mimicking those in which the two agents inhibit mast cell secretion [46]. Nemeth and Douglas have reported that quinacrine induces mast cell secretion [44]. N-7-nitro-2,1,3-benzoxadiazol-4-yl phosphatidylserine (NBD-PS) is a derivatized phospholipid that inhibits mast cell phospholipase A_2 in the same concentration range that it inhibits mast cell secretion [46, 47], but the coincidence of the concentration-response curves for this single compound is insufficient to more than implicate phospholipase A_2 in the secretory behavior of mast cells.

Since phospholipase A_2 releases lysophospholipids there is a possible role for the lysophospholipids in the histamine release reaction. A number of secretagogues, especially those dependent on IgE receptors require phosphatidylserine for optimal histamine release in vitro. No other natural phospholipid is as active as phosphatidylserine [47].

Lyso-PS is also effective [48] and recently lyso-phosphatidyl-2-carboxyethanol has been demonstrated to enhance concanavalin A-induced secretion [49]. A phospholipase A_2 that acts on phosphatidylserine has been described in rat mast cells [46]. Lysophospholipids are membrane fusogens, and it has been proposed that they may be required for the fusion of mast cell granules with each other and with the plasma membrane to elicit histamine release. Diacylglycerol is also fusogenic [50] and activates protein kinase C [51].

3.2 Arachidonic acid

Several pathways exist in membranes for the production of arachidonic acid: phospholipase A_2 action on phosphatidylcholine, release by diacylglycerol lipase from the 1,2, diacylglycerol formed by the action of phospholipase C on phosphatidylinositol 4,5 diphosphate [52], or hydrolysis of its substrate by a phosphatidic acid specific phospholipase A_2 [53]. Arachidonic acid in turn is metabolized by two major enzyme pathways: cyclooxygenase and lipoxygenase. Different cells preferentially generate specific metabolites by the cyclooxygenase or lipoxygenase pathways [44, 45], and the same cell type from different species may preferentially generate different metabolites. Furthermore agents which block one pathway of arachidonic acid may enhance products of the alternative pathway.

Indomethacin and aspirin, inhibitors of the cyclooxygenase system, weakly inhibit histamine release from rat peritoneal mast cells induced by a number of secretagogues [28, 56–58] but at considerably higher concentrations than those required to inhibit the formation of prostaglandins [28, 56]. Conversely, histamine release from human basophils is enhanced by indomethacin [59–61]. Dapsone (diaminodiphenylsulfone), another cyclooxygenase inhibitor, does not inhibit histamine at concentrations which inhibit the antigen-induced generation of PGD_2 from rat mast cells [62].

Eicosatetraynoic acid (ETYA), an analogue of arachidonic acid which inhibits both the lipoxygenase and cyclooxygenase pathways inhibits histamine release induced by antigen, anti-IgE, Con A, and the ionophore A23187 by both rat mast cells [28, 56, 63–65] and basophils [66]. The mechanism by which ETYA inhibits histamine release is unclear. It has been shown to inhibit the increased phosphate labelling of phosphatidylinositol, phosphatidylcholine and phosphatidic acid normally observed after stimulation by anti-IgE [66, 67]. However 48/80, although it stimulates phosphate incorporation in the manner of the other secretagogues, elicits mast cell secretion in the presence of ETYA [64]. It is unusual that an inhibitor of A23187-induced release does not similarly inhibit release stimulated by one of the polyamines. Consequently Nemeth and Douglas [64] suggest that ETYA may be acting directly on A23187 and not the mast cell. If this proposal is correct then the findings would be consistent with an action of ETYA on an early step common to IgE receptor aggregation perhaps mediated by lipoxygenase metabolites.

Arachidonic acid itself has opposing effects upon mediator release from mast cells. Preincubation of mast cells with arachidonic acid has an inhibitory effect on subsequent stimulation by anti-IgE, Con A, A23187 and 48/80 [28, 66–68]. However, if arachidonic acid is added to cells after stimulation by anti-IgE or Con A, an enhancement of release is observed [67]. Low concentrations of aspirin or indomethacin which inhibit prostaglandin formation but do not inhibit histamine release, were able to suppress the inhibition of release caused by preincubation with arachidonic acid, when the secretagogue was anti-IgE or Con A [28, 67] implicating cyclooxygenase metabolites in the effect of exogenous arachidonic acid.

3.3 Phospholipid methylation inhibition

Phospholipid methylation has received considerable attention in recent years. Evidence has been presented for methylation of phospholipids of rat mast cells [69], human mast cells [70] and RBL cells [71, 72], but a recent, seemingly careful study has failed to confirm the phospholipid methylation previously reported to be associated with stimulation [49]. Inhibitors of N-methylation, 3-deazaadenosine [70, 72–74], and 5′-deoxy-5′-isobutylthiol-3-deazaadenosine [75, 76] have been shown to block secretion; homocysteine which can reverse the inhibition of phospholipid methylation releases the block to secretion. Homocysteine thiolactone potentiates the inhibitory effect of the adenosine analogues on both N-methylation and histamine release. Mutants of the RBL-2H3 cell line which have lost the ability to transmethylate phospholipids are incapable of secreting in response to IgE receptor cross-linking; secretion can be restored by fusion between two mutants complementary with respect to transmethylation [77].

A23187-induced secretion proceeds normally in the methylation deficient basophils [77], and 3-deazaadenosine has been reported not to inhibit secretion by the ionophore [74]. In the case of human mast cells, 3-deazaadenosine inhibited both IgE receptor and A23187-dependent histamine release but at considerably different concentrations; almost 20 × the concentration of the inhibitor was required to achieve equivalent inhibition of A23187-stimulated secretion as required for IgE-receptor dependent secretion [78]. 5′-Deoxy-5′isobutylthiol-3-deaza-adenosine in contrast is equally effective in inhibiting secretion induced by anti-IgE, A23187, polymyxin B and 48/80 [76].

3.4 Phosphatidylinositol

Phosphatidylinositol metabolism has recently become a major focus in the pursuit of an explanation of cell secretory events. Early studies indicated an increased metabolism of phosphoinositol [79–81]. In RBL-2H3 cells after stimulation by IgE dependent secretagogues but not by A23187 or ionomycin, the metabolic events involving phosphoinositol described in other secreting cells have been found [82]. Thus phosphoinositol metabolism is implicated in initiating an early rise in cytoplasmic calcium but not in those events subsequent to the increase in cytoplasmic calcium [82].

Table 2
Drugs that Alter Phospholipid Metabolism

Inhibitor	Secretagogue	Type of Cell	IC_{50} (mM)	Reference
Eicosatetraynoic Acid	Antigen	Rat Mast Cells	0.2	B
	Antigen	Rat Mast Cells	0.06	67
	Antigen	Rat Mast Cells	0.065	56
	Con A (100 ug/ml)	Rat Mast Cells	0.050	56
	A23187 (0.57 uM)	Rat Mast Cells	0.017	56
	Antigen	Human Basophils	0.03*	66
	Antigen	Human Basophils	0.01*	60
Eicosatriynoic Acid	Antigen	Human Basophils	0.008*	60
**NBD–PS	Con A (100 ug/ml)	Rat Mast Cells	0.01*	47
Arachidonic Acid	48/80 (0.4 ug/ml)	Rat Mast Cells	0.01*	68
Nordihydro-guaiaretic Acid	Antigen	Rat Mast Cells	0.01	B
	Antigen	Human Basophils	0.007	66

* Indicates estimated value for IC_{50} (mM)
** N-(7-nitro-2,1,3-benzoxadiazol-4-yl) phosphatidylserine
(B) A. M. Magro and M. Brai: Immunology *49,* 1 (1983)

4 Calcium and calmodulin antagonists

4.1 The role of calcium

Calcium's role as a key element in the stimulus-secretion coupling and subsequent release of inflammatory mediators has been firmly established. Micro-injection of calcium directly into a mast cell produced a degranulation comparable to that induced by the antigen-antibody reaction [83], whereas magnesium or potassium were without effect. Phospholipid vesicles containing calcium when fused with mast cells induced degranulation and histamine release [84]; again, magnesium and potassium could not mimic this effect. Ionophores facilitate the passive diffusion of calcium into mast cells, producing a concentration-dependent release of histamine [85–88]. External radiolabelled calcium ^{45}[Ca] is taken up by mast cells when stimulated by a number of secretagogues [89–91]. Whereas it has generally been found that there is a correlation between the ^{45}[Ca] uptake and histamine release with uptake of ^{45}[Ca] preceding release [89, 91–93], significant discrepancies between uptake and release with respect to timing have been observed [94].
Calcium does not have to enter from an extracellular pool to initiate exocytosis; antigen [91, 95], compound 48/80 [96], A23187 [97], and polylysine [98] may under special circumstances all induce some suboptimal histamine release in the absence of extracellular calcium. Dextran is the only secretagogue reported to have an absolute requirement for extracellular calcium in order to initiate exocytosis [99]. When internal stores of calcium are depleted with EDTA or EGTA [86, 91, 97, 98, 100] release is prevented in the absence of extracellar calcium.

4.2 Calcium channel blockers

The effects of calcium channel blockers on mast cells is not clear; both lack of effect [101] and inhibition [102, 103] including blocking ^{45}[Ca] uptake [103] have been described. The high concentrations of nifedipine or verapamil required to inhibit rat mast cell secretion led to the suggestion that non-specific actions of these drugs might be involved [104]. Human basophils have also been variously reported to be resistant to the inhibitory effects of these agents when induced to secrete by antigen [103, 105] or to exhibit inhibition by nifedipine, verapamil and nimodipine when induced by a variety of secretagogues [106]. In this latter case, an increase in the external calcium concentration overcame the inhibition by nifedipine and verapamil but not nimodipine, suggesting that they may not act by simply blocking calcium channels. Consistent with the latter proposal is the observation that verapamil inhibits A23187-induced histamine release from mast cells [107], since calcium influx under the influence of the ionophore evades the need for calcium channels.

4.3 Cations

A number of ions like strontium and barium mimic the calcium effect on histamine release, but higher concentrations are required [108, 109]. Lanthanum and other lanthanide ions appear to act as calcium antagonists and inhibit histamine release induced by antigen, Con A, 48/80 and peptide 401 [110–112]. This inhibitory effect occurs at low concentrations (10^{-9} M), whereas higher concentrations (10^{-4} M) potentiate the antigen-induced release, as well as the spontaneous release [111]. Yet higher concentrations (10^{-3} M) again resulted in inhibition [112]. Lanthanum is believed to competitively displace calcium from its binding sites, including internal stores, as it binds with greater affinity [93, 110, 112].

4.4 Measurement of cytoplasmic calcium with quin-2

The development of quin-2 as a reagent for assessing the free cytoplasmic calcium ion concentration in intact cells provides a major new probe for the evaluation of mechanisms of action of inhibitors of histamine secretion. Not only does it allow study of rapid changes of calcium in real time, it also directly measures cytoplasmic calcium ion concentration, the pertinent variable, in contrast to measurements of calcium influx and efflux. Published studies of mast cells using quin-2 at this time are still limited. Beaven et al. [113] have reported on increases in free cytoplasmic calcium in stimulated RBL-2H3 cells, and White and co-workers [114] have presented preliminary findings for histamine release from rat mast cells induced by polyamine, IgE-dependent secretagogues and A23187. With IgE-primed RBL-2H3 cells the increase in cell calcium induced by antigen or Con A was prevented by depressing cell ATP or blocking calcium channels with lanthanum. In rat mast cells the antigen-induced rise in cytoplasmic calcium was blocked by 3-deazaadenosine with homocysteine thiolactone, cromoglycate, and theophylline. When the cells were induced to secrete with A23187, cromoglycate inhibited histamine release without diminishing the increase in cytoplasmic calcium concentration, so that cromoglycate can not simply act to block calcium increases.

4.5 Calmodulin antagonists

Calmodulin is an intracellular regulatory protein mediating a number of calcium-dependent enzyme pathways [91, 115]. It affects cyclic nucleotide levels by stimulating both phosphodiesterase and adenylate cyclase and regulates a variety of kinases. Several classes of drugs including the phenothiazines and naphthalenesulfonamides turn out to be potent inhibitors of calmodulin [116]. W-7, a member of the latter class, inhibited the histamine release induced by a variety of secretagogues from human basophils [117] and rat mast cells [118], whereas an analogue, W-5, which binds only weakly to calmodulin, was unable to inhibit histamine release with either group of cells. In addition, a number of phenothiazines have been shown to inhibit histamine release from rat mast cells induced by antigen or A23187 [90]. There was a positive correlation between calmodulin-inhibiting capacity tested with isolated calmodulin and inhibition of histamine release with A23187 but not with antigen, suggesting different actions upon different secretagogues [90].
The possibility that inhibition by the phenothiazines and other calmodulin-binding drugs is in fact mediated through an action on calmodulin can be questioned, since these agents demonstrate considerable inhibitory action on several enzymes directly, including protein kinase C [119, 120], and also exhibit direct effects on membrane stabilization paralleling their anti-calmodulin effects [121]. It is interesting to note, although probably not pertinent, that the classic mast cell secretagogue, 48/80, is a potent calmodulin antagonist [122, 123].

Table 3
Calcium and Calmodulin Antagonists

Inhibitor	Secretagogue	Type of Cell	IC_{50} (mM)	Reference
Verapamil	Antigen	Rat Mast Cells	0.38*	104
	Antigen	Rat Mast Cells	0.03*	107
	Con A (20 ug/ml)	Rat Mast Cells	0.34*	104
	A23187 (0.5 uM)	Rat Mast Cells	0.5*	104
	A23187 (0.56 uM)	Rat Mast Cells	0.03*	107
	Antigenic	Human Basophils	0.07*	106
Nifedipine	Antigen	Rat Mast Cells	0.002*	104
	48/80 (0.1 ug/ml)	Rat Mast Cells	0.002*	104
	A23187 (0.25 uM)	Rat Mast Cells	0.003*	104
	A23187 (0.5 uM)	Human Basophils	0.006*	106
	Antigen	Human Basophils	0.006*	106
Nimodipine	Antigen	Human Basophils	0.006*	106
Chlorpromazine	48/80 (0.2 ug/ml)	Rat Mast Cells	0.084	90
	A23187 (0.38 uM)	Rat Mast Cells	0.094	90
Imipramine	48/80 (0.2 ug/ml)	Rat Mast Cells	0.218	90
	A23187 (0.38 uM)	Rat Mast Cells	0.264	90
Trifluoperazine	48/80 (0.2 ug/ml)	Rat Mast Cells	0.047	90
	A23187 (0.38 uM)	Rat Mast Cells	0.046	90
**W-7	48/80 (0.2 ug/ml)	Rat Mast Cells	0.122	90
	A23187 (0.38 uM)	Rat Mast Cells	0.105	90
	Antigen	Human Basophils	0.07	117

* Indicates estimated value for IC_{50} (mM)
** N-(6-aminohexyl)-5-chloro-1-naphthalenesulphonamide

5 Cromoglycate and flavonoids

Because of its clinical effectiveness in treating asthma, cromoglycate has been studied extensively for its effect on mast cells. The related flavonoids have recently also attracted considerable attention [124–126]. The disodium salt of cromoglycate (DSCG) has been shown to inhibit histamine release from rat peritoneal mast cells induced by a variety of secretagogues including antigen [127, 128], polylysine [98], dextran [129, 130] and neurotensin [131, 132]. Release induced by calcium ionophores is considered to be resistant to cromoglycate inhibition [130, 131]. However, when a suboptimal concentration of ionophore is used with consequent low histamine release, DSCG has been found to effectively inhibit histamine release [109, 127, 133]. Quercetin, the best studied of the flavonoids is capable of inhibiting the histamine release induced by antigen and A23187 from rat mast cells [107,

124, 127] and from human basophils [105, 124], whereas DSCG had no effect upon antigen-induced histamine release from human basophils [134, 135].

Both drugs apparently can inhibit histamine release induced by a number of agents in the absence of calcium [127], whereas release by substance P could only be blocked by DSCG in the presence of calcium [136]. The inhibitory effect of DSCG declines rapidly during preincubation with the cells before addition of the secretory stimulus; quercetin shows no such reduction in inhibitory ability after preincubation [137]. Quercetin in fact probably inhibits only after cell activation has occurred [125].

Although DSCG does not inhibit antigen-induced release from human basophils, it does inhibit the release from human chopped lung fragments [134, 135]. Mucosal mast cells are also unaffected by DSCG [138, 139], but DSCG strongly inhibited Con A-induced histamine release from peritoneal mast cells of the rat, was less active against those of the hamster and totally ineffective against the mouse [119]. Quercetin, was essentially equiactive against cells of all three species [119].

A structurally related compound, doxantrazole (3-(5-tetrazolyl)-thioxanthone-10,10-dioxide), is an active inhibitor of histamine release from the peritoneal cells of mouse, rat and hamster [140] as well as rat mucosal mast cells [138].

DSCG, quercetin, and a number of other unrelated inhibitors of secretion promote the incorporation of radioactive phosphate into a protein of molecular weight 70000 daltons [141]. Fluorescently labelled beads bound to DSCG were able to bind to mast cells and to RBL-2H3 cells and inhibit antigen-induced histamine release [142], suggesting the existence of a specific DSCG receptor. A variant of the RBL-2H3 cell line deficient in binding sites for DSCG was found to be unable to release histamine when induced by antigen, although A23187 was still active, suggesting that the cromolyn binding protein (CBP) was in some fashion involved in antigen-induced secretion [143]. This was confirmed when the extracted, purified CBP was implanted into the variant cell line, with restoration of the capacity to release histamine on exposure to antigen [144]; this release was inhibitable by DSCG.

Table 4
Cromoglycate and Flavonoids

Inhibitor	Secretagogue	Type of Cell	IC_{50} (mM)	Reference
Disodium Cromoglycate	Antigen	Rat Mast Cells	0.01*	129
	Antigen	Rat Mast Cells	0.001	138
	Antigen	Rat Mast Cells	0.01*	127
	Con A (10 ug/ml)	Rat Mast Cells	0.1*	140
	Con A (20 ug/ml)	Rat Mast Cells	0.1*	127
	A23187 (0.15 uM)	Rat Mast Cells	0.25*	127
	A23187 (0.15 uM)	Rat Mast Cells	0.5*	133
	Chlortetracycline (50 uM)	Rat Mast Cells	0.05	109
	Substance P (10^{-5} M)	Rat Mast Cells	0.006*	136
Quercetin	Antigen	Rat Mast Cells	0.02*	137
	Antigen	Rat Mast Cells	0.01*	107
	Antigen	Rat Mast Cells	0.032	139
	Con A (10 ug/ml)	Rat Mast Cells	0.014*	140
	Con A (10 ug/ml)	Rat Mast Cells	0.020*	137
	A23187 (0.2 uM)	Rat Mast Cells	0.003*	133
	A23187 (0.15 uM)	Rat Mast Cells	0.002*	127
	A23187 (6 uM)	Rat Mast Cells	0.1*	137
	Chlortetracycline (50 uM)	Rat Mast Cells	0.02	109
	Antigen	Rat Mucosal Mast Cell	0.03	139
	Antigen	Human Basophils	0.009	124
	Antigen	Human Basophils	0.01*	105
	Con A (0.8 to) (3.1 ug/ml)	Human Basophils	0.004	124
	A23187 (0.96 uM)	Human Basophils	0.044	124
	f-Met Leu Phe (0.5 uM)	Human Basophils	0.024	124
Doxantrazole	Antigen	Rat Mast Cells	0.25	138
	Con A (10 ug/ml)	Rat Mast Cells	0.1*	140
	A23187 (0.2 uM)	Rat Mast Cells	0.75*	133
	Chlortetracycline (50 uM)	Rat Mast Cells	0.1	109
	Antigen	Rat Mucosal Mast Cell	0.1	138

* Indicates estimated value for IC_{50} (mM)

6 Protease inhibitors

Austen and Brocklehurst [145], and Keller and Beeger [146] showed that a range of inhibitors of proteases could interfere with histamine release by rat mast cells. An additional group of agents have been examined in T. Ishizaka's laboratory, and the original findings confirmed and extended to the RBL-2H3 cell [147, 148]. The range and variety of effective agents and several examples of considerable apparent functional specificity among structurally closely related compounds make a compelling argument for the involvement of an activatable ester protease in mast cell secretion. However, as yet no enzyme has actually been identified to support the evidence from the inhibitor studies. One peculiar feature of the inhibition of mast cell secretion by diisopropylfluorophosphate (DFP) is its effectiveness. The rate constants for inhibition of proteases such as chymotrypsin and trypsin are quite low and since secretion is a rapid reaction, it is surprising that DFP can so thoroughly inhibit the putatively critical protease in the initiation of secretion. Without being able to assay the enzyme and its inactivation, it is impossible to correlate the effects of the inhibitors on the enzyme with effects on secretion.

Table 5
Protease Inhibitors

Inhibitor	Secretagogue	Type of Cell	IC_{50} (mM)	Reference
Diisopropyl-fluoro-phosphate	Antigen	Guinea Pig Lung	1.0	C
	Anti-IgE	Human Basophils	5	148
	Anti-IgE	Rat Baso. Leuc.	1.0	E
	Anti-Ig	Rat Mast Cells	1.0	D
N-acetyltrypto-phan ethyl ester	Anti-Ig	Rat Mast Cells	2.5	D
Tryptophan ethyl ester	Antigen	Guinea Pig Lung	2.0	C

(C) E. L. Becker and K. F. Austen: J. Exp. Med. *120,* 491 (1964)
(D) R. Keller: Int. Arch. Allergy *23,* 315 (1963)
(E) J. D. Taurog, C. Fewtrell and E. L. Becker: J. Immunol. *122,* 2150 (1979)

7 Drugs acting on cytoskeletal elements

7.1 Microtubules

Inhibition of secretion by colchicine and other agents capable of depolymerizing microtubules has been widely used to invoke the participation of these cytoskeletal elements in the secretory process. Histamine release was one of the first secretory systems to be shown to be inhibitable by colchicine [149]. However, peculiarly high concentrations are required to achieve substantial inhibition, and in a study correlating the loss of microtubules with the inhibition of histamine release from rat mast cells, marked dissociation between the two effects of colchicine was found [150]. Subsequent quantitative measurements of mast cell microtubule content indicated that as much as 90% of the microtubules could be depolymerized with less than 10% inhibition of histamine release [151]. The use of vinblastine to study the relationship of microtubule integrity to cell secretion was not successful because of the requirement for concentrations that significantly reduced mast cell ATP. Nocodazole, a relatively new microtubule depolymerizing agent, has been shown to be without inhibitory effect on secretion by interphase RBL-2H3 cells at concentrations sufficient to prevent mitosis [152]. Cells actually trapped in mitosis did exhibit significant inhibition of histamine release.

Stimulation of histamine release by deuterium oxide [153, 154] has generally been assumed to support a role for microtubules in histamine secretion. This reasoning is based, as far as we can tell, solely on the fact that deuterium oxide has been shown to stabilize microtubules. The probability that deuterium oxide will significantly affect a range of protein functions in addition to microtubule stability not to mention the structure and function of other molecules in an aqueous environment seems not to have been sufficiently considered in interpreting the effects of deuterium oxide.

7.2 Actin

The microfilamentous component of the cytoskeleton is sensitive to disruption by cytochalasin B. Again in this instance although very high concentrations of cytochalasin B will partially inhibit histamine release from rat mast cells [155], the concentrations required appear to far exceed those required to deplete the mast cell of the preponderance of its functional microfilaments [156]. This conclusion has not yet been bolstered by quantitiative measurements of microfilament integrity.

Table 6
Drugs Acting on Cytoskeletal Elements

Inhibitor	Secretagogue	Type of Cell	IC_{50} (mM)	Reference
Colchicine	Polymyxin B (2 ug/ml)	Rat Mast Cells	0.5*	150
	Antigen	Human Lung	0.075*	F
	Antigen	Human Basophils	0.5	G

* Indicates estimated value for IC_{50} (mM)
(F) M. Kaliner: J. Clin. Invest. *60,* 951 (1977)
(G) D. Levy and J. A. Carlton: Proc. Soc. Exp. Biol. Med. *130,* 1333 (1969)

8 Steroid hormones

Glucocorticosteroids are widely utilized to relieve the symptoms of allergic hypersensitivity. The mechanisms of their anti-inflammatory action have remained elusive despite extensive investigations. Dexamethasone or hydrocortisone in micromolar concentrations when preincubated with mouse or rat peritoneal mast cells for 16–24 hours inhibited the release of histamine induced by antigenic stimulation [157, 158]. A23187 and 48/80-induced release was unaffected by this steroid treatment. However, if higher concentrations of steroids are used ($2–4 \times 10^{-4}$ M), inhibition of histamine release can be achieved after a 20-minute preincubation, whether induced by antigen, 48/80 [159, 160] or A23187 [159]. Furthermore, chronic administration of dexamethasone (100 µg/day intramuscularly for 4 days) to rats also inhibited mediator release from peritoneal cells when subsequently induced by either antigen or A23187 [161].

Antigen-induced histamine release from human lung fragments or purified human lung mast cells was not inhibited by a 24-hour preincubation with micromolar concentrations of dexamethasone [162]. However, incubation for the same period with dexamethasone or hydrocortisone does inhibit histamine release from human leucocytes at concentrations of 10^{-7}–10^{-8} and 10^{-6}–10^{-7} M, respectively [163]. This inhibitory effect begins after 2 hours and increases up to 24 hours [164].

An innovative proposal to explain the effects of steroids holds that they can combine with a cytoplasmic receptor protein which is translocated to the nucleus, leading to the synthesis of a phospholipase inhibitory protein (lipomodulin [165] or macrocortin [166]) that inhibits arachidonic acid release by inhibition of phospholipase A_2 or C [167].

When exogenous lipomodulin was added to mouse mast cells no inhibition of secretion was observed [157].
The inhibition of antigen, 48/80 or A23187-induced histamine release from rat peritoneal mast cells by high concentrations (10^{-4} M) of steroids can be reversed by the addition of glucose [159, 160] therefore another proposal for the action is the impairment of ATP production and/or intracellular calcium distribution by mitochondria [159].

Table 7
Steroid Hormones

Inhibitor	Secretagogue	Type of Cell	IC_{50} (mM)	Reference
Dexamethasone	Antigen	Human Basophils	1.0×10^{-5}*	60
(24 hours)	Antigen	Human Basophils	7.8×10^{-6}	164
Hydrocortisone	Antigen	Human Basophils	2.0×10^{-4}*	60
(24 hours)	Antigen	Human Basophils	1.5×10^{-4}	164
(20 mins)	Antigen	Rat Mast Cells	0.36	159
	A23187 (1 uM)	Rat Mast Cells	0.2	159
Prednisolone				
(24 hours)	Antigen	Human Basophils	2.2×10^{-5}	164
(20 mins)	Antigen	Rat Mast Cells	0.23	159
	A23187 (1 uM)	Rat Mast Cells	0.14	159
9α–Fluoro-cortisone	Antigen	Human Basophils	4.5×10^{-5}	164
(24 hours)	Antigen	Human Basophils	7.0×10^{-5}*	60

* Indicates estimated value for IC_{50} (mM)

9 Arylalkylamines

A substantial number of drugs with varying specific pharmacological action fall into the class of arylalkylamines. These include antihistamines, tranquilizers, anti-depressants, catecholamines, anti-malarials and local anesthetics. Read and Lagunoff (unpublished) have studied a series of these drugs and related compounds under standardized conditions with secretion stimulated by each of three secretagogues, 48/80, A23187, and concanavalin A with phosphatidylserine. The 50% inhibitory concentrations range from 0.02 mM to 20 mM. A few agents like benzalkonium chloride [168] and norepinephrine exhibited considerable degrees of specificity with respect to secretagogue, but the IC_{50} for most of the drugs did not vary more than a factor of 5 when compared with respect to their effects on secretion induced by each of the 3 secretagogues. The group of anti-psychotic agents whose members have been found to strongly bind to and inhibit calmodulin [169] contains some of the most active componds in inhibiting histamine release. The evidence that these compounds are directly effective in inhibiting protein kinase C [120] indicates the lack of specificity of calmodulin. Read and Lagunoff argue for the hypothesis that these amphipaths disorder membrane structure [170, 171] rather than bind to any selective functional receptor site. Quantitative measurements of membrane perturbations by the drugs and correlation with inhibitory effects are essential to substantiate this speculation.

Table 8
Arylalkylamines

Inhibitor	Secretagogue	Type of Cell	IC_{50} (mM)	Reference
Diphenhydra-mine	Con A (30 ug/ml)	Rat Mast Cells	0.27	H
	48/80 (1 ug/ml)	Rat Mast Cells	0.52	H
	A23187 (1.9 uM)	Rat Mast Cells	0.56	H
Propranolol	Con A (30 ug/ml)	Rat Mast Cells	0.15	H
	48/80 (1 ug/ml)	Rat Mast Cells	0.30	H
	A23187 (1.9 uM)	Rat Mast Cells	0.34	H
Tripelennamine	Con A (30 ug/ml)	Rat Mast Cells	0.20	H
	48/80 (1 ug/ml)	Rat Mast Cells	1.18	H
	A23187 (1.9 uM)	Rat Mast Cells	0.99	H
Amphetamine	Con A (30 ug/ml)	Rat Mast Cells	0.46	H
	48/80 (1 ug/ml)	Rat Mast Cells	0.79	H
	A23187 (1.9 uM)	Rat Mast Cells	2.86	H
Epinephrine	Con A (30 ug/ml)	Rat Mast Cells	3.00	H
	48/80 (1 ug/ml)	Rat Mast Cells	2.81	H
	A23187 (1.9 uM)	Rat Mast Cells	4.96	H
Ephedrine	Con A (30 ug/ml)	Rat Mast Cells	1.44	H
	48/80 (1 ug/ml)	Rat Mast Cells	6.75	H
	A23187 (1.9 uM)	Rat Mast Cells	5.99	H
Alprenolol	Con A (30 ug/ml)	Rat Mast Cells	0.67	H
	48/80 (1 ug/ml)	Rat Mast Cells	1.17	H
	A23187 (1.9 uM)	Rat Mast Cells	1.05	H

(H) G. W. Read: Unpublished

10 Sulfhydryl reagents

The first studies of the inhibitory actions of sulfhydryl reagents on mast cells were performed in the course of screening a variety of agents for their effects on mesenteric mast cell degranulation [172] and subsequently histamine secretion by isolated mast cells [173]. Ethacrynic acid was studied with basophils and mast cells by Magro [174, 175] who proposed that the inhibition of histamine release produced by the drug was related to the inhibitory effect on a Ca^{++}, Mg^{++}-dependent ATPase present in mast cells. Cytochalasin A was originally found to be far more effective than cytochalasin B [155] and later shown to be a sulfhydryl reagent [176]. Cytochalasin A and dithiodipyridine were equally effective in inhibiting rat mast cell histamine release by polymyxin B or A23187 [177]. N-ethyl maleimide and the fluorescent N-(7-dimethyl-amino-4-methylcoumarin) maleimide and p-mercuribenzene sulfonate (p-chloromercuribenzoate) were significantly more effective against secretion induced by polymyxin B than A23187 [177]. Based on the pattern of inhibition it was proposed that there are at least two SH groups in the mast cell critical for secretion; one not accessible to p-mercuribenzene sulfonate is required in the secretion elicited by polymyxin B or A23187, the other accessible to p-mercuribenzene sulfonate is required when the cells are stimulated by polymyxin B but not A23187. A different interpretation has been offered by Nemeth and Douglas [178]. Neither the proteins nor the functions involved in inhibition by the SH reagents have been identified beyond the possible involvement of a membrane ATPase.

Table 9
Sulfhydryl Reagents

Inhibitor	Secretagogue	Type of Cell	IC_{50} (mM)	Reference
Cytochalasin A	A23187 (1 uM)	Rat Mast Cells	1.7×10^{-4}	177
	Polymyxin B (1.4 uM)	Rat Mast Cells	2.4×10^{-4}	177
2,2'-Dithiodipyridine	A23187 (1 uM)	Rat Mast Cells	0.002	177
	Polymyxin B (1.4 uM)	Rat Mast Cells	0.001	177
N-Ethylmaleimide	A23187 (1 uM)	Rat Mast Cells	0.044	177
	Polymyxin B (1.4 uM)	Rat Mast Cells	0.003	177
	Antigen	Rat Mast Cells	0.5	J
Bromophenacyl bromide	Antigen	Human Basophils	0.001*	60
	Antigen	Human Basophils	0.02*	174
	A23187 (1.9 uM)	Human Basophils	0.02*	174
	A23187 (1.9 uM)	Rat Mast Cells	0.025*	178
Ethacrynic acid	Antigen	Rat Mast Cells	0.2*	175
	A23187 (1.9 uM)	Rat Mast Cells	0.1*	175

* Indicates estimated value for IC_{50} (mM)

(J) R. Keller: Monographs in Allergy 2, Elsevier, New York, (1966)

References

1 L. M. Lichtenstein and S. Margolis: Science **161,** 902 (1968).
2 E. S. K Assem and H. O. Schild: Int. Arch. Allergy appl. Immun. **40,** 576 (1971).
3 L. J. Loeffler, W. Lovenberg and A. Sjoerdsma: Biochem. Pharmac. **20,** 2287 (1971).
4 T. Ishizaka, K. Ishizaka, R. P. Orange and K. F. Austen: J. Immun. **106,** 1267 (1971).
5 R. P. Orange, M. A. Kaliner, P. J. Laraia and K. F. Austen: Fed. Proc. **30,** 1725 (1971).
6 H. R. Bourne, L. M. Lichtenstein and K. L Melmon: J. Immun. **108,** 695 (1972).
7 A. I. Tauber, M. Kaliner, D. J. Stechschulte and K. F. Austen: J. Immun. **111,** 27 (1973).
8 M. Kaliner and K. F. Austen: J. Immun. **112,** 664 (1974).
9 Y. Kimura, Y. Inoue and H. Honda: Immun. **26,** 983 (1974).
10 H. Hayashi, A. Ichikawa, T. Saito and K. Tomita: Biochem. Pharmac. **25,** 1907 (1976).
11 H. R. Bourne, K. L. Melmon and L. M. Lichtenstein: Science **173,** 745 (1971).
12 R. Kerouac, S. St-Pierre and F. Rioux: Res. Commun. Chem. Pathol. Pharmacol. **45,** 309 (1984).
13 L. M. Lichtenstein, C. S. Henney, H. R. Bourne and W. B. Greenough (III): J. Clin. Invest. **52,** 691 (1973).
14 B. B. Fredholm, I. Guschin, K. Elwin, G. Schwab and B. Uvnas: Biochem. Pharmac. **25,** 1583 (1976).
15 P. S. Skov, A. Geisler, R. Klysner and S. Norn: Experientia **33,** 965 (1977).
16 S. T. Holgate, R. A. Lewis, J. F. Maguire, L. J. Roberts (II), J. A. Oates and K. F. Austen: J. Immun. **125,** 1367 (1980).
17 A. Sydbom, B. Fredholm and B. Uvnas: Acta Physiol. Scand. **112,** 47 (1981).
18 J. A. Chabot, P. S. Riback and D. E. Cochrane: Life Sci. **28,** 1155 (1981).
19 A. Sydbom and B. B. Fredholm: Acta Physiol. Scand. **114,** 243 (1982).
20 P. E. Alm and G. D. Bloom: Life Sci. **30,** 213 (1982).
21 A. Khandwala, R. van Inwegen, S. Coutts, V. Dally-Meade, N. Jariwala, F. Huang, J. Musser, R. Brown, B. Loev, I. Weinry, S. Henney, H. R. Bourne and W. B. Greenough (III): J. Clin. Invest. **52,** 691 (1973).
22 E. Gillespie: Experientia **29,** 447 (1973).
23 T. J. Sullivan, K. L. Parker, S. A. Eisen and C. W. Parker: J. Immun. **114,** 1480 (1975).
24 A. R. Johnson, N. C. Moran and S. E. Mayer: J. Immun. **112,** 511 (1974).
25 D. S. Burt and D. R. Stanworth: Biochem. biophys. Acta **762,** 458 (1983).
26 T. J. Sullivan and C. W. Parker: Am. J. Path. **85,** 437 (1976).
27 T. J. Sullivan, K. L Parker, A. Kulczycki and C. W. Parker: J. Immun. **117,** 713 (1976).
28 R. A. Lewis, S. T. Holgate, L. J. Roberts (II), J. F. Macguire, J. A. Oates and K. F. Austen: J. Immun. **123,** 1663 (1979).
29 T. Ishizaka, F. Hirata, A. R. Sterk, K. Ishizaka and J. A. Axelrod: Proc. natn. Acad. Sci. **78,** 6812 (1981).
30 B. Fredholm and A. Sydbom: Agent Action **10,** 145 (1980).
31 D. L. Marquardt, C. W. Parker and T. J. Sullivan: J. Immun. **120,** 871 (1978).
32 G. Marone, S. R. Findlay and L. M. Lichtenstein: J. Immun. **123,** 1473 (1979).
33 M. Nishibori, K. Shimamura, H. Yokoyama, K. Tsutsumi and K. Saeki: Arch. int. Pharmacodyn. Thér. **265,** 17 (1983).

34 S. T. Holgate, R. A. Lewis and K. F. Austen: Proc. natn. Acad. Sci. **77,** 6800 (1980).
35 C. M. Winslow, R. A. Lewis and K. F. Austen: J. exp. Med. **154,** 1125 (1981).
36 D. S. Burt and D. R. Stanworth: Biochem. Pharmac. **32,** 2729 (1983).
37 B. Uvnas, B. Diamant and B. Hogberg: Arch. int. Pharmacodyn. Thér. **140,** 577 (1962).
38 E. Y. Chi, W. R. Henderson and S. J. Klebanoff: Lab. Invest. **47,** 579 (1982).
39 J. P. Durkin and W. T. Shier: Biochem. biophys. Acta **663,** 467 (1981).
40 R. L. Heinrikson, E. T. Krueger and P. S. Keim: J. Biol. Chem. **252,** 4913 (1977).
41 J. J. Volwerk, W. A. Pieterson and G. H. deHaas: Biochemistry **13,** 1446 (1974).
42 G. Marone: IRCS Med. Sci. **8,** 802 (1980).
43 A. McGivney et al.: Arch. Biochem. biophys. **212,** 572 (1981).
44 E. F. Nemeth and W. W. Douglas: Naunyn-Schmiedeberg's Arch. Pharmacol. **324,** 38 (1983).
45 M. Volpi, R. I. Sha'Afi and M. B. Feinstein: Molec. Pharmacol. **20,** 263 (1981).
46 T. W. Martin and D. Lagunoff: Biochemistry **21,** 1254 (1982).
47 T. W. Martin and D. Lagunoff: Science **204,** 631 (1979).
48 T. W. Martin and D. Lagunoff: Nature **279,** 250 (1979).
49 J. P. Moore, A. Johannson, T. R. Hesketh, G. A. Smith and J. C. Metcalfe: Biochem. J. **221,** 675 (1984).
50 D. A. Kennerly, T. J. Sullivan and C. W. Parker: J. Immun. **122,** 152 (1979).
51 Y. Nishizuka: Science **225,** 1365 (1984).
52 C. W. Parker: J. Allergy Clin. Immunol. **74,** 343 (1984).
53 M. M. Billah, E. G. Lapetina and P. Cuatrecasas: J. Biol. Chem. **256,** 5399 (1981).
54 E. A. Higgs and S. Moncada: Gen. Pharmac. **14,** 7 (1983).
55 H. Bisgaard: Allergy **39,** 413 (1984).
56 T. J. Sullivan and C. W. Parker: J. Immun. **122,** 431 (1979).
57 G. D. Champion, R. O. Day, J. E. Ray and D. N. Wade: Br. J. Pharmacol. **59,** 29 (1977).
58 G. P. Lewis and B. J. R. Whittle: Br. J. Pharmacol. **61,** 229 (1977).
59 G. Marone, A. Kagey-Sobotka and L. M. Lichtenstein: J. Immun. **123,** 1669 (1979).
60 S. P. Peters, R. P. Schleimer, G. Marone, A. Kagey- Sobotka, M. I. Siegel and L. M. Lichtenstein: Leukotrienes and other Lipoxygenase Products, p. 315. Raven Press, New York 1982.
61 S. P. Peters, A. Kagey-Sobotka, D. W. MacGlashan, Jr., and L. M. Lichtenstein: J. Pharmac. exp. Thér. **228,** 400 (1984).
62 T. Ruzicka, S. I. Wasserman, N. A. Soter and M. P. Printz: J. Allergy Clin. Immun. **72,** 365 (1983).
63 W. Konig, F. Pfeiffer and H. W. Kunau: Int. Arch. Allergy appl. Immun. **66,** 149 (1981).
64 E. F. Nemeth and W. W. Douglas: Eur. J. Pharmacol. **67,** 439 (1980).
65 A. Nakao, A. M. Buchanan and D. S. Potokar: Int. Arch. Allergy appl. Immun. **63,** 30 (1980).
66 W. Dorsch, J. Ring and H. Riepel: Int. Arch. Allergy appl. Immun. **73.** 274 (1984).
67 D. L. Marquardt, R. A. Nicolotti, D. A. Kennerly and T. J. Sullivan: J. Immun. **127,** 845 (1981).
68 J. Dainaka, A. Ichikawa, M. Okada and K. Tomita: Biochem. pharmac. **33,** 1653 (1984).
69 F. Hirata, J. Axelrod and F. T. Crews: Proc. natn. Acad. Sci. **76,** 4813 (1979).
70 T. Ishizaka, D. H. Conrad, F. S. Schulman, A. R. Sterk and K. Ishizaka: J.

Immun. **130,** 2357 (1983).
71 F. T. Crews, Y. Morita, A. McGivney, F. Hirata, R. P. Siraganian and J. Axelrod: J. Immun. **130,** 2357 (1983).
72 F. T. Crews, Y. Morita, F. Hirata, J. Axelrod and R. P. Siraganian: Biochem. biophys. Res. Commun. **93,** 42 (1980).
73 Y. Morita, P. K. Chiang and R. P. Siraganian: Biochem. Pharmac. **30,** 785 (1981).
74 Y. Morita and R. P. Siraganian: J. Immun. **127,** 1339 (1981).
75 T. Ishizaka, F. Hirata, K. Ishizaka and J. Axelrod: Proc. natn. Acad. Sci. **77,** 1903 (1980).
76 Y. Morita, R. P. Siraganian, C. K. Tang and P. K. Chiang: Biochem. Pharmac. **31,** 2111 (1982).
77 A. McGivney, F. T. Crews, F. Hirata, J. Axelrod and R. P. Siraganian: Proc. natn. Acad. Sci. **78,** 6176 (1981).
78 R. C. Benyon, M. K. Church and S. T. Holgate: Biochem. Pharmac. **33,** 2881 (1984).
79 D. A. Kennerly, T. J. Sullivan and C. W. Parker: J. Immun. **122,** 152 (1979).
80 D. A. Kennerly, C. J. Secosan, C. W. Parker and T. J. Sullivan: J. Immun. **123,** 1519 (1979).
81 S. Cockcroft and B. D. Gomperts: Biochem. J. **178,** 681 (1979).
82 M. A. Beaven, J. P. Moore, G. A. Smith, T. R. Hesketh and J. C. Metcalfe: J. Biol. Chem. **259,** 7137 (1984).
83 T. Kanno, D. E. Cochrane and W. W. Douglas: Can. J. Physiol. Pharmacol. **51,** 1001 (1973).
84 T. C. Theoharides and W. W. Douglas: Science **201,** 1143 (1978).
85 J. C. Foreman, J. L. Mongar and B. D. Gomperts: Nature **245,** 249 (1973).
86 D. E. Cochrane and W. W. Douglas: Proc. natn. Acad. Sci. **71,** 408 (1974).
87 R. P. Siraganian, A. Kulczycki, Jr., G. Mendoza and H. Metzger: J. Immun. **115,** 1599 (1975).
88 J. P. Bennett, S. Cockcroft and B. D. Gomperts: J. Physiol. **317,** 335 (1981).
89 J. C. Foreman, M. B. Hallett and J. L. Mongar: J. Physiol. **271,** 193 (1977).
90 W. W. Douglas and E. F. Nemeth: J. Physiol. **323,** 229 (1982).
91 F. L. Pearce: Progress in Medicinal Chemistry, p. 59. Elsevier Biomedical Press 1982.
92 T. Ishizaka and K. Ishizaka: Prog. Allergy **34,** 188 (1984).
93 N. R. Ranadive and N. Dhanani: Int. Arch. Allergy appl. Immun. **61,** 9 (1980).
94 N. Grossman and B. Diamant: Agent Action **8,** 338 (1978).
95 J. R. White and F. L. Pearce: Immunology **46,** 361 (1982).
96 T. Johansen: Life Sci. **27,** 369 (1980).
97 T. Johansen: Eur. J. Pharmacol. **62,** 329 (1980).
98 M. Ennis, F. L. Pearce and P. M. Weston: Br. J. Pharmacol. **70,** 329 (1980).
99 J. R. White and F. L. Pearce: Agent Action **11,** 4 (1981).
100 B. Diamant and S. A. Patkar: Int. Arch. Allergy appl. Immun. **49,** 183 (1975).
101 D. M. Ritchie, J. N. Sierchio, C. M. Bishop, C. C. Hedli, S. L. Levinson and R. J. Capetola: J. Pharmac. exp. Ther. **229,** 690 (1984).
102 Y. Tanizaki, H. Komagoe, M. Sudo, J. Ohtani, I. Kimura, K. Akagi and R. G. Townley: Acta Med. Okayama **37,** 207 (1983).
103 Y. Tanizaki, K. Akagi, K. N. Lee and R. G. Townley: Int. Arch. Allergy appl. Immun. **72,** 102 (1983).
104 M. Ennis, P. W. Ind, F. L. Pearce and C. T. Dollery: Agent Action **13,** 144 (1983).
105 E. Middleton, Jr., G. Drzewiecki and D. Triggle: Biochem. Pharmac. **30,** 2867 (1981).
106 C. Jensen, P. S. Skov and S. Norn: Allergy **38,** 233 (1983).

107 W. L. Parker and E. Martz: Agent Action **12,** 276 (1982).
108 J. C. Foreman and J. L. Mongar: J. Physiol. **224,** 753 (1972).
109 F. L. Pearce, K. E. Barrett and J. R. White: Agent Action **13,** 117 (1983).
110 J. C. Foreman and J. L. Mongar: Br. J. Pharmacol. **48,** 572 (1973).
111 J. C. Foreman and J. L. Mongar: Nature New Biol. **240,** 255 (1972).
112 F. L. Pearce and J. R. White: Br. J. Pharmacol. **72,** *341 (1981).*
113 M. A. Beaven, J. Rogers, J. P. Moore, T. R. Hesketh, G. A. Smith and J. C. Metcalfe: J. Biol. Chem. **259,** 7129 (1984).
114 J. R. White, T. Ishizaka, K. Ishizaka and R. I. Sha'Afi: Proc. natn. Acad. Sci. **81,** 3978 (1984).
115 A. R. Means, J. S. Tash and J. G. Chafouleas: Physiol. Rev. **62,** 1 (1982).
116 T. Tanaka, T. Ohmura and H. Hidaka: Pharmacology **26,** 249 (1983).
117 G. Marone, M. Columbo, S. Poto, P. Bianco and M. Condorelli: IRCS **11,** 97 (1983).
118 T. Suzuki, K. Ohishi and M. Uchida: Gen. Pharmac. **14,** 273 (1983).
119 B. D. Roufogalis, A. E. V. M. Minocherhomjee and A. Al-Jobore: Can. J. Biochem. Cell Biol. **61,** 927 (1983).
120 Y. Nishizuka: Phil. Trans. R. Soc. Lond. **302,** 101 (1983).
121 Y. Landry, M. Amellal and M. Ruckstuhl: Biochem. Pharmacol. **30,** 2031 (1981).
122 K. Gietzen: Biochem. J. **216,** 611 (1983).
123 K. Gietzen, E. Sanchez-Delgado and H. Bader: IRCS **11,** 12 (1983).
124 E. Middleton, Jr., and G. Drzewiecki: Biochem. Pharmacol. **33,** 3333 (1984).
125 E. Middleton, Jr.: Trend Pharmacol. Sci. **5,** 335 (1984).
126 J. C. Foreman: J. Allergy Clin. Immun. **74,** 769 (1984).
127 M. Ennis, A. Truneh, J. R. White and F. L. Pearce: Nature **289,** 186 (1981).
128 F. L. Pearce: Hospital Update **10,** 25 (1984).
129 L. G. Garland and J. L. Mongar: Br. J. Pharmac. **50,** 137 (1974).
130 J. C. Foreman, J. L. Mongar, B. D. Gomperts and L. G. Garland: Biochem. Pharmacol. **24,** 538 (1975).
131 M. Kurose and K. Saeki: Eur. J. Pharmacol. **76,** 129 (1981).
132 R. Carraway, D. E. Cochrane, J. B. Lansman, S. E. Leeman, B. M. Paterson and H. J. Welch: J. Physiol. **323,** 403 (1982).
133 F. L. Pearce and A. Truneh: Agent Action **11,** 44 (1981).
134 M. Radermecker: Respiration **41,** 45 (1981).
135 B. A. Spicer, J. W. Ross, G. D. Clarke, E. J. Harling, P. A. Hassall, H. Smith and J. F. Taylor: Agent Action **13,** 301 (1983).
136 M. Nishibori and K. Saeki: Jap. J. Pharmac. **33,** 1255 (1983).
137 C. M. S. Fewtrell and G. D. Gomperts: Biochem. biophys. Acta **469,** 52 (1977).
138 F. L. Pearce, A. D. Befus, J. Gauldie and J. Bienenstock: J. Immun. **128,** 2481 (1982).
139 F. L. Pearce, A. D. Befus and J. Bienenstock: J. Allergy Clin. Immun. **73,** 819 (1984).
140 K. B. P. Leung, K. E. Barrett and F. L. Pearce: Agent Action **14,** 461 (1984).
141 W. Sieghart, T. C. Theoharides, W. W. Douglas and P. Greengard: Biochem. Pharmac. **30** (1981).
142 N. Mazurek, G. Berger and I. Pecht: Nature **286,** 722 (1980).
143 N. Mazurek, P. Bashkin, A. Petrank and I. Pecht: Nature **303,** 528 (1983).
144 N. Mazurek, P. Bashkin, A. Loyter and I. Pecht: Proc. natn. Acad. Sci. **80,** 6014 (1983).
145 K. F. Austen and W. E. Brocklehurst: J. exp. Med. **113,** 521 (1961).
146 R. Keller and I. Beeger: Int. Arch. Allergy **22,** 31 (1963).
147 T. Ishizaka: J. Allergy Clin. Immun. **67,** 90 (1981).
148 T. Ishizaka and K. Ishizaka: Prog. Allergy **34,** 188 (1984).

149 E. Gillespie, R. J. Levine and S. E. Malawista: J. Pharmac. exp. Ther. **164,** 158 (1968).
150 D. Lagunoff and E. Y. Chi: J. Cell Biol. **71,** 182 (1976).
151 D. Lagunoff and E. Y. Chi: The Cell Biology of Inflammation, p. 217, Elsevier, Amsterdam 1980.
152 T. R. Hesketh, M. A. Beaven, J. Rogers, B. Burke and G. B. Warren: J. Cell Biol. **98,** 2250 (1984).
153 E. Gillespie and L. M. Lichtenstein: J. Clin. Invest. **51,** 2941 (1972).
154 E. Gillespie and L. M. Lichtenstein: Proc. Soc. exp. Biol. Med. **140,** 1228 (1972).
155 T. S. C. Orr, D. E. Hall and A. C. Allison: Nature **236,** 350 (1972).
156 D. Lagunoff and E. Y. Chi, unpublished experiments.
157 M. Daeron, A. R. Sterk, F. Hirata and T. Ishizaka: J. Immun. **129,** 1212 (1982)
158 A. S. Heiman and F. T. Crews: J. Pharmac. exp. Ther. **230,** 175 (1984).
159 N. Grosman and S. M. Jensen: Agent Action **14,** 21 (1984).
160 G. P. Lewis and B. J. R. Whittle: Br. J. Pharmacol. **61,** 229 (1977).
161 D. L. Marquardt and S. I. Wasserman: J. Immun. **131,** 934 (1983).
162 R. P. Schleimer, E. S. Schulman, D. W. MacGlashan, jr., S. P. Peters, E. C. Hayes, G. K. Adams (III), L. M. Lichtenstein and N. F. Adkinson, jr.: J. Clin Invest. **71,** 1830 (1983).
163 H. Bergstrand, A. Bjornsson, B. Lundquist, A. Nilsson and R. Brattsand: Allergy **39,** 217 (1984).
164 R. P. Schleimer, L. M. Lichtenstein and E. Gillespie: Nature **292,** 454 (1981).
165 F. Hirata, E. Schiffmann, K. Venkatasubramanian, D. Salomon and J. Axelrod: Proc. natn. Acad. Sci. **77,** 2533 (1980).
166 R. J. Flower and G. J. Blackwell: Nature **278,** 456 (1979).
167 F. Hirata: J. Biol. Chem. **256,** 7730 (1981).
168 G. W. Read and E. F. Kiefer: J. Pharmac. exp. Ther. **211,** 711 (1979).
169 B. Weiss, W. C. Prozialeck and T. L. Wallace: Biochem. Pharmac. **31,** 2217 (1982).
170 M. P. Sheetz and S. J. Singer: Proc. natn. Acad. Sci. **71,** 4457 (1974).
171 M. R. Lieber, Y. Lange, R. S. Weinstein and T. L. Steck: J. Biol. Chem. **259,** 9225 (1984).
172 B. Hogberg and B. Uvnas: Acta Physiol. Scand. **41,** 345 (1957).
173 B. Uvnas and I.-L. Thon: Exp. Cell Res. **23,** 45 (1961).
174 A. M. Magro: Clin. exp. Immun. **30,** 160 (1977).
175 A. M. Magro: Clin. exp. Immun. **29,** 436 (1977).
176 D. Lagunoff: Biochem. biophys. Res. Commun. **73,** 727 (1976).
177 D. Lagunoff and H. Wan: Biochem. Pharmac. **28,** 1765 (1979).
178 E. F. Nemeth and W. W. Douglas: Naunyn-Schmiedeberg's Arch. Pharmacol. **302,** 153, 1978.

Dopamine agonists: Structure-activity relationships

By Joseph G. Cannon
The University of Iowa, Iowa City, Iowa 52242

Introduction

In the several years since Blaschko [1] first suggested that dopamine might be a neurotransmitter substance, many chemical compounds have been found to possess dopamine-like actions. This rcview surveys classes of structures for which putative dopaminergic agonism has been reported, and cites structure-activity correlations. Some aspects of metabolism of the agents are addressed, which seem to have relevance to structure-activity considerations and to strategy of drug design.

It is well established and widely accepted that dopamine functions as a neurotransmitter in the central nervous system. Lackovic and Neff [2] and Relja and Neff [3] have cited considerable evidence that dopamine is also utilized physiologically as a neurotransmitter and/or a cotransmitter by many peripheral nerves.

Although dopamine possesses no center of asymmetry, the in vivo dopamine receptors seem to possess a high degree of chirality, and they clearly discriminate between enantiomers of chiral dopaminergic agonists. In this respect, the dopaminergic nervous system resembles the cholinergic receptor system. Dopamine receptors also recognize and discriminate between geometric isomers of certain dopamine analogs.

A serious problem in dopamine structure-activity studies arises from the large variety of animal models and in vivo and in vitro pharmacological assays utilized to assess dopaminergic effects. Comparable test data in the same animal species using the same biological end point and the same criteria for assessment of potency/activity are not available for many agents. Thus, it is frequently not possible to make valid comparisons of actions and potencies among compounds described in the literature, and a *caveat* must be expressed with respect to (at least) the quantitative validity of many attempted structure-activity correlations.

Considerable biochemical and physiological evidence indicates that multiple classes of dopamine receptors are present in vertebrate and invertebrate systems. Leff and Creese [4] pointed out that classification of dopamine-sensitive receptors must integrate pharmacological criteria, e. g., how different drugs interact with each discriminating response or putative structural subtype of receptor mediating response. Creese [5] and Offermeier and Van Rooyen [6] have described the difficulties attendant to subcategorization of dopamine receptors, and they have presented possible strategies for bringing order to this chaotic scientific area. Cavero et al. [7] suggested that *peripheral* dopamine receptors be designated as DA_1 (which are located postjunctionally in renal and mesenteric arterial beds. Stimulation produces direct smooth muscle relaxation); and DA_2 (which are located on postganglionic sympathetic neurons. Stimulation produces a reduction of neural release of norepinephrine). Leff and Creese [4] have concluded that the preponderance of pharmacological data can be encompassed by a model defined by Kebabian and Calne [8]. These workers classified dopamine receptors (central and peripheral) on the basis of regulation of activity of adenylate cyclase: *D-1* receptors mediate stimulation of adenylate cyclase activity (resulting in increased production of adenosine 3′,5′-monophosphate, 'cyclic AMP'), while *D-2* receptors are either unassociated with this enzyme, or mediate its inhibition. Kebabian et al. [9] provided further experimental support for their *D-1, D-2* classification, and Stoof and Kebabian [10] have presented a review of the biochemistry, physiology, and pharmacology of *D-1* and *D-2* receptors. Wreggett and Seeman [11] made a compelling argument for subdivision of *D-2* receptors into high and low affinity (for dopamine) ones, and they also proposed usage of *D-3* and *D-4* receptor terminology. Repeated attempts to classify dopamine receptors solely on the basis of structural features of agonist molecules have been unsuccessful.

2 β-Phenethylamines and Related Systems

2.1 Stereochemical and physical chemical aspects

Quantum mechanical calculations led to the conclusion [12,13] that there are three favored conformations for the dopamine molecule: *trans* (antiperiplanar), illustrated by the Newman projection I, with the plane of the ring perpendicular to the plane of the side chain; and two unfolded (*gauche*, synclinal) forms, Newman projections II and III.

An X-ray crystallographic study [41] of dopamine hydrochloride revealed that the molecule exists in the conformation I in the crystalline state. Rotman et al. [14], proposed that a *gauche* form of dopamine is permitted, if not preferred, in the reuptake process in nervous tissue. However, Granot [15] concluded that when dopamine interacts with ATP (a component of some dopamine receptors) in aqueous solution,

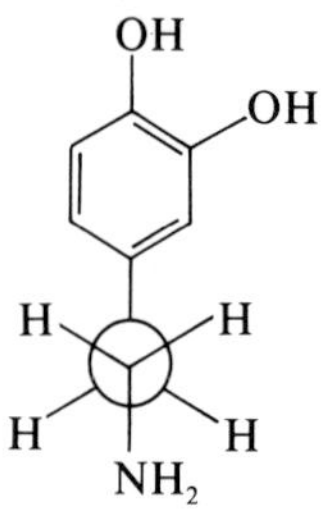

I

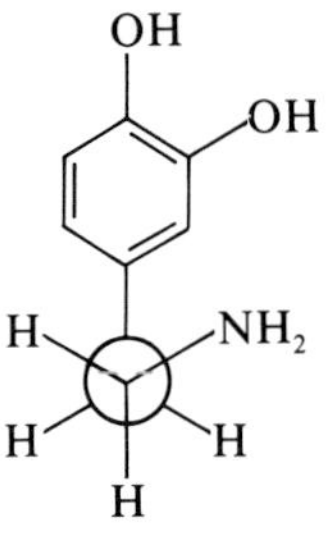

II

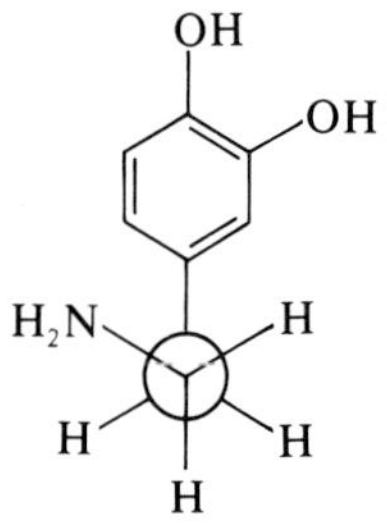

III

there is a significant preference for a *trans* conformation of the dopamine molecule. The dopamine-ATP complex is stabilized, inter alia, by hydrogen bonding between the catechol hydroxyls and purine nitrogens, and by electrostatic interaction between the protonated ammonium group of the dopamine and a negative phosphate group. Cannon [16,17] defined α- and β-conformers of dopamine (structures IV and V), in which the catechol ring is coplanar with the plane of the ethylamine side chain, and he proposed that these conformations are significant in agonist-receptor interactions.

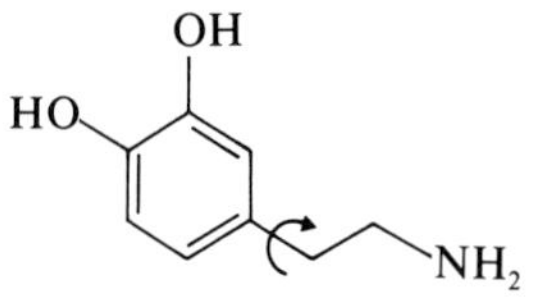

IV α-conformer

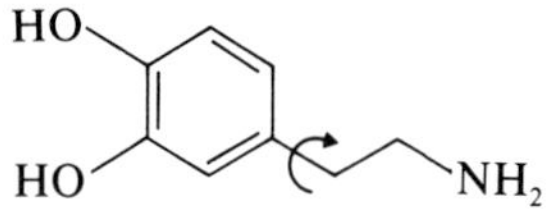

V β-conformer

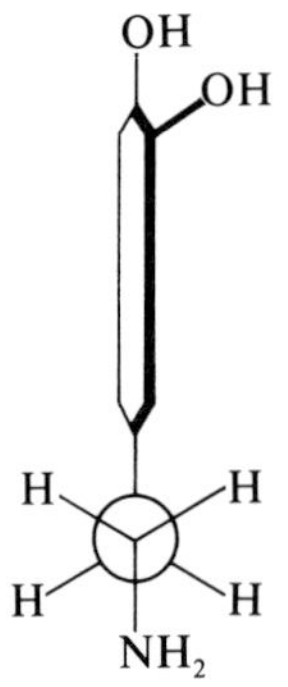

VI

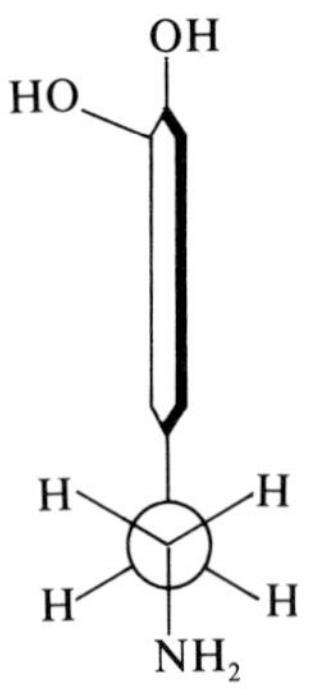

VII

In the α-conformer (Newman projection VI), the *m*-OH is projected over the ethylamine side chain; the nitrogen-to-*p*-OH distance is 7.8 A, and the nitrogen-to-*m*-OH distance is approximately 6.2 A. In the β-conformer (Newman projection VII), the *m*-OH is directed away from the ethylamine side chain; the nitrogen-to-*p*-OH distance is the same as in the α-conformer (7.8 A), but the nitrogen-to-*m*-OH distance is greater (7.3 A). Other structural features of a dopaminergic agonist may modify the ability of the molecule to interact with dopamine receptors or with specific populations of dopamine receptors, or these structural features may destroy agonist effects. However, the achievement of a conformation that corresponds to or approximates the α- or the β-conformation of dopamine seems to be a sine qua non for dopamine agonist activity in many types of chemical structures.

Katz et al. [13] concluded that the *meta*-phenolic monoanion of dopamine free base is more stable than the *para*-monoanion. 2- and 5-Fluorodopamines (structure VIII) are equipotent to dopamine in a dog renal vascular assay, whereas the 6-fluoro isomer is four times less potent than dopamine [18].

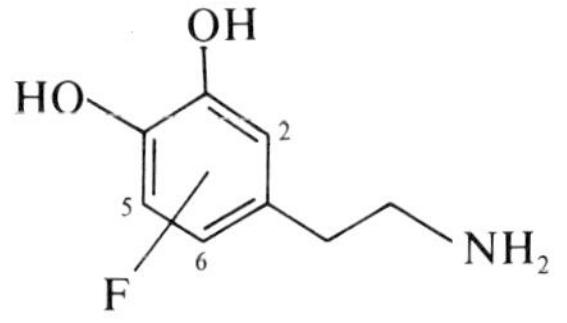

VIII

All of the monofluorodopamines were at least equipotent to dopamine in assays for α- and β-adrenergic activity. In binding studies on rat striatal tissue, the three monofluorodopamines are equipotent in displacement of ^{3}H-spiroperidol, but the 2- and 6- fluoro systems are less potent than the 5-fluoro- or dopamine in displacing ^{3}H-apomorphine [19]. Kirk [20] suggested that a role of fluorine in these molecules is to change the acidity of the phenolic groups, which may modify the character and/or strength of the agonist-receptor interaction. Fluorination of the ring of dopa influences the site of methylation by catechol-O-methyltransferase, due to alteration of the acid strength of the phenolic OH [21].

Armstrong and Barlow [46], on the basis of pK_a studies in a short series of β-phenethylamine derivatives, concluded that the active form of dopamine at its in vivo receptors may be the uncharged molecule rather than the ammonium cation. However, Anderson et al. [47] found that the S,S-dimethylsulfonium analog IX of dopamine demonstrates pharmacological properties consistent with its being a dopaminergic agonist, albeit of relatively low potency.

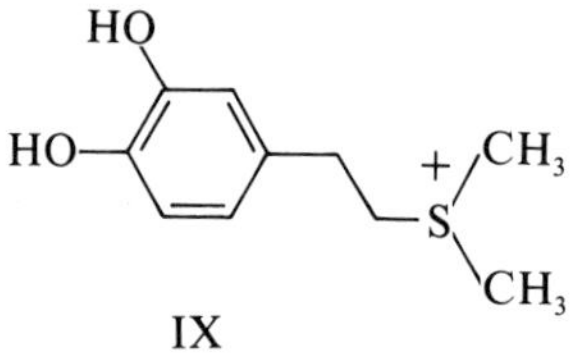

IX

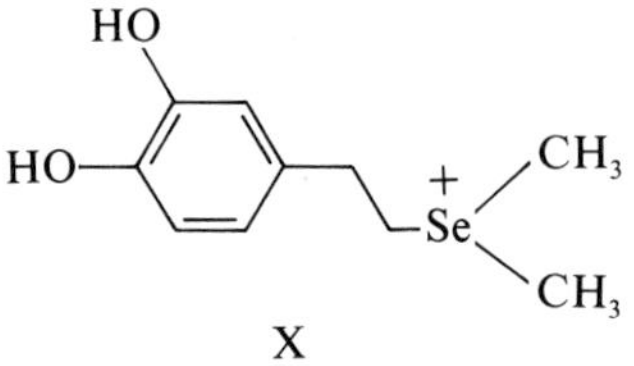

X

The selenonium analog X, like dopamine, is taken up by heart and adrenal medulla tissue [48]. However, no studies of pharmacological actions of this selenonium congener were reported. From these preceding results, it might be inferred that the dopamine receptor(s) involved accept a positively charged ion as a part of the dopamine molecule. The activity of the sulfonium derivatives is intriguing, in that the literature has not revealed any example of a quaternary ammonium derivative which displays dopaminergic agonist effects. The N-trimethyl quaternary derivative of dopamine is inactive in cyclic AMP production in rat striatal homogenate [40]. In addition, Hamada et al. [49] observed that there are significant differences in the structure-activity relationships of sulfonium and amine analogs of dopamine, and that the permanent charge on the sulfonium moiety seems to minimize the otherwise important role played by the other structural features. Camerman and Camerman [50] have pointed out that the spatial orientation of the nitrogen lone pair of electrons on dopamine and/or dopaminergic agonist molecules has not been clearly defined and related to phenomena incident to agonist-receptor interactions. The role of atomic inversion [51] about the basic nitrogen in agonist-receptor interactions seems as yet unclear.

2.2 Structural modifications of the dopamine molecule

Laduron and co-workers [23,24] demonstrated in vitro N-methylation of dopamine by a rat brain enzyme preparation. 5-Methyltetrahydrofolic acid was the source of methyl. The product of this reaction, N-methyldopamine (epinine: XI, R = CH_3), is a potent dopaminergic agonist in a variety of assays [25-27].

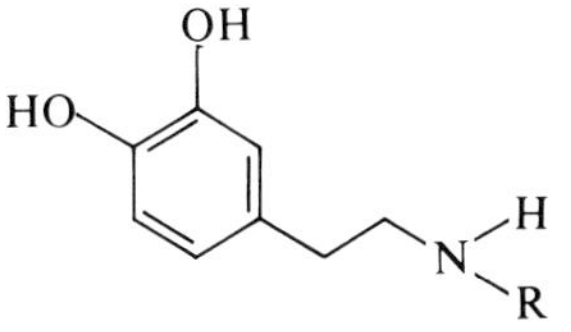

XI

It still seems uncertain whether epinine is an endogenous neurotransmitter in mammals. Other N-alkyl secondary amine homologs of dopamine (XI: R = C_2H_5; *n*-C_3H_7; 2-C_3H_7) showed extremely weak activity compared to dopamine and epinine in a rat caudate nucleus adenylate cyclase assay [25].
Studies [28-39] of N,N-disubstituted dopamines reveal that the combinations of alkyl groups in XII a-d impart unusually high central and/or peripheral dopamine agonist effects.

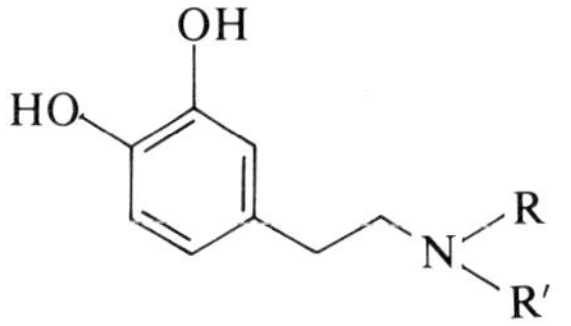

XII

XIIa R = R′ = CH_3
b R = R′ = C_2H_5
c R = R′ = n-C_3H_7
d R = n-C_3H_7; R′ = n-C_4H_9

N,N-Dialkyldopamines frequently show greater selectivity of pharmacological effect than dopamine itself, or than epinine (XI: R = CH_3), or apomorphine XXVIII a prototypical dopaminergic agonist [31,40]. N,N-Dimethyldopamine XII a has a marked effect in the cat cardioaccelerator nerve assay [31], but it elicits no emetic effect in dogs [31] and it is inactive in production of hyperactivity or stereotypic behavioral response by direct intracerebral administration in rats [26]. Remarkably, whereas N,N-di-*n*-propyldopamine XII c [31] and N-*n*-propyl, N-*n*-butyldopamine XII d [29] are potent dopaminergic agonists, N,N-di-*n*-butyldopamine (XII: R = R′ = *n*-C_4H_9) is inert [31]. It has been suggested [17,42] that enhanced dopaminergic effects frequently conferred by N-*n*-propyl group(s) are not related merely to the

effect of the alkyl chain on solubility or partitioning phenomena, but rather that certain dopamine receptors have a positive affinity for the N-*n*-propyl group. Longer chains (e. g., *n*-butyl) do not fit the receptor subsite; shorter chains fit, but not optimally. Repeatedly, it will be seen that the most active members of a series of dopaminergic agonists bear one or more *n*-propyl groups on the nitrogen. Goldman and Kebabian [43] have termed this the 'N-propyl phenomenon'. Bradbury et al. [44], found that in several chemical types of dopaminergic agonist molecules, the N-*n*-propyl substitution generally conferred the greatest selectivity of motor inhibitory action in mice.
Attempts have been made [45] to relate N-alkyl substituents of β-phenethylamines with partition coefficients, but no attempt was made to seek biological correlations. It was concluded that diastereoisomerism had no detectable effect on partitioning behavior in the series of compounds studied.
Incorporation of the nitrogen of dopamine into a piperidine ring (structure XIII) resulted in complete loss of dopaminergic effect [30].

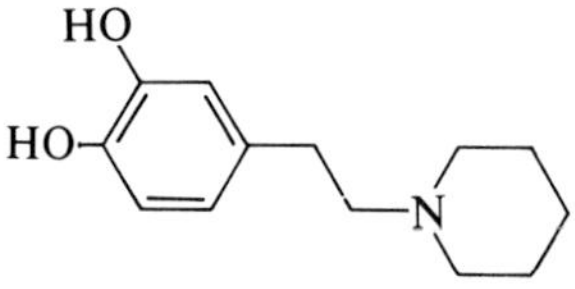

XIII

Geissler [52] cited literature evidence that *m*-tyramine XIVa is a central dopaminergic agonist.

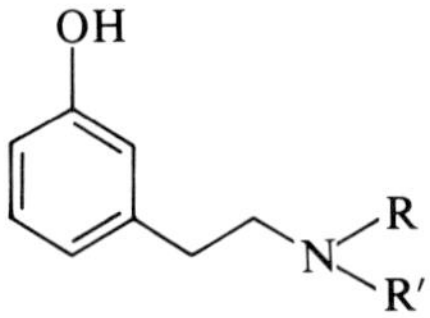

XIVa R = R' = H
b R = R' = n-C_3H_7

However, both *m*-tyramine and *p*-tyramine are inert in hyperactivity/stereotypy assays in rodents [26] and in stimulation of cyclic AMP production [40]. N,N-di-*n*-Propyl *m*-tyramine XIVb is a relatively selective central and peripheral dopaminergic agonist [52,53]. N,N-di-*n*-propyl *p*-tyramine has apparently not been evaluated for dopamine-like effects. α-Methyl *m*-tyramine XV has been described as a 'false dopaminergic transmitter' [54].

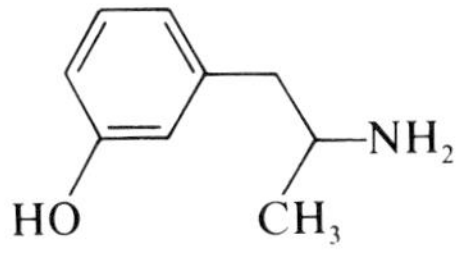

XV

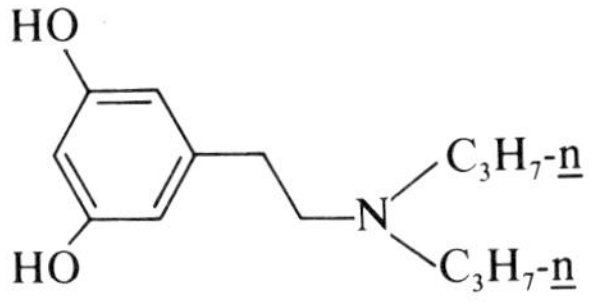

XVI

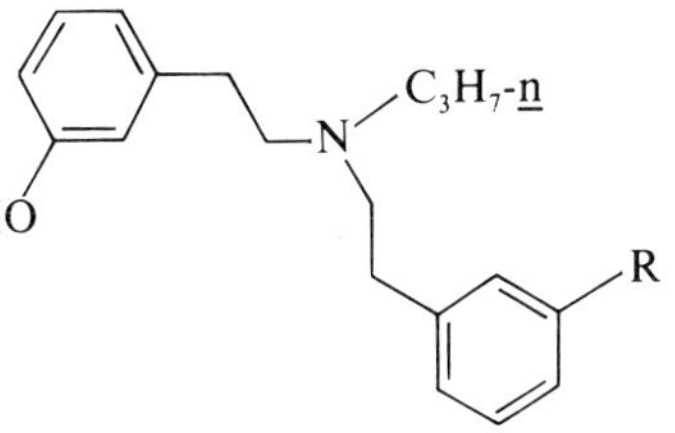

XVIIa R = H
b R = OH

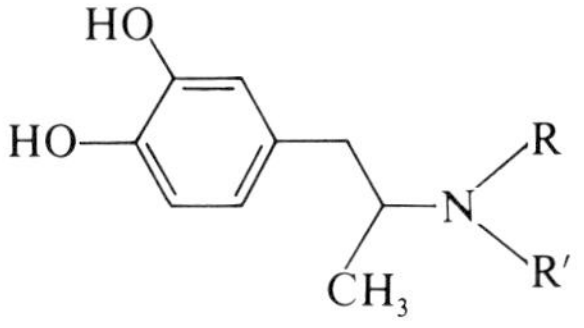

XVIII

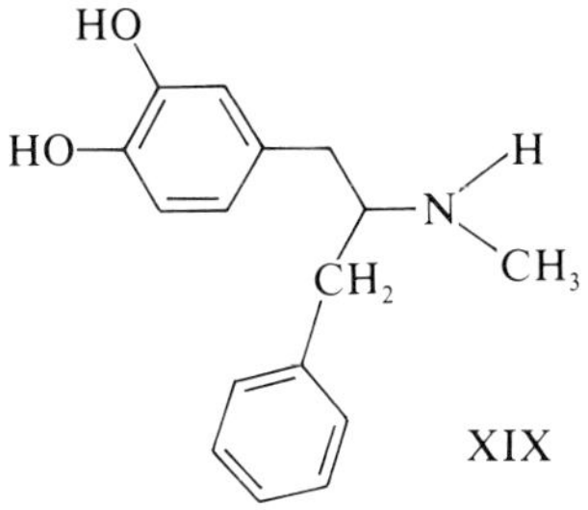

XIX

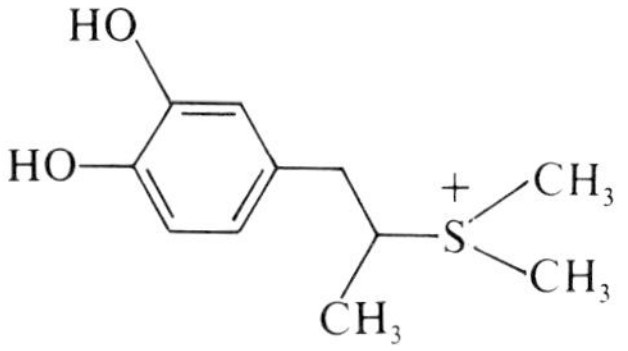

XX

An important role has been suggested [16,17] for the 'meta' OH of dopamine in agonist-receptor interactions, and it is unexpected that the 'di-meta'-OH system XVI is inert in a variety of assays for central dopaminergic effects [53]. Two compounds (XVIIa, b), representing a variation of the *m*-tyramine structure, exhibited a variety of dopamine agonist effects [53,55-57]. The added β-phenethylamine moiety does not bestow unique potency nor activity, but the compounds have a longer duration of action in a rodent stereotypy assay [53] and in affecting prolactin release from the anterior pituitary [57]. Stoof and Kebabian [10] have classed XVIIa and b as selective *D-2* agonists.

Introduction of a methyl group into the α-position of the dopamine side chain (structure XVIII: R,R′ = combinations of H and alkyl) greatly decreases dopaminergic agonist effects in some assays [40,58] and abolishes effects in others [59]. This loss of activity was explained [59] by a steric effect of the α-methyl group in forcing the catechol ring to lie perpendicular to the plane of the ethylamine side chain, a deviation of the dopamine moiety from either the α- or the β-rotameric disposition. However, Erhardt [60] has rejected this explanation and has suggested that the molecular bulk added by the α-methyl group interferes with receptor interaction because a steric protrusion resides above the general plane of the receptor in a proposed [61] topographical model. An α-benzyl epinine XIX has been reported [62] to display 'a clear dopaminomimetic character', but details of pharmacologic effects were not provided.

The S-enantiomer of α-methyldopamine (XVIII: R = R′ = H) is equipotent to dopamine in production of a series of cardiovascular effects, and is approximately twice as potent as the R-enantiomer [63]. However, these pharmacological effects probably do not reflect direct actions upon the dopaminergic nervous system. The sulfonium analog XX of α-methyl N,N-di-methyldopamine was reported to demonstrate dopamine agonist effects of the same magnitude as those produced by the dopamine isostere IX and the aminotetralin isostere LXXI (vide infra) [49]. These similarities contrast with the marked potency/activity differences shown by the nitrogen analogs dopamine, XVIII and XLVI (R = R′ = CH_3, 'TL-99').

Attempts to improve dopamine's relatively poor, capricious absorption and distribution through the body and its failure of passage through the blood brain barrier have involved latentiation of the molecule by esterification of the phenolic groups, to increase the overall lipophilicity of the molecule.

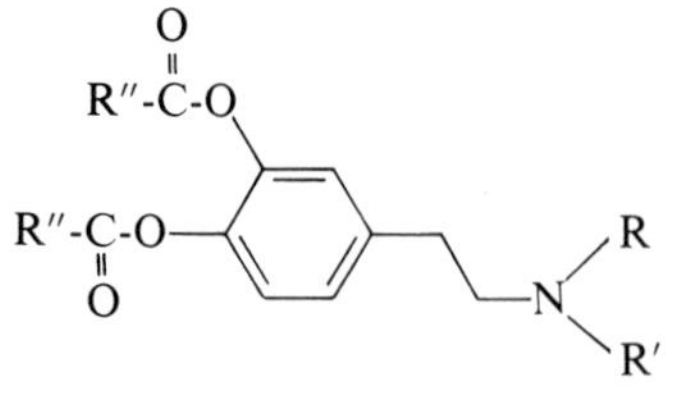

XXI

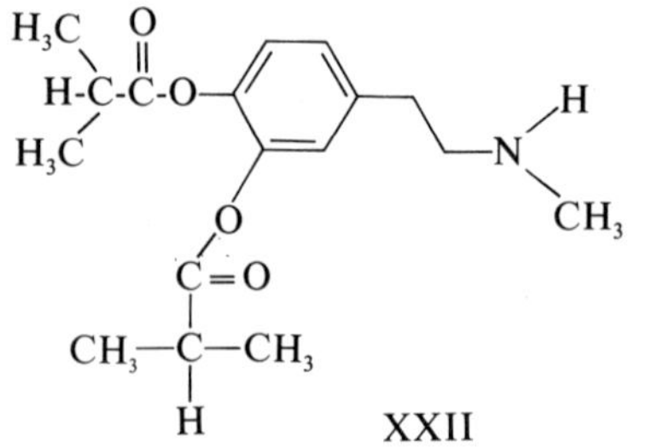

XXII

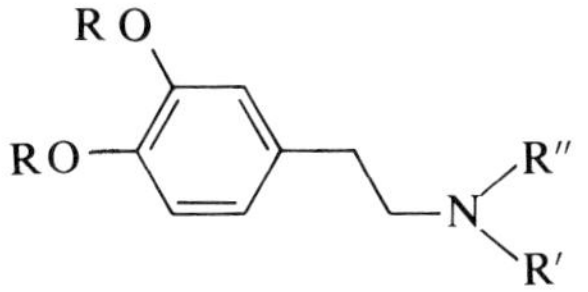

XXIIIa R = $(CH_3)_3Si$; R′ = R″ = H
b R = R′ = $(CH_3)_3Si$; R″ = H
c R = R′ = R″ = $(CH_3)_3Si$

Casagrande and Ferrari [64] prepared the acetate, pivalate, and benzoate esters of dopamine (XXI: R = R′ = H; R″ = CH_3; $C(CH_3)_3$; C_6H_5) but these workers reported no pharmacological data on their latentiated molecules. From studies of the diacetate esters of dopamine (XXI: R = R′ = H; R″ = CH_3), epinine (XXI: R = H; R′ = R″ CH_3), and N,N-dimethyldopamine (XXI: R = R′ = R″ = CH_3), Borgman et al. [65] concluded that both O-acetylation and N-diakylation are required to provide sufficient lipophilicity to penetrate the blood brain barrier. Trimethylsilylation of dopamine (structures XXIIIa-c) enhances lipophilic character of dopamine [66]. Preliminary investigation showed the desired central effects of these compounds. Compounds XXIII b and c, upon intraperitoneal injection in the rat, influenced positively the rigid akinetic syndrome which was induced by reserpine. In this experiment, dopamine itself had no effect. Randolph et al. [67] showed in studies in rats, dogs, and man that orally administered ibopamine XXII is de-esterified to give measurable circulating levels of epinine. In the dog and man, plasma levels of epinine and its conjugates accounted for 50% of the total drug, and in the rat, 75%. Casagrande and co-workers [68] found that the principal urinary metabolites of ibopamine in rat, dog, and man are dihydroxphenylacetic acid and homovanillic acid; N-methyldopamine-3-O-sulfate in the dog and man; and N-methyldopamine-4-O-glucuronide in the rat. N-Methyldopamine-3-O-sulfate appeared to be excreted without being reconverted to N-methyldopamine (by the action of aryl sulfatases). However, Harvey, et al. [69], found that large (600 mg) oral doses of ibopamine in normal humans elicit no change in urine flow, sodium, potassium, or creatinine excretion, and no change in blood pressure or heart rate, except for a transient rise in systolic pressure. The lack of effects suggested that the epinine, freed from the ester, undergoes first pass metabolism resulting in conjugation to form biologically inactive derivative(s). Diacetyl N,N-dimethyldopamine (XXI: R = R′ = R″ = CH_3) demonstrated dopaminergic actions in vivo but not in vitro [70]. It has been concluded that the labile ester groups are cleaved after penetration of the blood brain barrier, and prior to the molecule's interaction with dopamine receptors. The benzoate ester XXIV of the *m*-tyramine derivative XVII a is a potent dopaminergic, showing a prolonged duration of action in CNS assays [53].

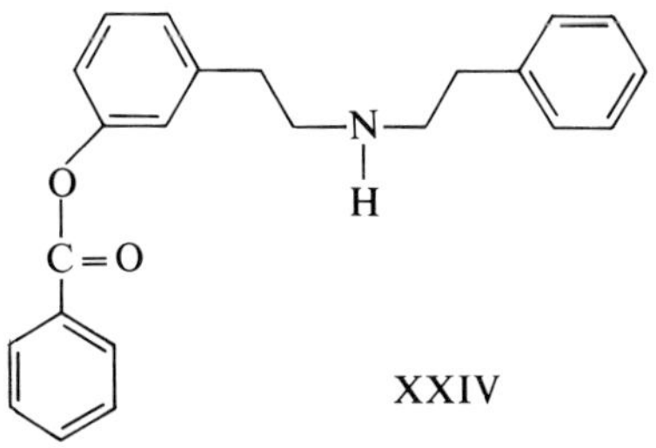

XXIV

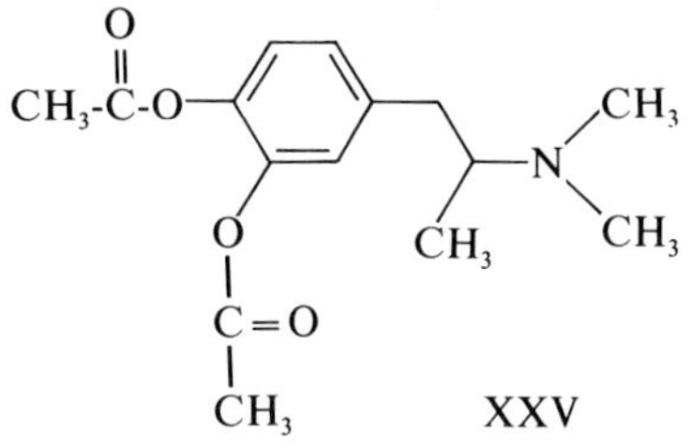

XXV

The likelihood of this compound's conversion to the free phenolic system XVII a in the brain was cited.

The diacetate ester of N,N-dimethyl α-methyldopamine XXV penetrates the blood brain barrier to exert central dopamine-related effects, although the introduction of the α-methyl group decreases activity as compared with the diacetate ester of N,N-dimethyldopamine [58].

2.3 Metabolic fate of dopamine and related β-phenethylamines

Dopamine is metabolized by a number of competing pathways, including β-hydroxylation (to form norepinephrine), O-methylation, oxidative deamination, and conjugation with sulfuric or glucuronic acid [71]. N-Acetyldopamine occurs in human urine, primarily in a conjugated form [22]. A compound tentatively identified as N-acetyldopamine has been found in the human kidney. It appears that conjugated N-acetyldopamine might be an endogenous metabolite of dopamine in man [22]. It has been speculated that this compound is formed peripherally, and not in the brain. Whether N-acetyldopamine has some role in mammals analogous to its sclerotization process in insects has not been investigated, nor has it been established whether dopamine is first N-acetylated, then conjugated, or vice versa. In vivo, dopamine is a substrate for catechol-O-methyltransferase ('COMT'), and the product is almost exclusively the 3-methyl ether XXVI a [72], which is inert as a neurotransmitter. Indeed, both the 3- and the 4-monomethyl ethers of dopamine are inert in the assay for elevation of cyclic AMP production [27].

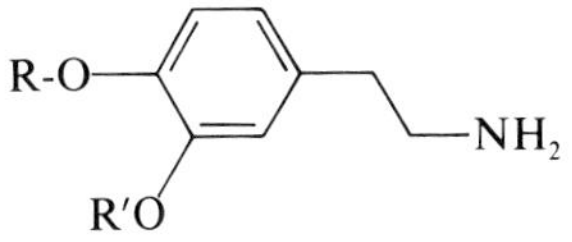

XXVIa R = H; R′ = CH_3
b R = CH_3; R′ = H
c R = R′ = CH_3

However, Smith and Hartley [73] noted 'abnormal' in vivo metabolites of dopamine: the 4-methyl ether XXVIb and the dimethyl ether XXVI c, which they suggested may be involved in pathogenesis of Parkinson's disease and/or schizophrenia. It was further speculated that dopa-induced psychotomimetic side effects may be caused by these metabolites. Feenstra et al. [74], in a study of a series of N,N-dialkyldopamines, found that maximal brain concentrations occurred approximately 10 minutes after peripheral injection, and there followed a rapid fall in concentration. O-Methylation was concluded to be a major factor in this rapid fall in brain concentration. 3-O-Methyl N,N-di-*n*-propyldopamine was cited as being especially rapidly formed in vivo, and it was eliminated more slowly than the parent non-etherified compound.

In vitro enzymatic (via COMT) O-methylation of dopamine affords mixtures of 3- and 4-monomethyl ethers [72,75], the relative amounts of each depending upon the biological source of the COMT used and the experimental conditions. A kinetic study [76] led to the conclusion that O-methylation of catecholamines by COMT is first order in concentration of catecholamine. Polarity of the substrate seems to contribute to the reaction rate at low concentrations of the substrate.

Dopamine and its 3-O-methyl derivative are substrates for monoamine oxidase ('MAO') [77], being oxidatively deaminated in the process. Glover et al. [78] found that in the rat, the 'A' type of monoamine oxidase (which, by definition, exhibits a preference for 5-hydroxytryptamine substrate) attacks dopamine, whereas in man, dopamine is attacked by the 'B' type of MAO (which, by definition, exhibits a preference for β-phenethylamines). Suzuki et al. [79] noted that 4-methoxyphenethylamines seem to be specific substrates for type 'B' MAO, whereas, 3, 4-dimethoxyphenethylamines are substrates for both the type 'A' and 'B' types. These workers suggested that 4-O-methylation of dopamine enhances the preference of the substrate for type 'B', whereas, 3-O-methylation contributes to a substance's being a type 'A' substrate.

Sulfate ester(s) is (are) among the metabolites of dopamine in humans [80]. Normal subjects and Parkinsonian patients on high oral doses of

L-dopa excrete up to 80% of dopamine as sulfo conjugates. Dopamine-3-O-sulfate XXVII a has been described as the predominant sulfate ester isomer in man, and 3-O-sulfo conjugation probably competes with 3-O-methylation [71]. Unger et al. [81] supported the concept that dopamine-O-sulfate may represent: (1) a peripheral transport form of dopamine; and/or (2) a norepinephrine precursor in peripheral organs under conditions of increased sympathetic nervous activity.

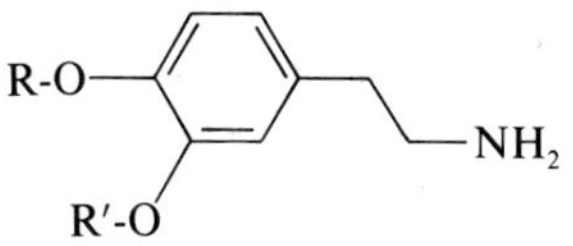

XXVIIa R = H; R′ = SO_3H
b R = SO_3H; R′ = H

Qu et al. [82,83] have suggested that dopamine-3-O-sulfate XXVII a and dopamine-4-O-sulfate XXVII b can act as substrates for catechol-O-methyl-transferase. Racz and co-workers [84] determined that 74% of the dopamine in bovine plasma is present as sulfo conjugate(s). Aldosterone secretion is inhibited in cultured bovine adrenal cells by dopamine and by dopamine-3-O-sulfate XXVII a, but the 4-O-sulfate XXVII b was inert. These workers suggested an in vivo physiological role for dopamine-3-O-sulfate in dopaminergic inhibition of aldosterone secretion at the adrenal level.

Buu et al. [85] found that stereotaxic implantation of 'dopamine sulfate' into the left lateral ventricle of the rat led to severe and generalized convulsions. These seizures were not blocked by phenoxybenzamine, metoclopramide, or by haloperidol, but they were reduced by propranolol, and they were suppressed by diazepam. The 'dopamine sulfate' used in this study was described as a mixture of 3- and 4-O-sulfate esters. Buu et al. [86] found that 'dopamine-3-sulfate' and 'dopamine-4-sulfate' produce convulsions in the rat similar to those produced by bicuculline; they were blocked by muscimol and by diazepam, but not by haloperidol. It was concluded that dopamine sulfate interacts with GABA receptors. The 3- and 4-sulfate esters, prepared by a method of Jenner and Rose [87], were equally potent. Rivett et al. [88] suggested that sulfate conjugation of biogenic amines and their metabolites may occur predominantly within neurons. Phenolsulfotransferase activity varies over a 17-fold range in different regions of rat brain, and it is possible that the enzymes involved may be restricted to specific types of neurons.

Bronaugh et al. [80] and Jenner and Rose [89] suggested that dopamine sulfate is the end product of dopamine metabolism in Parkinsonian patients. Merits [90] noted that very small doses of exogenous ^{14}C-dopamine-3-O-sulfate in the guinea pig are almost completely desulfated and metabolized according to the pattern characteristic for orally administered dopamine in this species. In the rat, approximately 40% of larger doses was totally desulfated and metabolized. In the dog, 80% of the dose of labelled dopamine-3-O-sulfate was excreted unchanged. However, the question of sulfate conjugation of dopamine in normal human metabolism still seems unsettled [91].
A significant factor in the confusion related to the biochemical/physiological/pharmacological role of 'dopamine sulfate' is the uncertainty about the chemical nature of the material which has been utilized in biological studies under the designation 'dopamine sulfate'. Idle et al. [92] have reviewed the 'dopamine sulfate' synthetic method of Jenner and Rose [87] (which is the preparative method employed by most of the earlier workers and by some contemporary workers), and they have concluded that this method gives rise to *mixtures* of the 3-O-sulfate XXVII a, the 4-O-sulfate XXVII b, and /or a sulfonic acid (ring sulfonation). The relative amounts of these products are highly dependent upon a number of experimental conditions, including the reaction temperature and the concentration of the sulfuric acid used. An unequivocal synthesis and characterization of dopamine-3- and 4-sulfates have been published [93], and NMR spectral analysis verified the assigned structures. Ackerman et al. [94] have reported preparation of dopamine-4-O-sulfate, and they have presented NMR spectral data supporting the assigned structure. The 4-O-sulfate ester elicited much weaker pharmacological effects than dopamine in a series of assays, and it was concluded that it possesses little dopaminergic or adrenergic activity. These workers [94] cited an earlier report of Scott and Elchisak [95], to conclude that dopamine-3-O-sulfate is the more active sulfate ester.

3 Aporphines and Related Systems

3.1 Apomorphine structural modifications

The prototypical dopaminergic agonist is a derivative of the aporphine ring system, R-(–)-apomorphine XXVIII ($R = CH_3$), which is derived from acid-catalyzed rearrangement of naturally occurring morphine.

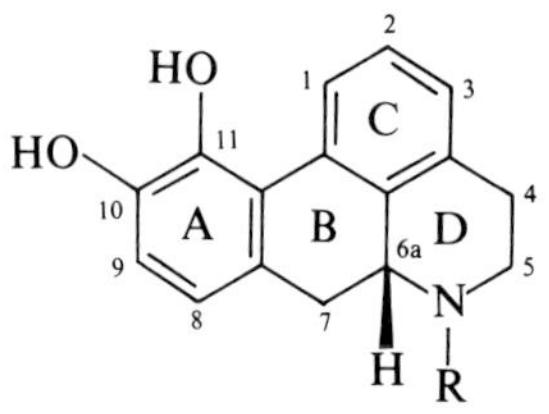

XXVIII

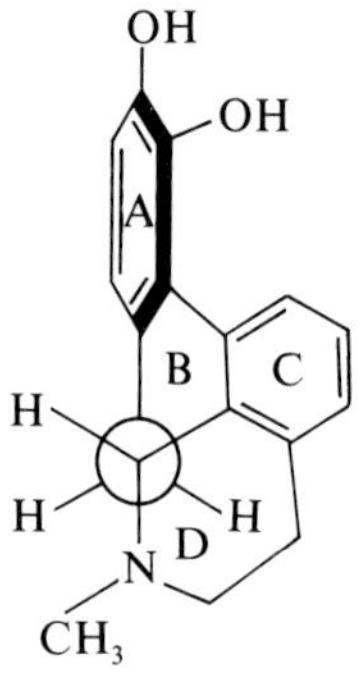

XXIX

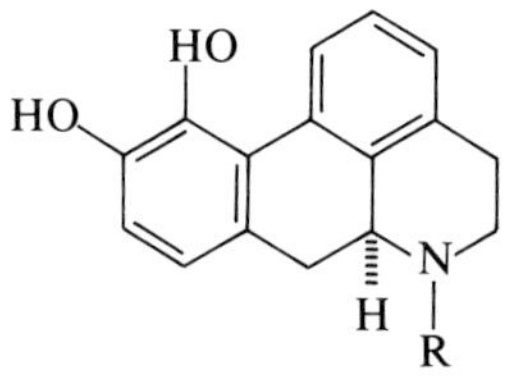

XXX

The absolute configuration (6a-R) of this apomorphine enantiomer was determined by chemical degradation means [96,97] and was verified by optical rotatory dispersion technique [98] and an X-ray crystallographic study [99]. The 6a-S-enantiomer (XXX: R = CH_3) and its dimethyl ether have been synthesized [100]. 6a-S-Apomorphine was ineffective in production of postural asymmetries in mice, emesis in dogs, and in increasing renal blood flow in dogs, assays in which 6a-R-apomorphine (XXVIII: R = CH_3) is highly active/potent. Riffee et al. [101] investigated the ability of 6a-S-apomorphine XXX to antagonize cage climbing induced by 6a-R-apomorphine XXVIII in mice. Initial studies using a racemic mixture had indicated that the presence of the S-enantiomer XXX produced a depression of cage climbing behavior. In subsequent experiments, it was demonstrated that XXX has no agonist activity in the cage climbing model, but that it can act as an antagonist to stereotypic activity induced by 6a-R-apomorphine XXVIII. Lehmann et al. [102] found that S-(+)-apomorphine (XXX) antagonized inhibition of uptake of ^{3}H-dopamine and of acetylcholine from cat caudate homogenate by R-(–)-apomorphine (XXVIII), to produce a parallel 5-fold shift in the dose-response curve of R-(–)-apomorphine to the right. Goldman and Kebabian [43], on the basis of their findings that S-(+)-apomorphine acts as an antagonist at D-1 and D-2 receptors, hypothesized that an N-methyl tertiary amine moiety in a molecule of appropriate configuration can confer dopaminergic antagonist activity.

Dopaminergic potency und activity in N-substituted homologs of apomorphine (structure XXVIII) vary with the substituent, the highest in most assays residing in *n*-propyl [103,104]. The *n*-butyl homolog shows almost no activity [103]. N-Allylnorapomorphine (XXVIII: R $= CH_2\,CH = CH_2$) has emetic activity in the dog comparable to that of apomorphine, but it is less potent than apomorphine in production of stereotypy in mice and compulsive pecking in pigeons [105]. Schoenfeld and co-workers [106] noted that of the two enantiomers of N-*n*-propylnorapomorphine, potency in the rat sniffing assays resides in the R-(–)-isomer (XXVIII: R= *n*-C_3H_7). A later communication [107] indicated that this R-(–)-enantiomer possessed all of the activity in rat stereotypy production; in an assay for effect on dopamine-sensitive adenylate cyclase; and in binding assays vs. ^{3}H-apomorphine and ^{3}H-2-amino-6,7-dihydroxy-1,2,3,4-tetrahydronaphthalene. The S-(+)-enantiomer did not antagonize the action of the R-(–)-enantiomer in the production of stereotypy in rats. However, the S-enantiomer was described as 'the weakly preferred configuration' for substrate activity for rat liver catechol-O-methyltransferase.

Molecular models reveal that the dopamine moiety within the apomorphine molecule is held rigidly in the α-conformation, with the catechol ring deviating from coplanarity with the ethylamine side chain by approximately 30°, as illustrated in the Newman projection XXIX. X-ray crystallographic data [108] support this conformational analysis.

(RS)-C-8-, 9-, 10-, and 11-monohydroxyapomorphines (structure XXXI) have been investigated [109-111] and of these, the 11-hydroxy isomer, a 'meta-hydroxy' system, is most potent and active [110], although less so than apomorphine. On the basis of rat rotation data, it was concluded that 10-hydroxy-N-*n*-propylnorapomorphine (XXXI: R = *n*-C_3H_7) is not a direct acting dopaminergic agonist [112].

RS-1,2-Dihydroxyapomorphine XXXII was reported to exhibit little or no dopamine-like actions in the dog [113] and in the pigeon [114]. The (+)- and (–)-enantiomers were similarly inactive in the dog [113]. This inactivity can be rationalized on the basis that the aporphine ring structure imposes upon the dopamine moiety of XXXII a conformation in which the catechol ring and the amino nitrogen are in a gauche (synclinal) disposition, Newman projection XXXIII, rather than the

XXXI

XXXII

XXXIII

antiperiplanar disposition which has been proposed to be essential for dopaminergic agonist-receptor interactions (structures VI, VII) and which occurs in apomorphine.

RS-'Isoapomorphine' XXXIV, which bears the dopamine moiety in a β-conformation, is inert as a central [109,113,115] and a peripheral [116] dopaminergic agonist.

XXXIV

XXXV

XXXVI

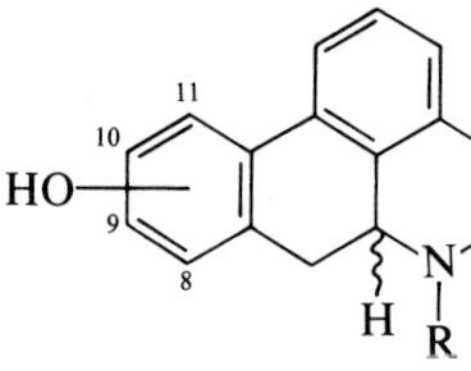

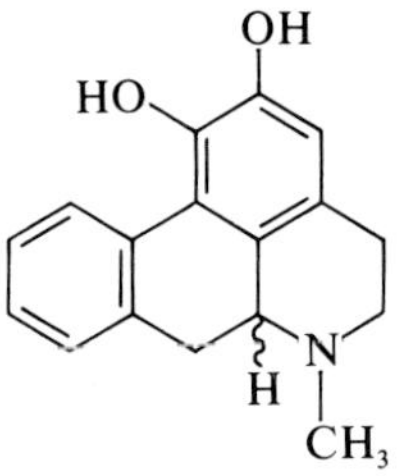

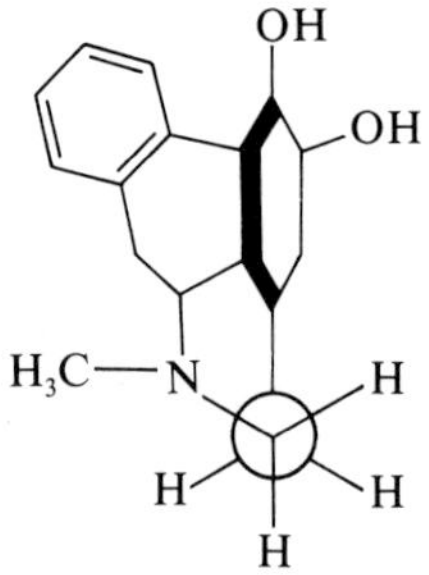

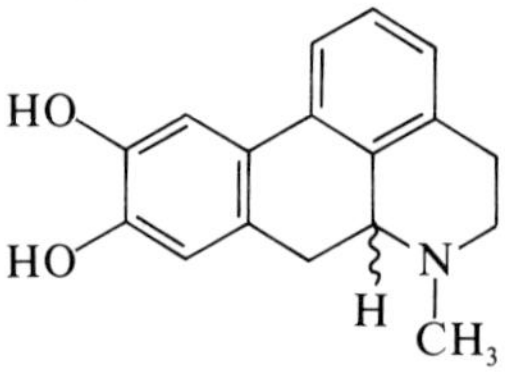

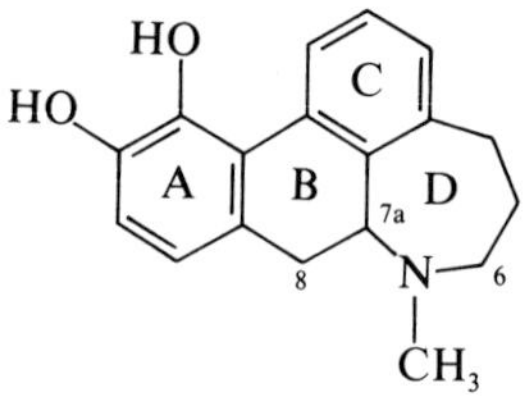

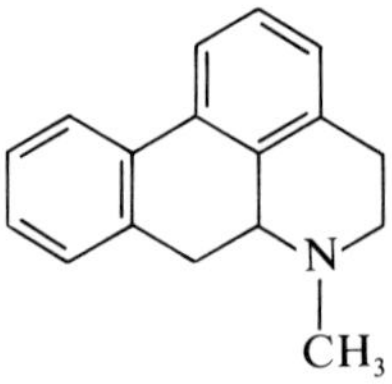

On the basis of the prominent agonist effects of congeners of the β-conformer of dopamine derived from 2-aminotetralin (vide infra), some agonist activity might have been predicted for XXXIV. Expansion of the D ring of apomorphine into a seven-membered ring (structure XXXV) destroys all dopamine-like effects [117]. An X-ray crystallographic study [117] revealed that the seven-membered ring distorts the dopamine moiety in structure XXXV (N-C_{7a}-C_8-ring A) away from the α-conformation. The ring nitrogen is gauche (synclinal) to the catechol ring, and this was concluded to be unfavorable for agonist actions.
The non-oxygenated aporphine system XXXVI is inactive [27].

3.2 Disruption of aporphine ring systems

In general, disruption of the ring system of the apomorphine molecule results in profound loss of dopaminergic agonist actions, even though a dopamine moiety be retained in the molecule. Examples of 'open ring' congeners are the phenanthrene derivatives XXXVII [114], XXXVIII [118], XXXIX [119], and the 1-benzyltetrahydroisoquinoline derivative XL [30].

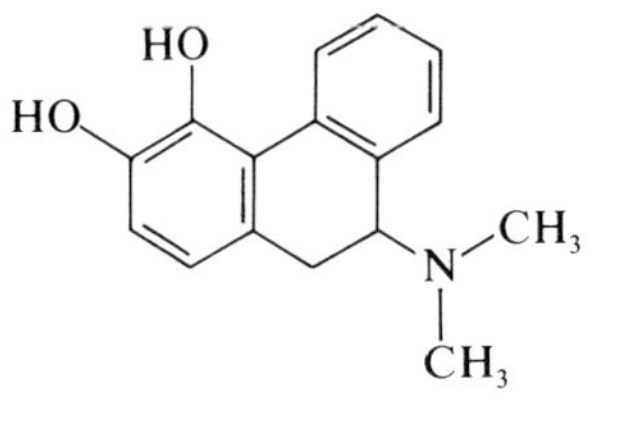

XXXVII

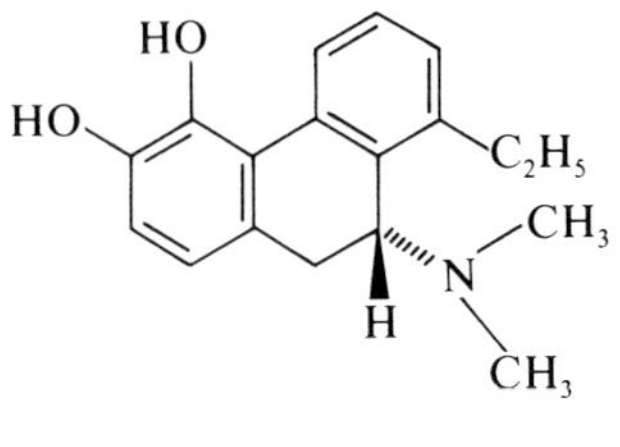

XXXVIII

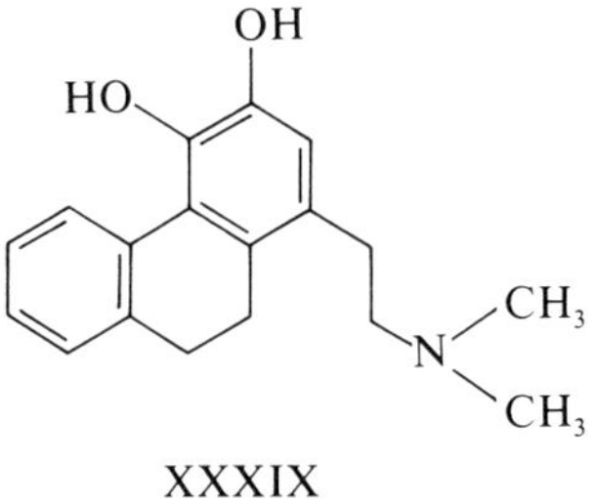

XXXIX

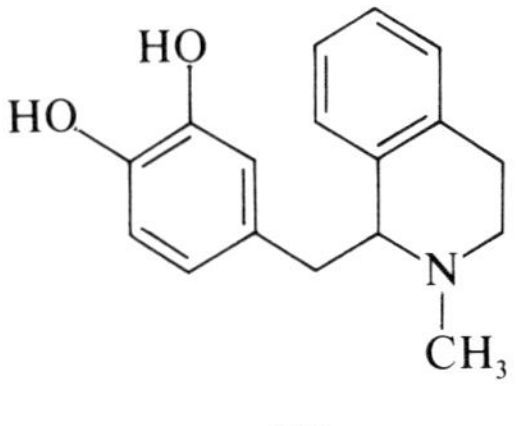

XL

The dopaminergic inactivity of the phenanthrene derivatives XXXVII and XXXVIII was rationalized [114,118] on the basis that the more stable steric disposition for these molecules is that 'flip' conformation in which the amino group is attached to the ring by a *pseudo*-axial bond, in which case the dopamine moiety is held with the catechol ring and the amino group *gauche* (Newman projection XLI), which conformation is not complimentary to the dopamine receptor(s).

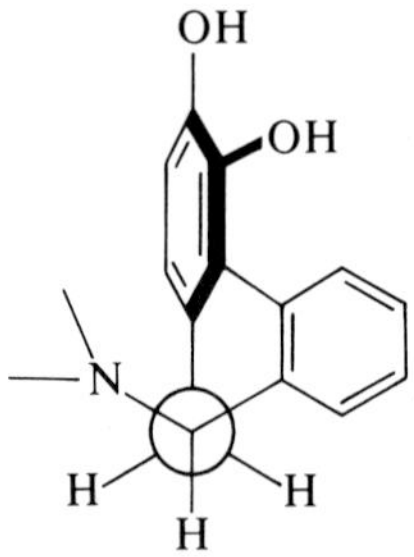

XLI

This conformational analysis of 9-substituted, 9,10-dihydrophenanthrenes has been supported by an NMR study of a series of 9-substituted, 9,10-dihydrophenanthrenes [120].

3.3 Latentiated aporphines and prodrugs

Apomorphine has been latentiated by conversion to its 10,11-diesters with a series of carboxylic acids XLII [121].

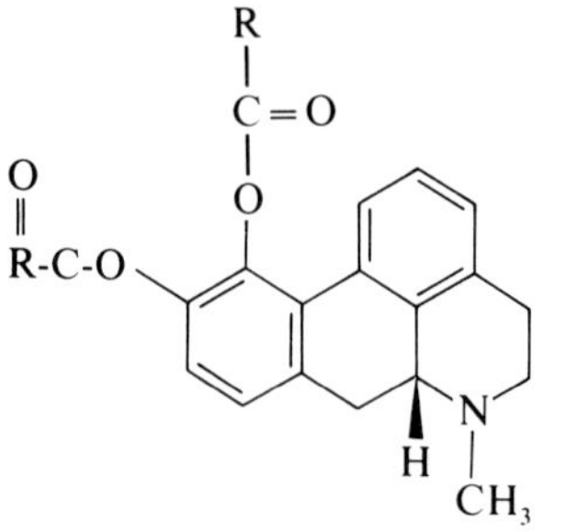

XLIIa R = CH_3
b R = C_2H_5
c R = $(CH_3)_2CH$-
d R = $(CH_3)_3C$-
e R = C_6H_5

All of the esters induced compulsive gnawing and turning in lesioned rats. All esters were less potent than apomorphine; they demonstrated a lag period before eliciting a pharmacological response; and they provided a prolonged duration of response compared to apomorphine. Diacetylapomorphine XLIIa appeared to be identical to apomorphine in regard to minimal emetic dose following subcutaneous administration in the dog. Likewise, the time of onset of action was similar to that of apomorphine, although the severity of vomiting was of a lower order with the diacetate ester. The ester was a potent pecking syndrome stimulant in the pigeon [122]. In a more detailed study of the diisobutyrate ester XLIIc, Baldessarini and co-workers [123] found that apomorphine, but not the ester, stimulated adenylate cyclase activity in mouse striatal homogenates. The total behavioral response in the mouse to larger doses of this ester was greater than to apomorphine. Systemic injection of the ester provided detectable brain levels of free apomorphine in the mouse, and the compound was advanced as a possibly clinically useful depot dopaminergic agonist prodrug. (–)-10,11-Methylenedioxy-N-*n*-propylnorapomorphine XLIII has been described [124,125] as a long-acting dopaminergic prodrug.

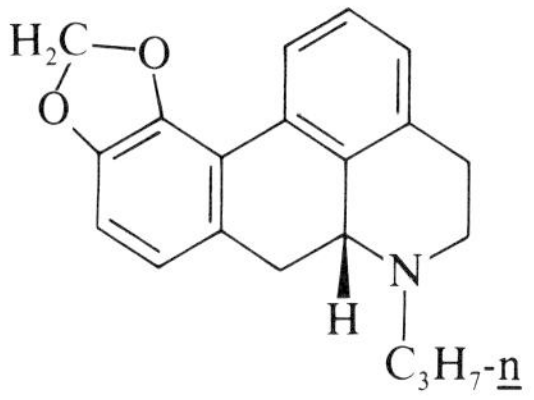

XLIII

The drug's stereotypic effects in rodents were blocked by haloperidol but not by reserpine pretreatment. Effects of large or small doses of XLIII were blocked by a microsomal enzyme inhibitor which did not interfere with the actions of N-*n*-propylnorapomorphine.

3.4 Metabolism of apomorphine and derivatives

Kaul et al. [126-128] reported that the rabbit, the rat, and the horse excrete mixtures of apomorphine O-glucuronides, with the 10-glucuronide predominating. The fraction of the total dose of apomorphine accounted for as excreted 'bound' apomorphine in these studies ranged from 13.6 to 71.9%. Under certain special experimental conditions, small percentages of unaltered apomorphine could be detected in rabbit urine. In vitro incubation of apomorphine with rat liver catechol-O-methyltransferase and S-adenosylmethionine iodide produced a mixture of monomethyl ethers, with the 10-methyl isomer XLIVa greatly predominating [129]. McKenzie and White [130] concluded that apomorphine is metabolized by COMT both in vitro and in vivo, but these workers made no attempt to isolate nor to identify the metabolite(s). Sourkes and co-workers [131,132] found that inhibitors of COMT (pyrogallol and tropolone) prolong the stereotypic behavioral response in rats produced by apomorphine. It was concluded that apomorphine is a substrate for rat brain COMT, and that COMT inactivates apomorphine in rat brain.

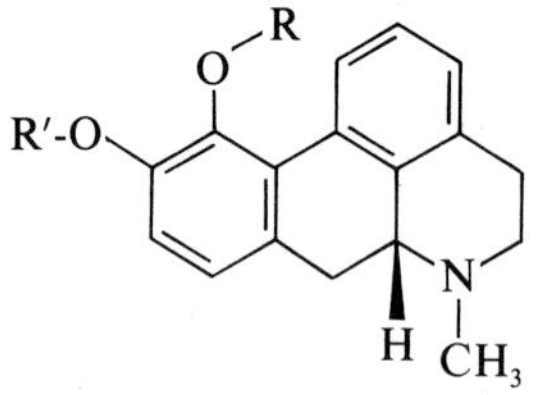

XLIVa R = H; R′ = CH_3
b R = CH_3; R′ = H
c R = R′ = CH_3

Both of the isomeric monomethyl ethers of apomorphine XLIVa, b are devoid of significant dopaminergic activity, as are the dimethyl ethers of apomorphine XLIVc and of N-*n*-propylnorapomorphine [27,129,133].

4 Aminotetralins and Aminoindans

4.1 Aminotetralins

5,6-Dihydroxy-2-aminotetralin derivatives XLV ('A-5,6-DTN') represent a fragment of the apomorphine molecule.

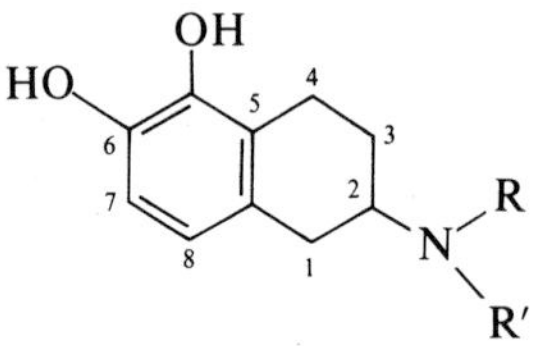

XLV

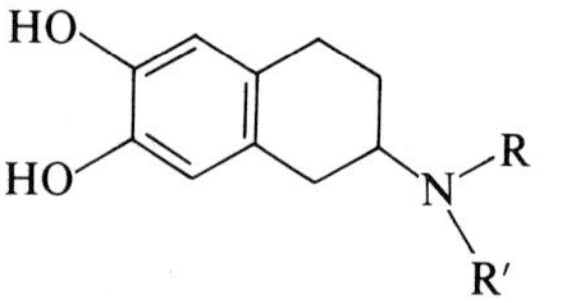

XLVI

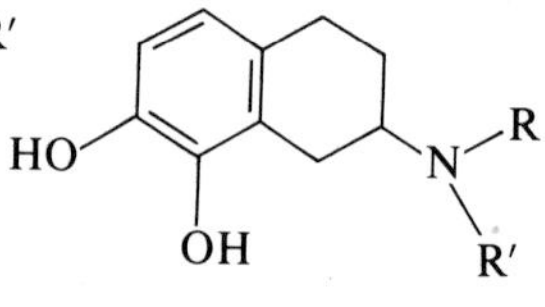

XLVII

The tetralin ring system is much less flexible than the dopamine molecule, but nevertheless it can exist in two 'flip' conformations: XLIX a (amino group pseudoequatorial) and XLIX b (amino group pseudoaxial), illustrated also by the Newman projections XLIX c and XLIX d, respectively.

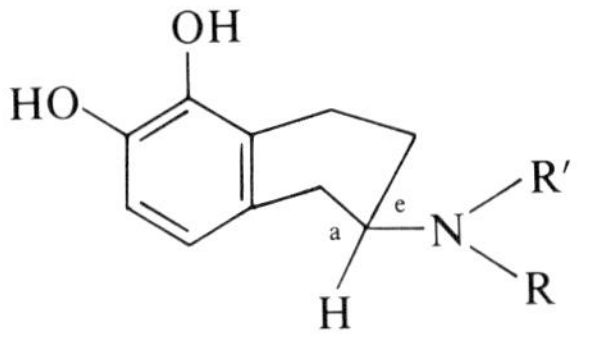

XLIXa

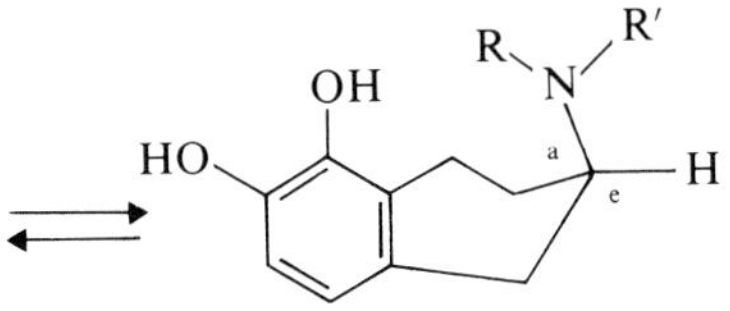

XLIXb

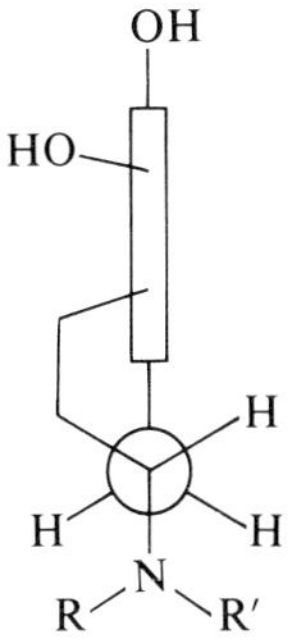

XLIXc

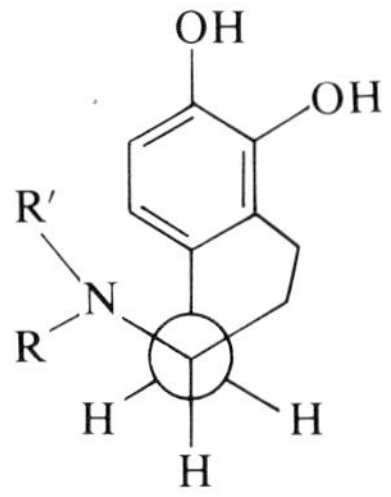

XLIXd

It would be predicted that the pseudoequatorial amino group conformer XLIX a would be the energetically preferred one; X-ray studies [134] have indicated this to be the case in the crystalline state. This geometry is parallel to that of the analogous portion of the apomorphine molecule. The α-conformation of the dopamine moiety within XLV bears the catechol ring very nearly coplanar with the plane of the ethylamine side chain, as shown in the Newman projection XLIX c. 6,7-Dihydroxy-2-aminotetralin XLVI ('A-6, 7-DTN') represents a fragment of the isoapomorphine XXXIV molecule, and it contains a dopamine β-conformer, Newman projection XLVIII.
In contrast with the dopaminergic inactivity of isoapomorphine, derivatives of XLVI as well as those of XLV (where R and R′ represent combinations of H, methyl, n-propyl, and 2-propyl) display prominent

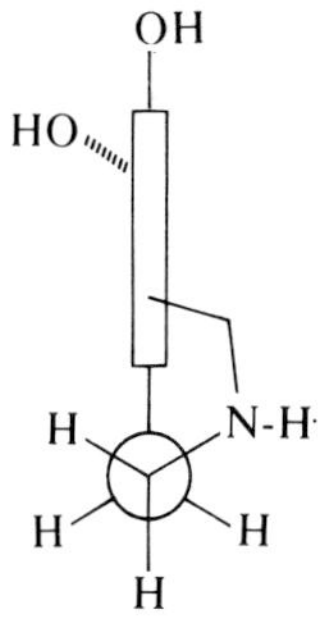

XLVIII

central and peripheral dopaminergic agonist effects [115,135-137], but the two isomeric systems show different spectra of activities. The N,N-di-*n*-propyl homolog of XLV displays high potency/activity, but the di-*n*-butyl homolog is virtually inert [136]. 5,6-Dihydroxy-2-di-*n*-propylaminotetralin (XLV: R = R′ = *n*-C_3H_7) is equipotent to dopamine itself as a D-1 agonist [138]. The dimethyl ethers of the 5,6-dihydroxy system XLV (where R = R′ = H; R = H, R′ = CH_3; and R = R′ = CH_3) were inert in assays for compulsive gnawing in mice and compulsive pecking in pigeons, and in production of emesis in dogs [139]. However, the dimethyl ethers of the N-methyl and of the N,N-dimethyl derivatives of XLV were far more potent/active emetics in pigeons than was apomorphine [139]. The dimethyl ether of A-6,7-DTN (XLVI: R = R′ = H) is dopaminergically inert in a guinea pig blood pressure model and in the mouse locomotor assay [137]. Kohli et al. [140] found that N-ethyl-, *n*-propyl- and *n*-butyl secondary amine derivatives of XLVI had low activity in the canine renal blood flow assay. The N,N-di-*n*-propyl and N-*n*-propyl, N-*n*-butyl tertiary amines of XLVI were active in this assay, but were of lower potency than dopamine.

In comparison studies of the 5,6- and 6,7-dihydroxy-2-aminotetralin derivatives, Kohli and co-workers [141] noted that 2-di-*n*-propylamino-5,6-dihydroxytetralin (XLV: R = R′ = *n*-C_3H_7) is a DA-1 agonist at least as potent as dopamine. This finding contrasts with previous observations of these workers that the 6,7-rather than the 5,6-dihydroxy pattern is required for DA-1 agonist effects. In the 5,6-dihydroxy-2-aminotetralin series, both the primary amine (XLV: R = R′ = H) and the N,N-dimethyl (XLV: R = R′ = CH_3) are devoid of

DA-1 effects on the canine renal vascular bed. The fact that di-*n*-propyl substitution can convert an otherwise inactive conformation into an active DA-1 agonist suggested [141] that N-substitution is more important for this pharmacologic response than is the disposition of the OH groups. However, two hydroxyl groups seem to be required. Seiler and Markstein [142] speculated that these changes in preferred rotameric form upon N-propylation might be due to changes in acid strength of the phenolic OH group(s) depending upon its (their) position on the ring and the subtitution pattern of the amino group. However, this is not the case, as evidenced by measurements of pK_a values of the phenolic OH group of 5-hydroxy-2-aminotetralin, 5-hydroxy-N,N-di-*n*-propyl-2-aminotetralin, 7-hydroxy-2-aminotetralin, and 7-hydroxy-N,N-di-*n*-propyl-2-aminotetralin, all of which gave pK_a approximately 12.6. In this series of 2-aminotetralins, the ionization constants of the protonated amino groups showed, as expected, higher values for the more basic primary amines than for the tertiary amines, but the values did not vary considerably between the OH-positional isomers, which rules out differences in ionization behavior between the 5-OH and the 7-OH series as a determinant of pharmacological difference(s).

Initial studies [143] led to the conclusion that 2-dimethylamino-6,7-dihydroxytetralin ('TL-99', XLVI: R = R′ = CH_3) is a selective presynaptic dopaminergic agonist. However, more recent studies [144–146] reveal stimulation of postjunctional receptors by this compound, and Williams et al. [147] have suggested that the apparent dopaminergic autoreceptor selectivity noted for TL-99 in vivo may be due at least in part to its α_2-adrenoceptor agonist activity.

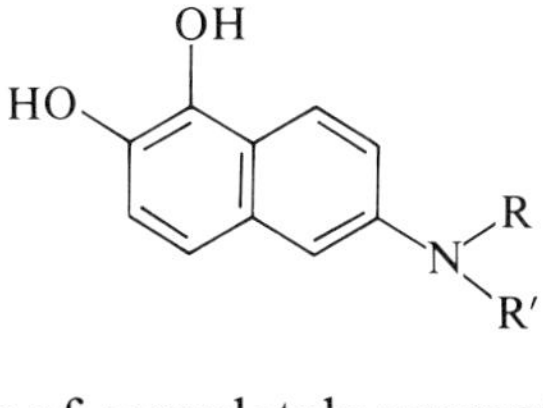

La R = R′ = H
b R = R′ = CH_3
c R = H; R′ = CH_3

The series of completely aromatic congeners La–c of A-5'6-DTN were inert in assays for dopaminergic effects [139].

Two 7,8-dihydroxy systems (XLVII: R = R′ = CH_3 and R = R′ = *n*-C_3H_7) exhibited dopaminergic agonist effects in three assay models [136], albeit of lower potency than the 5,6-dihydroxy congeners.

In monophenolic 2-aminotetralins, the most prominent central dopaminergic agonist effects and highest potency reside in the 5-hydroxy isomer LI, a 'meta-OH' system [149–152], although the potency of the 5-monohydroxy-N,N-di-*n*-propyl derivative was less than that of the analogous 5,6-dihydroxy compound [149]. The N,N-di-*n*-propyl-6- and 7-hydroxy systems display only very modest dopamine-like actions [149].

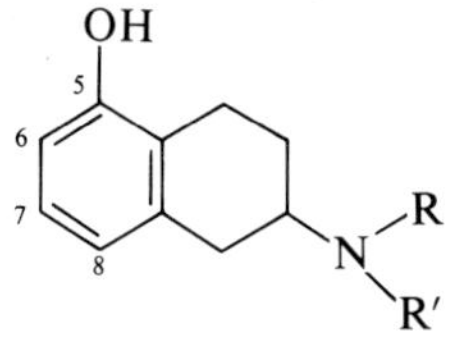

LI

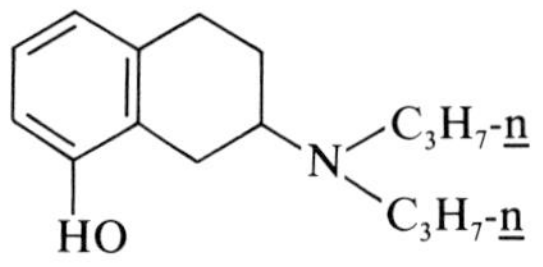

LII

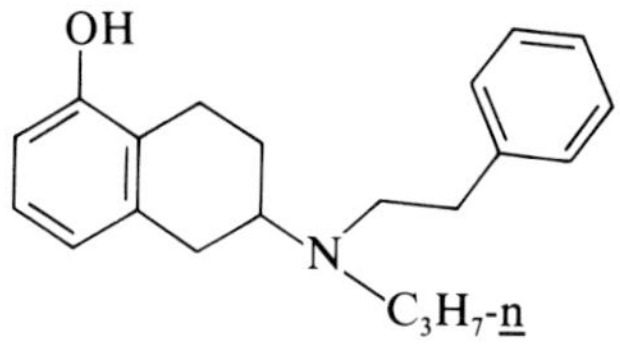

LIII

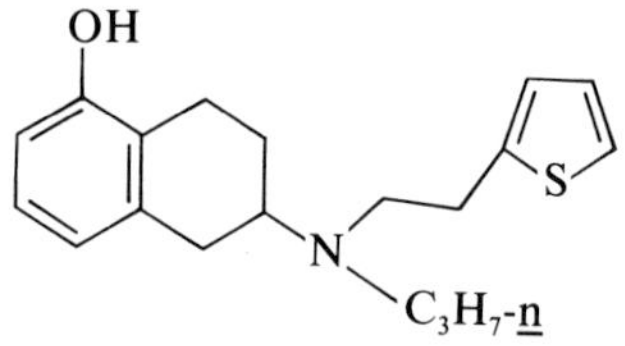

LIV

The dopaminergic potency of the 5-hydroxy-2-aminotetralins reflects some consistency with the relatively high potency/activity of certain *m*-tyramine derivatives and of 11-hydroxyaporphine derivatives, as described previously. Those 5-hydroxy-2-aminotetralins bearing at least one *n*-propyl group on the nitrogen have the highest potency/activity, and the di-*n*-propyl homolog is the most potent and active [42,151], and it is a selective D-2 agonist [138]. The di-*n*-butyl homolog (LI: R = R′ *n*-C_4H_9) is inactive. The N,N-diallyl homolog (LI: R = R′ = $CH_2CH{=}CH_2$) has decidedly less dopaminergic effect than the N,N-di-*n*-propyl congener [153]. Compound LIII, a tetralin congener of the *m*-tyramine derivative XVII a (vide supra) has been described as a specific D-2 agonist [154]. The 2-thienyl isostere of LIII (compound LIV) has been described [138] as the most potent D-2 agonist encountered to date. Beaulieu et al. [138] have stated that a selection of N-substituted 5-hydroxy-2-aminotetralins studied by them only marginally stimulated the D-1 receptor.

On the basis of a multiple regression analysis study of twenty-eight N-alkylated 2-aminotetralin derivatives, Lien and Nilsson [155] concluded that for central dopamine receptor stimulant activity, the most important structural feature appears to be the 5-hydroxy group. Symmetrical N-substituents with chain length no more than three carbons also contribute positively to receptor binding. Hydrophobicity was concluded to have only a minor influence on the activity in this series. The 8-hydroxy compound LII has been described [151,156] as a pre- and postsynaptic serotonergic agonist, devoid of dopaminergic agonist properties. Data have now appeared in the literature [157] showing that this compound has a component of action on in vitro suppression of prolactin release from anterior pituitary tissue supportive of the view that is has dopamine agonist properties at some populations of dopamine receptors.

In assessing overall pharmacological effects of 5-hydroxy-, 5,6-dihydroxy-, 7-hydroxy-, and 6,7-dihydroxy 2-aminotetralins, Van Oene and co-workers [153] concluded that high postsynaptic dopaminergic agonist effectiveness was achieved with those 2-aminotetralins bearing a hydroxy group at position 5, whereas those systems lacking the 5-OH possessed high dopaminergic autoreceptor specificity.

2-aminotetralin systems have one chiral center which is analogous to the 6a-chiral center of apomorphine (structure XXVIII). In both of these ring systems, the chiral center is a part of the dopamine moiety. It should be remembered that the Cahn-Ingold-Prelog RS rules assign different letters to the same absolute configuration at the chiral center in 2-aminotetralins and in apomorphine. McDermed et al. [158] reported separation of the enantiomers of the potent dopaminergic agonists 2-di-*n*-propylamino-5-hydroxytetralin (LI: R = R′ = *n*-C_3H_7), and 2-amino-6,7-dihydroxytetralin (XLVI: R = R′ = H), and determination of their absolute configurations by chemical degradative procedures. Giesecke [134] has verified the assignments by X-ray crystallographic analysis.

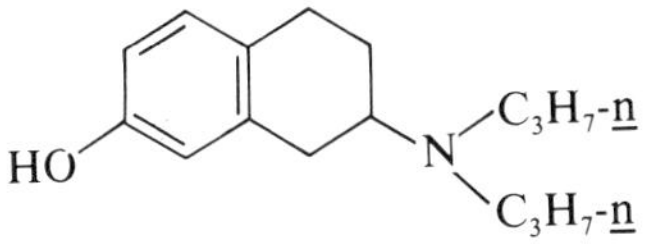

LV

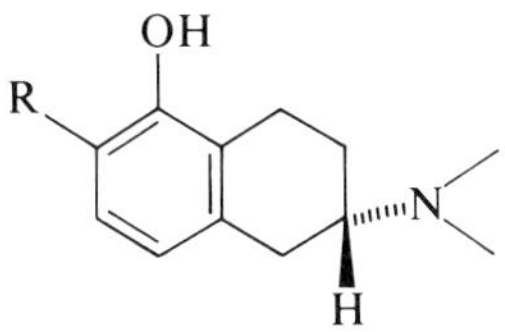

LVI R = H or OH

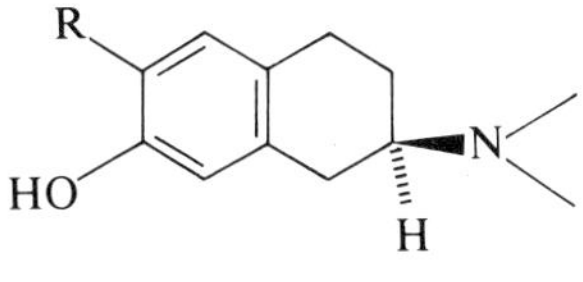

LVII R = H or OH

Subsequently [159], 7-hydroxy-2-di-*n*-propylaminotetralin LV and 5,6-dihydroxy-2-di-*n*-propylaminotetralin (XLV: R = R′ = *n*-C_3H_7) were resolved and absolute configurations of the enantiomers were established. In all cases (+)-enantiomers (HBr or HI salts in methanol) have the 2-R configuration. A considerable number of pharmacological studies [149,158,160–163) establish that the more active enantiomer in the 5-hydroxy and the 5,6-dihydroxy analogs is the S-(–)- (structure LVI), which is consistent with the absolute configuration (but not the letter designation) of the active enantiomer of apomorphine (structure XXVIII). However, in the 7-hydroxy and the 6,7-dihydroxy series, the more active enantiomer is the R-(+)- (structure LVII) [158,164].
McDermed et al. [158] provided a rationalization for the apparent inconsistency of absolute configuration for dopaminergically active 2-aminotetralins: these compounds are all analogs of dopamine constrained in an extended conformation so that the two 'meta' positions (5- and 7-OH) are distinguished: the *α*- and the *β*-rotamers. The pharmacological data were interpreted by the McDermed group to imply that when these and related agonists bind to dopamine receptors, the steric orientation required of the amino group of the agonist is controlled primarily by the location of a hydroxyl group in one of the two possible 'meta' positions. This, in turn, implies that, in order to achieve a proper orientation of the amino group and the 'meta' hydroxyl group with respect to their presumed complimentary binding subsites on the receptor, the R-LVII must be rotated with respect to S-LVI and to R-apomorphine, as illustrated in figure 1.

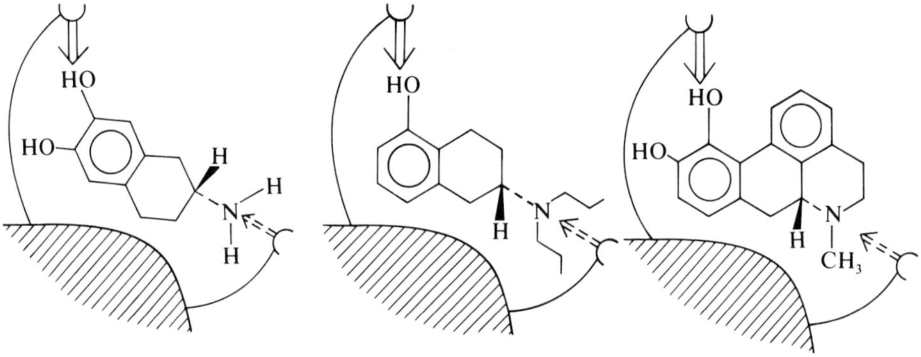

R-LVII S-LVI R-Apomorphine

Figure 1
Attachment of R- and S-2-aminotetralins to dopamine receptor: the McDermed model.

Grol and Rollema [165] and Seiler and Markstein [142] have proposed graphic representations of a dopamine receptor which in general agree with the McDermed model. Wikström et al. [166,167] integrated these preceding representations of the ('a'??) dopamine receptor into a modification which permitted rationalization of the dopaminergic agonist activity or inactivity of a variety of other molecules, and which provides an explanation for the seemingly inconsistent biological effects noted for various N-alkyl groups in a variety of molecules.

From studies on isomeric monohydroxy- and dihydroxy-2-aminotetralins, Seiler and Markstein [384] obtained data which suggested that there are two similar major binding sites in D-1 and D-2 receptors, but that there are differences between the two with respect to additional binding sites. According to this model, dopamine itself would react with D-1 and D-2 receptors in its β-conformation. However, N,N-dipropylation should cause a change in preferred conformation toward the α-rotamer.

The marked difference in spectrum of dopaminergic activity cited for the 5,6- as compared to the 6,7-dihydroxy 2-aminotetralins can in part be explained on the basis of metabolic differences. The aminotetralins are not substrates for monoamine oxidases [168], but they are affected by catechol-O-methyltransferase ('COMT'). The lower in vivo central nervous system potency of A-6,7-DTN (XLVI: R = R′ = H) as compared with A-5,6-DTN (XLV: R = R′ = H) was ascribed [169] to different rates of O-methylation by COMT. Inhibition of this enzyme in rodents produced almost equal brain concentrations of A-5,6-DTN and of A-6,7-DTN. In vitro incubation of A-5,6-DTN with COMT produced only very small amounts of a single methyl ether (thc 5-), and this compound was biologically inert at dopamine receptors [170]. In contrast, in vitro incubation of A-6,7-DTN with COMT resulted in extensive conversion to two isomeric monomethyl ethers, with the 7-ether greatly predominating. In vivo, A-6,7-DTN is converted exclusively into the 7-methyl ether which is inactive as a dopaminergic agonist [168,170]. Rollema et al [171] in studies (in the rat in vivo and in rat liver in vitro) on dopaminergic activity, brain concentrations, and metabolism of (+)-(2-R)- and (–)-(2-S)-N,N-di-*n*-propyl-A-5,6-DTN LVIII and LIX found that the brain concentrations of the (less biologically active) (+)-(2-R)-enantiomer are much lower than those of the dopaminergically more potent (–)-(2-S)-enantiomer after equimolar doses, and this was due to a remarkable difference in the susceptibility of the two enantiomers to COMT.

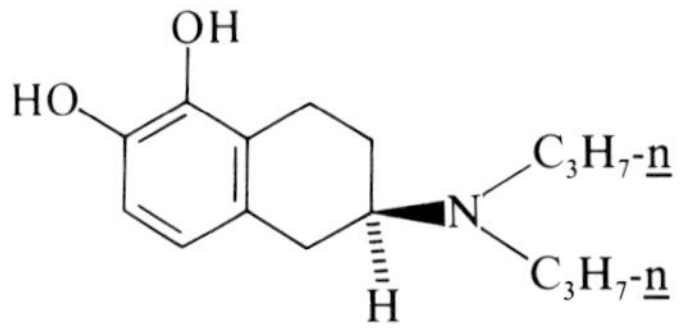

LVIII (2-R)

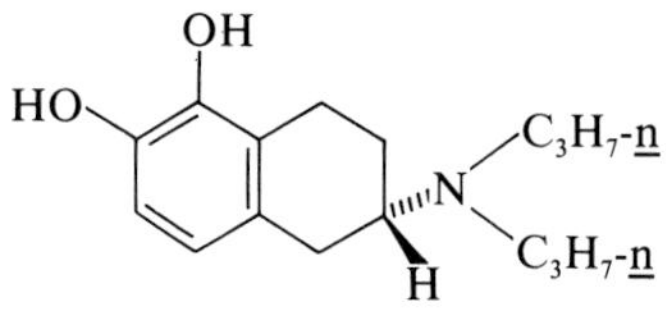

LIX (2-S)

In vitro, the less potent (+)-(2-R)-enantiomer was an excellent substrate for COMT, forming the 5-methyl ether exclusively [172]. Youde et al [173] found that in series of N,N-dialkylated A-5,6- and 6,7-DTN derivatives, increasing the alkyl chain length increases the susceptibility of both systems to COMT, and series of N,N-dialkylated dopamines showed the same trend.

McDermed et al [136] reported that a 5-methyl-6-hydroxytetralin derivative LX had dopamine agonist effects, although its potency in three assay models was approximately 1/100 that of the 5,6-dihydroxy congener.

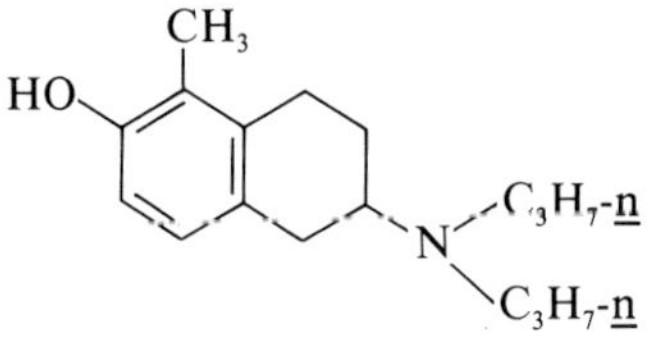

LX

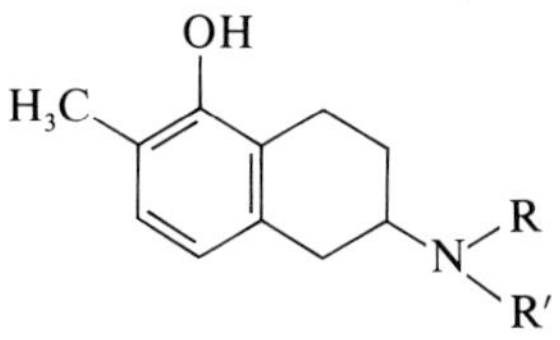

LXIa R = R′ = C_2H_5
b R = R′ = n-C_3H_7

Derivatives LXI a–b of the hydroxy-methyl group positional isomeric system were found [174,175] to display a variety of central and peripheral dopaminergic effects, qualitatively somewhat different from those reported for the 5-hydroxy-2-aminotetralins LI, and manifesting maximal response only after a 25–30-minute lag period following intravenous administration. The compounds showed only weak emetic activity in the dog [175], and in this respect, they differed from the corresponding, 5,6-dihydroxy congeners which were violent emetics [136]. The primary amine and the N,N-dimethyl homologs of LXI a–b demonstrated little or no dopamine receptor agonist activities [174]. The N,N-diethyl homolog LXI a did not appear to stimulate postsynaptic dopamine receptors in the central nervous system. Verimer et al. [175] suggested that LXI a and b do not directly activate dopamine receptors, but rather that they are metabolized to dopaminergically active molecule(s). Koons and co-workers [176,177] presented a pharmacological profile and data on changes in pharmacological effects of LXI b following pretreatment with inhibitors of drug metabolism, which are consistent with the proposal of in vivo metabolic activation. To evaluate the speculation that the ring methyl of LXI a–b is metabolically hydroxylated, and further that this alcohol might be further oxidized in vivo, the derivatives LXII a-c were prepared [178].

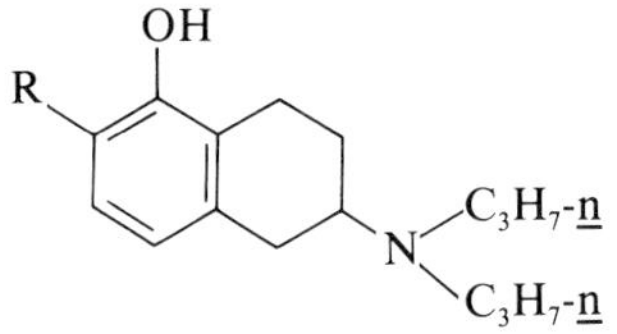

LXIIa R = CH_2OH
b R = CHO
c R = COOH

Both LXI a and b were highly potent in the cat cardioaccelerator nerve assay. The hydroxymethyl congener LXII a was more potent in this assay and in the rat rotation model than the 6-methyl derivative LXI b, and it was approximately equipotent to apomorphine. The lag period between administration of the drug and observation of maximal pharmacological effects in vivo was decidedly less for the 6-hydroxymethyl compound than for the 6-methyl [179]. Administration of ^{14}C-labelled 5-hydroxy-6-methyl derivative LXI b to rats permitted recovery of approximately 45% of the dose from the urine in the form of the tetralin-6-carboxylic acid LXII c [180]. This metabolite was inert as a dopaminergic agent.

No accounts of metabolism studies of the 5-methyl-6-hydroxytetralin system LX have appeared in the literature, and a possible role of metabolic activation of this molecule cannot be assessed.

Resorcinol-derived 2-aminotetralins LXIII which have the 'meta-OH' pattern of both the α- and the β-rotamers of dopamine, were less po-

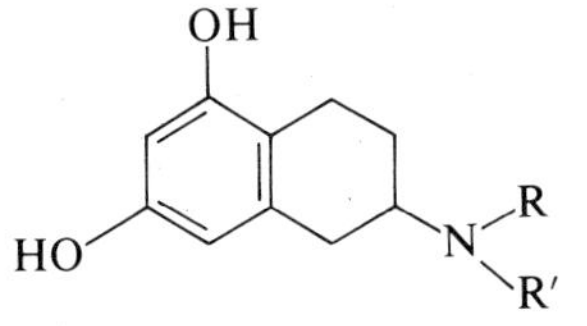

LXIIIa R = R' = C_2H_5
b R = R' = n-C_3H_7

tent and less active than their catechol-derived isomers XLV and XLVI [181], but unlike the *β*-phenethylamine derivative XVI, they were not dopaminergically inert.

The two racemic stereoisomeric 1-methyl-2-aminotetralin derivatives LXIV and LXV are equipotent at central dopamine receptors, and both compounds are less potent than the non-C-methylated system LXVI [182].

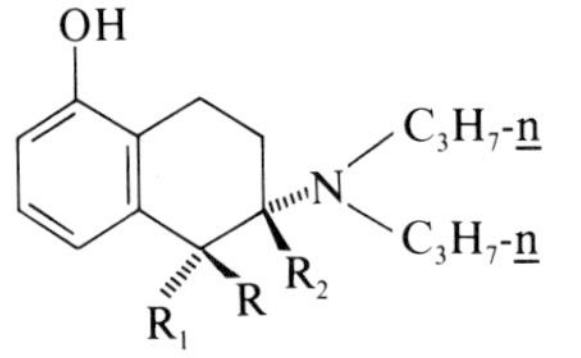

	R	R_1	R_2
LXIV	CH_3	H	H (trans)
LXV	H	CH_3	H (cis)
LXVI	H	H	H
LXVII	CH_3	CH_3	H
LXVIII	H	H	CH_3

Both the *trans*- (LXIV) and the *cis*-(LXV)isomers displayed dopaminergic autoreceptor stimulant capacity, but only the *trans*-system LXIV elicited clear-cut postsynaptic agonist actions at larger doses. The 1,1-dimethyl homolog LXVII was inactive at all dopaminergic receptors studied [182], as was the 2-methyl derivative LXVIII [183]. Nichols et al. [184] reported the dopaminergic inactivity of the 2-methyl analog LXIX of A-6,7-DTN.

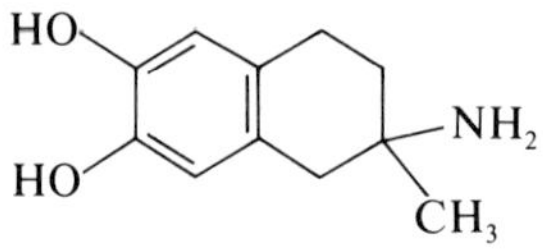

LXIX

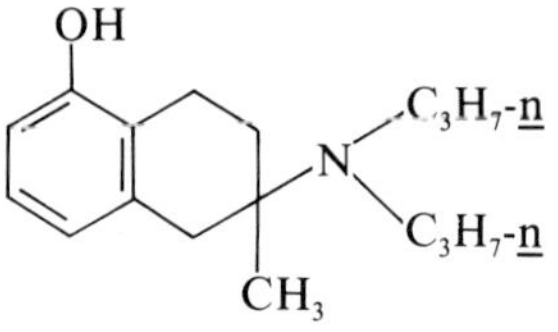

LXX

High field ^{1}H-NMR studies led to the conclusion that LXIX probably exists as a rapidly equilibrating mixture of conformers, and it seemed likely that it can adopt the active conformation proposed to be required for dopamine receptor interactions. The lack of pharmacological activity was therefore attributed to the steric effect of the 2-methyl group, consistent with the concept that the dopamine receptor(s) cannot tolerate alkylation at the side chain *α*-carbon. Andersson et al. [185], reported effects on serotonin receptors produced by an enantiomer of the 2-methyl-2-aminotetralin derivative LXX, but no dopamine-like effects were cited. Isolation of the enantiomers of the *cis*- and *trans*-isomers LXV and LXIV has been reported [186], and rat locomotor activity and dopamine accumulation data in reserpinized and non-reserpinized rats have been reported [187] for the *cis*-enantiomers. *cis*-(1-R), (2-S)-LXV and its methyl ether are central dopaminergic agonists, whereas *cis*-(1-S), (2-R)-LXV and its methyl ether are central dopamine receptor antagonists [186].

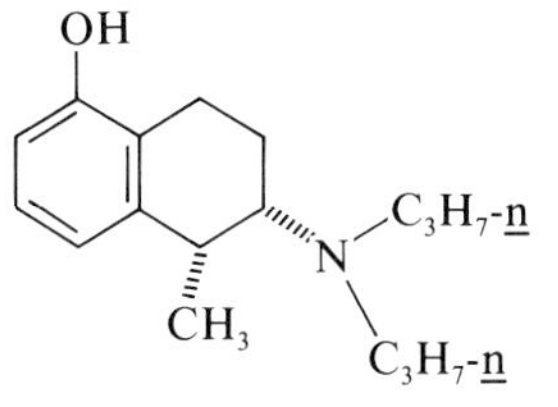

(1R, 2S)-LXV

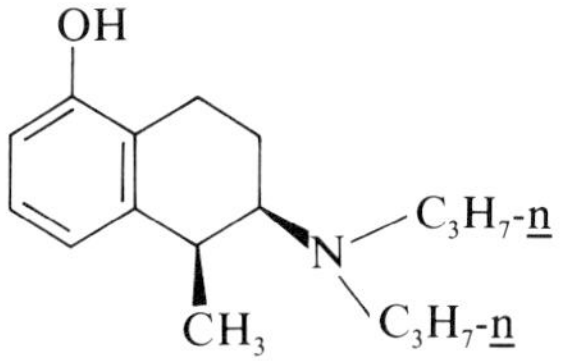

(1S, 2R)-LXV

Thus, in LXV, the absolute configuration at position 2 is consistent with that in the more pharmacologically potent enantiomers of the C-desmethyl 5-hydroxy- and 5,6-dihydroxy 2-aminotetralins. The equipotency of the *cis*- and the *trans*-isomers LXV and LXIV is surprising, in view of the inactivity of the 1,1-dimethyl derivative LXVII, and the marked difference in potency between *cis*- and *trans*-isomers in the octahydrobenzo(f)quinoline series (vide infra).
The sulfur isostere LXXI of 'TL-99' (XLVI: R = R' = CH_3) exhibited dopaminergic agonist effects in an assay of ability to inhibit potassium-induced release of ^{3}H-acetylcholine from mouse striatal slices [49]. In contrast to their amine analogs, chemical modifications of the sulfonium compounds produced little change in their dopamine agonist activity.

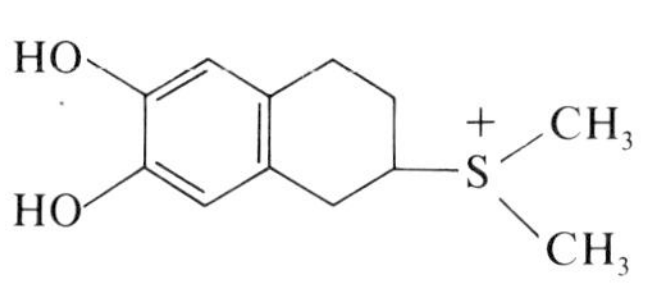

LXXI

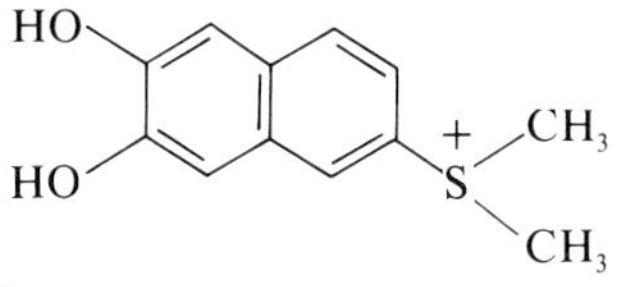

LXXII

The completely aromatic naphthalene derivative LXXII produced predominantly indirect action in this assay.
A variety of dopamine agonist effects has been reported [115, 136,188,189] for non-hydroxylated 2-aminotetralins LXXIII, which results may be interpreted as reinforcing the view that a catechol moiety is not essential for dopaminergic agonist effect.
However, the possibility that these compounds undergo in vivo metabolic hydroxylation has been cited [188,189,190], and this point seems

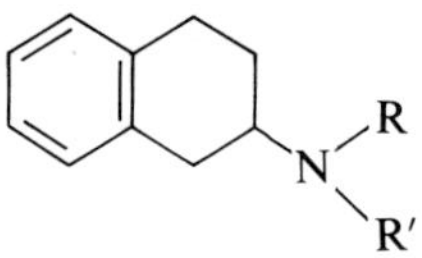

LXXIII

R, R′ = combinations of H, CH_3, C_2H_5, and n-C_3H_7

as yet unresolved in the literature. The finding that N,N-di-*n*-propylaminotetralin (LXXIII: R = R′ = *n*-C_3H_7) is active in both in vivo and in vitro assays for dopaminergic effects has suggested [191] that this compound does not require metabolic activation before interaction with dopaminergic receptors. This compound has been described [138] as a selective D-2 agonist, albeit of rather low potency. The primary amine homolog of LXXIII (R = R′ = H) [188], as well as secondary amines and tertiary amines bearing N-substituents larger than C_3H_7 [136] were reported to be inert as dopaminergics, although the primary amine and the N-ethyl- and *n*-propyl secondary amine homologs of LXXIII were reported [192] to induce ipsilateral or contralateral rotational behavior in rats.

Further variations in the tetralin-derived dopamine structure have involved placing the amino group one carbon from the ring, as illustrated in LXXIV [193,194].

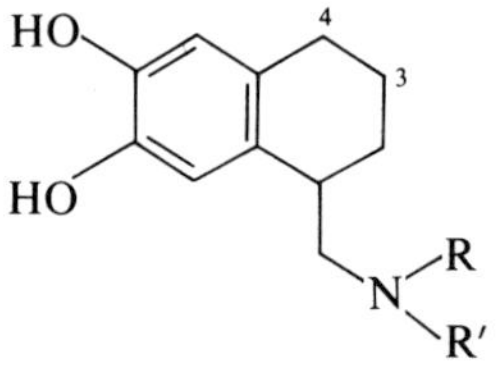

LXXIV

R, R′ = primary, secondary, or tertiary amines. combinations of H, CH_3, C_2H_5, n-C_3H_7

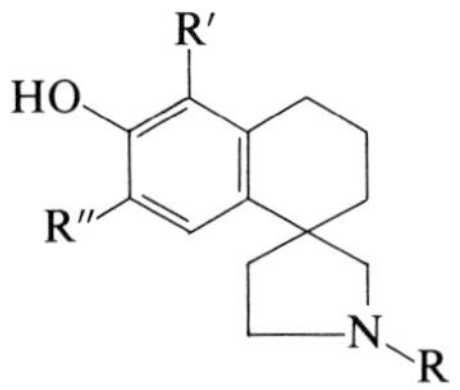

LXXVa R′ = OH; R″ = H

b R′ = H; R″ = OH

In these systems, the dopamine moiety can assume the α-conformation, with the catechol ring deviating from coplanarity with the ethylamine side chain by only a relatively small angle, all of which would seem to favor dopaminergic agonist effect. All of the derivatives of LXXIV showed extremely low activity or no activity in the cat cardioaccelerator nerve model [194] and/or in the canine renal vascular assay [193]. This inactivity was rationalized [194] on the basis that proper interaction of the catechol moiety and the amino function of LXXIV derivatives with the dopamine receptor(s) is inhibited the physical bulk in the molecule represented by carbons 3 and 4 of the tetralin ring. This rationalization seems applicable also to the *trans*-octahydrobenzo(h)isoquinoline system CLXXXVIII (vide infra).

Crooks et al. [195] reported that the spirotetralin systems LXXV a–b (R = H, CH_3) showed no or only weak dopamine agonist/antagonist effects. Inspection of molecular models of LXXV revealed [194] that the dopamine moiety cannot attain the catechol ring-amino nitrogen antiperiplanar disposition characteristic of the biologically significant α-rotamer of dopamine.

A naphthalene-derived dopamine system LXXVI [196] was found [179] to be inert as a dopaminergic agent, as was the N,N-di-*n*-propyl homolog.

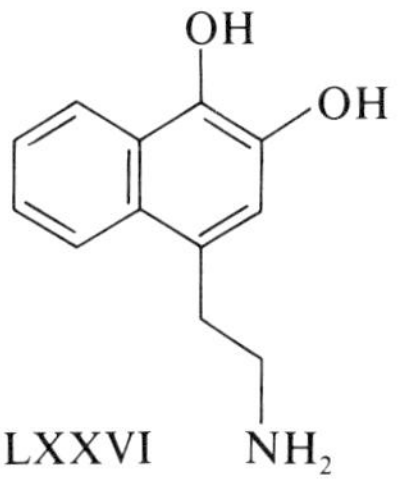

LXXVI

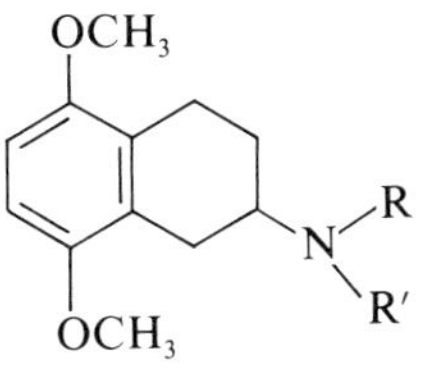

LXXVII

α_1Adrenoceptor agonist action has been reported [197] for certain derivatives of 5,8-dimethoxy-2-aminotetralin LXXVII. A derivative of LXXVII where R = R′ = *n*-C_3H_7 exhibits some dopamine receptor agonist activity in some assays [179,198].

The seven-membered ring benzocycloheptene congeners of most of the aminotetralin structural variants described above have been prepared, bearing a variety of N-substituents (H, CH_3, C_2H_5, *n*-C_3H_7).

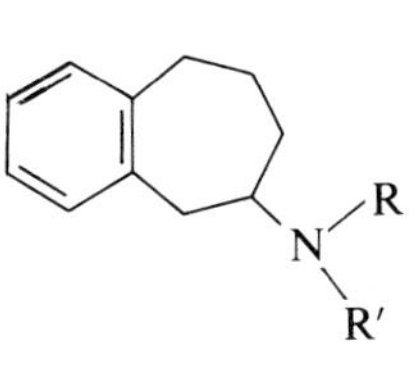

LXXVIII (192, 200)

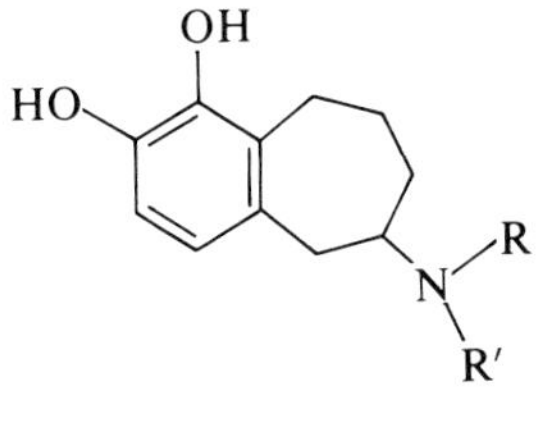

LXXIX (200)

[Structures LXXX (201) and LXXXI (202)]

Some pharmacological effects have been reported [199] for the non-oxygenated system LXXVIII R = R' = H), but these did not seem to involve a dopaminergic action. No significant dopaminergic effects were elicited by any of the derivatives LXXVIII-LXXXI. It was concluded [192,201] that the structural components of the benzocycloheptene ring system that maintain the plane of the aryl ring of the dopamine moiety in a steric disposition approaching perpendicularity with the ethylamine side chain (Newman projection LXXXII) are detrimental to dopamine-like activity.

[Structures LXXXII and LXXXIII]

6,7-Dihydroxy-3-chromanamine LXXXIII, an oxygen isostere of A-6,7-DTN (XLVI: R = R' = H), demonstrated a spectrum of dopaminergic agonist effects parallel to A-6,7-DTN, although the potency of the oxygen isostere was less [203]. Molecular models indicate that the dopamine moiety in LXXXIII assumes a β-conformation, as in A-6,7-DTN.

Dihydroquinazoline congeners LXXXIV and LXXXV of the 2-aminotetralin systems XLV and XLVI were inactive as dopaminergic agonists in the canine renal artery model [204].

[Structures LXXXIV and LXXXV]

Measured pK_a values revealed that these molecules were stronger bases than dopamine, and this was advanced as one possible reason for the lack of dopamine-like action.

Horn et al. [205], prepared dibenzoate esters LXXXVI and LXXXVII of A-5,6-DTN and of A-6,7-DTN, and noted that these penetrate the blood brain barrier of the rat much more easily than do the free catechol systems, to provide stable drug levels over many hours.

LXXXVI

LXXXVII

Constant brain levels in rats for at least 10 hours indicated that A-5,6-DTN leaves the brain only with great difficulty [206]. Accumulation of A-5,6-DTN in dopamine-rich areas of rat brain was observed after administration of the dibenzoyl derivative LXXXVI. Brain levels of both A-5,6- and 6,7-DTN following administration of the dibenzoyl prodrug were five times those when the free dihydroxy system was given, and the concentration of the 5,6-dihydroxy isomer was five to seven times that of the 6,7-dihydroxy isomer. By comparing striatal levels of both ADTN isomers after administration of the dibenzoates with data on their potencies to decrease homovanillic acid concentrations in the brain area, Westerink et al. [206], quantitatively determined the in vivo activity of the two isomers. A-6,7-DTN was twelve to fourteen times *more potent* in this assay than was A-5,6-DTN. As has been cited and discussed previously, these workers [206] ascribed to metabolic actions (via COMT) on the ADTN isomers a critical role in determining in vivo CNS activity and variations in brain concentrations.

4.2 Aminoindans

LXXXVIIIa R = R' = H
b R = R' = CH_3
c R = R' = C_2H_5
d R = R' = n-C_3H_7
e R = H; R' = CH_3
f R = H; R' = C_2H_5
g R = H; R' = n-C_3H_7
h R = H; R' = 2-C_3H_7

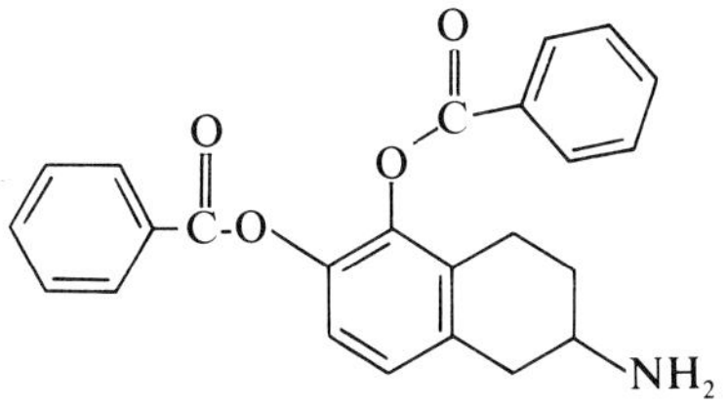

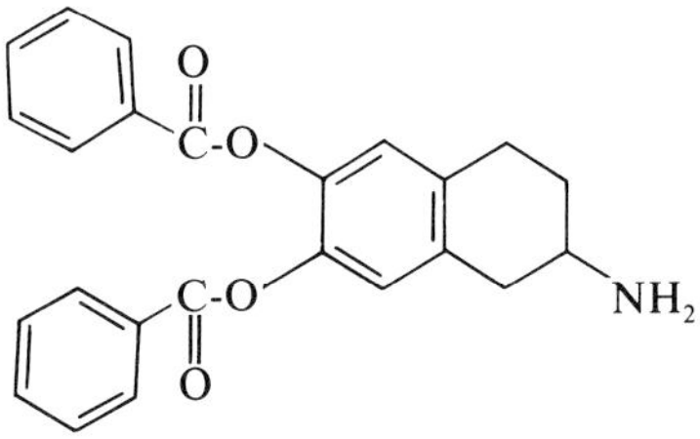

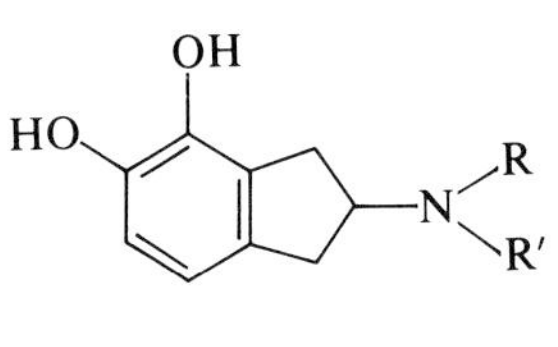

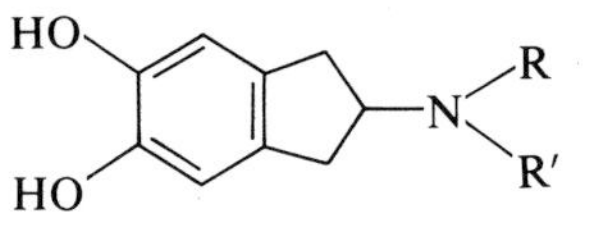

HO-[ring]-N(R)(R') R, R' = combinations of H, CH$_3$, C$_2$H$_5$, n-C$_3$H$_7$
HO

LXXXIX

4,5-Dihydroxy-2-dimethylaminoindan LXXXVIII b was approximately four times as potent an emetic as apomorphine in the pigeon, but in the dog emesis model it was 0.008 times as potent as apomorphine [135]. Compound LXXXVIII b resembled apomorphine and the analogous aminotetralin in a rat circling assay, although the indan derivative was less potent [207]. The primary amine LXXXVIII a was inert in these assays, unlike the aminotetralin analog. Homologs LXXXVII c and d (N,N-diethyl- and di-n-propyl) were approximately equipotent to apomorphine in blocking response to stimulation of the cat cardioaccelerator nerve, an index of peripheral presynaptic dopaminergic effect [208]. All other members of the series LXXXVIII, including the secondary amines e–h and the dimethyl ethers of all of the derivatives, were inert. Compounds LXXXVIII c and d were potent emetics in the dog. None of the compounds in the series LXXXVIII were effective in displacement of ^{3}H-spiperone or ^{3}H-A-6,7-DTN from calf striatal tissue homogenate. No members of the hydroxyl group positional isomers LXXXIX showed any dopamine-like activity in the dog emesis assay or the cat cardioaccelerator nerve model, nor were they effective in binding studies [208]. The inactivity of this series LXXXIX contrasts with the high potency/activity described for the structurally analogous 6,7-dihydroxy-2-aminotetralins, but it is consistent with the inactivity of isoapomorphine XXXIV. A resorcinol congener XC displayed only very weak dopamine-like effects [201].

OH
[resorcinol-indan]-N(C$_3$H$_7$-n)(C$_3$H$_7$-n)
HO

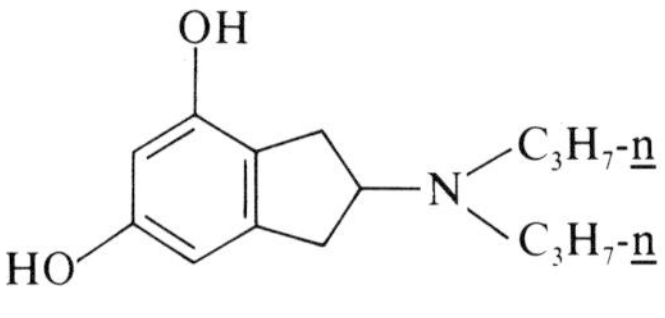

XC

Hacksell et al. [202] found that 2-di-n-propylamino-4-hydroxyindan XCI is much more potent and active in a series of CNS dopaminergic assays than is the isomeric-5-hydroxyindan XCII.

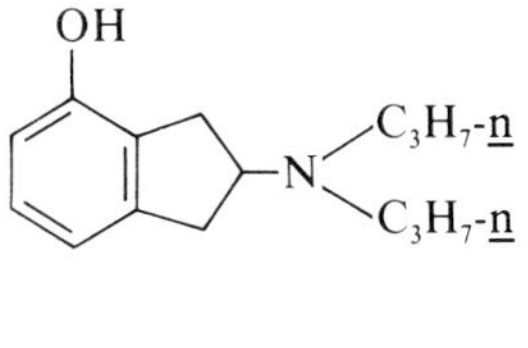

XCI

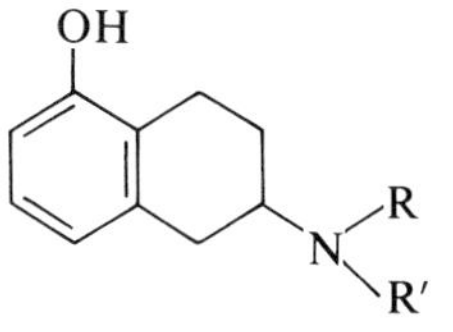

LI

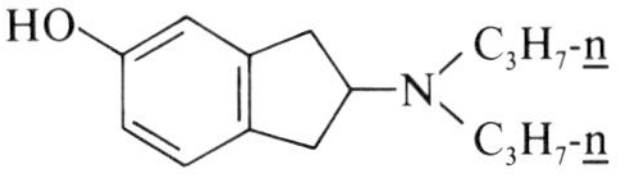

XCII

These results are consistent with the observation of higher potency residing in the analogously substituted 5-hydroxytetralin LI, and they demonstrate the importance of the meta-OH in the α-conformer of dopamine. Compound XCI was resolved, and the R-enantiomer XCIII was approximately 100 times as potent as the S- in an assay for dopamine agonist effect in the isolated cat atrium [209].

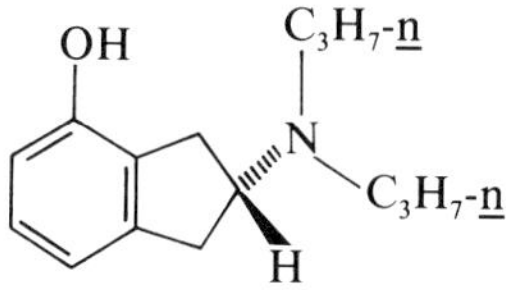

XCIII

The stereochemistry of XCIII coincides with that of the more active enantiomers of apomorphine XXVIII, 5-hydroxy- and 5,6-dihydroxy-2-aminotetralins LVI, and 7-hydroxyoctahydrobenzo(f)quinoline CLXXVIII (vide infra).

A critical factor in assessment of structure-activity relationships of 2-aminoindans is (as was presented for 2-aminotetralins, aporphines, and β-phcncthylamines) a conformational analysis of the system. NMR studies [210,211] have led to the conclusion that the cyclopentene ring of 2-substituted indan systems is not planar, and that the 2-substituent adopts a *pseudo*-equatorial disposition, possibly in an attempt of the molecule to relieve the extensive eclipsing of bonds extant in the all-planar conformation of the molecule. Molecular models suggested [208] that the torsion angle N-C_1-C_2-C_3 (structure XCIV) could approach 140–150°. The Newman projection XCV shows that in the 2-aminoindan system (where the torsion angle N-C_1-C_2-C_3 is 140-150°) the β-phenethylamine moiety approaches the antiperiplanar disposition, and the benzene ring is approximately coplanar with the ethylamine side chain, which conformation has been proposed [16,17] to be required for interaction with the dopamine receptor(s). Thus, the dopamine moieties of the two hydroxyl group positional isomers LXXXVIII and LXXXIX, as illustrated in the Newman projections XCVI and XCVII, approximate, respectively, the α- and β-conformers of dopamine, although the degree of correspondence of the indans is not as close as is the case of the 2-aminotetralins, whose analogous torsion angle is 180° (structures XCVIII, XCIX).

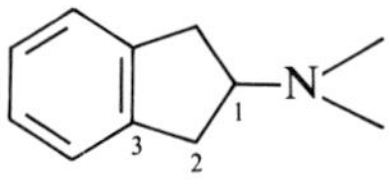

XCIV

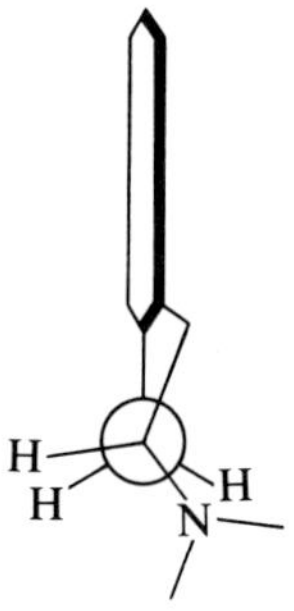

XCV

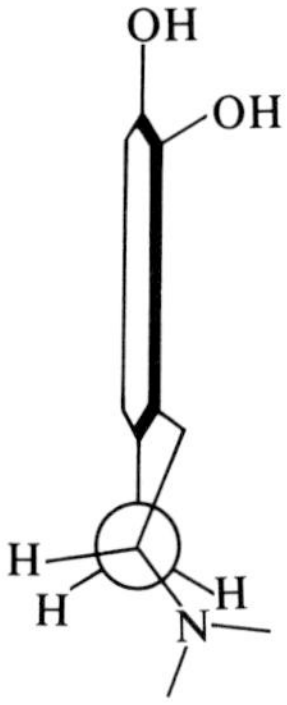

XCVI

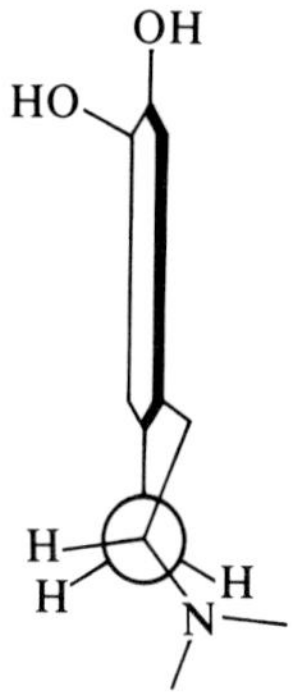

XCVII

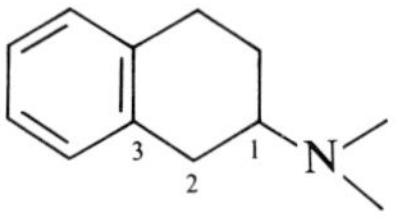

XCVIII

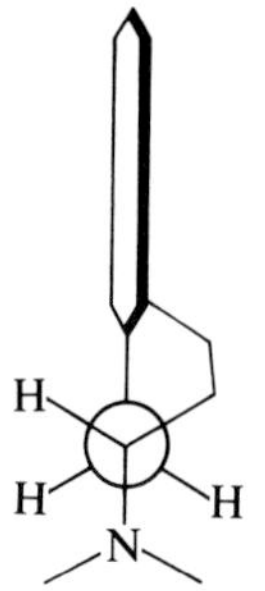

XCIX

In sharp contradiction to this molecular model-based conformational analysis are the results of an X-ray crystallographic study [209] of the *d*-tartrate salt of C, in which the torsion angle N-C_1-C_2-C_3 was determined to be –90°.

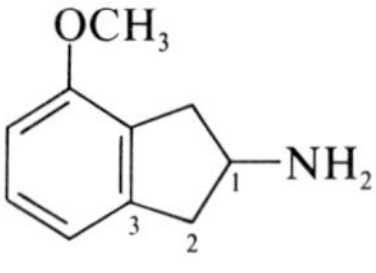

C

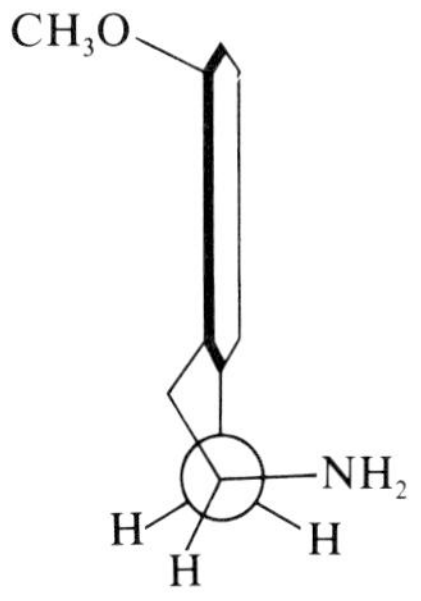

CI

Thus, the amino group is in a *pseudo*-axial disposition with respect to the 5-membered ring (structure CI), and the N-C_1 bond is nearly perpendicular to the plane of the bezene ring (*gauche* conformation). This steric disposition of the β-phenethylamine structure seems completely detrimental to dopaminergic agonist activity, on the basis of conformational analyses of other dopaminergically active and inactive molecules. However, it cannot be assessed whether the conformation observed in the crystalline state for the D-(–)-tartrate salt of a methyl ether of a primary amine has any relevance to the solution conformation of the very potent free phenolic N,N-di-n-propyl tertiary amine XCI when it interacts with in vivo dopamine receptor(s).
4,7-Dimethoxy-2-di-n-propylaminoindan CII demonstrated equal activity with apomorphine in activation of peripheral presynaptic dopamine receptors [212].

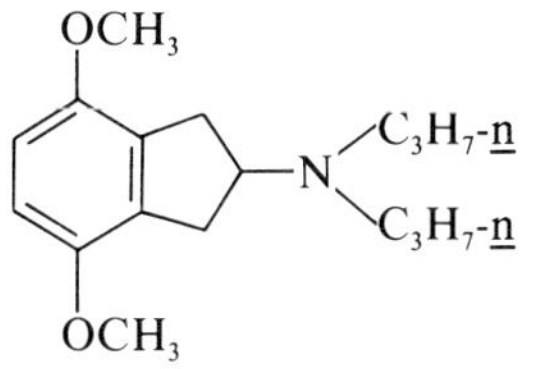

CII

The compound also activated central pre- and postsynaptic receptors, but it elicited only weak, transient effects in heart rate and blood pressure assays which would be indices of sympathomimetic effects. Evidence has been presented [213] suggesting that CII activates central dopamine receptors to produce hypotension and bradycardia in rats. Compound CII was significantly more selective than apomorphine in activating dopamine autoreceptors (as opposed to postsynaptic receptors) in the rat nigrostriatal pathway [198]. In most assays, the indan derivative CII was more potent/active than its tetralin congener LXXVII. Primary and secondary amino homologs of CII elicited little or no dopaminergic pre- or postsynaptic actions. Compound CII and its tetralin congener LXXVII represent a non-classical dopaminergic agonist moiety, '*p*-dimethoxy systems', for which a chemical structural rationalization of agonist activity is not readily apparent. Compound CII elicits potent effects on sexual behavior in male rats [214–216], for which a dopamine-related spinal mechanism has been suggested [215]. Gaino et al. [217] claimed adrenergic β_2 activity for the 1-aminomethylindan derivatives CIII, but no pharmacological data were provided. Nichols et al. [193], have reported that the primary amino analog (CIII: R = H) is, like its tetralin-derived congener LXXIV, devoid of dopamine-like effects in the canine renal artery assay.

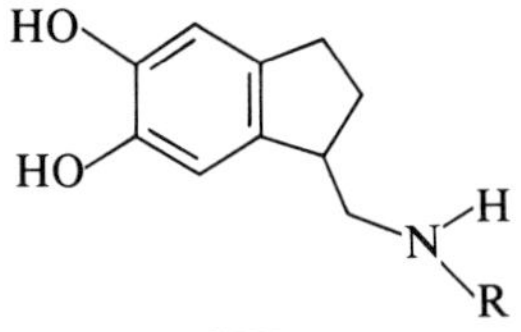

CIII

R = H; CH_3; 2-C_3H_7

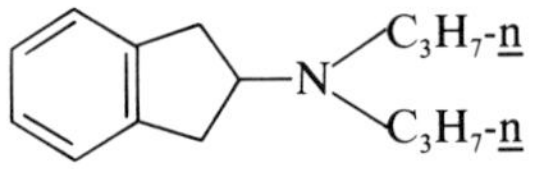

CIV

A non-oxygenated indan derivative CIV elicited a variety of dopaminergic effects [188]. The N,N-di-*n*-propyl substitution pattern was the sole active one in the brief series studied.

5 Ergot alkaloid derivatives, congeners, and fragments (partial structures)

5.1 Ergoline derivatives

The ergoline ring system CV, upon which the dopaminergically significant ergot alkaloids are based, contains a β-phenethylamine moiety.

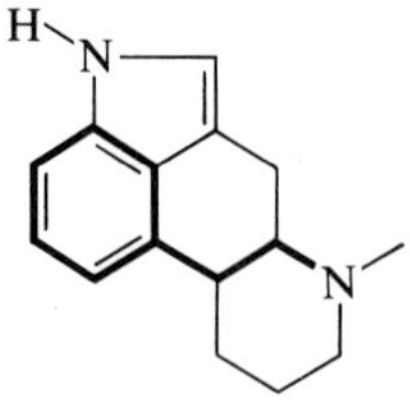

CV

CVI   R = CH$_3$; R' = Cl; R'' = –CH$_2$CN
CVII R = n-C$_3$H$_7$; R' = H; R'' = –CH$_2$–S–CH$_3$
CVIII R = CH$_3$; R' = Br; R'' = –C = O; $\Delta^{9,10}$

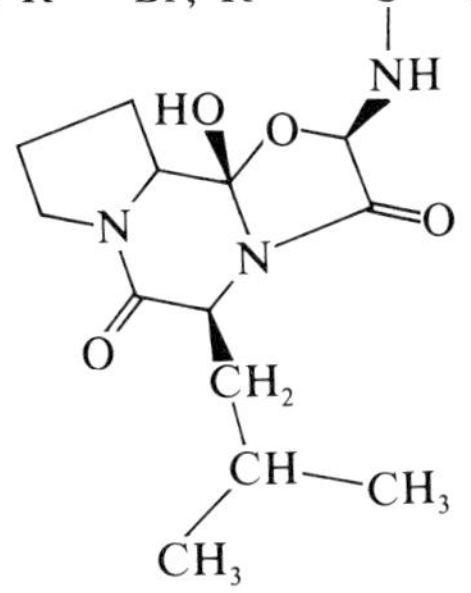

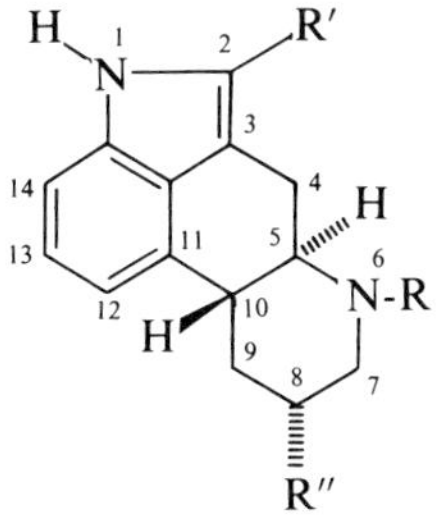

CIX $\Delta^{9,10}$; R = CH$_3$; R' = H; R'' = H;

$$C_8 = \begin{matrix} \text{N-H} \\ | \\ \text{C=O} \\ | \\ \text{N} \\ \diagup \quad \diagdown \\ \text{H}_5\text{C}_2 \quad \text{C}_2\text{H}_5 \end{matrix}$$

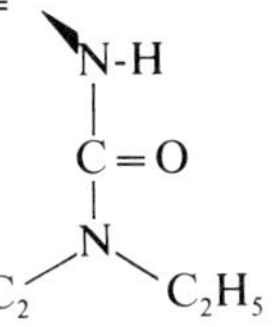

Central dopaminergic agonist properties of the semisynthetic ergoline derivatives lergotrile CVI, pergolide CVII, bromocriptine, CVIII, and lisuride CIX are established [218–222], and CNS dopaminergic activity in man has been described for some of these [223–225]. Pergolide CVII has been described [226] as having the highest relative selectivity for presynaptic dopamine receptors (as compared to postsynaptic receptors) in a heterogeneous series containing ergoline derivatives, oxaergoline derivatives, aporphines, aminotetralins, and β-phenethylamines. However, there was no correlation between overall potency and selectivity for the dopamine autoreceptor.

A considerable body of studies [227–231] has led many investigators to suggest that the ergoline derivatives CVI–CIX are not pure dopaminergic agonists, but rather that they are partial agonists or mixed agonists-antagonists. In addition to their effects on dopaminergic mechanisms, some ergolines have a high affinity for α-adrenoceptors and for serotonin receptors [232]. Lisuride CIX promotes mounting behavior in rats [222], an effect previously cited for the 4,7-dimethoxyindan derivative CII, and this action of lisuride has been ascribed, in part, to 'catecholaminergic activation'.

Wong and Bymaster [233] and Rubin et al. [234] reported that 13-hydroxylergotrile CX, a metabolite of lergotrile, had greater affinity than lergotrile in a dopamine agonist binding assay, and that CX was more potent than lergotrile in inhibiting prolactin release from the anterior pituitary in infrahuman species. Parli et al. [235] reported that 13-hydroxylergotrile was 100 times more active than lergotrile in vitro in inhibiting prolactin release from the anterior pituitary.

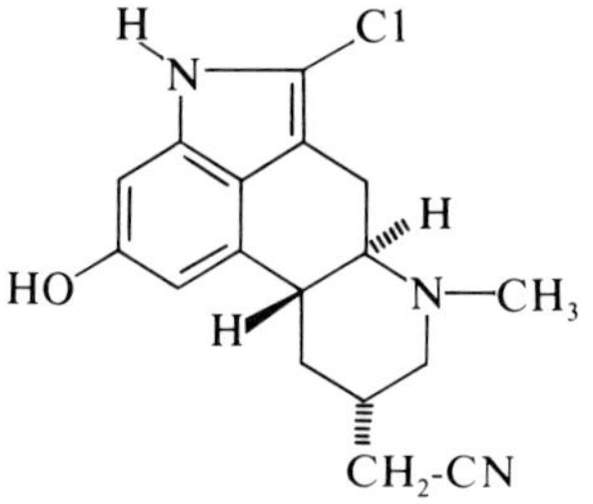

CX

However, it seems that in vivo metabolic hydroxylation of ergoline derivatives does not always occur, nor is it a prerequisite for dopaminergic agonist activity in all of the ergoline derivatives.

Kehr et al. [236] described powerful dopaminergic agonist properties in a spectrum of assay models for N-*n*-propylnorlisuride CXI, *trans*-dihydrolisuride CXII, and N-*n*-propyl *trans*-dihydronorlisuride CXIII.

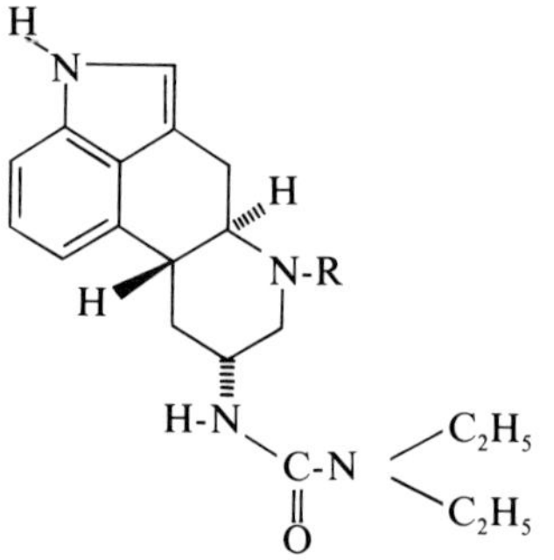

CXII $R = CH_3$
CXIII $R = n\text{-}C_3H_7$

The N-*n*-propyl congeners were, in general, more potent agonists than the N-methyl homologs. It was concluded that probably, *trans*-dihydrolusuride CXII combines both agonistic and antagonistic properties, and it acts as a partial agonist at least at some dopamine autoreceptors. Kehr et al. [236] noted that hydrogenation of lisuride at position 9,10 shifted the dopaminergic profile to an antidopaminergic one. The intent of this comment is not apparent, in that lisuride has been reported [231] to have mixed agonist-antagonist properties, and the Kehr et al. group have described agonist actions for the *trans*-dihydrolisurides. Stütz et al. [237] have presented a brief historical development of use of ergolines as central dopaminergics, and they have cited six of the most promising candidates (CXIV-CXIX) resulting from their studies.

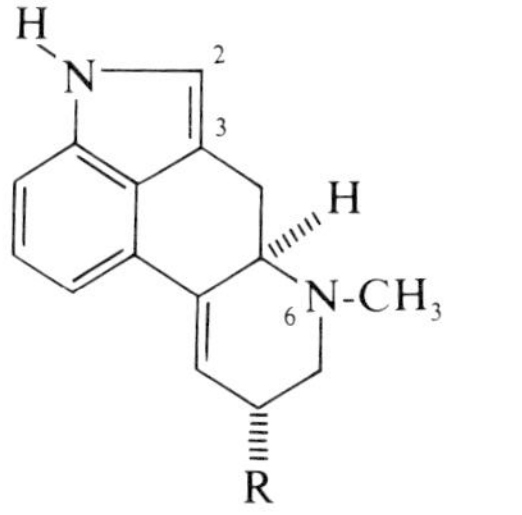

CXIV R = CH_2–S–CH_3
CXV R = CH_2–S

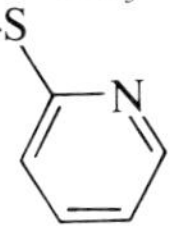

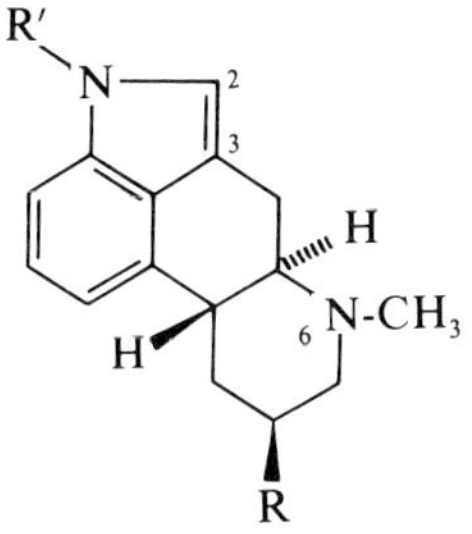

CXVI R = CH_2CN; R′ = H
CXVII R = $NHSO_2N(CH_3)_2$; R′ = H
CXVIII R = $NHSO_2N(C_2H_5)_2$; R′ = H
CXIX R = $NHSO_2N(CH_3)_2$; R′ = CH_3

A considerable number of structural variations was made on these compounds (e. g., halogenation at position 2; variation of the N_6 substituent; reduction of the $\Delta^{2,3}$ double bond). Compounds CXVII and CXIX inhibited prolactin secretion in the rat [380]. It was reported that the N_1-methyl compound CXIX is a partial agonist which becomes fully active in vivo only after metabolic conversion (presumably, to the N_1-desmethyl system CXVII). The N_1-methyl compound CXIX showed hypotensive action, like bromocriptine, mainly due to an action on dopaminergic receptors [381]. The fact that a 2,3-dihydro system CXX and its N_1 formyl derivative CXXI displayed dopaminergic effects suggested [237] that an indole ring might not be absolute prerequisite for high central dopaminergic activity. However, an alternate explanation is that both CXX and CXXI are prodrugs to the 2,3-unsaturated molecule CXXII.

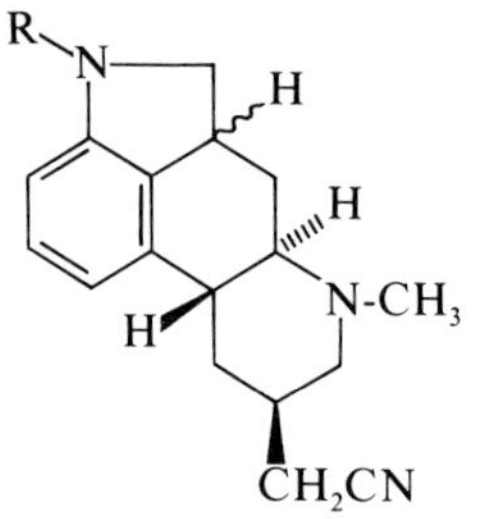

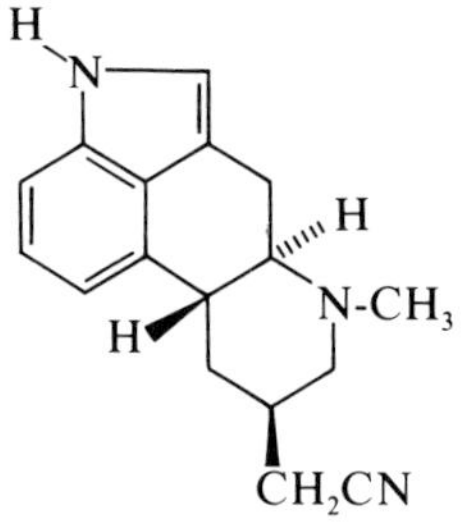

CXX R = H
CXXI R = CHO

CXXII

A prerequisite for central dopaminergic activity in 9,10-dihydroergolines is a *trans*-junction of the C/D rings, as in CXII, CXIII, CXVI-CXIX, CXX and CXXI. The effects of variation of the alkyl substituents on N_6 (the basic nitrogen) resemble those produced on apomorphine: activity increases from methyl to ethyl to *n*-propyl, but diminishes with *n*-butyl, and isopropyl reduces activity to zero [237]. It has been stated [237] that a hydrogen atom in position 1 (indole ring nitrogen) is not essential, as evidenced by the retention of activity in rat rotation and stereotypy assays in a series of N_1-methyl ergolines [238]. However, this series seemed to demonstrate a trend of lower potency/activity in the N_1-methyl homologs, which was also noted [239] in the prolactin release assay. An N_1-methyl ergoline derivative was reported [240] to have initial dopamine antagonist actions, but it underwent in vivo metabolic N_1-demethylation and it then displayed agonist actions. Assessment of the possible role of metabolic N_1-demethylation in the dopaminergic effects of other N_1-methyl ergolines has not appeared in the literature.

Halogenation of ergoline derivatives seems to produce no consistent effects. Lergotrile CVI and bromocriptine CVIII bearing, respectively, chlorine and bromine atoms at position 2, are highly potent/active compounds. However, Stütz et al. [237] found that 2-chlorine substitution of the position 8 epimer of lergotrile results in clearly reduced potency, and Wachtel and co-workers [241] observed that 2-bromolisuride showed central pre- and postsynaptic dopamine receptor blocking action of the same order of potency as haloperidol.

In a series of 6-methylergolines CXXIII, a 2-formyl subtituent resulted in 'not significant' prolactin-inhibiting action [242].

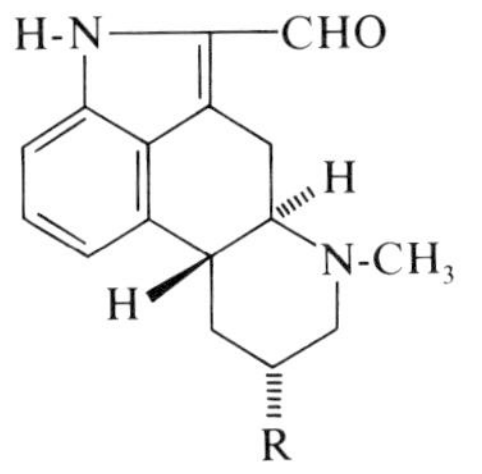

CXXIIIa R = $COOCH_3$
b R = CH_2CN
c R = CH_2CONH_2
d R = CH_3
e R = CH_2Cl

The absolute configuration at C-5 of the ergoline ring system (structure CXXIV) of naturally occurring derivatives is R [243, 244]; and, where comparison has been possible, molecules with the opposite absolute configuration (5-S) are biologically inactive [245].

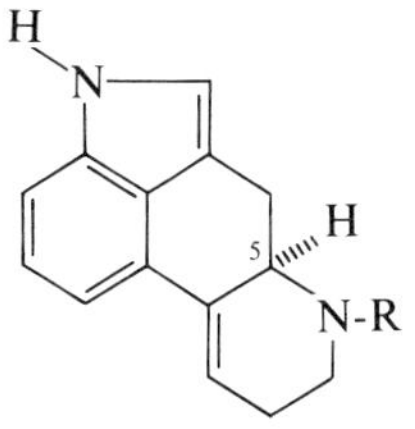

CXXIV

As has been illustrated in structural drawings, the dopaminergically active semisynthetic ergot derivatives lergotrile CVI, pergolide CVII, bromocriptine CVIII, and lisuride CIX also have the R-configuration at C-5 [219, 243, 246]. The enantiomers of CVI–CIX have apparently not been reported in the literature. However, Burt et al. [247] found that the ability of LSD CXXV to displace ^{3}H-dopamine and ^{3}H-haloperidol from dopamine binding sites in calf caudate homogenates resides mainly in the (+)-enantiomer which has the 5-R, 8-R configuration, as illustrated.

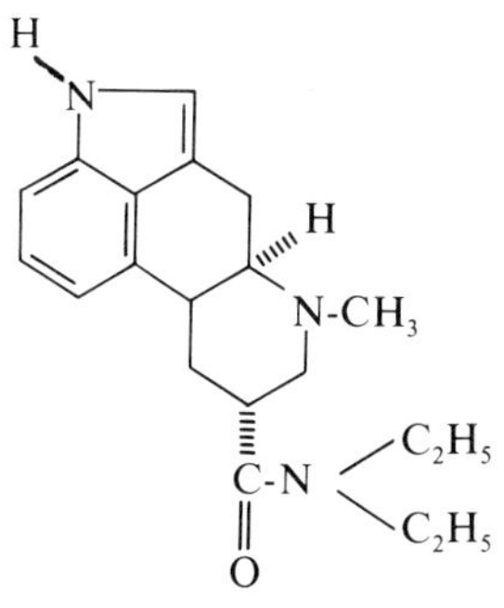

CXXV

Hofmann [248] has stated that the 5-R, 8-S diastereomer of LSD is devoid of ^{3}H-dopamine- and ^{3}H-spiroperidol blocking activity. Flückiger et al. [249] have reported that C-8 *iso*bromocriptine (8-S absolute configuration) is inert as a dopamine receptor agonist. It might seem reasonable to predict that the 8S-diastereomers of the other ergoline-derived systems would similarly be inert as dopaminergics, and it also seems reasonable to extrapolate that enantiomers of the dopaminergically active ergoline derivatives would be devoid of agonist effects. However, Stütz et al. [250] noted that one C-8 epimer of CXXVI was significantly the more active in a rat turning model, whereas the other epimer was the more active in inducing apomorphine-like stereotypy.

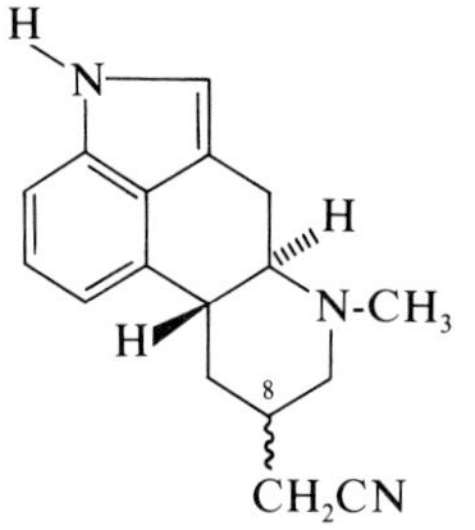

CXXVI

Stütz and co-workers [237] have cited X-ray crystallographic studies of ergotamine CXXVII and bromocriptine CVIII, showing that in the crystal state, the preferred conformation of the C-8 substituent (carboxamido group) in ergotamine is *equatorial*, but in bromocriptine the preferred conformation is *axial.*

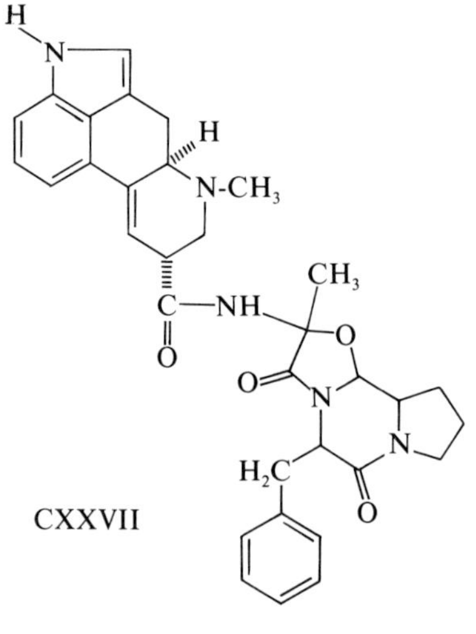

CXXVII

The C-8 substituents on the dopaminergically active ergoline derivatives cited previously are not a part of the β-phenethylamine or of the pyrrole-3-ethylamine moiety, nor of the indole-4-ethylamine moiety, which have been proposed by many workers to be the pharmacophoric portion(s) of the molecules. Nevertheless, it is appealing to speculate that the steric disposition of the C-8 substituent in the ergoline system is an important factor in agonist-receptor interactions.
As was described previously for the 7-hydroxy- and 6,7-dihydroxy-2-aminotetralins, the dopaminergically active ergoline derivatives appear to have the 'wrong' absolute configuration at the chiral center (position 5) which forms a part of the β-phenethylamine moiety, in comparison with the analogous carbon (6 a) in the R-(–)-apomorphine molecule (structure XXVIII). It will be noted that application of the Cahn-Ingold-Prelog rules to ergoline and to apomorphine requires that the analogous chiral centers in the two molecules, which have the *opposite* spatial orientations, be both designated as R. Nichols [251] rationalized the differences in absolute configurations of apomorphine and the ergolines upon the premise that the rigidly held pyrrole ethylamine moiety of the ergoline system corresponds to the β-3,4-dihydroxyphenethylamine system within the apomorphine molecule (structures CXXVIII and CXXIX):

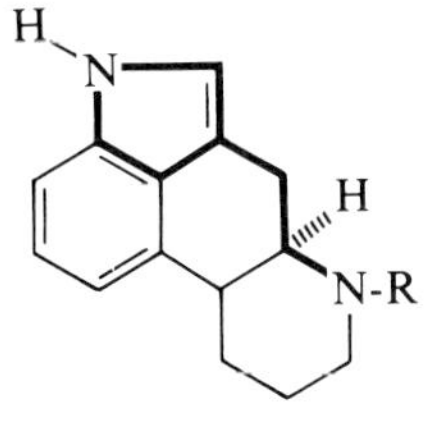

CXXVIII

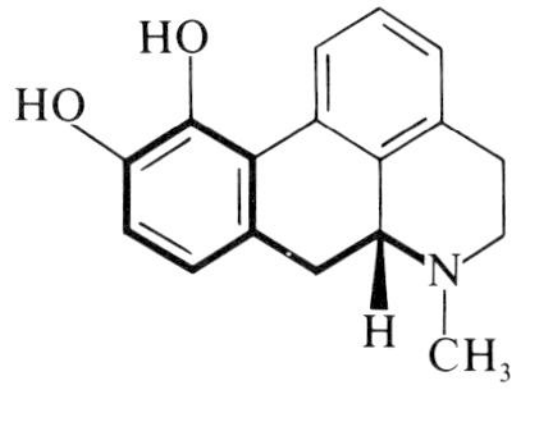

CXXIX

Priority was placed on correspondence at the chiral carbons and the aliphatic nitrogens, as illustrated in structures CXXX and CXXXI:

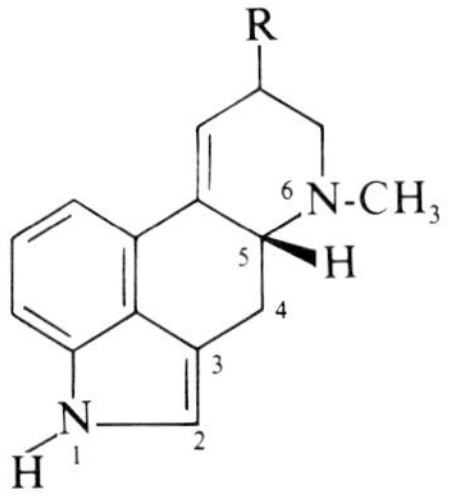

CXXX

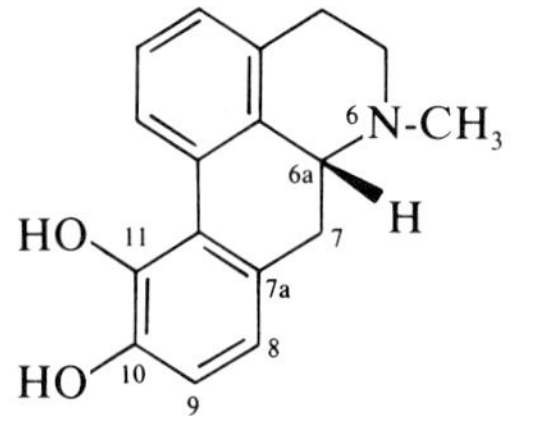

CXXXI

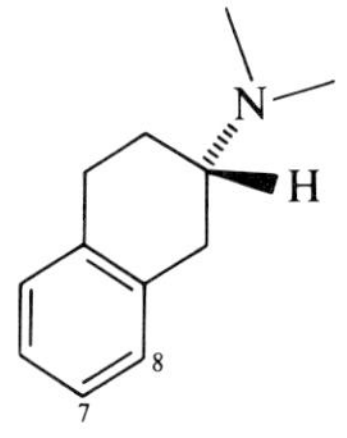

CXXXII

The 2,- 3,- 4,- 5,- and 6-positions of the ergoline system CXXX were proposed to correspond to the 6-, 6a-, 7-, 7a-, and 8-position of apomorphine CXXXI. Nichols suggested that the C-10 hydroxy of apomorphine is in about the same spatial position as the indole N-H of the ergoline (N-1). The 2-position of the ergolines corresponds to the 8-position of apomorphine and of the 2-aminotetralins CXXXII. Molecular models reveal that the distance between N-6 and the indole nitrogen (N-1) in the ergoline system is approximately 6.5 A, and the nitrogen-to-C-10 distance in apomorphine is approximately 7.8 A. It seems to follow from Nichols' proposals that the indole N-H of the ergolines and the C-10 ('para') hydroxyl of apomorphine are biologically equivalent for interaction with the same subsite on the dopamine receptor(s). However, consideration of pharmacological data on the isomeric monohydroxyaporphines (vide supra) indicates that the 11-hydroxy derivative is decidely more potent than the 10-hydroxy isomer [110]. The dopamine receptor map described by Wikström [166] seems to accommodate the structural features of the 5-R ergolines as well as their biologically active 13-hydroxy metabolites, although Wikström has proposed that the hydroxylated ergoline molecule binds in a somewhat different manner.

Camerman et al. [246] provided a structural comparison of the ergoline portion of bromocriptine CVIII with apomorphine, taken as a prototypical dopaminergic agonist. Their demonstration (illustrated in figure 2) involved superimposition of the ergoline ring system of bromocriptine upon 6a-R-apomorphine (the biologically active enantiomer) and upon 6a-S-apomorphine (the biologically inactive enantiomer), so that the 'dopamine-like' features of bromocriptine (illustrated in CXXXIII: ring A, ring B, C-10, C-5, N-6) most closely overlap the corresponding entities on the aporphine molecule.

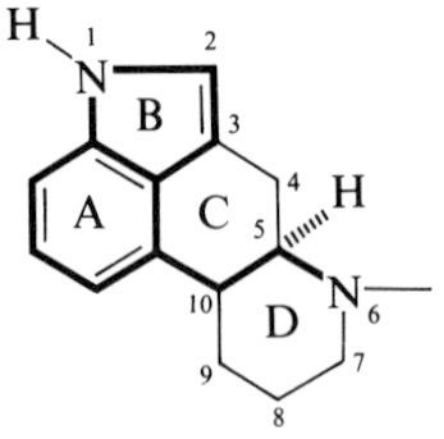

CXXXIII

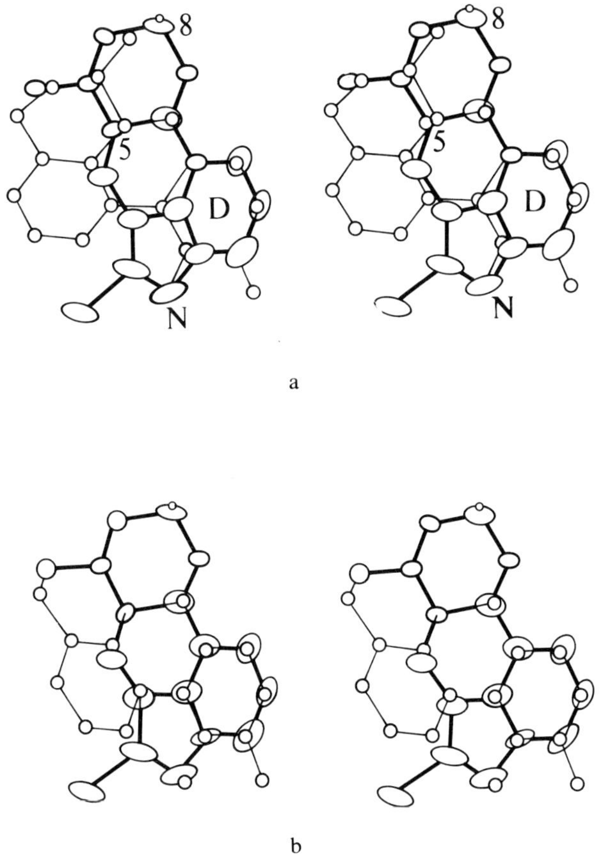

Figure 2
Structural relationships of ergoline ring and apomorphine.

The Camerman group concluded that the degree of fit for bromocriptine is better with the pharmacologically inactive enantiomer of apomorphine. It seemed obvious that dopaminergic activity could not be rationalized on the basis of 'goodness of fit'. The Camerman group suggested that the main difference in the conformations of the two enantiomers of apomorphine involves the two rings *not* containing the dopamine moiety (rings C and D, left edge of the molecules, fig. 2). In R-(–)-apomorphine the conformation is such that these rings bend away from the viewer, while in S-(+)-apomorphine, they are oriented sharply up toward the viewer. It was concluded that the approach of these agonists to the dopamine receptor must be from the side of the molecules closest to the viewer. The differing activities of the apomorphine enantiomers can then be rationalized by proposing that the upward extension of the two rings *not* containing the dopamine moiety in S-(+)-apomorphine blocks the close approach of the dopaminergic portion of the molecule to the receptor. Thus, the critical factor seems to be the change in spatial disposition of the entire molecule, which is controlled by the nature of the chiral center. Inspection of molecular models of the 8-R- and 8-S-diastereomers of an ergoline system (CXXXIV and CXXXV) suggests that if it is assumed that the approach of the ergoline molecules to the dopamine receptor(s) occurs at the side of the molecule closest to the viewer, the change in stereochemistry at position 8 to the S (CXXXV) will cause the 'R' substituent to interfere sterically with close approach of the molecule to the receptor from the front side. Thus, a rationalization is provided for the observed stereospecificity of the C-8 position of some of the ergoline-derived dopaminergics.

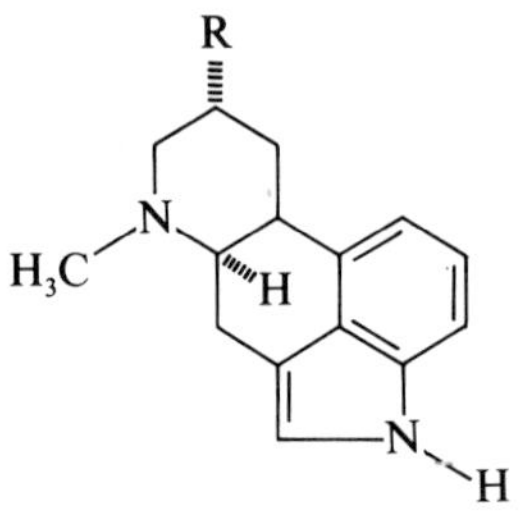

CXXXIV (5R, 8R)

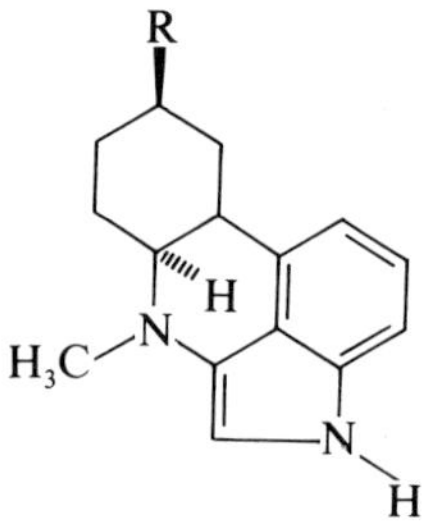

CXXXV (5R, 8S)

On the basis of X-ray crystallographic studies of bromocriptine free base, bromocriptine hydrochloride, and apomorphine hydrochloride, Camerman and Camerman [50] noted that despite the differing chiralities in bromocriptine and apomorphine, the direction of orientation of the nitrogen lone electron pair is the same for both. These workers concluded that the direction of the nitrogen atom lone pair electrons is dependent upon environment, and the absolute configuration at the asymmetric carbon atom does not definitely fix the nitrogen lone pair orientation.

A new chemical class, *cis*- and *trans*-6-ethyl-9-oxaergolines CXXXVI represents a modification in which the π-electron center present in the $\Delta^{9(10)}$ ergolines (e. g., bromocriptine CVIII or lisuride CIX has been replaced by the unoccupied *p* orbitals of oxygen [252, 253]. In dopamine binding assays vs. ^{3}H-apomorphine and in a rat rotation model, activity peaked with the N-*n*-propyl substituent, and dropped with *n*-butyl [252].

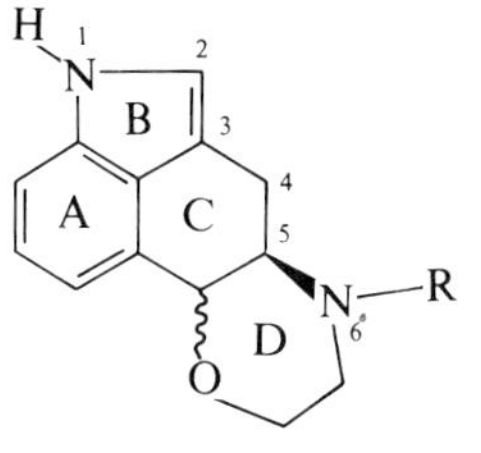

CXXXVI

R = H; CH_3; C_2H_5; n-C_3H_7; allyl; n-C_4H_9

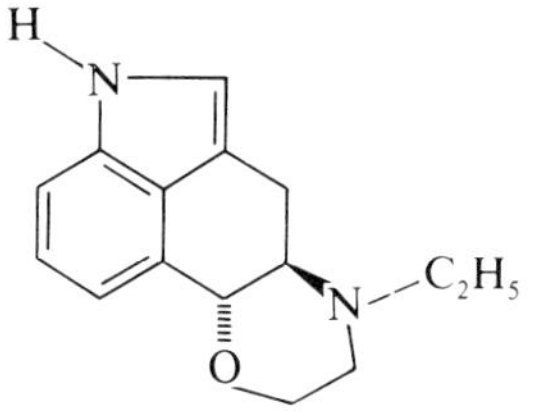

CXXXVII

Activity resided almost in toto in the *trans*-fused C/D system, and resolution of the *trans*-N-ethyl derivative showed an R,R-configurational requirement (structure CXXXVII) for dopaminergic actions [254, 255], which is consistent with the absolute configurations of the active ergoline derivatives. R,R-CXXXVII is also more potent at 5-HT_1 and 5-HT_2 receptors [256]. In a dopamine-sensitive adenylate cyclase assay, CXXXVII was a partial agonist, whereas its S,S-enantiomer was inert. However , at α_1 and α_2 adrenoceptors, the S,S-enantiomer was more potent [256]. Boissier et al. [257] found that the (±)-*trans*-N-*n*-propyl homolog (CXXXVI: R = *n*-C_3H_7) had a similar dopaminergic agonist profile to pergolide CVII at striatal and anterior pituitary sites. Derivatives CXXXVIII of the *trans*-oxacrgoline system lacking the pyrrole ring have been reported by two groups [258–260].

The (+)-enantiomer of the 9-hydroxy derivative CXXXVIIIb is a di-

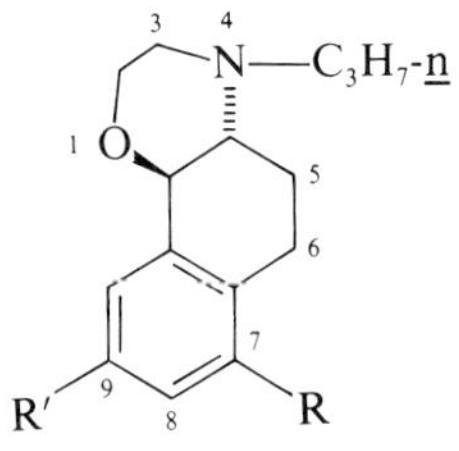

CXXXVIIIa R′ = H; R = OH
b R′ = OH; R = H

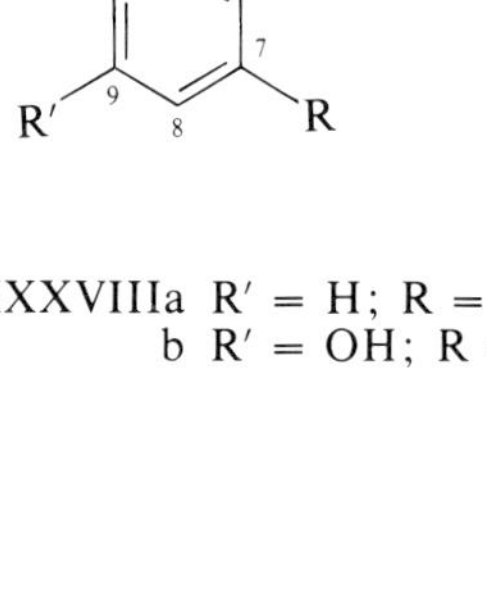

rect dopaminergic agonist in several in vivo and in vitro models, and it has been described [258] as one of the most potent dopaminergic agonists yet discovered. Horn et al. [259] found that the hydroxyl group positional isomers CXXXVIII a–b exhibit greater selectivity of pharmacological action than do the oxaergoline systems CXXXVI, due to a lower affinity of CXXXVIII a-b for adrenergic and for 5-HT receptors. Unexpectedly, the 9-hydroxy isomer CXXXVIII b, rather than the 7-hydroxy isomer CXXXVIII a, displayed the same high potency as the oxaergoline CXXXVII. It was suggested that the hypothesis (vide infra) that the N-H of the pyrrole ring of ergolines is equivalent pharmacologically to the 'meta-OH' of dopamine may be an oversimplification.

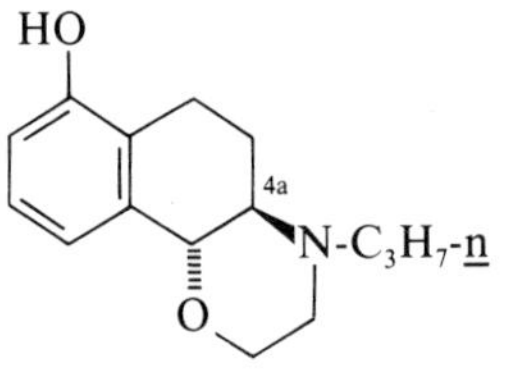

CXXXVIIIa

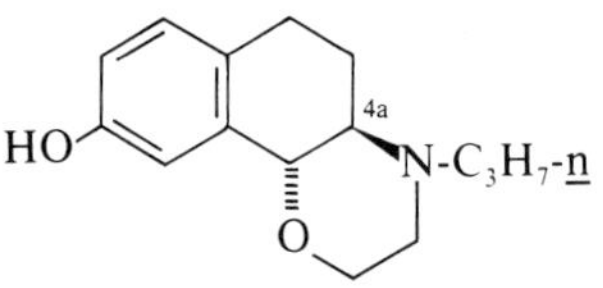

CXXXVIIIb

Compounds CXXXVIII a–b can be viewed as 2-aminotetralin derivatives; the more active hydroxyl group positional isomer CXXXVIII b is a congener of 7-hydroxy-2-aminotetralin, and the less active positional isomer CXXXVIII a is a congener of 5-hydroxy-2-aminotetralin, for which data have been cited (vide supra) demonstrating that the 5-hydroxy-2-aminotetralin is consistently the more potent/active dopaminergic agonist as compared to the 7-hydroxy. It is noted that the absolute configuration of carbon 4a in CXXXVIII b is the same as that of the analogous carbon in the more active enantiomer of 2-amino-7-hydroxytetralin [158, 164]. Horn and co-workers [259] noted that the position of hydroxylation in CXXXVIII b corresponds to that in the hydroxy substituted metabolite of lergotrile (structure CX), which is an extremely potent dopaminergic agonist. The non-hydroxylated congener CXXXIX and the catechol derivative CXL have been cited [261], but no biological data were specified for them.

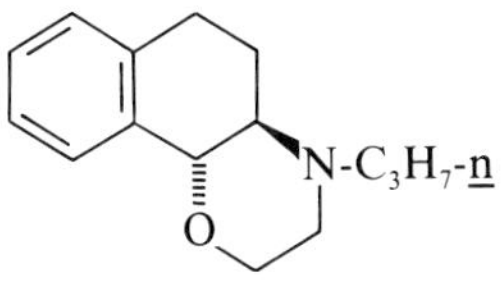

CXXXIX

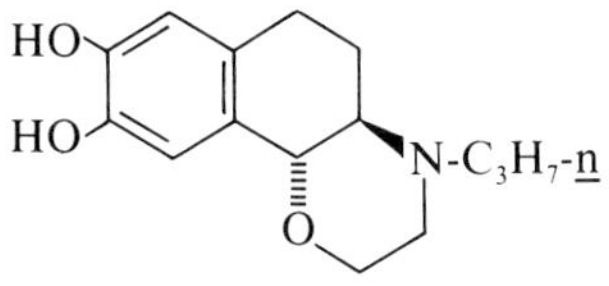

CXL

Further 1,4-oxazine congeners CXLI, CXLII of ergoline ring fragments have appeared in the literature [262].

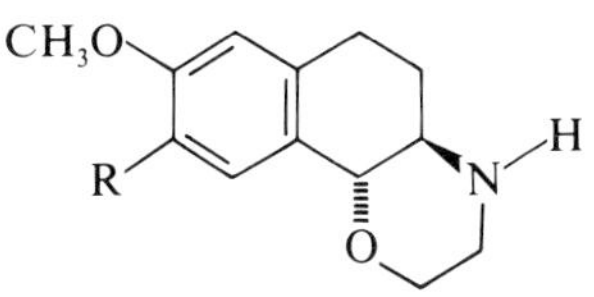

CXLI R = H or OCH_3

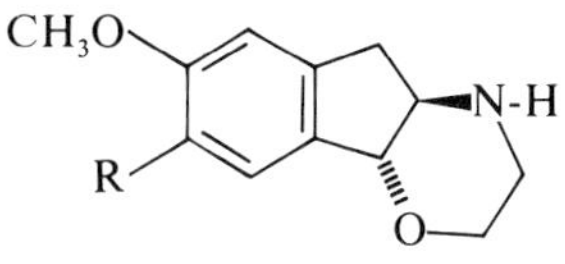

CXLII R = H or OCH_3

However, this communication did not describe preparation of the free phenolic systems, nor did it report any pharmacological data.
In 9-oxa-C-homoergolines CXLIII, the N-*n*-propyl homolog (R = *n*-C_3H_7) was the most effective of the N-alkyl substituents studied, but the compounds in this seven-membered ring series were generally less active than the corresponding oxaergolines CXXXVI [263].

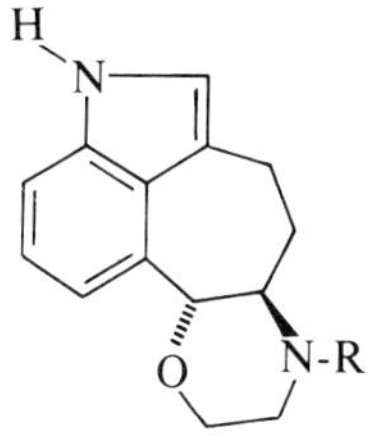

CXLIII

5.2 Ergoline partial structures and fragments

Despyrrolopergolide CXLIV showed a 'dramatic' diminution of activity (although not a total loss), compared to pergolide CVII in inhibition of prolactin secretion and in a rat rotation model [264].

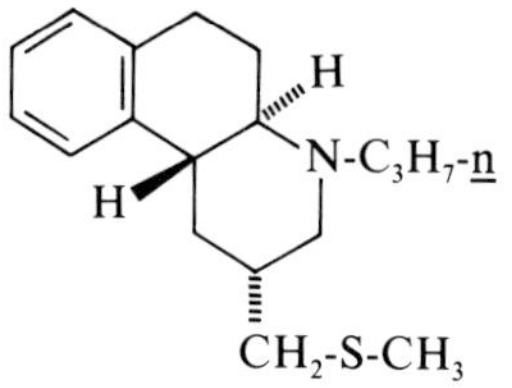

CXLIV

These data were invoked in support of the idea that the pyrroleethylamine moiety of ergolines is highly important for dopaminergic activity.

Parallel to the proposals of Nichols [251] concerning the pharmacophore in the ergoline-derived dopaminergics, Kornfeld et al. [265] as a working hypothesis, suggested that in the ergoline class, the rigid pyrroleethylamine moiety (rather than the β-phenethylamine) is responsible for dopamine agonist properties. To test this hypothesis, they have designed and synthesized ergoline partial structures, and have evaluated them in standard dopaminergic tests. The ABC rings partial structure CXLV was active, albeit less potent than pergolide [266]. These results do not advance the pyrroleethylamine hypothesis, however, since CXLV, like the ergolines, contains a pyrroleethylamine and a β-phenethylamine moiety.

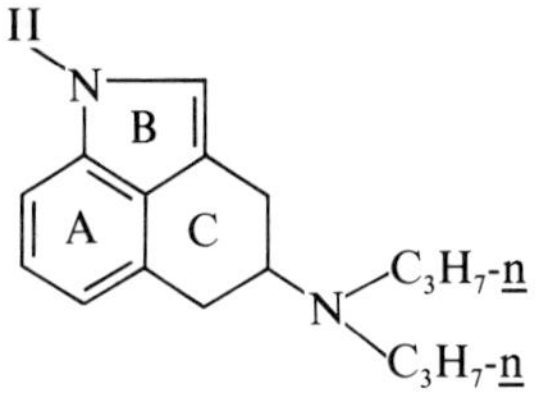

CXLV

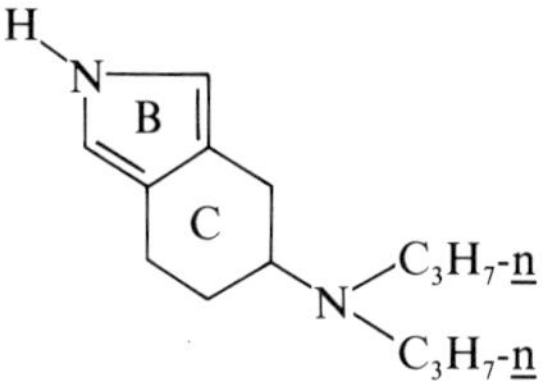

CXLVI

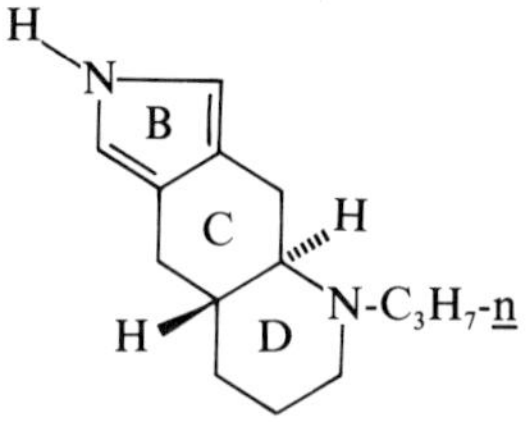

CXLVII

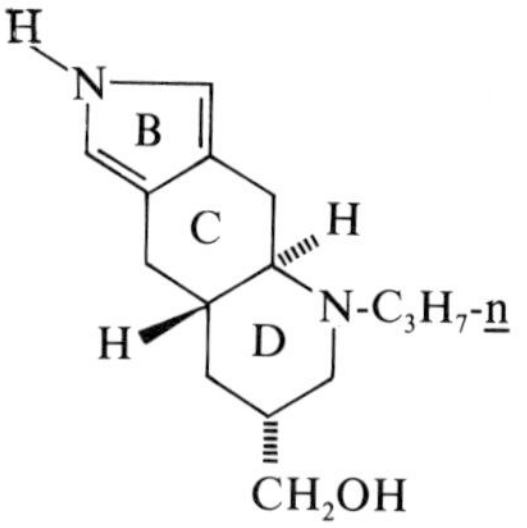

CXLVIII

The BC partial structure CXLVI, the tricyclic BCD partial structure CXLVII, and the hydroxymethyl derivative CXLVIII (all lacking the benzene ring of the ergolines) showed dopamine-like effects, but the potency in all of these rigid systems was inferior to that of pergolide [266]. However, these results were taken as strong support for the hypothesis that the dopaminergic moiety within the ergolines is indeed the pyrroleethylamine [266].

Compounds based upon the seven-membered ring congener CXLIX were practically devoid of dopaminergic activity [263].

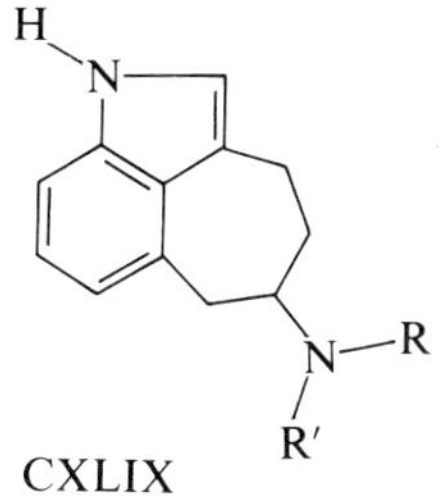

CXLIX

The pyrazole isosteres CL, CLI of ergoline BCD ring fragments were at least as active as the related pyrroles [265].

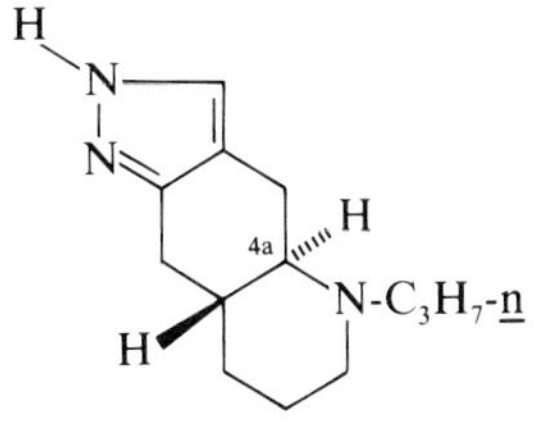

CL

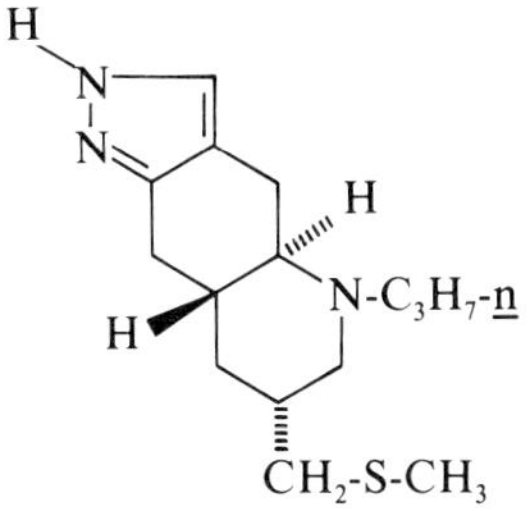

CLI

The compound CL has been described [265] as a more 'pure' dopaminergic agonist than a typical ergoline. It stimulates the D-2 receptor, but not the D-1 receptor [10]. It is devoid of α-adrenoceptor and of serotonin agonist effects, and it is a selective DA-2 agonist in the cardiovascular system [267]. Fuxe and co-workers [268], on the basis of binding studies, concluded that CL has low affinity for central dopaminergic receptors, and they discounted the possibility of metabolic activation of the molecule. Their data suggested that the compound reacts with the 'high affinity' state of the dopamine receptor, but not with the 'low affinity' state. Titus et al. [269] determined the absolute configuration

(4a-R) of the biologically active enantiomer of CL to be as shown in the structure, which is the same as that of the natural ergolines. Stoof and Kebabian [10] have commented that the reported ability of (±)- and of 4a-R-(–)-CL selectively to activate peripheral presynaptic receptors, an example of the DA-2 receptor, suggests a pharmacological similarity between D-2 and DA-2 receptors. Armstrong and co-workers [270] described histamine H_2-agonist actions for (±)-CL, but it is not established whether this activity resides in the (–)-enantiomer as is the case for D-2 agonist activity.
Two new BCD ergoline ring fragments CLII, CLIII involve yet other heterocyclic rings [271].

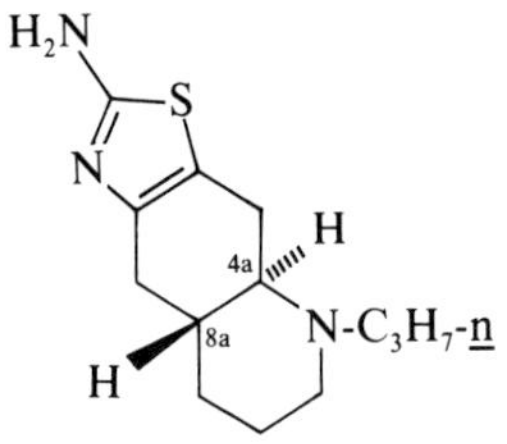

CLII

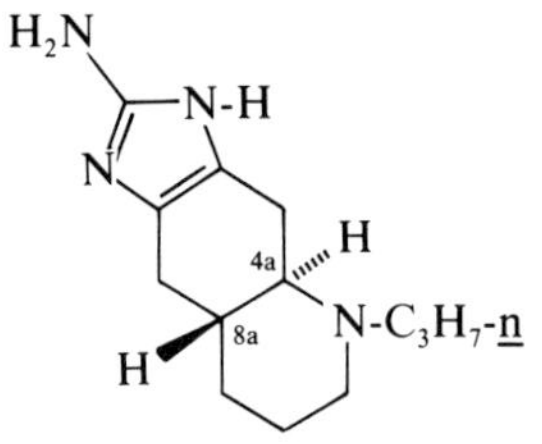

CLIII

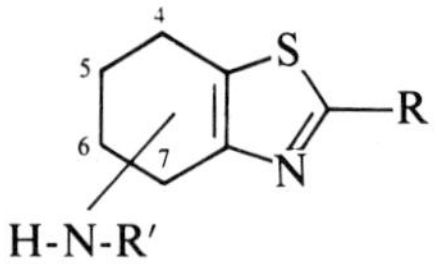

CLIV

These were described as possessing 'potent dopaminergic activity'. In each case the 4a-R, 8a-R antipode (illustrated) possessed all of the dopaminergic activity of the racemate. Derivatives of CLIV where R = H, methyl, or *n*-butyl and which bore a primary amino group or an N-methyl secondary amino group at position 6 had central dopaminergic activity [382], but isomers bearing the amino function at 4, 5, or 7 were inactive in rat stereotypy and rotation assays. Thus, the active isomer in this series is an amino group positional isomer of CLII.
A patent disclosure [272] described the ring system (CLV), where R, R_1, R_2 = H or alkyl; R_3, R_4 = H and the other = NR_5R_6; and R_5, R_6 = H or alkyl.

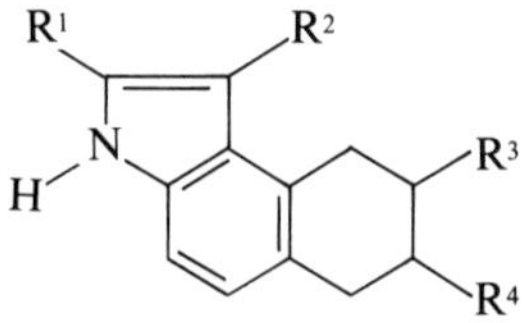

CLV

Certain of these compounds (not specified) showed dopaminergic agonist activity comparable to apomorphine and bromocriptine. The literature has not revealed pharmacological data on these derivatives. Ring systems CLVI, CLVII were designed [265] as hybrids of the apomorphine ring system or apomorphine and the ergoline partial structure congeners CL.

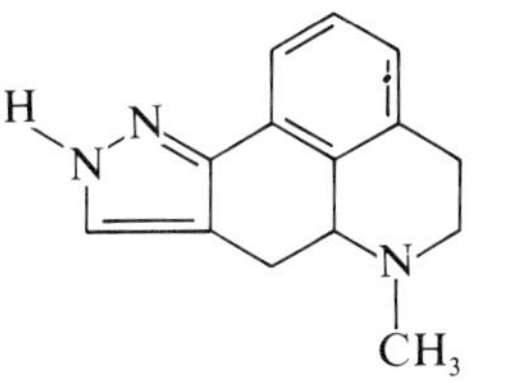

CLVI

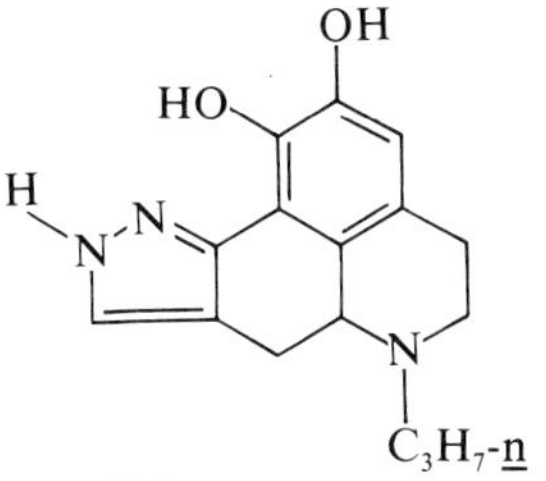

CLVII

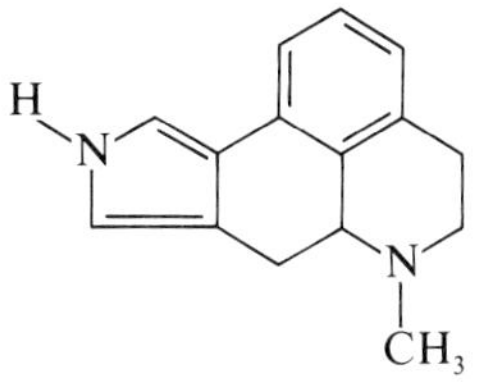

CLVIII

Compound CLVI was a dopamine *antagonist* in the rat stereotypy model and in the rat rotation model. In contrast, CLVII was a dopaminergic agonist, which seems remarkable, in that the molecule is related structurally to the 1,2-dihydroxy-apomorphine XXXII which demonstrated no dopaminergic actions. Berney and Schuh [273] have reported synthesis of racemic CLVI and of a related system CLVIII, and they stated that these compounds 'have shown dopaminergic activity like apomorphine'. Thus, currently, the status of CLVI is unclear.

In contrast to the Nichols [251] interpretation of the dopaminergic pharmacophoric portion of the ergolines, Cannon and co-workers [274] offered the following conformational analysis/interpretation: Inspection of a molecular model of legotrile CVI reveals that it is a very rigid molecule. If a sight be taken down the ethylamine side chain of the β-phenethylamine moiety of the molecule with the nitrogen atom (N-6) nearest the eye, the molecule appears as shown in the Newman projection CLIX.

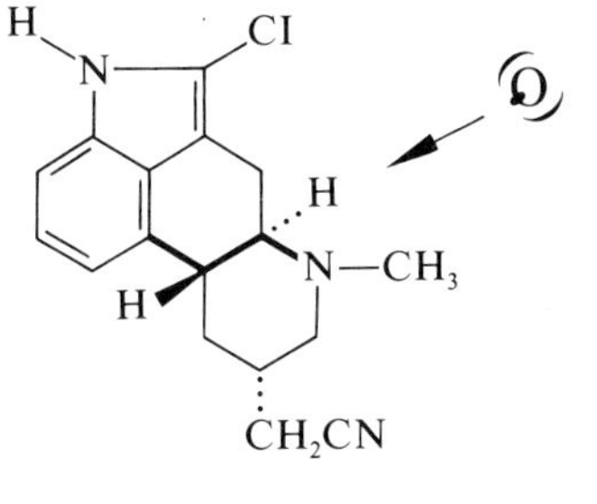

CVI

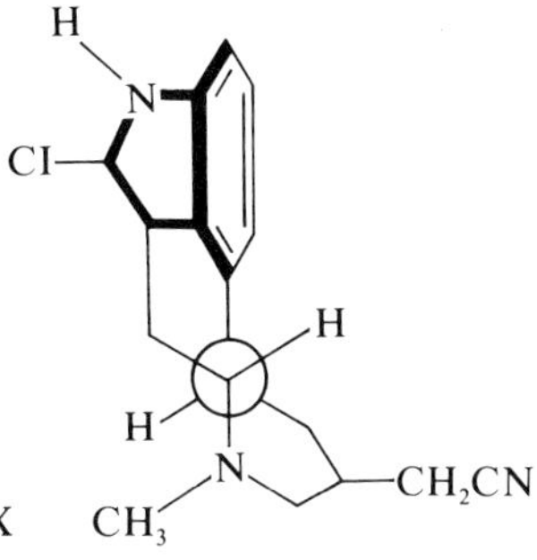

CLIX

It is difficult to represent this Newman projection accurately, to illustrate the significant aspects of the structure in proper perspective. The entire indole ring system of the molecule is planar and, as drawn, the benzene ring and the pyrrole ring project out toward the viewer. The indole ring system is *anti-* to the amino nitrogen (N-6) and, according to analysis of the model, is approximately 20° out of coplanarity with the ethylamine side chain. This is an excellent overall approximation of the α-rotamer of dopamine (structure VI), in which the indole nitrogen of lergotrile (which nitrogen is weakly acidic and is considered to be isosteric with a phenolic OH group) occupies the same position in space as does the 'meta' OH of dopamine in its α-conformation. Measurements of molecular models reveal that the distance from the amino nitrogen to the 'meta'-OH oxygen in the α-conformer of dopamine is 6.4 A, and in the lergotrile molecule the distance between N-6 and the indole nitrogen (N-1) is approximately the same. Models of the tricyclic ergoline fragment CXLV, for which some dopaminergic activity was noted, suggest that the benzene ring is skewed approximately 30° out of coplanarity with the ethylamine side chain. Nevertheless, this structure (Newman projection CLX) superimposes very well, atom-for-atom, upon the α-rotamer of dopamine.

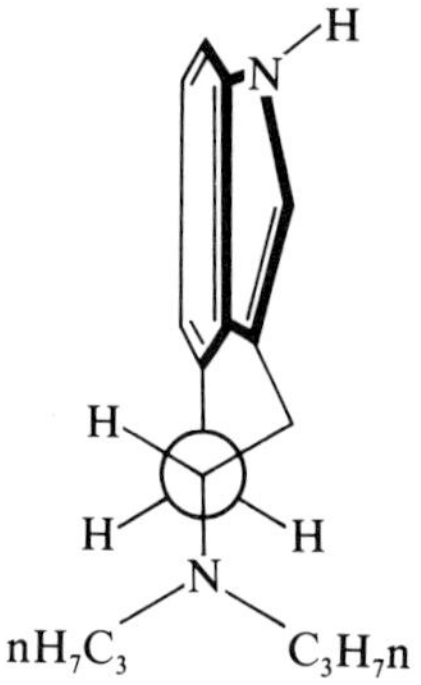

CLX

The dopaminergically active ergoline fragments lacking the benzene ring (e. g., CXLVI, CXLVII, CXLVIII) are rigid or semi-rigid systems, and molecular models (illustrated for the BC fragment CXLVI) reveal that in a reasonable conformation of the molecule where the amino group is attached to the ring by a *pseudo*-equatorial bond, the pyrrole nitrogen (which, like the ring nitrogen of indole, is weakly acidic and may be viewed as being isosteric with the phenolic OH group) is in almost the same position in space as is the 'meta' OH of the α-rotamer of dopamine (Newman projection CLXI).

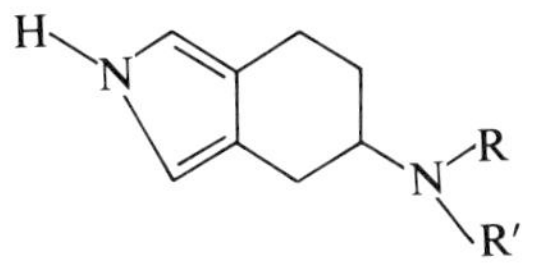

CXLVI

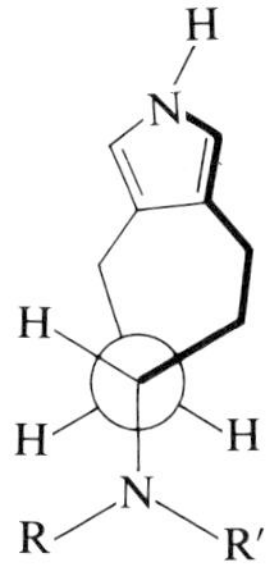

CLXI

A degree of *anti*-arrangement of the amino N and the pyrrole ring N-H is maintained, and the internitrogen distance is approximately 6.2 A, very near to the N-'meta'-OH distance in the α-rotamer of dopamine. It is therefore appealing to speculate that the significant components of the ergoline system for dopaminergic effects are the basic nitrogen (N-6) and the indole nitrogen (N-1), and that a molecule incorporating these groups in an appropriate steric relationship with the appropriate internitrogen distance (approximately 6.4 A) may be expected to demonstrate dopamine-like effects.

The biological inactivity of the pyrrole-3-ethylamine CLXII was rationalized [266] (without any apparent experimental confirmation) on the basis of the possibility of rapid in vivo inactivation by monoamine-oxidases.

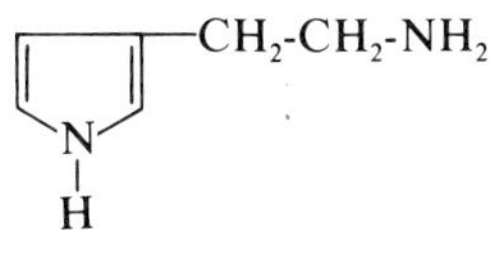

CLXII

On the basis of the foregoing discussion, it would be predicted that the N,N-di-*n*-propyl homolog of CLXII (which would not be a substrate for monoamine oxidase enzymes) should manifest dopaminergic agonist effects. However, this derivative seems never to have evaluated or prepared.

Based on the premise that the dopaminergic pharmacophore of the ergolines is a 4-(2′-aminoethyl) indole system, Cannon et al. [274, 275] investigated a series CLXIII a–c.

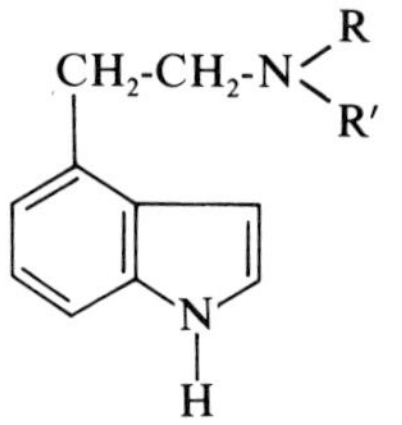

CLXIIIa R = R′ = CH_3
b R = R′ = C_2H_5
c R = R′ = n-C_3H_7

Analysis of molecular models of CLXIII shows that this molecule can assume both the α- and the β-conformations of dopamine, with the assumption that the indole N-H is isosteric with the ‘meta’ OH of dopamine, as shown in Newman projections CLXIV and CLXV.

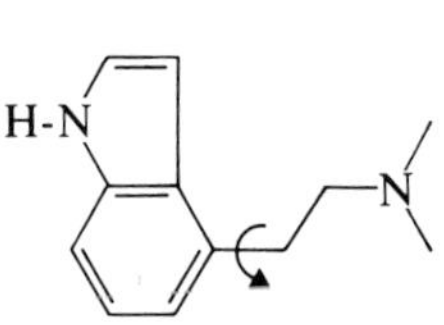

CLXIII

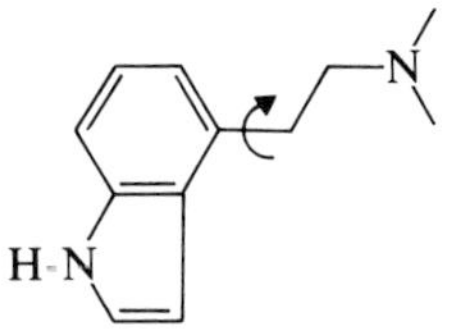

CLXIII

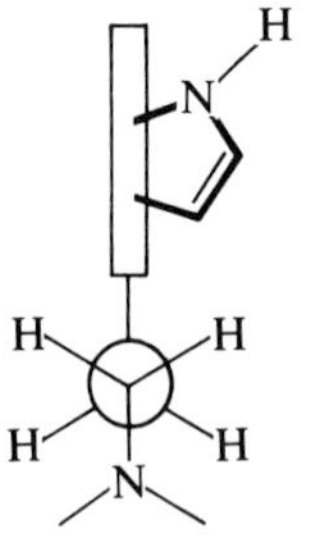

Alpha

CLXIV

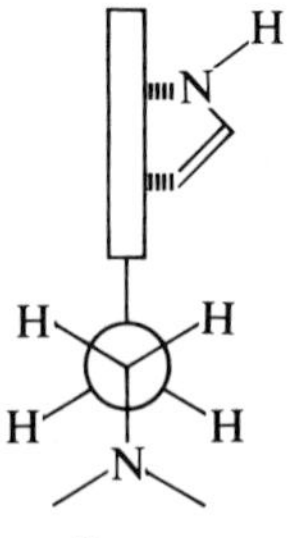

Beta

CLXV

In these conformations, the indole ring is coplanar with the ethylamine side chain. In the β-conformer CLXV, the pyrrole ring portion of the molecule is projected away from the ehtylamine side chain, behind the plane of the page, and in the α-conformer CLXIV, the pyrrole ring is projected out over the ethylamine side chain. Inspection of molecular models suggests that the α-conformer CLXIV may be less favored energetically than the β- because in the α-conformer, the projection of the pyrrole ring over the ethylamine side chain seems likely to result in undesirable destabilizing non-bonded interactions which are not present in the β-conformer. Nevertheless, since the pharmacological effects of members of the series CLXIII (vide infra) so closely mimic those of pergolide and lergotrile, and since the rigidity inherent in the ergoline system of these latter molecules maintains the 4-aminoethyl moiety within them in an α-like conformation, it has been speculated [274] that the simple 4-(2′-aminoethyl) indoles CLXIII can indeed assume the α-like rotameric disposition for interaction with dopamine receptor(s).

The N,N-dimethyl homolog CLXIII a was inert in the cat cardioaccelerator nerve assay, but the N,N-diethyl- and di-*n*-propyl systems CLXIII b and c were active, the potency of the latter derivative approaching that of lergotrile CVI. Both CLXIII b and c exhibited a delay of onset of action in vivo and they were nearly inactive in vitro [276]. Binding studies showed only weak binding abilities for CLXIII b–c and almost none for CLXIII a. On the basis that a portion of the dopaminergic effect of lergotrile CVI has been ascribed [233–235] to in vivo formation of a 13-hydroxy metabolite CX, it was speculated [277] that an analogous metabolic hydroxylation occurs on the simple indoles CLXIII to form a 6-hydroxyindol system CLXVI.

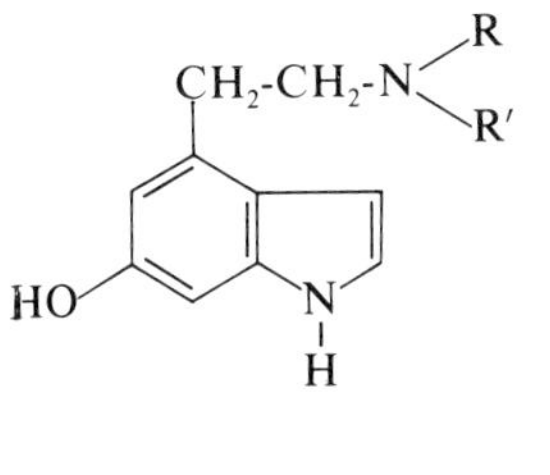

CLXVI

The derivative of CLXVI where R = R′ = n-C_3H_7 demonstrated potent dopaminergic effects in a battery of assays [276, 277] both in vivo and in vitro, but it lacked α- and β-adrenoceptor stimulating activity.

It has been suggested [276] that this 6-hydroxy derivative may be the active in vivo metabolite of CLXIII c.

Boissier and co-workers [263] prepared derivatives of 4-piperidinyl- and 4-tetrahydropyridinyl indole CLXVII, CLXVIII as fragments of the ergoline ring system.

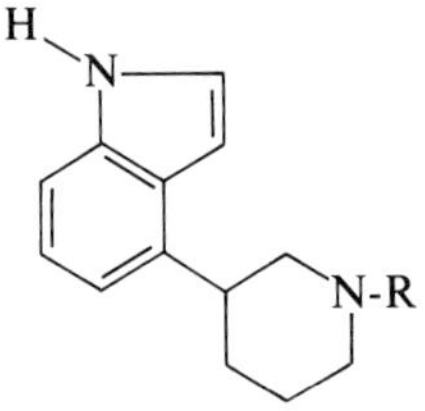

CLXVII

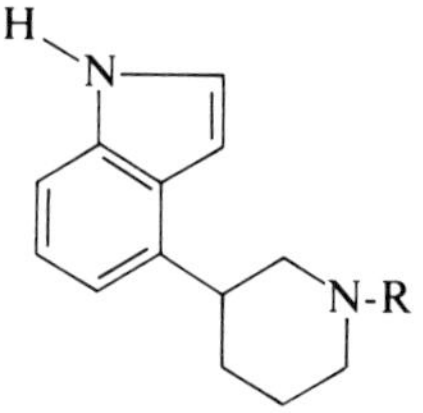

CLXVIII

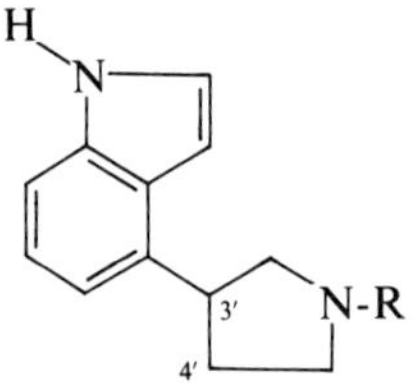

CLXIX

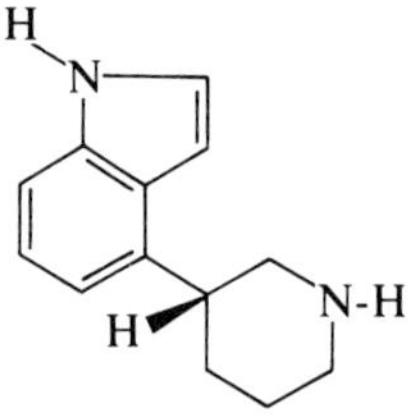

CLXX

In CLXVII and CLXVIII, when R = CH_3, C_2H_5, or n-C_3H_7, potent in vivo dopaminergic agonist effects were noted [278]. There was no clear pharmacological difference between the saturated series CLXVII and the unsaturated series CLXVIII. None of the active compunds of either series had more than a weak affinity for dopamine receptors in binding studies. In the two series, there was no clear correlation between the nature of the N-alkyl substituent on the piperidine or the tetra-hydropyridine ring and the pharmacological activity. It has been suggested [263] that in vivo, CLXVII and CLXVIII are converted into pharmacologically active metabolites, although the nature of these metabolites was not addressed. The secondary amine analog of CLXVII (R = H) was resolved, and the pharmacological activity was found [263] to reside in the R-(–)-enantiomer CLXX, whose absolute configuration is the same as that of the analogous chiral center (C-10) in the naturally occurring ergolines.

The pyrrolidine congeners CLXIX and derivatives having a double bond between 3′ and 4′ were inactive in all tests for dopaminergic effect [263].
Boissier et al. [263] interpreted the data on their series of compounds, especially the piperidinyl indoles CLXVII, the tetrahydropyridinyl indoles CLXVIII, and the 9-oxa-homoergolines CXXXVI, as suggesting that the β-phenethylamine moiety in the ergolines, in addition to the pyrroleethylamine moiety, may play an important role in dopaminergic agonism. Boissier et al. [263] cited conformational studies [279] which show some structural similarities between the piperidinyl indoles, the 9-oxa-C-homoergolines, and the ergoline structure. However, the inactivity of the pyrrolidinyl indole series CLXIX remains to be explained.
Huffman et al. [280] described indolones CLXXI a–b as 'catechol replacement analogues' of dopamine and N,N-di-*n*-propyldopamine.

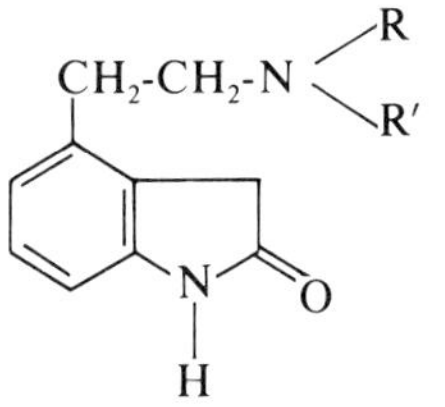

CLXXIa $R = R' = H$
b $R = R' = n\text{-}C_3H_7$

The tertiary amine CLXXI b was described as one of the most potent presynaptic dopamine receptor agonists reported to date.
Pharmacological studies of an extended series of 3-{ω-(1-piperidyl) alkyl}-indoles [281] revealed that the most promising compound was CLXXII.

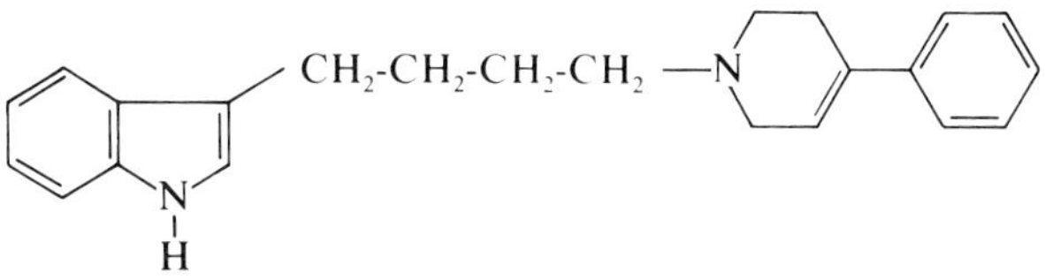

CLXXII

The pharmacology of this molecule seems complex [282–284]; the compound has actions at both postsynaptic supersensitive and intact presynaptic striatal dopaminergic receptors, and it has neither marked agonist nor marked antagonist actions in intact postsynaptic striatal receptors [285]. It has been suggested [283] that, because of the unusual

structure of CLXXII (described as a 'non-rigid-ergoline type' of dopamine agonist), it might offer a new approach to understanding the steric requirements and conformational changes necessary for dopamine agonism at various receptor site subpopulations.
A tetracyclic indole derivative CLXXIII has been stated [286] to have preponderant dopaminergic agonist action, although it displays multiple and complex CNS actions.

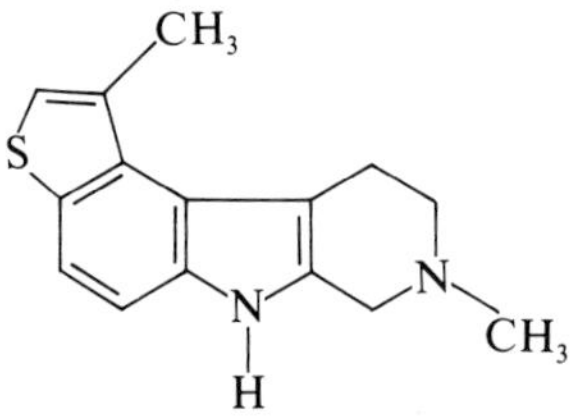

CLXXIII

6 Benzoquinoline derivatives and related molecules

6.1 Benzo(f)quinoline derivatives

The octahydrobenzo(f)quinolines CLXXIV and CLXXV represent a strutural bridge between the 2-aminotetralins and the ergolines.

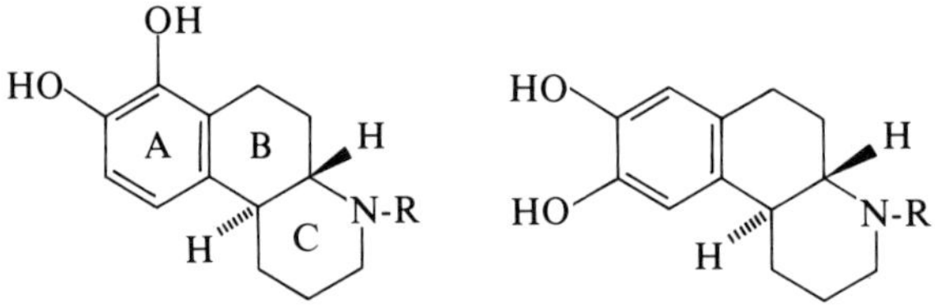

CLXXIV CLXXV

a R = H
b R = CH_3
c R = C_2H_5
d R = n-C_3H_7

The ring systems CLXXIV and CLXXV can exist as *trans*-(illustrated) or *cis*-fused BC ring isomers. In a variety of assays for dopaminergic effects [287–289], the *trans*-fused systems having the hydroxylation pattern CLXXIV were more potent/active than the *cis*-. The dimethyl ethers of both the *cis*- and the *trans*-CLXXIV were dopaminergically inert. In most tests, the N-*n*-propyl substituent CLXXIVd conferred the highest dopaminergic effect [288].

The series of hydroxyl group positional isomers CLXXV also showed biological dependence upon the stereochemistry of BC ring fusion [290], the *trans*-isomers showing higher activity/potency. The *cis*- and *trans*-secondary amines CLXXVa were sympathomimetics rather than dopaminergics. For the *trans*-tertiary amines CLXXVb–d, high dopaminergic activities were manifested only in assays for assessment of peripheral dopaminergic effects. These compounds were inactive in assays for CNS dopaminergic actions, and they thus present a clear separation of CNS effects from very potent peripheral ones.

The resorcinol-derived (±)-*trans*-octahydrobenzo(f)quinolines CLXXVI a–b displayed a very high degree of dopamine-like effects in a series of assays [291].

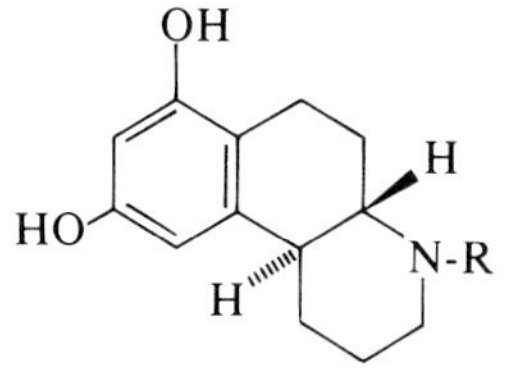

CLXXVIa R = C_2H_5
b R = n-C_3H_7

These compounds were more potent/active than their tetralin-derived congeners (compounds LXIII a–b). The secondary amine homolog (CLXXVI: R = H) was virtually inert. The *cis*-fused BC ring isomers of CLXXVI a–b were inert.

The isomeric monophenolic (±)-*cis*- and *trans*-octahydrobenzo(f)-quinolines CLXXVII a–d have been studied in some detail [287, 289, 292].

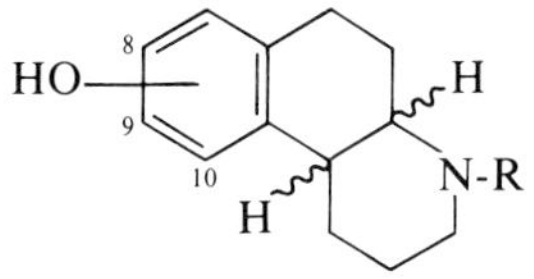

CLXXVIIa 7-OH
b 8-OH
c 9-OH
d 10-OH

R = H; CH_3; n-C_3H_7; n-C_4H_9

The 7-, 8-, and 9-hydroxy isomers (CLXXVII a–c) exhibited dopaminergic activity in a variety of assays, parallel to effects of analogous monohydroxy 2-aminotetralins. The 10-hydroxy isomers CLXXVII d showed central serotonergic effects [292]. This latter compound type has a hydroxylation pattern analogous to the 8-hydroxytetralin LII which also exhibited serotonergic effects. The *trans*-isomers CLXXVII a–d consistently showed higher potency than the *cis*- in dopaminergic or the serotonergic assays [292]. Remarkably, the *trans*-7-hydroxy-N-*n*-butyl homolog CLXXVII a (R = *n*-C_4H_9) was more active than the *n*-propyl homolog in a biochemical model and in a behavioral model for dopamine-like effect [293 a]. This result contrasts with the data cited previously, indicating low or no dopaminergic activity for N,N-di-*n*-butyl-2-aminotetralins, N,N-di-*n*-butyl-dopamine, and N-*n*-butylnorapomorphine. The N-*n*-butyl-9-hydroxy system CLXXVII c (R = *n*-C_4H_9) was dopaminergically inert, whereas the *n*-propyl homolog was very potent [294].
The *trans*-7-hydroxy secondary amine (CLXXVII: R = H) was unique in the series, in that is showed essentially no classical postsynaptic dopamine receptor agonist effects, even at high doses, but it exhibited agonist properties at presynaptic dopamine receptors [294].
Wikström [293 b] found high dopaminergic activity only in the 4 a-S, 10 b-S enantiomer CLXXVIII of *trans*-7-hydroxy-4-*n*-propyloctahydrobenzo(f)quinoline, which is the same absolute configuration for position 4 a as in the analogous chiral center in the dopaminergically active enantiomers of apomorphine and 5-hydroxy- and 5,6-dihydroxy-2-aminotetralins.

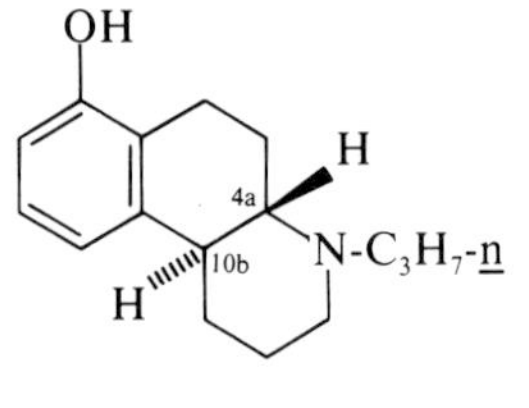

CLXXVIII

Weak dopaminergic effects noted for the 4 a-R, 10 b-R enantiomer of CLXXVIII were rationalized [293 b] by contamination with the more active enantiomer. Compound CLXXVIII non-selectively stimulated both dopaminergic autoreceptors and postsynaptic receptors in the CNS.

The marked differences in dopaminergic activity seen between the *cis*- and *trans*-octahydrobenzo(f)quinolines can be rationalized on conformational grounds [288]. The *trans*-ring system is a highly rigid molecule and, as shown in the conformational representation CLXXIX for the *trans*-7,8-dihydroxy system, it is overall a planar molecule, like apomorphine and the ergolines.

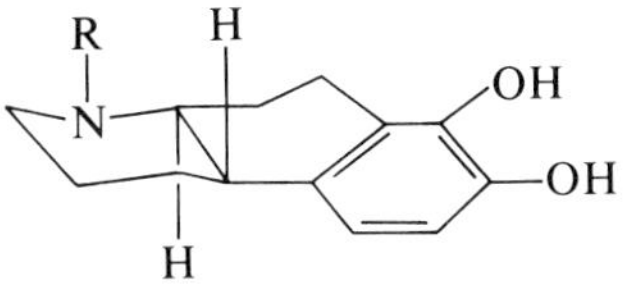

CLXXIX

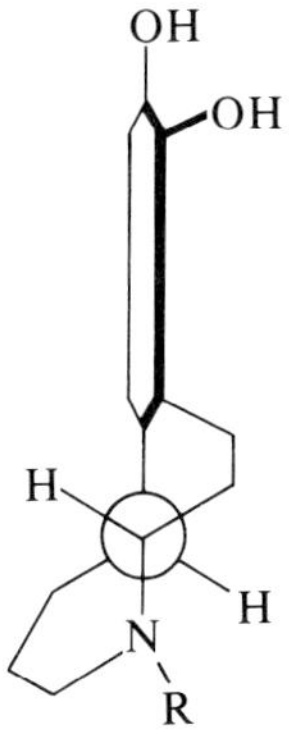

CLXXX

As illustrated in the Newman projection CLXXX, the dopamine moiety in CLXXIX is held firmly with the catechol ring and the amino group in an antiperiplanar conformation, and with the plane of the catechol ring deviating from coplanarity with the ethylamine side chain by only 15–20°. This molecule is thus held in the α-conformation of dopamine, and the OH-positional isomeric system, *trans*-CLXXV, is firmly held in the β-conformation of dopamine, as illustrated in the Newman projection CLXXXI.

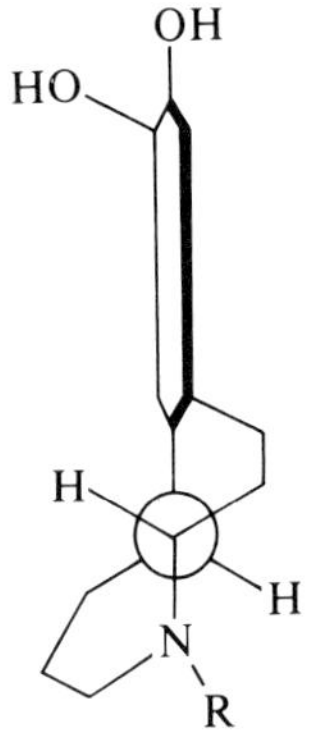

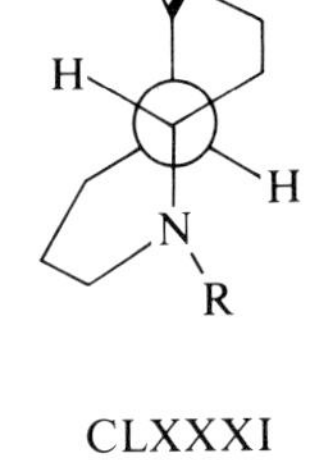

CLXXXI

Molecular models reveal that the *cis*-fused systems CLXXXII are not completely rigid, and they can exist in two interconvertible 'flip' conformations, in one of which the amino group and the catechol ring are synclinal (Newman projection CLXXXIII), and one in which the dopamine moiety approaches the antiperiplanar disposition and the catechol ring approaches coplanarity with the ethylamine side chain (Newman projection CLXXXIV).

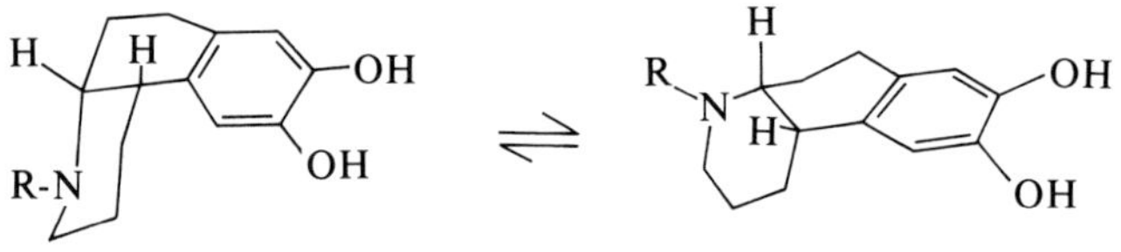

CLXXXIIa CLXXXIIb

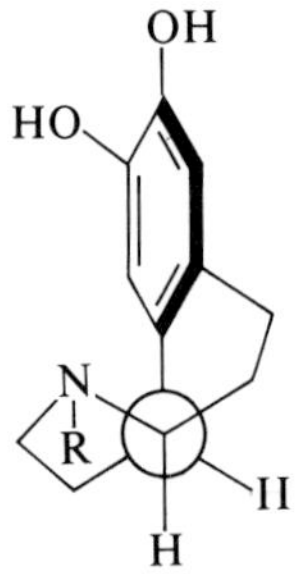

CLXXXIII

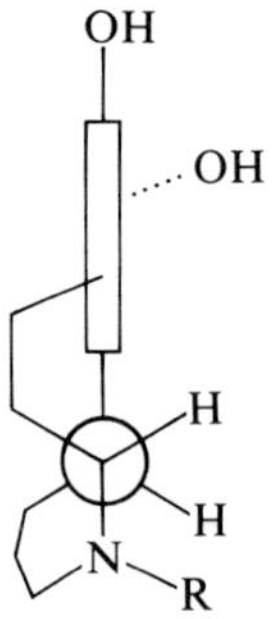

CLXXXIV

It might be speculated that the *cis*-'flip' conformer CLXXXIV would be capable of interaction with dopamine receptor(s) to produce a response. However, overall, neither conformation of the *cis*-fused molecule is planar; the C-ring (heterocyclic ring) makes almost a right angle with the plane of the A and B rings, and this deviation of the molecule from overall planarity was proposed [288] to be detrimental to proper interaction with the dopamine receptor(s).

Cannon et al. [287] found that the *cis*- and *trans*-non-oxygenated derivatives CLXXXV (where R = CH_3) lacked dopaminergic effects in a variety of assays, but the compounds antagonized apomorphine-induced emesis in dogs, and they potentiated stimulation of the cardioaccelerator nerve in the cat in vitro but not in vivo.

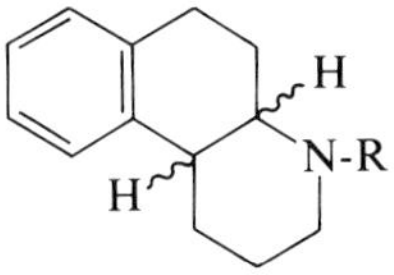

CLXXXV
(cis- or trans-)

Bach and co-workers [264] found that CLXXXVI exhibits no dopamine-like effects in the prolactin release assay, nor in the rat rotation model.

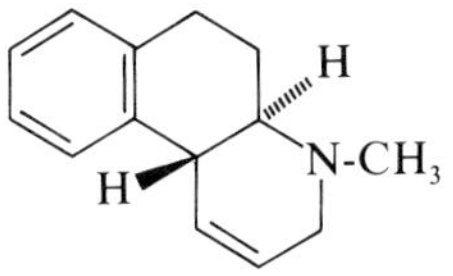

CLXXXVI

However, *trans*-CLXXXV where R = *n*-C_3H_7 exhibited marked dopamine-like effects in the cat cardioaccelerator nerve assay and in a rat rotation model [295]. Further studies on this N-*n*-propyl homolog [296] revealed other dopaminergic effects, in addition to α-adrenoceptor blocking activity. These divergent properties provide direct evidence that presynaptic α_2-adrenoceptors and dopamine receptors are different entities on the sympathetic nerve terminal. The pharmacological effects described for the ring system *trans*-CLXXXV provide yet another example of the remarkable ability of the N-*n*-propyl group to bestow dopamine-like actions upon a molecule. The *cis*-fused isomer CLXXXV (R = *n*-C_3H_7) was inactive [295].

6.2 Benzo(h)isoquinoline derivatives

A series of isomeric tricyclic congeners CLXXXVII of dopamine derived from octahydrobenzo(f)quinoline has been investigated [297].

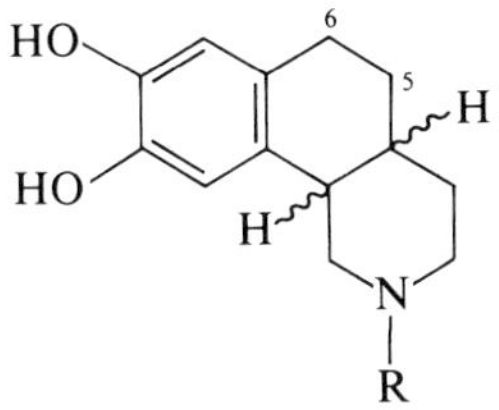

CLXXXVIIa R = H
b R = n-C_3H_7

In this series, the (±)-*cis*- and *trans*-derivatives CLXXXVII a–b all produced inhibition of the cat cardioaccelerator nerve of the same order of potency, which was very low, and *very large* intravenous doses of all members of the series produced hypotensive and negative chronotropic effects in cats, which were blocked by haloperidol. One 'flip' conformer of the *cis*-fused ring system holds the amino nitrogen synclinal (*'gauche'*) to the catechol ring, and the other holds the catechol ring perpendicular to the plane of the ethylamine side chain. Neither of these conformations is conducive to dopamine agonist activity. The *trans*-fused ring system provides a rigidly held dopamine moiety in the α-rotameric disposition (Newman projection CLXXXVIII), and the low order of activity of members of this ring system was unexpected.

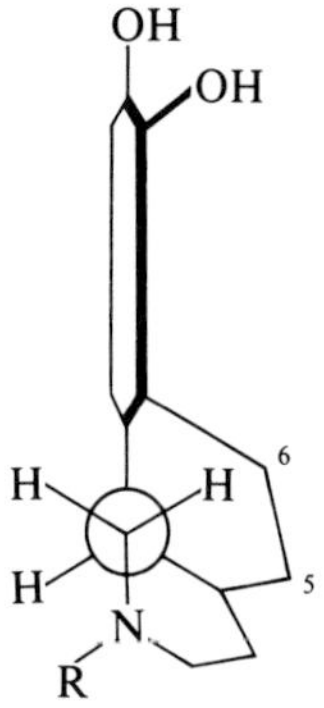

CLXXXVIII

Study of molecular models [297] led to the proposal that carbons 5- and 6-(structures CLXXXVII and CLXXXVIII) represent a bulky region in the octahydrobenzo(h)isoquinoline molecule not present in that spatial region in the molecules of the potent dopaminergic agonists (dopamine itself, apomorphine, 2-aminotetralins, ergolines, and octahydrobenzo(f)quinolines). It is possible that this region of molecular bulk prevents optimal interaction of CLXXXVIII with dopamine receptor(s).

6.3 Benzo(g)quinoline derivatives

The linearly annulated octahydrobenzo(f)quinoline CLXXXIX is an apomorphine molecule lacking the non-oxygenated benzene ring and, like apomorphine, it bears the dopamine moiety in an α-conformer when (as illustrated) the stereochemistry of BC ring fusion is *trans.*

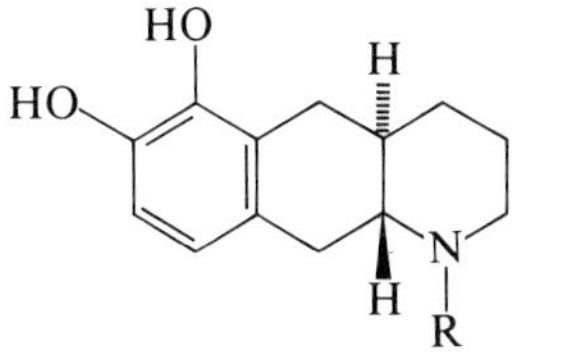

CLXXXIXa R = H
b R = CH_3
c R = C_2H_5
d R = n-C_3H_7

The *cis*-fused isomers of CLXXXIX have not been reported in the literature. The (±)-secondary amine CLXXXIX a exhibited low potency and activity [298]. However, the (±)-tertiary amines CLXXXIX b–d were somewhat more potent than apomorphine in inhibition of striatal dopa accumulation in the rat, and they also produced stereotypy in mice; N-ethyl CLXXXIX c and *n*-propyl CLXXXIX d homologs were at least equipotent to apomorphine, and the N-methyl CLXXXIX b was less potent. The N-methyl and *n*-propyl homologs induced a circling response in unilaterally lesioned rats, but the N-ethyl homolog produced no rotational response at doses ten times those required for good response to (±)-CLXXXIX d and apomorphine. The rotational inactivity of the N-ethyl derivative is difficult to explain on chemical grounds. Costall et al. [299, 300] have suggested that gross, centrally mediated rotational effects produced by dopaminergic drugs may result from more than one physiological mechanism, involving more than one population of dopamine receptors, having different agonist structural requirements. Only the *n*-propyl homolog CLXXXIX d exhibited a marked dopamine-like effect in the canine renal blood flow assay [298], having a potency ratio to dopamine of 0.2. This is a tenfold increase in potency over that reported [33] for N,N-di-*n*-propyldopamine XII c.

Kocjan and Hadzi [301] have reported a molecular mechanics study of dopaminergically active 2-aminotetralins CXC, apomorphines CXCI, octahydrobenzo(f)quinolines CXCII, and octahydrobenzo(f)quinolines CXCIII.

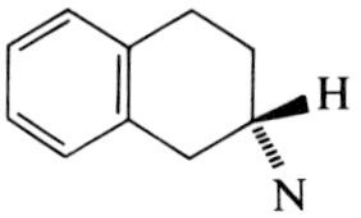

CXC

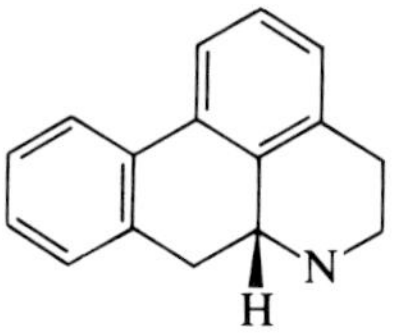

CXCI

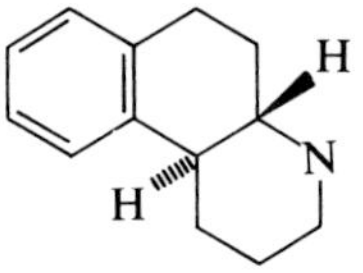

CXCII

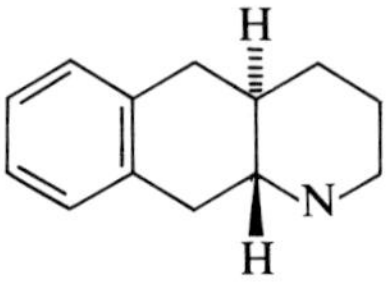

CXCIII

The most stable conformations for these systems were computed, and the values of the torsion angles τ_1, τ_2, and τ_3 were compared with the analogous torsion angles in the dopamine molecule CXCIV.

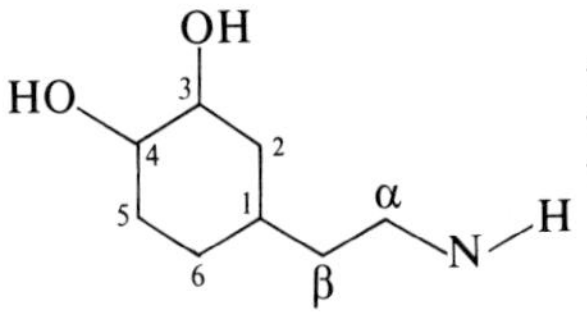

$\tau_1 = C_6C_1C_\beta C_\alpha$
$\tau_2 = C_1C_\beta C_\alpha N$
$\tau_3 = C_\beta C_\alpha NH$

CXCIV

For all of the ring systems studied (CXC–CXCIII) the $\tau_1 \simeq 200°$; $\tau_2 \simeq 180°$; and $\tau_3 \simeq 60°$. It was postulated that the stereo structures calculated for CXC–CXCIII are those required by the brain and the cardioaccclerator nerve dopamine receptor(s). Attempts to exploit this molecular mechanics strategy in assessment of vascular bed dopamine receptors were not successful.

In the octahydrobenzo(f)quinoline ring system, all of the β-rotamer congeners (±)-CXCV a–d, like isoapomorphine XXXIV, were inert in central and peripheral assays in which A-6,7-DTN derivatives XLVI are active [290].

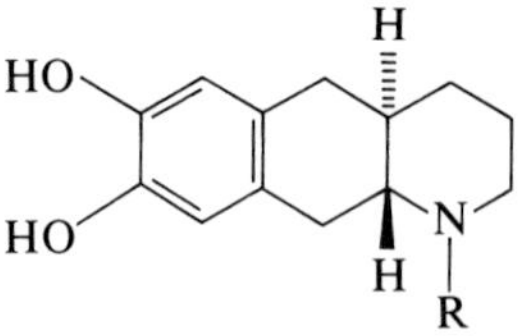

CXCVa R = H
b R = CH_3
c R = C_2H_5
d R = n-C_3H_7

Introduction of the rescorcinol hydroxylation pattern into the (±)-*trans*-linearly annulated benzoquinoline system CXCVI resulted in almost total loss of biological activity [291].

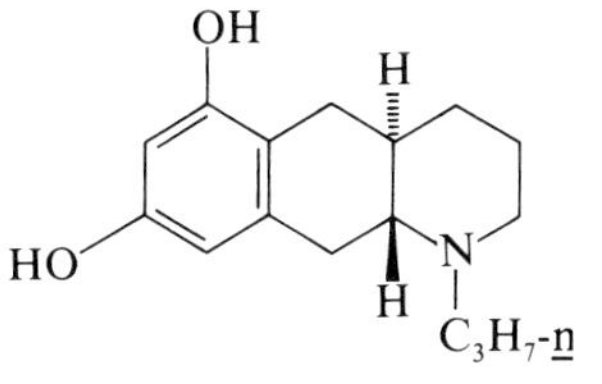

CXCVI

Thus, as can be noted from pharmacological data on the various resorcinol-substituted ring systems (XVI, LXIII, XC, CLXXVI, and CXCVI) there seems to be no consistency in the pharmacological effect of changing the catechol (1,2-dihydroxy) moiety of a dopaminergic agonist molecule to a resorcinol (1,3-dihydroxy) moiety. No obvious explanation appears for the dopaminergic agonism of the resorcinol-derived-2-aminotetralins LXIII and octahydrobenzo(f)quinolines CLXXVI and for the low or complete lack of activity of the β-phenethylamines XVI, 2-aminoindans XC, and octahydrobenzo(f)quinolines CXCVI.

The monohydroxylated octahydrobenzo(f)quinolines CXCVII, CXCVIII were designed as analogs of fragments of the potent dopaminomimetic ergolines CXVIII and pergolide CVII, respectively, in which the fused pyrrole ring is replaced by a benzene ring [302].

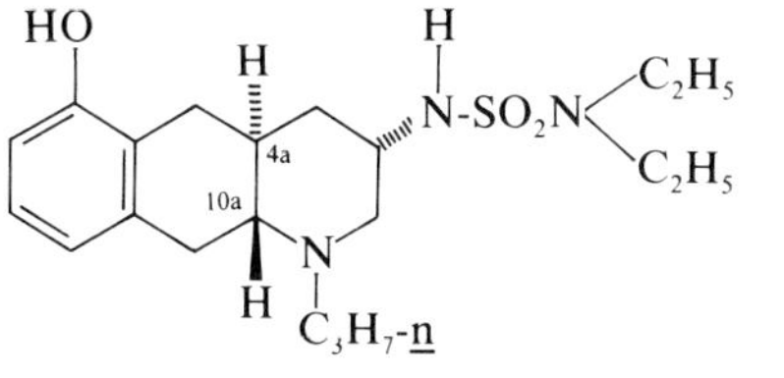

CXCVII

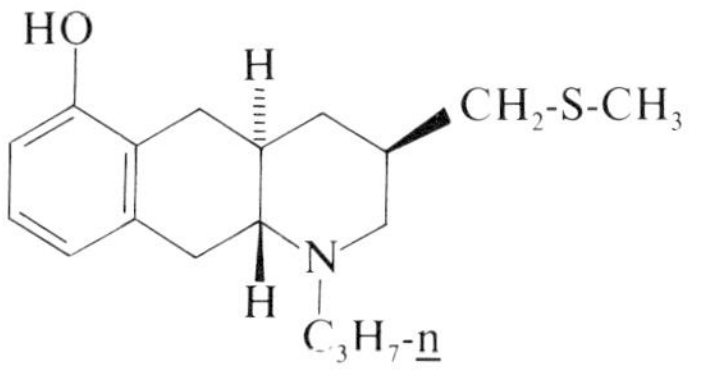

CXCVIII

Preliminary pharmacological evaluation (prolactin secretion in male rats, in vitro binding data) of (±)-CXCVII and (±)-CXCVIII suggested that they combine the specificity of apomorphine with the potency, long duration of action, and good oral activity of the ergolines. Compound CXCVII was resolved, and the absolute configurations at positions 10a and 4a of the biologically active enantiomer were found to correlate with the corresponding positions in the natural ergolines. These data were interpreted to support the Nichols [251] proposal that the dopaminomimetic pharmacophore of the ergolines is a rigid pyrroleethylamine [303].

6.4 'Open chain' analogs of benzoquinolines

Derivatives CXCIX a–c of an 'open chain' analog of the octahydrobenzo(g)quinoline ring system were inert in tests for dopaminergic activity [30].

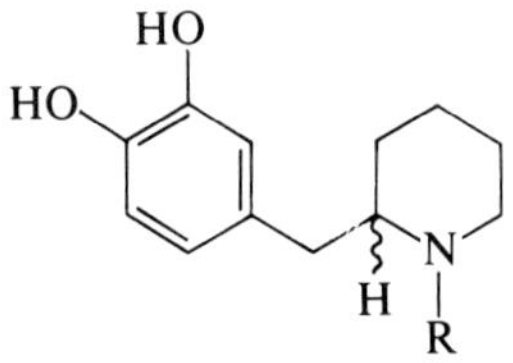

CXCIXa R = H
b R = CH_3
c R = n-C_3H_7

The 3-phenylpiperidine system CC could be viewed as an 'open chain' analog of the octahydrobenzo(f)quinoline ring system (e. g., CLXXIV, CLXXV).

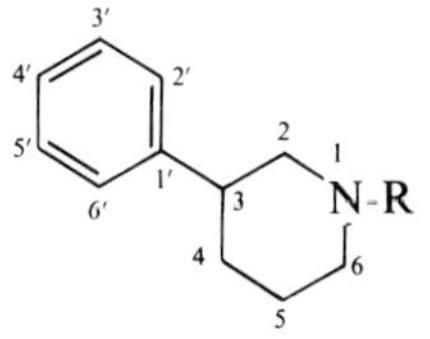

CC

The 3′-monohydroxy derivatives were the most potent hydroxylated members of the series [304]; remarkably, a 3′,4′-dihydroxy derivative was less potent. Molecular models reveal that in a stable conformation for 3-(3′-hydroxyphenyl)-piperidine CC (piperidine ring in a chair form; benzene ring attached to C-3 by an equatorial bond), the β-phenethylamine system (N-C_2-C_3-benzene ring) assumes a *trans* (antiperiplanar) conformation and there seems to be no steric impediment to the 3′-OH to assume either the α-(CCI) or the β-(CCII) conformation.

A 2′-*n*-propyl homolog CCIII is dopaminergically inert [294], but the isomeric 4′-*n*-propyl compound CCIV has some little dopaminergic action.

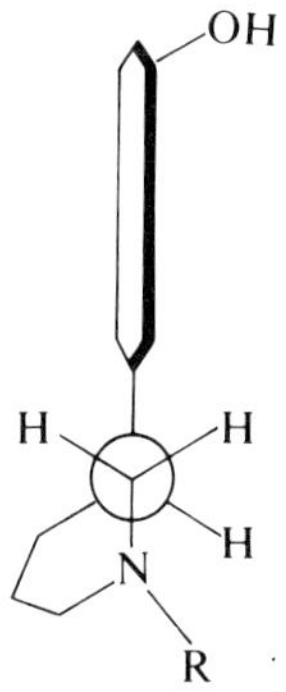

CCI

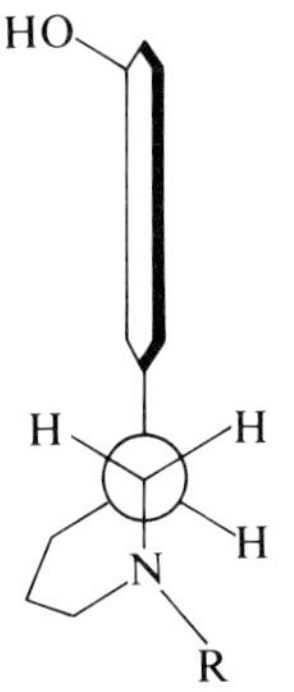

CCII

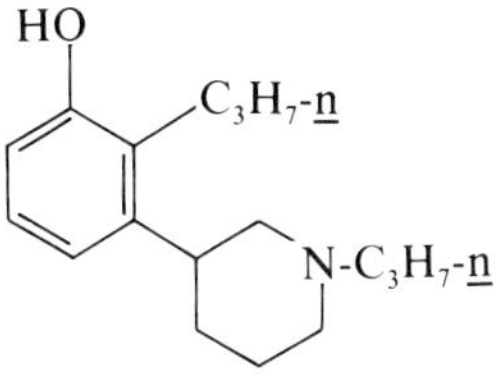

CCIII

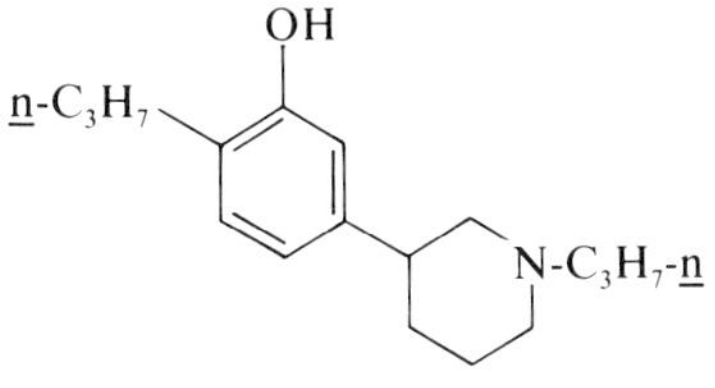

CCIV

The inactivity of CCIII can be rationalized on the basis of steric effect of the 2′-*n*-propyl group, the prevent the molecule's assuming the *α*-conformation CCI.

Movement of the 3′-OH in CC to the 2′- or the 4′-position resulted in diminution or complete loss of activity. Replacement of the piperidine ring in CC by pyridine CCV, N-*n*-propylpyrrolidine CCVI, N-*n*-propylperhydroazepine CCVII, quinuclidine CCVIII, or quinuclidinene CCIX resulted in loss of biological activity [304].

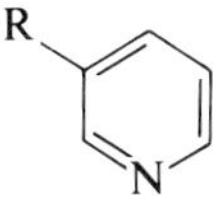

CCV

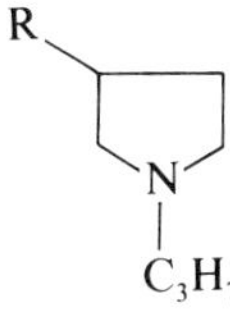

CCVI

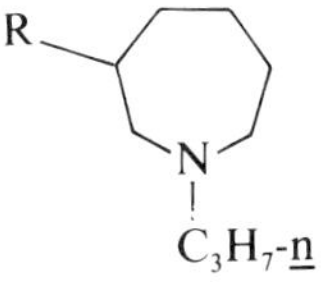

CCVII

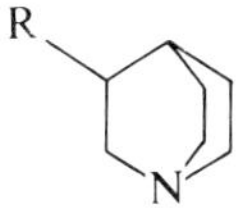

CCVIII

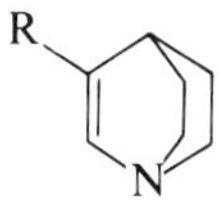

CCIX

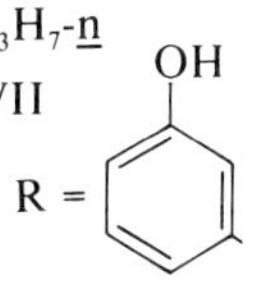

The inactivity of the pyrrolidine ring derivative CCVI was explained [304] on the basis of the inability of the five-membered ring system to place the β-phenethylamine moiety in the benzene ring-amino nitrogen antiperiplanar disposition. The critical torsion angle (N-C_2-C_3-$C_{1'}$) in CCX was stated [304] to be approximately 155° (Newman projection CCXI).

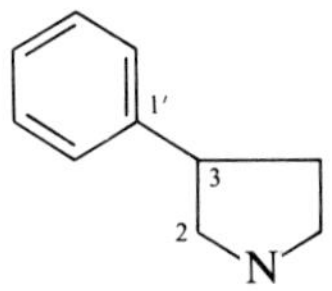

CCX

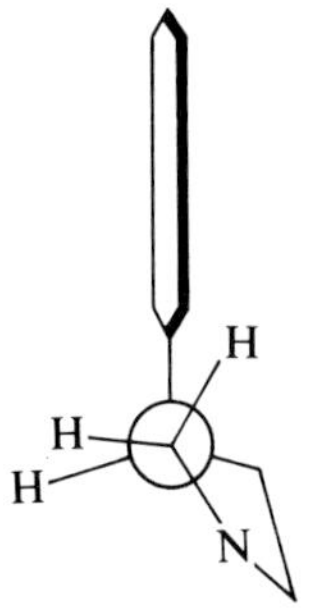

CCXI

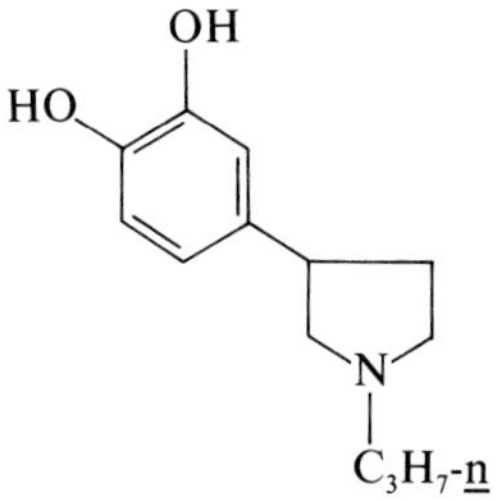

CCXII

However, as was described previously, the analogous torsion angle in biologically active 2-aminoindans (see structures XCIV and XCV) is *at most* 140–150°, which does not, in certain compounds, destroy dopaminergic effects. A catechol derivative CCXII of 3-phenylpyrrolidine showed actions in a series of assays in rats, consistent with its being a dopaminergic agonist [148]. It exhibited lower potency but slightly longer duration of action than apomorphine. However, the activity of the compound in the presence of haloperidol or other dopaminergic antagonist was not reported.

For the other non-aromatic systems CCVII–CCIX, the likelihood of unfavorable conformation(s) of the β-phenethylamine moiety, in addition to the large amount of mulecular bulk about the amino nitrogen, was invoked to rationalize the pharmacological inactivity. In the case of the pyridine ring derivative CCV, it may be that the lowering of the base strength of the ring nitrogen (in pyridine, as compared with piperidine) is detrimental to interaction with receptor subsites. It has been noted previously that the completely aromatic congeners L a–c of the highly potent and active 2-aminotetralins were inert in all assays for dopaminergic effects.

In the derivatives of 3′-hydroxy CC, the structure-activity relationships for the N-substituent were complex [304]. The N-isopropyl, *n*-butyl, *n*-pentyl, and 2-phenethyl derivatives were *more* potent compounds than the N-methyl, ethyl, or *n*-propyl, in contrast to results cited previously for β-phenethylamines, 2-aminotetralins, and apor-

phines, but consistent with observations cited for 7-hydroxybenzo(f)quinoline CLXXVII a.
Earlier reports (e. g., [304]) of studies on (±)-N-*n*-propyl-3-(3′-hydroxyphenyl)piperidine CCXIII ('3-PPP') described a high degree of specificity as a dopaminergic autoreceptor agonist. However, resolution of CCXIII and study of the enantiomers [305, 306] revealed that R-(+)-CCXIII is a dopaminergic agonist at postsynaptic sites as well as at autoreceptors, but in contrast, in vivo, S-(–)-CCXIII activates autoreceptors and acts concomitantly as an *antagonist* at postsynaptic receptors.

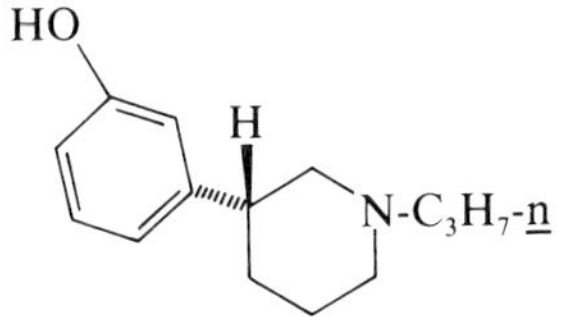

R-(+)-CCXIII

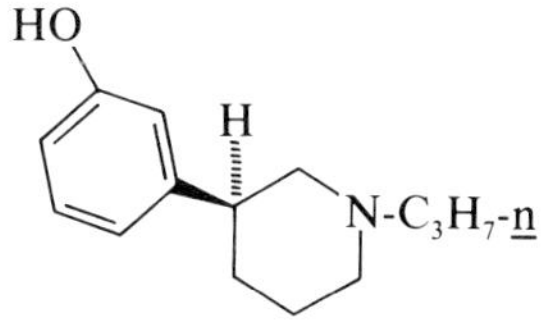

S-(–)-CCXIII

In vitro, neither (±)-CCXIII nor S-(–)-CCXIII appeared to be a presynaptic agonist in the assay on rat neostriatal slices for K^+-induced acetylcholine release, but rather these two compounds behaved as postsynaptic D-2 antagonists. R-(+)-CCXIII was a weak D-2 agonist in this assay. Neither the R-(+)- nor the S-(–)-enantiomer of CCXIII was a D-1 agonist in a rat neostriatum assay for efflux of cyclic AMP, but the S-(–)-enantiomer was an antagonist [307]. Other studies in other test systems [308, 309] were consistent in finding presynaptic agonist effects for R-(+)-CCXIII and S-(–)-CCXIII, but postsynaptic agonism for R-(+)-CCXIII and antagonistic action for S-(–)-CCXIII.
In a study of a series of R- and S-N-alkyl (methyl, ethyl, *n*-propyl, 2-propyl, *n*-butyl, *iso*-pentyl, and β-phenethyl) homologs of CCXIII, Wikström et al. [310] found that in the R-series, N-alkyl derivatives all behave as classical dopamine receptor agonists, with affinity and intrinsic activity for both pre- and postsynaptic receptors. The same bifunctional profile seems valid for the S-enantiomers with N-substituents larger or bulkier than *n*-propyl. However, the N-ethyl and *n*-propyl homologs in the S-series had affinity for both pre- and postsynaptic receptors, but showed intrinsic activity only at presynaptic receptors.

The chiral center in R-CCXIII has the same absolute configuration as the equivalent center (C-10) in the potent ergoline-derived dopaminergics, e. g., lergotrile CVI and pergolide CVII. But, the absolute configuration in S-CCXIII is the same as in the equivalent position (C-10b) in the pharmacologically potent enantiomer of the octahydrobenzo(f)quinoline CLXXVIII, which nonselectively stimulates pre- and postsynaptic CNS receptors. The chiral center in CCXIII has no equivalent chiral center in the aporphines, aminotetralins, or aminoindans.

Arnt and co-workers [311] proposed that a non-oxygenated derivative of 3-phenylpiperidine, S-(+)-1-β-phenethyl-3-phenylpiperidine CCXIV is a more selective model than 3-PPP CCXIII for differential studies of pre- and postsynaptic receptors.

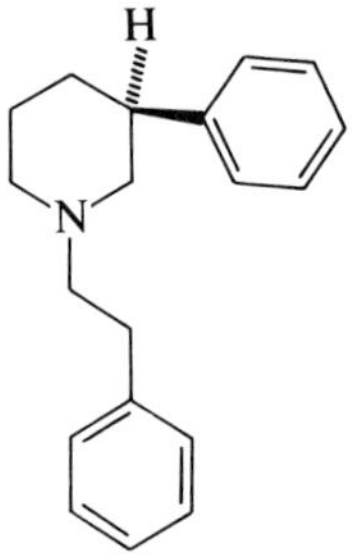

CCXIV

Rollema and Mastebroek [312] noted that discrepancies between in vivo and in vitro activities of 3-PPP CCXIII have been occasionally ascribed to the possibility of metabolic activation of 3-PPP in vivo. Accordingly, 3-PPP was incubated with rat liver microsomes in the presence of NADPH and Mg^{++}. In vivo formation of catecholamine was studied in rats following intraperitoneal injection of 3-PPP and dissection of various brain regions. In vitro, 3-PPP is hydroxylated at position 4′, forming CCXV. However, in vivo, the initially formed catechol system CCXV is rapidly converted (via COMT) into the 4-methyl ether CCXVI.

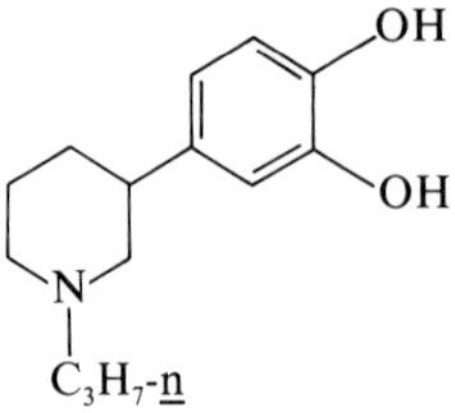

CCXV

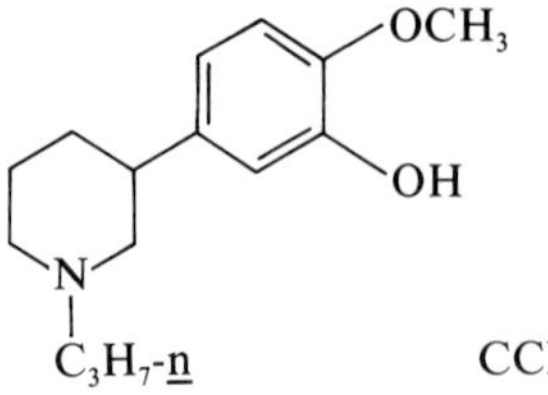

CCXVI

Similar results were obtained with both enantiomers of 3-PPP; no significant differences were noted between the in vivo or the in vitro hydroxylation of (+)- or (–)-3-PPP.
Several N-alkyl-3-(1,2,5,6-tetrahydro 3-pyridyl) phenols CCXVII were compared with the corresponding saturated analogs in assay for central dopaminergic effects, both in vivo and in vitro [313].

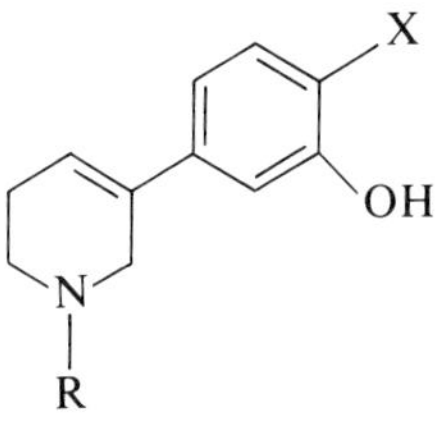

CCXVII
X = H, OH
R = H, n-C_3H_7, CH_2-c-C_3H_5

The new compounds CCXVII (with the exceptions of the N-*n*-propyl and N-cyclopropylmethyl monophenolic derivatives, X = H, which were antagonists) exhibited dopaminergic agonist properties. There were differences in action between the saturated system (typified by 3-PPP) and the unsaturated one CCXVII, which were discussed on the basis of conformational differences.

7 Azepine derivatives

The tetrahydrobenzazepine derivative CCXVIII dilates the canine renal vascular bed and has been descibed [314] as a partial agonist in this assay.

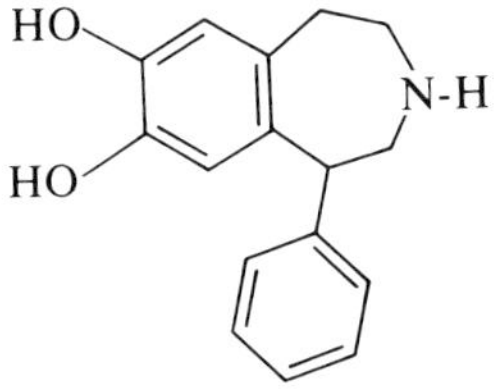

CCXVIII

It was an agonist in the rat rotation model [315, 316]. Compound CCXVIII did not affect prolactin levels nor dopamine turnover, nor was it an emetic in the dog. It did not induce stereotypy in normal rats. The biological profile of this compound suggests selectivity for D-1 receptors [317]. Stoof and Kebabian [10] have stated that although

CCXVIII is presumed not to stimulate D-2 receptors in vivo, in biochemical models of the D-2 receptor, higher concentrations of CCXVIII did interact with the receptor. The reason for this discrepancy between physiological and biochemical data was not evident.
Kaiser et al. [318] have noted that the tetrahydrobenzazepines (typified by CCXVIII) differ from most dopaminergic agonists in which the dopamine-like moiety within the molecule can attain the catechol ring-amino nitrogen antiperiplanar conformation. The dopamine moiety in structure CCXIX (ring A–C_1–C_2–N) can vary from a fully eclipsed N-catechol ring (synperiplanar) conformation CCXX to one (CCXXI) in which the catechol ring and the basic nitrogen are in a partially eclipsed (anticlinal) disposition. Moreover, in these conformations, the plane of the catechol ring system seems to deviate significantly from coplanarity with the ethylamine side chain of the dopamine moiety.

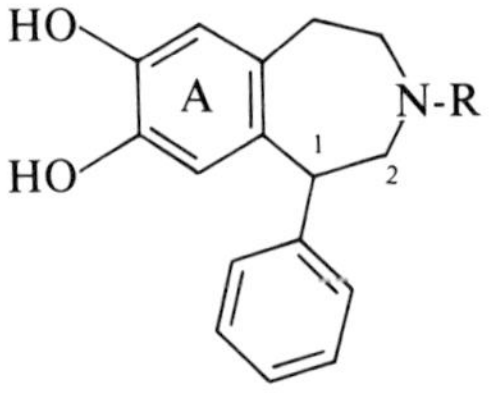

CCXIX

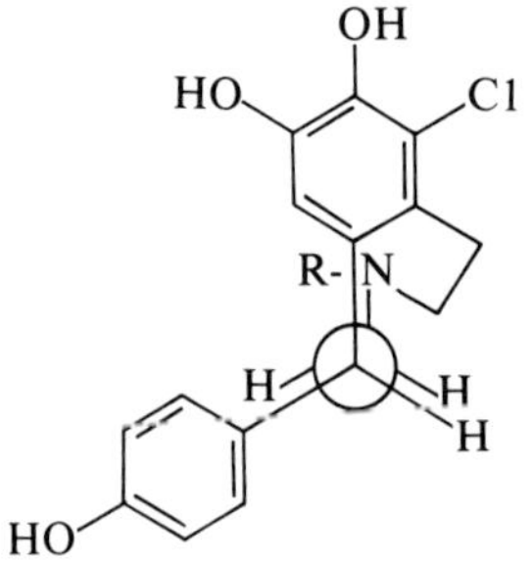

CCXX

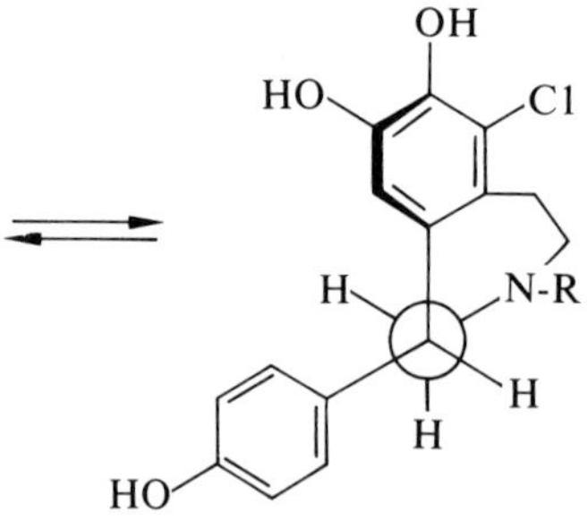

CCXXI

Kaiser et al. [318] stated that the preponderance of evidence suggests that these agonists interact with the dopamine receptor(s) in the conformation CCXXI, in which the distance between the nitrogen and either catecholic OH, both of which seem to be required for D-1 activity, is approximately 7.0 A. This distance falls between the values defined [16] for the N-to-*m*-OH distance for the dopamine α-rotamer (6.2 A) and the N-to-*m*-OH distance for the β-rotamer (7.3 A), and it is somewhat less than the N-to-*p*-OH distance in both rotamers (7.8 A). Resolution of CCXVIII [319] revealed that dopaminergic activity resides almost exclusively in the R-enantiomer CCXXII.

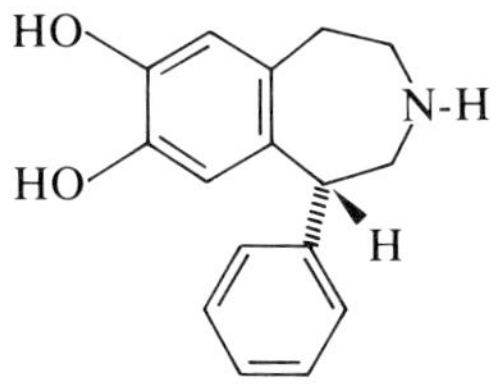

CCXXII

Molloy and Waddington [320] reported that the R-enantiomer CCXXII, but not the S-enantiomer, promoted grooming behavior in rats, which was considered to be a model specific for D-1 receptor effects.

When the benzazepine system bears a chlorine atom at position 6 (structure CCXXIII) the potency in the dog renal vascular assay increases greatly [314]; potency and activity in the rat rotation model is retained; and the chloro compound is somewhat more potent in the rat adenylate cyclase assay.

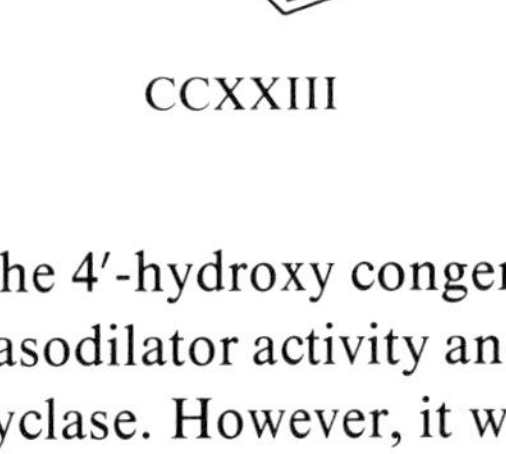

CCXXIII

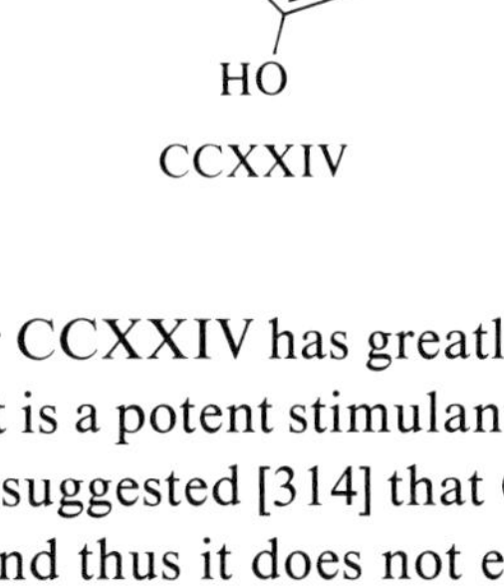

CCXXIV

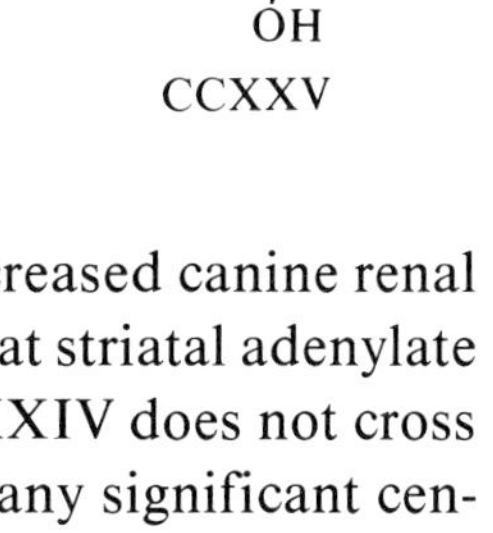

CCXXV

The 4′-hydroxy congener CCXXIV has greatly increased canine renal vasodilator activity and it is a potent stimulant of rat striatal adenylate cyclase. However, it was suggested [314] that CCXXIV does not cross the blood brain barrier and thus it does not exert any significant central dopaminergic activity. When the 6-chloro group was removed from CCXXIV (structure CCXXV), the renal vasodilator activity was lost. Compound CCXXV produced significant contralateral rotation in the lesioned rat when administered intracaudally but it was not active when given intraperitoneally. It was speculated [314] that the 6-chloro substituent enhances binding at the receptor in a conformation which induces maximum activation of the renal receptor. Alternatively, it was speculated that the 4′-hydroxy group enhances the polarity which decreases entry into lipophilic drug compartments and thus promotes higher concentration in the kidney.

Kaiser and co-workers [318, 319] have stated that replacement of the 1-phenyl group of CCXVIII by H (structure CCXXVI) results in 'only a weak dopaminergic receptor agonist'.

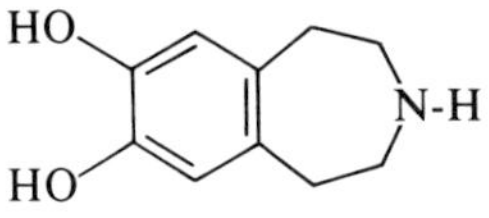

CCXXVI

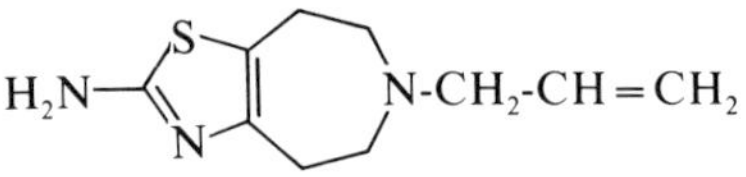

CCXXVII

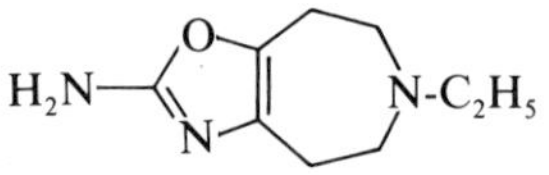

CCXXVIII

However, an aminothiazole congener CCXXVII bearing an N-allyl group has been reported [321] selectively and potently to stimulate dopamine autoreceptors in rat brain, based upon a motor activity assay. An N-ethyl oxazole isostere CCXXVIII seemed devoid of dopaminergic activity.

From a study of an extended series of 6-chloro-1-phenyltetrahydrobenzazepines based upon CCXXIX, Pfeiffer et al. [322] concluded that the most potent compounds in the canine renal vascular assay contained a hydroxyl group on the 1-phenyl moiety, or were substituted at the 3′-position with chloro, methyl, or trifluoromethyl.

The compounds with the best central dopaminergic activity were generally those which were the most lipophilic, were substituted on the 3′-position of the 1-phenyl ring, and contained either a 3-N-methyl or 3-N-allyl group. Two compounds, CCXXIV and CCXXX, were selected for more extensive examination, and a summary of the pharmacologi-

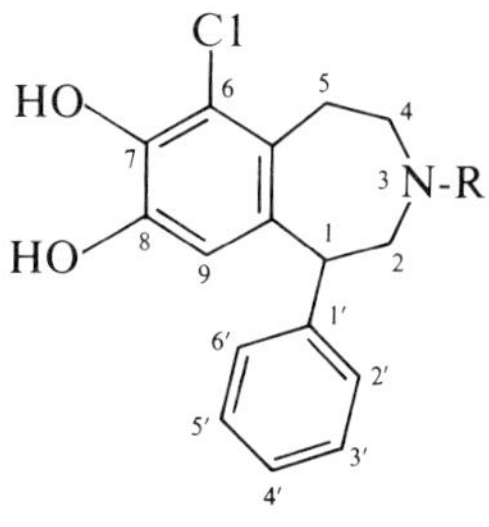

CCXXIX

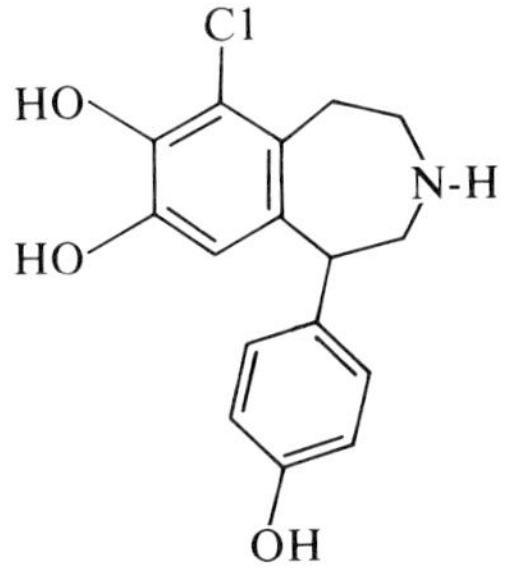

CCXXIV

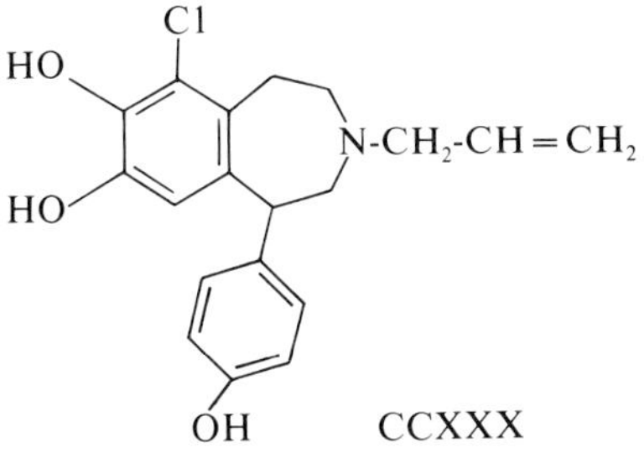

CCXXX

cal actions of the (±)-compounds has been presented [318]. Both compounds were resolved, and the absolute configurations of the enantiomers were determined. In agreement with the results obtained with the optical antipodes of CCXVIII, in vitro tests showed the potencies of the R-enantiomers of CCXXIV and CCXXX to be greater than those of either the corresponding racemates or of the S-enantiomers. These observations were interpreted as reinforcing the suggestion [319] that the 1-phenyl substituent is important for receptor activation, perhaps by interaction with a chirally defined accessory site.

In the 6-chlorobenzazepine series, N-allyl substitution enhanced potency 4- to 7-fold in the adenylate cyclase and the spiroperidol binding tests [318]. It was speculated that the significant potency of both enantiomers of CCXXIV and CCXXX in decreasing renal vascular resistance in dogs may reflect differing modes of action, perhaps as a consequence of N-allylation. The potency of R-CCXXX in the adenylate cyclase assay, coupled with its effectiveness in the canine renal vasodilator assay, suggested that it may have both D-1 and D-2 receptor agonist actions, as had been proposed previously [323] for the racemate. The comparatively weak potency of S-CCXXX relative to that of S-CCXXIV in these tests may reflect the receptor's spatial ability to accommodate both antipodes of CCXXIV, although S-CCXXIV seems to fit with less facility than its R-counterpart. N-allylation of the S-antipode interferes with receptor interaction. Kaiser et al. [318] noted that these structure-activity correlations must be considered in definitions of any model for the D-1 receptor. The literature does not reveal sufficient examples of N-allyl dopaminergic agonists to permit extensive conclusions about consistent effects (if any) of allyl groups on dopamine-like activity (cf. pharmacological data cited previously for N-allylnorapomorphine and the oxaergoline system CXXXVI). A 7-chloro-8-hydroxybenzazepine derivative CCXXXI was devoid of dopaminergic agonist effects in an assay for efflux of cyclic AMP from superfused rat neostriatal slices (D-1 receptor simulation) and in affecting K^+-evoked release of ^{3}H-labelled acetylcholine (D-2 receptor stimulation) [324].

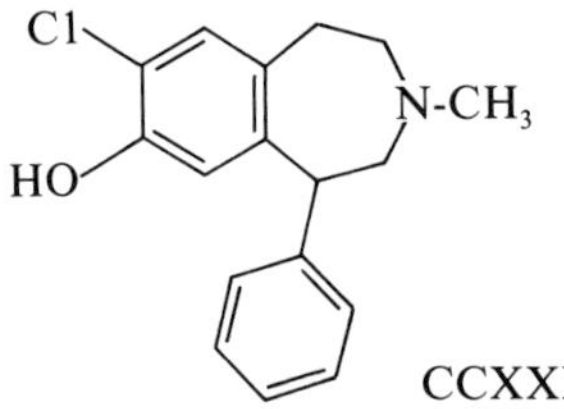

CCXXXI

Rather, this compound was an antagonist at both receptors [325]. Despite the appearance that the dopaminergic benzazepines represent an exception to the proposed [16, 17] requirement of an α- or a β-conformation for the dopamine moiety, Kaiser et al. [318] concluded that CCXXIV and CCXXX can be accommodated on proposed receptor models, and they illustrated possible agonist-receptor interactions using the McDermed model [158]. 2-Amino-6-(*p*-chlorobenzyl)-4H-5,6,7,8-tetrahydrothiazolo-5,4-*d*-azepine CCXXXII, a congener of a series of α-adrenoceptor agonists, was reported [383] to stimulate dopaminergic autoreceptors and to block noradrenergic autoreceptors in the mouse brain.

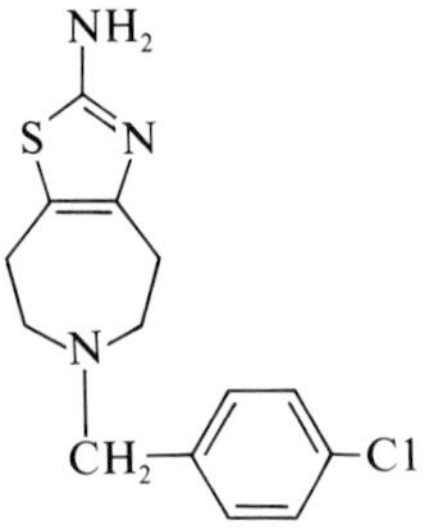

CCXXXII

8 Tetrahydroisoquinoline derivatives

The tetrahydroisoquinoline derivatives CCXXXIII and CCXXXIV are devoid of dopaminergic effects [27, 326].

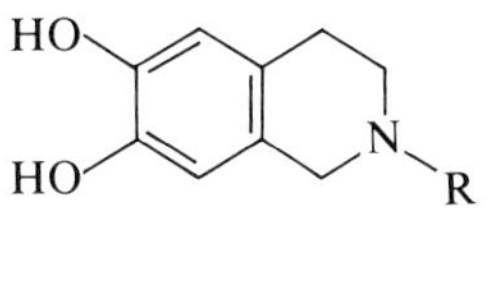

CCXXXIIIa R = H
b R = CH_3

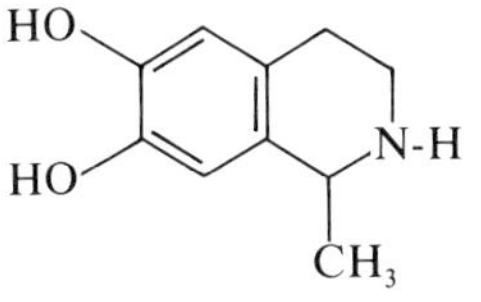

CCXXXIV

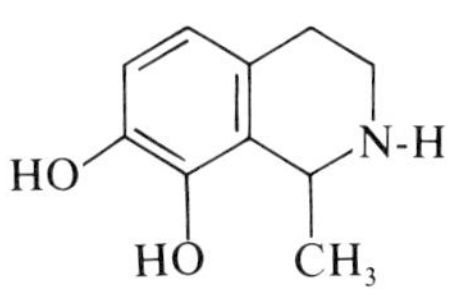

CCXXXV

This inactivity can be rationalized on the basis that the ring system holds the dopamine moiety with the catechol ring and the amino nitrogen *gauche* (synclinal range) rather than antiperiplanar, which has been proposed to be conducive to dopaminergic agonist effect. The OH group positional isomer CCXXXV is similarly inert [27].
Another isoquinoline derivative, nomifensine CCXXXVI, has shown dopaminergic agonist effects in a variety of animal studies [327–331].

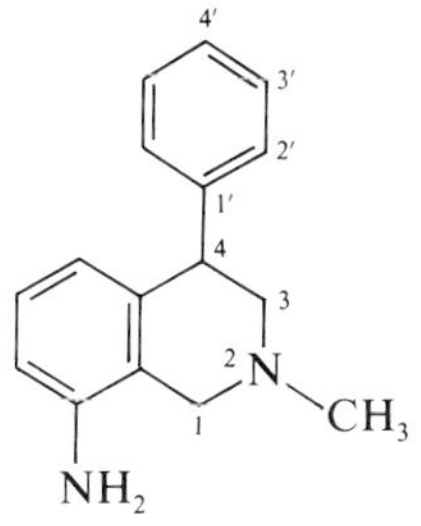

CCXXXVI

An extensive review clinical aspects of nomifensine has appeared [332].
Three major metabolites (CCXXXVII–CCXXXIX) of nomifensine have been found in human serum [333].
Although neither nomifensine CCXXXVI nor its 4′-hydroxy metabolite CCXXXVII had effect on rat striatal dopamine-sensitive adenylate cyclase [334, 335], the metabolite demonstrated dopamine-like effects in a variety of tests, albeit with lower potency than nomifensine [333]. In contrast to nomifensine, compound CCXXXVII also display-

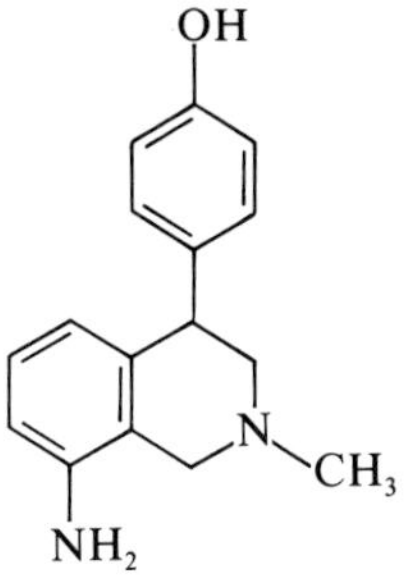

CCXXXVII

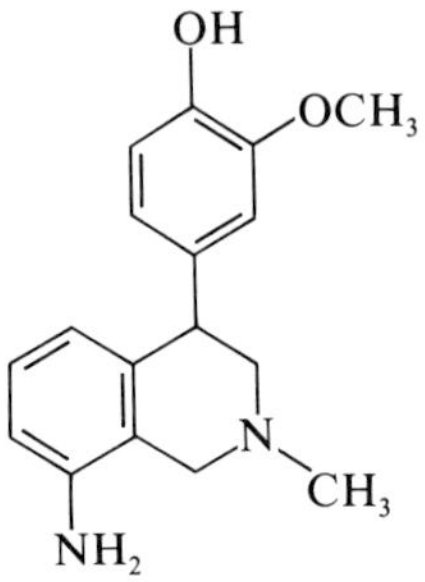

CCXXXVIII

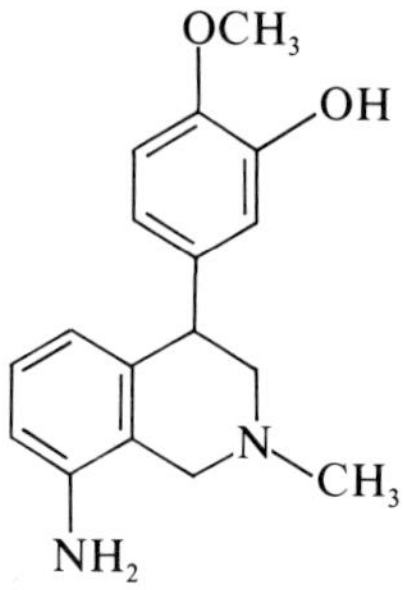

CCXXXIX

ed serotonergic activity and in general it had a wider spectrum of pharmacologic effect than nomifensine. However, Hoffmann [333] has commented that if the metabolite CCXXXVII contributes at all to the pharmacological effects seen with nomifensine, its quantitative share would be relatively small, since 'only about 7% of nomifensine are (sic) converted to this metabolite'. The catechol derivative CCXL was described [335] as a potent agonist in stimulation of adenylate cyclase in rat striatum, and it produced powerful dopamine-like effects upon injection into the nucleus accumbens of the mouse [336].

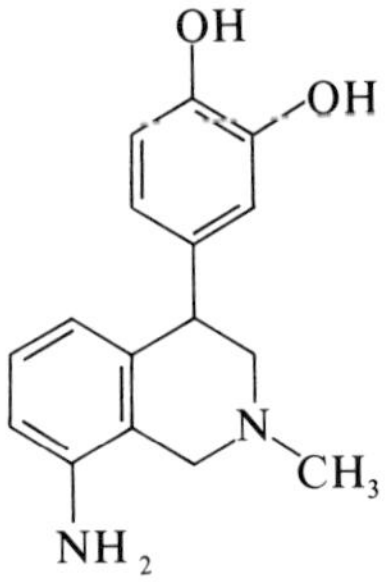

CCXL

This catechol derivative is a partial agonist on the renal vascular receptor, like apomorphine [337], and indeed, the structural similarity between CCXL and apomorphine XXVIII (shown below) was noted.

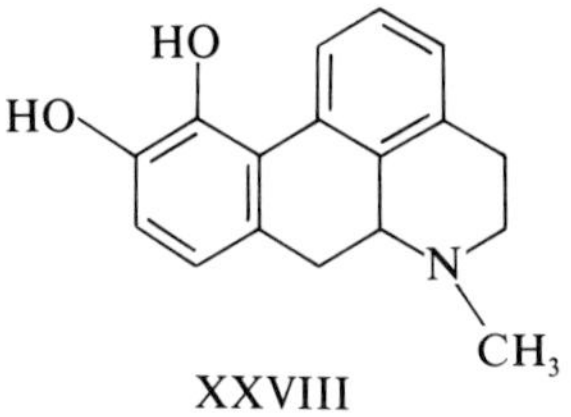

XXVIII

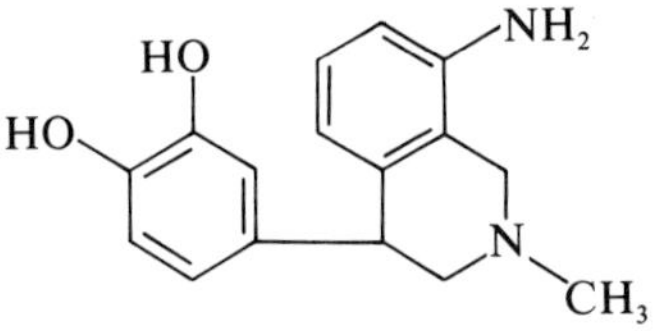

CCXL

It has been determined [338] that D-1 dopaminergic activity resides almost exclusively in the S-enantiomer of 3',4'-dihydroxynomifensine CCXLI.

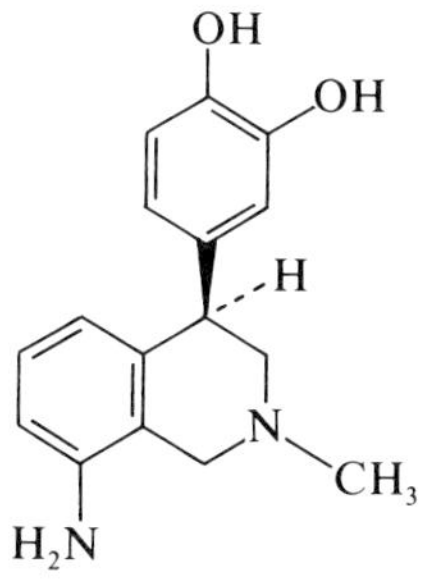

CCXLI

This absolute configuration was concluded [338] to fit the McDermed [158, 159] model of the dopamine receptor.
Both monomethyl ether metabolites CCXXXVIII, CCXXXIX of nomifensine are dopaminergically inert [333]. It may be concluded from the structures of CCXXXVIII and CCXXXIX that the catechol system CCXLI is formed from nomifensine in vivo. However, the free catechol system has apparently not been found in metabolism studies. It may be converted into the monomethyl ethers at a rate that precludes accumulation of significant amounts of the free catechol in the tissues. It therefore seems likely that the catechol derivative CCXLI does not play a significant role in the in vivo pharmacological actions of nomifensine.
Analysis of molecular models suggests that the torsion angle N-C_3-C_4-$C_{1'}$ in nomifensine (structure CCXXXVI) is in the antiperiplanar range (approximately 160°), which approximates the proposed pharmacologically optimal conformation for dopamine itself. The 4-phenyl group seems to be able to exist coplanar with the N-C_3-C_4 (ethylamine side chain) moiety, and it appears that the dopamine moiety in CCXL can assume either the α- or the β-conformation.
The 4-phenylisoquinoline derivative CCXLII was reported [339] to have dopamine-like properties, but improvement in Parkinsonian patients was small.

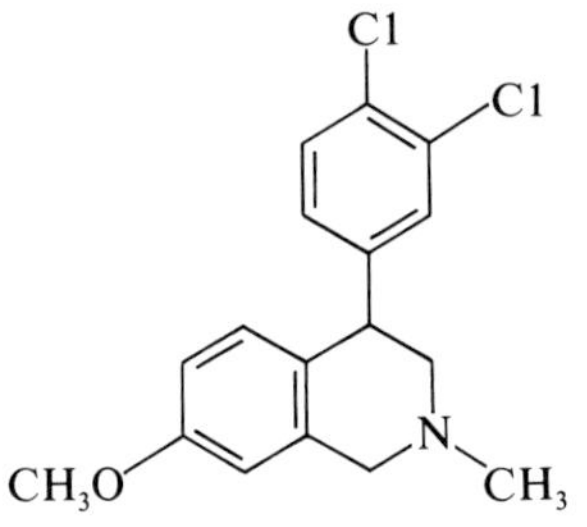

CCXLII

This compound was described [340] as probably the most potent inhibitor of dopamine uptake (rat brain synaptosomes) yet found.
A series of 8-acylamino and 8-acylhydrazino compounds CCXLIII has been reported [341].

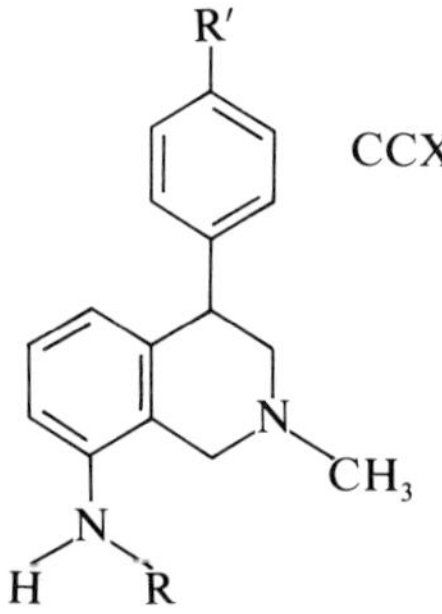

CCXLIIIa R = $CO(CH_2)_nNR^2R^3$
b R = $CONHR^2$,
c R = $COOR^2$
d R = $NHCOCH_3$
e R = $NHCOOC_2H_5$

The nature of the R′ group on the 4-phenyl moiety was not specified. It was stated that 'some of the new compounds have remarkable biological (antidepressant/antiparkinson) activity', but further details were not provided.

9 Miscellaneous Structures

Some aminomethylbenzocyclobutene derivatives CCXLIV have been identified as dopaminergic agonists [342].
Specifically, the N,N-di-*n*-propyl-3,4- and 4,5-dihydroxy derivatives (CCXLIVa and b) showed reasonably potent agonist activity at peripheral dopamine receptors. This result contrasts stongly with the lack of significant activity reported for the related aminotetralin derivatives (LXXIV) and the aminoindan CIII. The primary amines CCXLIVd

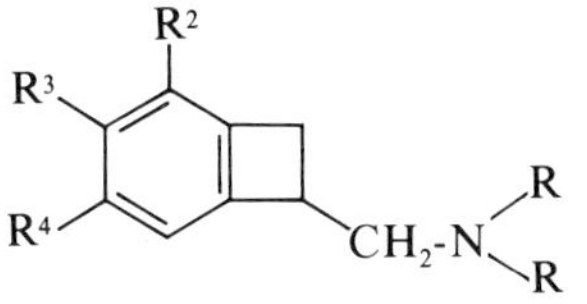

CCXLIVa R = n-C_3H_7; R^2 = R^3 = OH; R^4 = H
b R = n-C_3H_7; R^2 = H; R^3 = R^4 = OH
c R = n-C_3H_7; R^2 = R^3 = H; R^4 = OH
d R = H; R^2 = R^3 = OH; R^4 = H
e R = H; R^2 = H; R^3 = R^4 = OH
f R = CH_3; R^2 = R^3 = OH; R^4 = H

and e were devoid of activity at dopaminergic receptors, as was the 'meta' monohydroxy system CCXLIVc. Both compounds CCXLIVa and b showed considerable selectivity for the peripheral prejunctional (DA-2) receptor (rabbit isolated ear artery), compound CCXLIVa being about ten times more potent than CCXLIVb in vitro, and about three times more potent in vivo. No significant difference in selectivity was apparent between CCXLIVa and b [343].

Molecular models suggest that the β-phenethylamine moiety in CCXLIV can exist with the torsion angle N-C_1-C_2-$C_{1'}$ (structure CCXLV) approximately 180° (antiperiplanar).

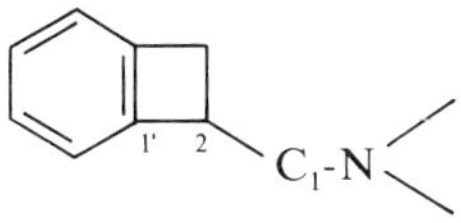

CCXLV

However, in this conformation, the plane of the benzene ring approaches perpendicularity (70–75°) with the plane of the ethylamine side chain moiety (N-C_1-C_2). Thus, this molecule seems to deviate from the definition for optimum dopamine conformation.

The *trans*-cyclobutane derivatives CCXLVI are more potent than their *cis*-isomers in binding studies on rat corpus striatum membrane tissue [344], but the potency is much lower than that of dopamine.

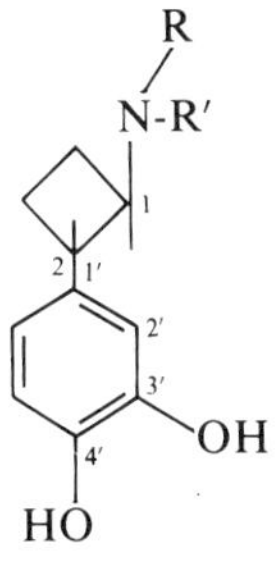

CCXLVIa R = R' = H
b R = R' = CH_3

The cyclobutane ring is not planar, but rather, exists in a puckered conformation [345]. Conformational analysis of molecular models of CCXLVI leads to the conclusion that the torsion angle N-C_1-C_2-$C_{1'}$ lies within the limits 125–170°, depending upon the direction and amount of puckering of the cyclobutane ring. Newman projection CCXLVII shows this torsion angle approximately 170°.

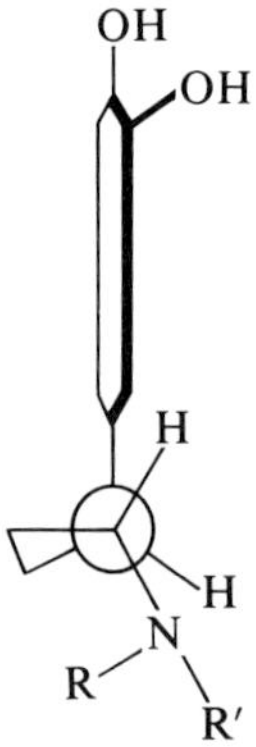

CCXLVII

Thus, the overall shape of the molecule seems to permit a close approximation of either the α-(illustrated in CCXLVII) or the β-rotamer of dopamine. The low potency of the *trans*-system CCXLVI may be a reflection of less-than-optimal correspondence of the dopamine moiety in the molecule with the topography of the dopamine receptor(s), due to lack of overall planarity of the cyclobutane-derived molecule. The *cis*-isomer of CCXLVI presents the dopamine moiety in an approximate *gauche* arrangement of the nitrogen and the catechol ring. The *cis*- and *trans*-cyclopropane derivatives CCXLVIII and CCXLIX are structurally similar to the *gauche*- and the *trans*-rotamers, respectively, of dopamine (structures I-III).

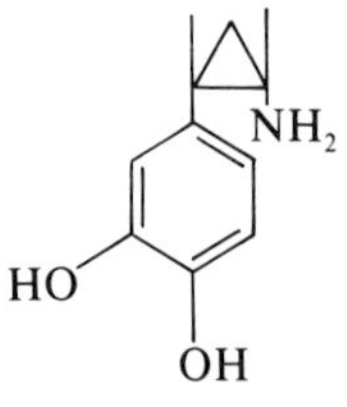

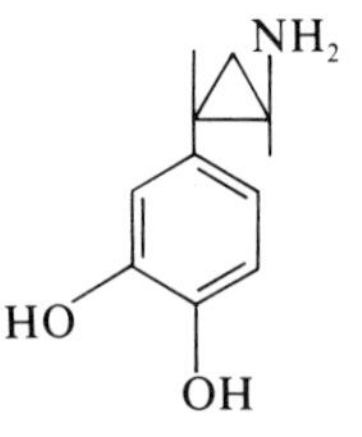

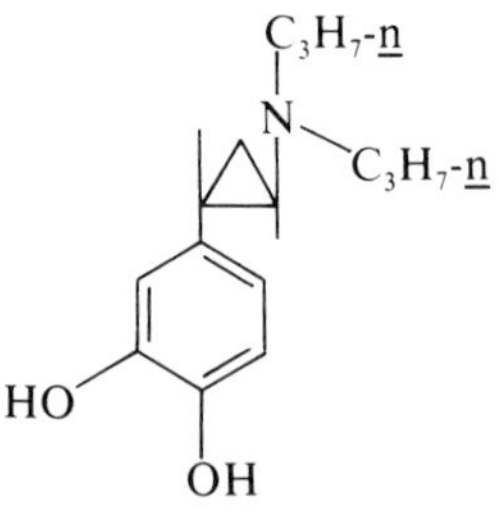

CCXLVIII CCXLIX CCL

Surprisingly, neither the *trans*-isomer CCXLIX nor the *cis*-isomer CCXLVIII showed dopaminergic agonist or antagonist effects [60, 61, 346–349]. It was concluded that the *trans*-isomer CCXLIX has a pharmacological profile similar to that of α-methyldopamine (structure XVIII: R = R' = H). Erhardt [60] rationalized the dopaminergic inactivity of the *trans*-system CCXLIX as being a reflection of the molecular bulk of the cyclopropane ring CH_2 group, which interferes with proper agonist-receptor interaction(s). The N,N-di-*n*-propyl homolog CCL of the *trans*-cyclopropane system exhibited weak dopamine agonist activity [350], yet another example of the 'N-propyl phenomenon'.

DPI, CCLI, has been described as a dopaminergic agonist in the brain of the snail, *Helix aspersa* [351], and in the cat caudate nucleus [352].

CCLI

However, more recent studies [353, 354] have lead to the conclusion that pharmacologically, DPI is a mixed α_1-, α_2-adrenoceptor agonist, and that evidence does not justify calling it a dopaminergic agonist. In a published polemic [385], these conclusions were challenged, and evidence was presented to support the contention that CCLI possesses adrenomimetic as well as dopamine-like effects.

An N_1-β-phenethyl-N_4-troponyl piperazine CCLII has been described [355, 356] as representing a novel class of central dopaminergic agonists, acting at supersensitive postsynaptic receptor sites.

CCLII

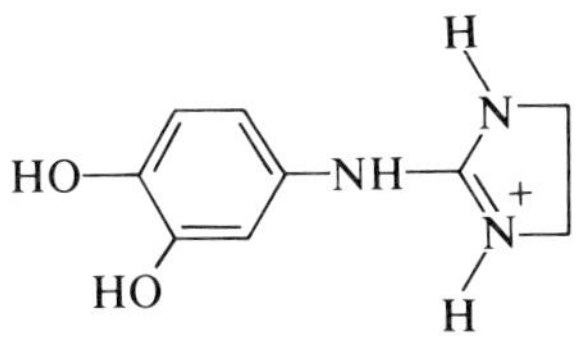

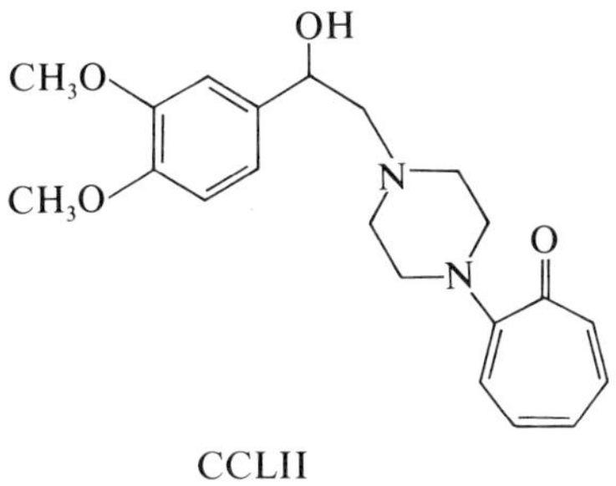

The behavioral effects of CCLII in 6-hydroxydopamine-lesioned rats reside in the S-(–)-enantiomer [355, 357, 358]. The R-(+)-enantiomer was essentially devoid of dopaminergic activity.
An extensive structure-activity study [359] of troponyl piperazines revealed that the N_1-β-phenethanol moiety of CCLII can be replaced by a variety of alkyl or substituted alkyl groups (structure CCLIII) with retention of dopaminergic effect in the rat hypokinesia model and the rat rotation model (6-hydroxydopamine lesion of the nigrostriatal dopamine pathway).

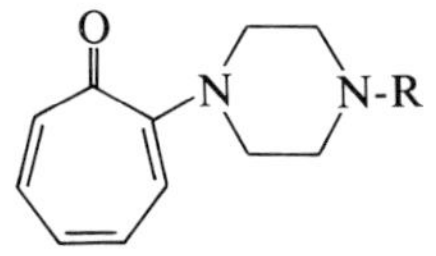

CCLIIIa R = H
b R = CH_3
c R = C_2H_5
d R = n-C_3H_7
e R = 2-C_3H_7
f R = t-C_4H_9
g R = $CH_2CH(CH_3)_2$
h R = n-C_4H_9

All members of this series except CCLIII f and h (R = *t*-C_4-H_9 or *n*-C_4H_9) were active in the rat hypokinesia assay, although the compounds were not highly potent. A series of N_1-hydroxyalkyl congeners CCLIV was studied.

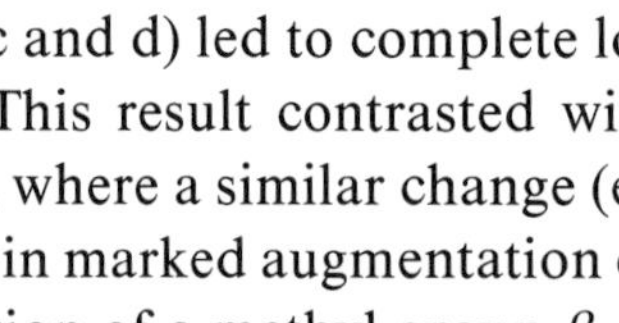

CCLIVa X = Y = H; R = OH
b X = H; Y = CH_3; R = OH
c X = CH_3; Y = H; R = OH
d X = CH_3; Y = CH_3; R = OH
e X = Y = H; R = CH_2OH
f X = H; Y = R = OH

Introduction of a methyl group onto the carbon α- to N_1 (as in CCLIV c and d) led to complete loss of activity in the rat hypokinesia model. This result contrasted with the results in the N-alkyl series CCLIII, where a similar change (e. g., the isopropyl system CCLIII e) resulted in marked augmentation of dopaminergic agonist activity. Incorporation of a methyl group β- to N_1 (as in CCLIV b) substantially increased biological activity. These findings were interpreted [359] to suggest that the alcohol function in CCLIV may be crucial for activity, and that any alteration by oxidative metabolism (e. g., $CH_2OH \rightarrow COOH$) may render the compound inactive. A methyl group α- to the OH group may prevent such oxidative deactivation. It was noted that compounds in which the Y group of CCLIV has a higher oxidation state (C = O) lacked biological activity. Esterification of the OH group in CCLIVa with acetic or pivalic acid permitted retention of a high level of activity. The ethyl ether of CCLIVa was active, but less so than the free OH derivative. The corresponding phenyl ether was inert. The 5'-chloro derivative of CCLIVa was inactive, but the 7'-bromo derivative was only slightly less active and potent than the parent compound CCLIVa.

Modification of the troponyl piperazine molecule by replacement of the piperazine ring by a 1-piperidinyl or a 4-piperindinyl system, or by a 1,4-diazepine ring resulted in complete loss of activity. Replacement of the troponyl ring by a benzene ring or by a 2-ketocycloheptene ring completely destroyed activity.

A piperonyl piperazine derivative, piribedil CCLV, was found by random screening to possess dopamine receptor agonist activity [360–365]. A review of the dopaminergic agonist effects of piribedil has appeared [366]. In addition to its effects on dopamine receptors, piribedil acts as a blocker on postsynaptic cholinergic receptors in the cockroach CNS [367].

CCLV

Jenner et al. [368] found compounds CCLVI–CCLVIII as metabolites of piribedil in the rat.

CCLVI   CCLVII

CCLVIII

There was also some indication of in vivo formation of a piperazine ring cleavage product.

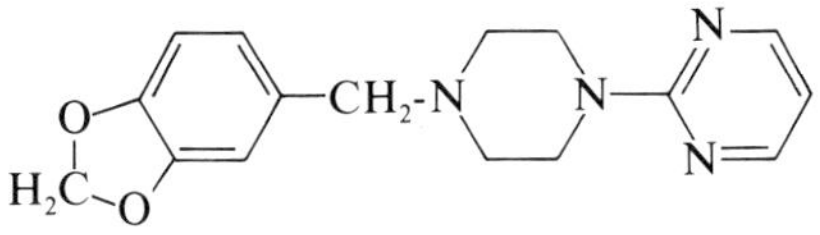

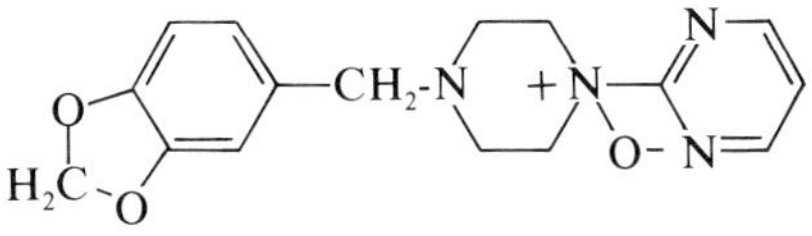

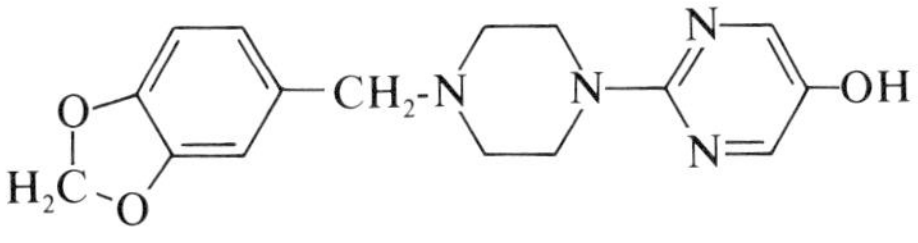

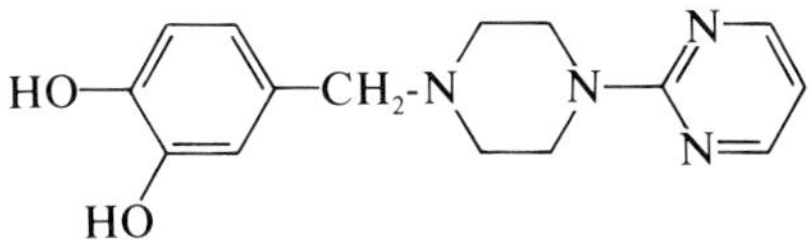

It was concluded that the pharmacological action of piribedil is due to a metabolite, and Creese [369] proposed that the catechol system CCLVIII (which has potent dopamine-like effects) is the active one. However, Schorderet [370] showed that, in the rabbit retinal adenylate cyclase assay, the effects of piribedil and of its catechol metabolite CCLVIII are *not* related to direct stimulation of dopamine receptors. Study of molecular models does not reveal close similarities of structure between α- or β-rotamers of dopamine or of the several rigid dopamine congeners (e. g., 2-aminotetralins, 2-aminoindans, aporphines, benzo(f)quinolines, or ergolines) and piribedil. The overall shape and reasonable conformations of the piribedil molecule, and pertinent interatomic distances (e. g., catechol OH-to-amino-N) seem completely at variance with those of the dopaminergically active systems cited above. The structural dimensions, parameters, and limitations which have been described for other dopaminergic agonists do not seem to apply to piribedil. As was stated ten years ago [16], piribedil is an enigma to the medicinal chemist.

Moragues and co-workers [371] prepared and studied a series of piribedil congeners CCLIX, based upon 4-aminopiperidine.

CCLIX

R, R_1 = H or CH_3
R_2, R_3 = combinations of H, CH_3, Cl, OCH_3, $-O-CH_2-O-$

Most of the members of the series showed some dopaminergic activity, although some exhibited primarily CNS effects and others exhibited primarily peripheral effects. No clear-cut structure-activity correlations were apparent. The closest pharmacological profile to that of piribedil was shown by compound CCLX.

CCLX

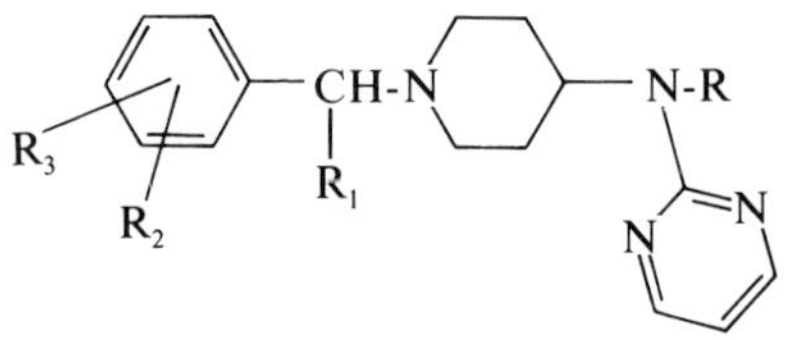

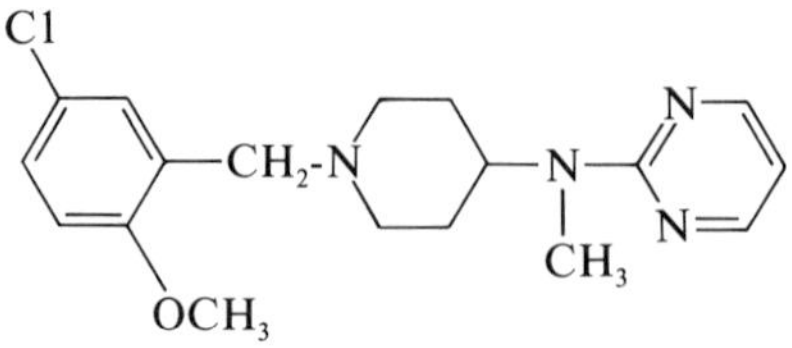

None of the members of the series were dopamine receptor blockers. Binding studies on another 2-pyrimidyl piperazine, buspirone (CCLXI), showed that it has affinity for dopamine receptors [372], and this compound has been termed a dopaminergic agonist [373].

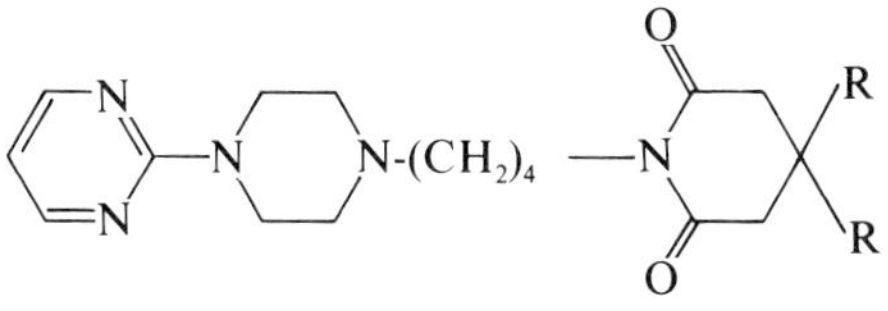

CCLXI R, R = $-(CH_2)_4-$
CCLXII R, R = CH_3

Yevich et al. [374] have cited results of several studies suggesting that buspirone may be a selective *antagonist* at presynaptic dopaminergic autoreceptors, and that its catalepsy-reversal effects may occur via a dopamine-independent mechanism. It was further stated [374] that the mechanistic implications of the findings of several other pharmacological studies of buspirone are as yet unclear. The analog CCLXII is equipotent to buspirone in conflict testing, but it is much weaker than buspirone in dopaminergic neurotransmission, thus suggesting that dopaminergic effects may not be important to the antianxiety actions of these drugs [375].

A structure-activity study of N-aryl or heteroaryl cyclic imide buspirone analogs CCLXIII has been reported [376].

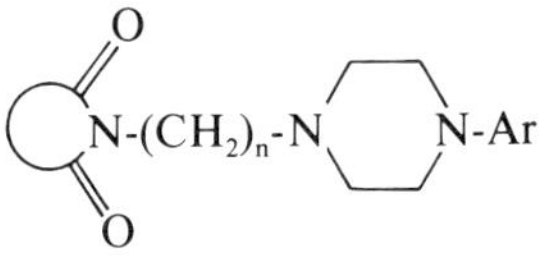

CCLXIII

However, it still seems unclear whether the spectrum of pharmacological effects reported for these compounds reflects dopaminergic agonism.

The question of whether the active conformation(s) of dopamine present(s) the benzene ring perpendicular to the plane of the ethylamine side chain (as in structure I) or coplanar with the ethylamine side chain (as in structures VI and VII) was addressed by Burn et al. [377] utilizing some derivatives CCLXIV, CCLXV of *exo*-2-aminobenzonorbornene.

I

VI

VII

CCLXIV

CCLXV

CCLXVI

These systems may be regarded as rigidly fixed, extreme conformations of 5,6- and 6,7-dihydroxy-2-aminotetralins (structures XLV and XLVI, respectively, and, as revealed by analysis of molecular models and as shown by the Newman projections below, the plane of the catechol ring makes an angle of approximately 70° with the plane of the ethylamine side chain. The two moieties cannot attain coplanarity in CCLXIV and CCLXV.

CCLXIV
Newman projection

CCLXV
Newman projection

CCLXVI
Newman projection

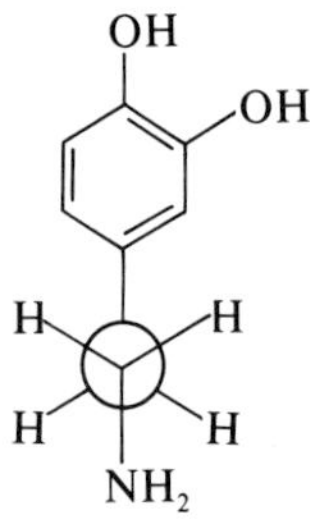

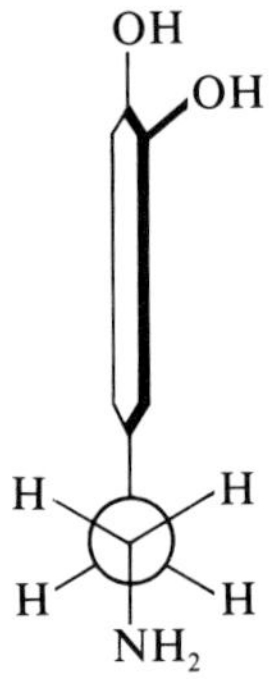

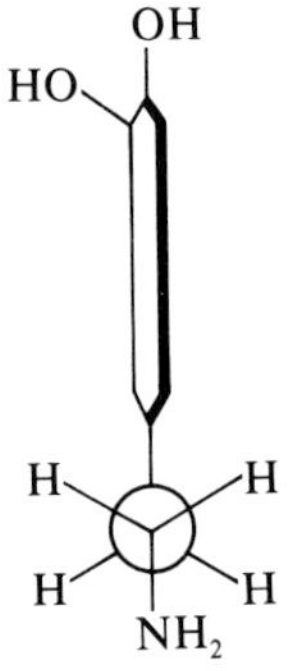

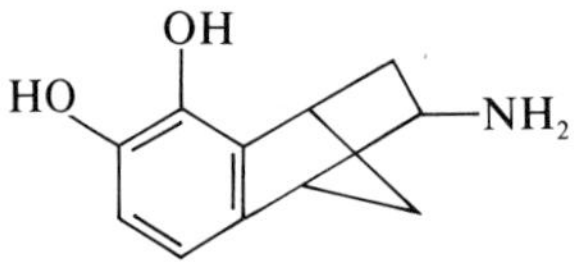

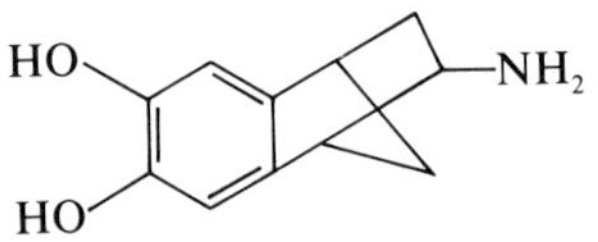

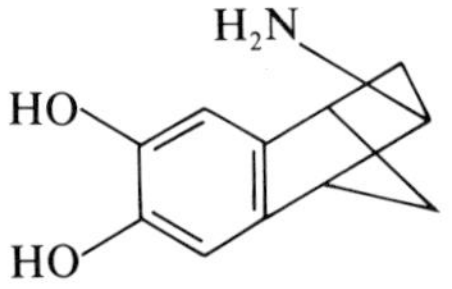

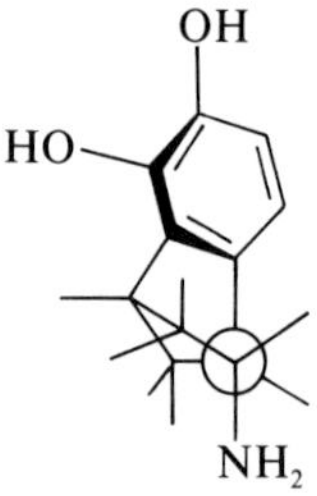

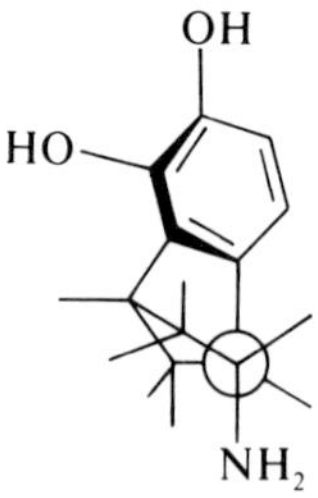

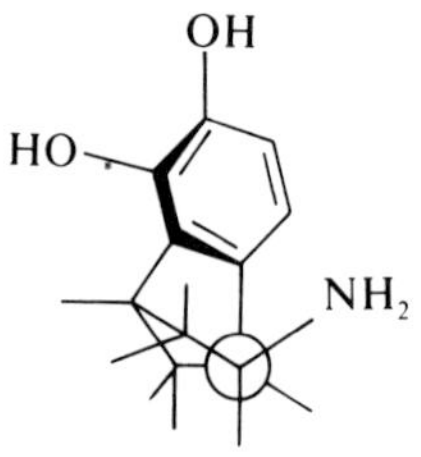

The isomeric *endo*-amino compound CCLXVI was prepared, a rigid analog related to the *gauche* conformers of dopamine (structures II, III). All three compounds CCLXIV–CCLXVI, together with the N-methyl- and N,N-dimethyl homologs of CCLXV and CCLXVI were inactive as dopaminergic agonists in the mouse stereotypy assay and the rat hyperactivity model. However, some of these compounds were effective in displacing ^{3}H-A-6,7-DTN XLVI ($R = R' = H$) and ^{3}H-N-*n*-propylnorapomorphine from rat striatal membranes. The negative behavioral data were interpreted as supporting the conformations VI and VII as the 'active' form of dopamine. Schuster and co-workers [378] have resolved the *exo*-system CCLXV and have reported that neither enantiomer was active in binding assays vs. ^{3}H-dopamine.
The *exo*-azabicyclo derivative CCLXVII maintains the dopamine moiety (N–C_1–C_2-catechol ring) in the *trans*-disposition (Newman projection CCLXVIII).

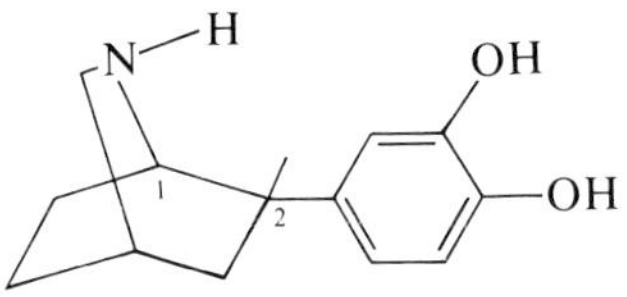

CCLXVII

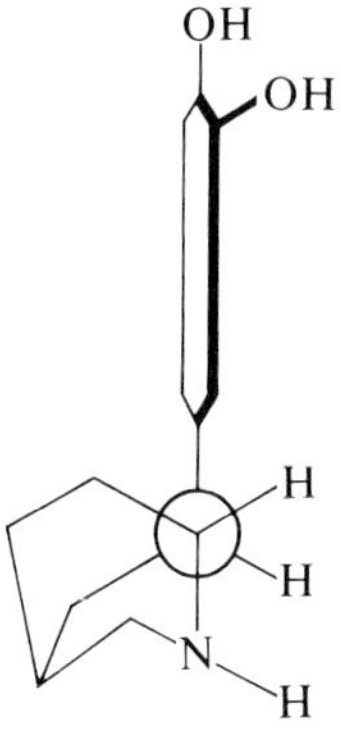

CCLXVIII

In this molecule, the catechol ring seems to be able to achieve either the α- (as illustrated) or the β-conformation without any apparent unfavorable non-bonded interactions. This compound is inert in several assays for dopamine-like effect [379], possibly a result of steric interference to receptor interaction caused by the bulky bicyclic ring. The molecule CCLXVII is typical of a number of compounds in which extraneous structural features predominate over the desirable steric disposition of the dopamine portion, and nullify dopaminergic actions. The achievement of conformational integrity of dopamine structure by incorporating it into a complex molecule is often at the expense of biological activity.

References

1 H. Blaschko: Experientia *13,* 9 (1957).
2 Z. Lackovic and N. H. Neff: Life Sci. *32,* 1665 (1983).
3 M. Relja and N. H. Neff: Fed. Proc., Fed. Am. Soc. exp. Biol. *42,* 2998 (1983).
4 S. E. Leff and I. Creese: Trends Pharmacol. Sci. *11,* 463 (1983).
5 I. Creese: Trends Neurosci. *5,* 40 (1982).
6 J. Offermeier and J. M. Van Rooyen: Trends Pharmacol. Sci. *3,* 326 (1982).
7 I. Cavero, R. Massingham and F. Lefèvre-Borg: Life Sci. *31,* 939 (1982).
8 J. W. Kebabian and D. B. Calne: Nature, Lond. *277,* 93 (1979).
9 J. W. Kebabian, M. Beaulieu and Y. Itoh: Can. J. Neurol. Sci. *11,* 1 suppl., 114 (1984).
10 J. C. Stoof and J. W. Kebabian: Life Sci. *35,* 2281 (1984).
11 K. A. Wreggett and P. Seeman: Acta Pharm. Suecica, suppl. *1983*: 1, 30.
12 B. Pullman, J.-L. Coubeils, Ph. Courrière and J.-P. Gervois: J. Med. Chem. *15,* 17 (1972).
13 R. Katz, S. R. Heller and A. E. Jacobson: Molec. Pharmacol. *9,* 486 (1973).
14 A. Rotman, J. Lundstrom, E. McNeal, J. Daly and C. R. Creveling: J. Med. Chem. *18,* 138 (1975).
15 J. Granot: J. Am. Chem. Soc. *100,* 1539 (1978).
16 J. G. Cannon: Adv. Neurol. *9,* 177 (1975).
17 J. G. Cannon: Adv. Biosci. *20,* 87 (1979).
18 L. I. Goldberg, J. D. Kohli, D. Cantacuzene, K. L. Kirk and C. R. Creveling: J. Pharmacol. exp. Ther. *213,* 509 (1980).
19 Y. Nimit, D. Cantacuzene, K. L. Kirk, C. R. Creveling and J. W. Daly: Life Sci. *27,* 1577 (1980)
20 K. L. Kirk: J. Org. Chem. *41,* 2373 (1976).
21 G. Firnau, S. Sood, R. Pantel and S. Garnett: Molec. Pharmacol. *19,* 130 (1981).
22 M. A. Elchisak and E. A. Hausner: Life Sci. *35,* 2561 (1984).
23 P. Laduron: Nature, Lond., New Biol. *238,* 212 (1972).
24 P. M. Laduron, W. R. Gommeren and J. E. Leysen: Biochem. Pharmacol. *23,* 1599 (1974).
25 H. Sheppard and C. R. Burghardt: Molec. Pharmacol. *10,* 721 (1974).
26 B. Costall, R. J. Naylor and R. M. Pinder: J. Pharm. Pharmacol. *26,* 753 (1974).
27 R. Miller, A. Horn, L. Iversen and R. M. Pinder: Nature, Lond. *250,* 238 (1974).
28 J. Z. Ginos, G. C. Cotzias and D. Doroski: J. Med. Chem. *21,* 160 (1978).
29 J. Z. Ginos and F. C. Brown: J. Med. Chem. *21,* 155 (1978).
30 J. Z. Ginos, G. C. Cotzias, E. Tolosa, L. C. Tang and A. J. Lo Monte: J. Med. Chem. *18,* 1194 (1975).
31 J. G. Cannon, F.-L. Hsu, J. P. Long, J. R. Flynn, B. Costall and R. J. Naylor: J. Med. Chem. *21,* 248 (1978).
32 J. D. Kohli, A. B. Weder, L. I. Goldberg and J. Z. Ginos: J. Pharmacol. exp. Ther. *213,* 370 (1980).
33 J. D. Kohli, L. I. Goldberg, P. H. Volkman and J. G. Cannon: J. Pharmacol. exp. Ther. *207,* 16 (1978).
34 M. Ilhan, J. P. Long and J. G. Cannon: Arch. int. Pharmacodyn. Thér. *212,* 247 (1974).
35 R. Massingham, M. L. Dubocovich and S. Z. Langer: Naunyn-Schmiedebergs Arch. Pharmacol. exp. Pathol. *314,* 17 (1980).
36 S. E. O'Conner, G. W. Smith and R. A. Brown: J. Cardiovasc. Pharmacol. *4,* 493 (1982).
37 I. Cavero, F. Lefèvre-Borg and R. Gomeni: J. Pharmacol. exp. Ther. *219,* 510 (1981).

38 W. H. Fennell, J. D. Kohli and L. I. Goldberg: J. Cardiovasc. Pharmacol. *2,* 247 (1980).
39 S. Z. Langer and R. Massingham: Br. J. Pharmacol. *69,* 297 P (1980).
40 R. Miller, A. Horn, L. Iversen and R. M. Pinder: Nature, Lond. *250,* 238 (1974).
41 D. Carlström, R. Bergin and G. Falkenberg: Q. Rev. Biophys. *6,* 257 (1973).
42 U. Hacksell, U. Svensson, J. L. G. Nilsson, S. Hjorth, A. Carlsson, H. Wikström, P. Lindberg and D. Sanchez: J. Med. Chem. *22,* 1469 (1979).
43 M. E. Goldman and J. W. Kebabian: Molec. Pharmacol. *25,* 18 (1984).
44 A. J. Bradbury, J. G. Cannon, B. Costall and R. J. Naylor: Eur. J. Pharmacol. *105,* 33 (1984).
45 D. Mihailova and B. Testa: Eur. J. Med. Chem. *13,* 49 (1978).
46 J. Armstrong and R. B. Barlow: Br. J. Pharmacol. *57,* 501 (1976).
47 K. Anderson, A. Kuruvilla, N. Uretsky and D. D. Miller: J. Med. Chem. *24,* 684 (1981).
48 S. A. Sadek, G. P. Basmadjian, P. M. Hsu and J. A. Rieger: J. Med. Chem. *26,* 947 (1984).
49 A. Hamada, Y. A. Chang, N. Uretsky and D. D. Miller: J. Med. Chem. *27,* 675 (1984).
50 N. Camerman and A. Camerman: Molec. Pharmacol. *19,* 517 (1981).
51 J. B. Lambert: Top. Stereochem. *6,* 20 (1971).
52 H. E. Geissler: Arch. Pharm., Weinheim *310,* 749 (1977).
53 C. Sumners, D. Dijkstra, J. B. de Vries and A. S. Horn: Naunyn-Schmiedebergs Arch. Pharmacol. exp. Pathol. *316,* 304 (1981).
54 R. L. Dorris and A. Parkhurst: Biochem. Pharmac. *23,* 867 (1974).
55 L. Nedelec, C. Dumont, C. Oberlander, D. Frechet, J. Laurent and J. R. Boissier: Eur. J. Med. Chem. *13,* 553 (1978).
56 J. R. Boissier, C. Dumont, J. Laurent and C. Oberlander: Psychopharmacology *68,* 15 (1980).
57 C. Euvrard, L. Ferland, T. Di Polo, M. Beaulieu, F. Labrie, C. Oberlander, J. P. Raynaud and J. R. Boissier: Neuropharmacology *19,* 379 (1980).
58 R. J. Borgman, M. R. Baylor, J. J. McPhillips and R. E. Stitzel: J. Med. Chem. *22,* 901 (1979).
59 J. G. Cannon, Z. Perez, J. P. Long, D. B. Rusterholz, J. R. Flynn, B. Costall, D. H. Fortune and R. J. Naylor: J. Med. Chem. *22,* 901 (1979).
60 P. W. Erhardt: Acta Pharm. Suecica, suppl. *1983,* 2, 56.
61 P. W. Erhardt: J. Pharm. Sci. *69,* 1059 (1980).
62 V. Valenta, A. Dlabac, M. Valchar and M. Protiva: Coll. Czech. Chem. Commun. *49,* 1002 (1984).
63 J. Wepierre, C. Doreau, A. Papin, C. Paultre and Y. Cohen: Arch. int. Pharmacodyn. Thér. *206,* 135 (1973).
64 C. Casagrande and G. Ferrari: Farmaco, Ed. Sci. *28,* 143 (1973).
65 R. J. Borgman, J. J. McPhillips, R. E. Stitzel and I. J. Goodman: J. Med. Chem. *16,* 630 (1973).
66 M. Gerlach, P. Jutzi, J.-P. Stasch and H. Przuntek: Z. Naturforsch., part. B *38,* 237 (1983).
67 W. C. Randolph, J. E. Swagzdis, G. L. Joseph and R. Gifford: Pharmacologist *25,* 117 (1983).
68 C. Casagrande, P. Castelnuovo, O. Cerri, R. Ferrini, L. Merlo, G. Miragoli, M. Paro, R. Pataccini, F. Pocchiari and F. Santangelo: Abstr. VIIIth Int. Symposium on Medicinal Chemistry, Uppsala, Sweden, August 27–31, 1984, p. 72.
69 J. N. Harvey, D. P. Worth, J. Brown and M. R. Lee: Br. J. Clin. Pharmacol. *17,* 671 (1984).
70 R. J. Baldessarini, K. G. Walton and R. J. Borgman: Neuropharmacology *14,* 725 (1975).

71 O. Kuchel, N. T. Buu and Th. Unger: Adv. Biosci. *20,* 15 (1979).
72 C. R. Creveling, N. Dalgard, H. Shimizu and J. W. Daly: Molec. Pharmacol. *6,* 691 (1970).
73 J. A. Smith and R. Hartley: J. Pharm. Pharmacol. *25,* 415 (1973).
74 M. G. Feenstra, J. W. Homan, R. Everts, H. Rollema and A. S. Horn: Naunyn-Schmiedebergs Arch. Pharmacol. exp. Pathol. *326,* 203 (1984).
75 F. A. Kuehl, M. Hichens, R. E. Ormond, M. A. P. Meisinger, P. H. Gale, V. J. Cirillo and N. J. Brink: Nature, Lond. *203,* 154 (1964).
76 M. T. I. W. Schüsler-Van Hees and G. M. J. Beijersbergen Van Henegouwen: Pharm. Weekbl. Sci. Ed. *5,* 291 (1983).
77 J. A. Roth and K. Feor: Biochem. Pharmacol. *27,* 1606 (1978).
78 V. Glover, M. Sandler, F. Owen and G. J. Riley: Nature, Lond. *265,* 80 (1977).
79 O. Suzuki, T. Matsumoto, Y. Katsumata and M. Oya: Experientia *36,* 895 (1980).
80 R. L. Bronaugh, S. E. Hattox, M. M. Hoehn, R. C. Murphy and C. O. Rutledge: J. Pharmacol. exp. Ther. *195,* 441 (1975).
81 Th. Unger, N. T. Buu and O. Kuchel: Adv. Biosci. *20,* 357 (1978).
82 Y.-Q. Qu, Y. Hashimoto and H. Miyazaki: J. Pharmacobiodyn. *5,* S-27 (1982).
83 Y.-Q. Qu: Life Sci. *32,* 1811 (1983).
84 K. Racz, N. T. Buu, O. Kuchel and A. De Leon: Am. J. Physiol. *247,* No. 4, part 1, E431 (1984).
85 N. T. Buu, J. Duchaime, O. Kuchel and J. Jenest: Life Sci. *29,* 2311 (1981).
86 N. T. Buu, J. Duchaime and O. Kuchel: Life Sci. *35,* 1083 (1984).
87 W. N. Jenner and F. A. Rose: Biochem. J. *135,* 109 (1973).
88 A. J. Rivett, A. Francis, R. Whittemore and J. A. Roth: J. Neurochem. *42,* 1444 (1984).
89 W. N. Jenner and F. A. Rose: Nature, Lond. *252,* 237 (1974).
90 I. Merits: Biochem. J. *25,* 829 (1976).
91 J. A. Roth and A. J. Rivett: Biochem. Pharmacol. *31,* 3017 (1982).
92 J. R. Idle, B. A. Osikowska, P. S. Sever and F. J. Swinbourne: Br. J. Pharmacol. *74,* 837 P (1981).
93 B. A. Osikowska, J. R. Idle, F. J. Swinbourne and P. S. Sever: Biochem. Pharmacol. *31,* 2279 (1982).
94 D. M. Ackerman, J. P. Hieble, H. M. Sarau and T. C. Jain: Arch. int. Pharmacodyn. Thér. *267,* 241 (1984).
95 M. C. Scott and M. A. Elchisak: Fed. Proc. Fed. Am. Soc. exp. Biol. *42,* 1363 (1983).
96 H. Corrodi and H. Hardegger: Helv. Chim. Acta *38,* 2038 (1955).
97 J. Kalvoda, P. Buchschacher and O. Jeger: Helv. Chim. Acta *38,* 1847 (1955).
98 J. C. Craig and S. K. Roy: Tetrahedron *21,* 395 (1965).
99 J. Giesecke: Acta Crystallogr. *33 B,* 302 (1977).
100 W. S. Saari, S. W. King and V. J. Lotti: J. Med. Chem. *16,* 171 (1973).
101 W. H. Riffee, R. E. Wilcox and R. V. Smith: Abstr. 8th Int. Congr. Pharmacol., Satellite Symposium on Dopamine, Okayama, Japan, July, 1981, p. 44.
102 J. Lehmann, R. V. Smith and S. Z. Langer: Eur. J. Pharmacol. *88,* 81 (1983).
103 E. R. Atkinson, F. J. Bullock, F. E. Granchelli, S. Archer, F. J. Rosenberg, P. G. Teiger and F. C. Nachod: J. Med. Chem. *18,* 1000 (1975).
104 M. V. Koch, J. G. Cannon and A. M. Burkman: J. Med. Chem. *11,* 977 (1968).
105 J. F. Hensiak, J. G. Cannon and A. M. Burkman: J. Med. Chem. *8,* 557 (1965).
106 R. I. Schoenfeld, J. L. Neumeyer, W. Dafeldecker and S. Roffler-Tarlov: Eur. J. Pharmacol. *30,* 63 (1975).

107 J. L. Neumeyer, D. Reischig, G. W. Arana, A. Campbell, R. J. Baldessarini, N. S. Kula and K. J. Watling: J. Med. Chem. *26,* 516 (1983).
108 J. Giesecke: Acta Crystallogr. *29 B,* 1785 (1973).
109 W. S. Saari, S. W. King, V. J. Lotti and A. Scriabine: J. Med. Chem. *17,* 1086 (1974).
110 J. L. Neumeyer, F. E. Granchelli, K. Fuxe, U. Ungersted and H. Corrodi: J. Med. Chem. *17,* 1090 (1974).
111 J. L. Neumeyer, W. P. Dafeldecker, B. Costall and R. J. Naylor: J. Med. Chem. *20,* 190 (1977).
112 R. J. Miller, P. H. Kelly and J. L. Neumeyer: Eur. J. Pharmacol. *35,* 77 (1976).
113 J. L. Neumeyer, M. M. McCarthy, S. M. Battista, F. J. Rosenberg and D. G. Teiger: J. Med. Chem. *16,* 1228 (1973).
114 J. G. Cannon, R. V. Smith, M. A. Aleem and J. P. Long: J. Med. Chem. *18,* 108 (1975).
115 J. G. Cannon, T. Lee, H. D. Goldman, B. Costall and R. J. Naylor: J. Med. Chem. *20,* 1111 (1977).
116 L. I. Goldberg, J. D. Kohli, A. N. Kotake and P. H. Volkman: Fed. Proc. Fed. Am. Soc. exp. Biol. *37,* 82 (1978).
117 D. Berney, T. J. Pechter, J. Schmutz, H. P. Weber and T. G. White: Experientia *31,* 1327 (1975).
118 J. G. Cannon, R. J. Borgman, M. A. Aleem and J. P. Long: J. Med. Chem. *16,* 219 (1973).
119 J. G. Cannon, P. R. Khonje and J. P. Long: J. Med. Chem. *18,* 110 (1975).
120 R. G. Harvey and P. P. Fu: J. Org. Chem. *41,* 3722 (1976).
121 R. J. Baldessarini, K. G. Walton and R. J. Borgman: Neuropharmacology *15,* 471 (1976).
122 J. G. Cannon, J. F. Hensiak and A. M. Burkman: J. Pharm. Sci. *52,* 1112 (1963).
123 R. J. Baldessarini, N. S. Kula, K. G. Walton and R. J. Borgman: Psychopharmacology *53,* 45 (1977).
124 R. J. Baldessarini, J. L. Neumeyer, A. Campbell, G. Sperk, V. Ram, G. W. Arana and N. S. Kula: Eur. J. Pharmacol. *77,* 87 (1982).
125 A. Campbell, R. J. Baldessarini, V. J. Ram and J. L. Neumeyer: Neuropharmacology *21,* 953 (1982).
126 P. N. Kaul, E. Brochmann-Hanssen and E. L. Way: J. Pharm. Sci. *50,* 244 (1961).
127 P. N. Kaul, E. Brochmann-Hanssen and E. L. Way: J. Pharm. Sci. *50,* 248 (1961).
128 P. N. Kaul and M. W. Conway: J. Pharm. Sci. *60,* 93 (1971).
129 J. G. Cannon, R. V. Smith, A. Modiri, S. P. Sood, R. J. Borgman, M. A. Aleem and J. P. Long: J. Med. Chem. *15,* 273 (1972).
130 G. M. McKenzie and H. L. White: Biochem. Pharmacol. *22,* 2329 (1973).
131 A. L. Symes, S. Lal and T. L. Sourkes: J. Pharm. Pharmacol. *27,* 947 (1975).
132 K. Missala, S. Lal and T. L. Sourkes: Eur. J. Pharmacol. *22,* 54 (1973).
133 J. L. Neumeyer, B. R. Neustadt, K. H. Oh, K. K. Weinhardt, C. B. Boyce, F. J. Rosenberg and D. G. Teiger: J. Med. Chem. *16,* 1223 (1973).
134 J. Giesecke: Acta Crystallogr. *36 B,* 110 (1980).
135 J. G. Cannon, J. C. Kim, M. A. Aleem and J. P. Long: J. Med. Chem. *15,* 348 (1972).
136 J. D. McDermed, G. M. McKenzie and A. P. Phillips: J. Med. Chem. *18,* 362 (1975).
137 G. N. Woodruff, A. O. Elkhawad and R. M. Pinder: Eur. J. Pharmacol. *25,* 80 (1974).
138 M. Beaulieu, Y. Itho, P. Tepper, A. S. Horn and J. W. Kebabian: Eur. J. Pharmacol. *105,* 15 (1984).

139 W. K. Sprenger, J. G. Cannon, B. K. Barman and A. M. Burkman: J. Med. Chem. *12*, 487 (1969).
140 J. D. Kohli, L. I. Goldberg and D. E. Nichols: Eur. J. Pharmacol. *56*, 39 (1979).
141 J. D. Kohli, L. I. Goldberg and J. D. McDermed: Eur. J. Pharmacol. *81*, 293 (1982).
142 M. P. Seiler and R. Markstein: Molec. Pharmacol. *22*, 281 (1982).
143 D. B. Goodale, D. B. Rusterholz, J. P. Long, J. R. Flynn, B. Walsh, J. G. Cannon and T. Lee: Science *210*, 1141 (1980).
144 G. E. Martin, D. R. Haubrich and M. Williams: Eur. J. Pharmacol. *76*, 15 (1981).
145 A. S. Horn, J. DeVries, D. Dijkstra and A. H. Mulder: Eur. J. Pharmacol. *83*, 35 (1982).
146 J. W. Kebabian, K. Miyazaki and C. W. Grewe: Neurochem. Int. *5*, 227 (1983).
147 M. Williams, G. E. Martin, D. E. McClure, J. J. Baldwin and K. J. Watling: Naunyn-Schmiedebergs Arch. Pharmacol. exp. Pathol. *324*, 275 (1983).
148 A. M. Crider, T. F. Hemdi, M. N. Hassan and S. Fahn: J. Pharm. Sci. *73*, 1585 (1984).
149 J. D. McDermed, G. M. McKenzie and H. S. Freeman: J. Med. Chem. *19*, 547 (1976).
150 J. L. Tedesco, P. Seeman and J. D. McDermed: Molec. Pharmacol. *16*, 369 (1979).
151 L.-E. Arvidsson, U. Hacksell, J. L. G. Nilsson, S. Hjorth, A. Carlsson, P. Lindberg, D. Sanchez and H. Wikström: J. Med. Chem. *24*, 921 (1981).
152 M. G. P. Feenstra, H. Rollema, D. Dijkstra, C. J. Grol, A. S. Horn and B. H. C. Westerink: Naunyn-Schmiedebergs Arch. Pharmacol. exp. Pathol. *313*, 213 (1980).
153 J. C. Van Oene, J. B. DeVries, D. Dijkstra, R. J. W. Renkema, P. G. Tepper and A. S. Horn: Eur. J. Pharmacol. *102*, 101 (1984).
154 A. S. Horn, P. Tepper, J. W. Kebabian and P. M. Beart: Eur. J. Pharmacol. *99*, 125 (1984).
155 E. J. Lien and J. L. G. Nilsson: Acta Pharm. Suecica *20*, 271 (1983).
156 M. Hamon, S. Bourgoin, H. Gozlau, M. D. Hall, C. Goetz, F. Artaud and A. S. Horn: Eur. J. Pharmacol. *100*, 263 (1984).
157 M. Simonovic, G. A. Gudelsky and H. Y. Meltzer: J. Neurol. Transm. *59*, 143 (1984) [C. A. *101*, 49142r (1984)].
158 J. D. McDermed, H. S. Freeman and R. M. Ferris, in: Catecholamines: Basic and Clinical Frontiers, vol. 1, p. 568–570. Eds. E. Usdin, I. J. Kopin and J. Barchas. Pergamon Press, Elmsford, New York 1979.
159 J. D. McDermed: Abstr. 8th Int. Congr. Pharmacol., Satellite Symposium on Dopamine, Okayama, Japan, July, 1981, p. 22.
160 M. Schorderet, J. D. McDermed and P. Magistretti: J. Physiol., Paris *74*, 509 (1978).
161 F. Ince, B. Springthorpe, R. A. Brown, J. C. Hall, S. E. O'Connor and G. W. Smith: Abstr. VIIIth Int. Symposium on Medicinal Chemistry, Uppsala, Sweden, August 27–31, 1984, p. 79.
162 M. Williams, G. E. Martin, D. E. McClure, J. J. Baldwin and K. J. Watling: Naunyn-Schmiedebergs Arch. Pharmacol. exp. Pathol. *324*, 275 (1983).
163 C. J. Grol, L. J. Jansen and H. Rollema: Abstr. VIIIth Int. Symposium on Medicinal Chemistry, Uppsala, Sweden, August 27–31, 1984, p. 74.
164 C. D. Andrews, J. D. McDermed and G. N. Woodruff: Br. J. Pharmacol. *64*, 433P (1978).
165 C. J. Grol and H. Rollema: J. Pharm. Pharmacol. *29*, 153 (1977).
166 H. Wikström: Ph. D. Thesis, Uppsala University, Sweden, 1983, p. 34–35.
167 H. Wikström, B. Andersson, D. Sanchez, P. Lindberg, K. Svensson, S.

Hjorth, A. Carlsson, L.-E. Arvidsson, A. M. Johansson and J. L. G. Nilsson: Abstr. VIIIth Int. Symposium on Medicinal Chemistry, Uppsala, Sweden, August 27–31, 1984, p. L43.
168 A. S. Horn, D. Dijkstra, T. B. A. Mulder, H. Rollema and B. H. C. Westerink: Eur. J. Med. Chem. *16*, 469 (1981).
169 M. G. P. Feenstra, H. Rollema, T. B. A. Mulder and A. S. Horn: Life Sci. *32*, 459 (1983).
170 H. Rollema, B. H. C. Westerink, T. B. A. Mulder, D. Dijkstra, M. G. P. Feenstra and A. S. Horn: Eur. J. Pharmacol. *64*, 313 (1980).
171 H. Rollema, C. J. Grol and M. G. P. Feenstra: Pharm. Weekbl. Sci. Ed. *4*, 205 (1982).
172 H. Rollema and C. J. Grol: Pharm. Weekbl. Sci. Ed. *5*, 159 (1983).
173 I. R. Youde, M. J. Raxworthy, P. A. Gulliver, D. Dijkstra and A. S. Horn: J. Pharm. Pharmacol. *36*, 309 (1984).
174 J. G. Cannon, D. L. Koble, J. P. Long and T. Verimer: J. Med. Chem. *23*, 750 (1980).
175 T. Verimer, J. P. Long, R. Bhatnagar, D. L. Koble, J. G. Cannon, J. R. Flynn, D. B. Goodale and S. P. Arneric: Arch. int. Pharmacodyn. Thér. *250*, 221 (1981).
176 J. C. Koons, L. J. Fischer, J. G. Cannon and J. P. Long: Pharmacologist *25*, 233 (1982).
177 J. C. Koons, J. R. Flynn, J. G. Cannon and J. P. Long: Fed. Proc. Fed. Am. Soc. exp. Biol. *41*, 1662 (1982).
178 J. G. Cannon, D. C. Furlano, D. L. Koble, J. C. Koons and J. P. Long: Life Sci. *34*, 1679 (1984).
179 J. P. Long: Univ. of Iowa, unpublished data (1984).
180 J. G. Cannon, D. C. Furlano, R. G. Dushin and J. P. Long: Presentation at VIIIth Int. Symposium on Medicinal Chemistry, Uppsala, Sweden, August 27–31, 1984.
181 J. G. Cannon, A. N. Brubaker, J. P. Long, J. R. Flynn and T. Verimer: J. Med. Chem. *24*, 149 (1981).
182 U. Hacksell, A. M. Johansson, L.-E. Arvidsson, J. L. G. Nilsson, S. Hjorth, A. Carlsson, H. Wikström, D. Sanchez and P. Lindberg: J. Med. Chem. *27*, 1003 (1984).
183 U. Hacksell: Ph. D. Thesis, Uppsala University, Sweden, 1981, p. 38.
184 D. E. Nichols, J. N. Jacob, A. J. Hoffman, J. D. Kohli and D. Glock: J. Med. Chem. *27*, 1701 (1984).
185 B. Andersson, D. Sanchez, P. Lindberg, H. Wikström, K. Svensson, S. Hjorth, A. Carlsson, L.-E. Arvidsson, A. Johansson, U. Hacksell and J. L. G. Nilsson: Abstr. VIIIth Int. Symposium on Medicinal Chemistry, Uppsala, Sweden, August 27–31, 1984, p. 173.
186 A. M. Johansson, L.-E. Arvidsson, J. L. G. Nilsson, D. Sanchez, B. Andersson, H. Wikström, K. Svensson, S. Hjorth and A. Carlsson: Abstr. VIIIth Int. Symposium on Medicinal Chemistry, Uppsala, Sweden, August 27–31, 1984, p. 80.
187 K. Svensson, S. Hjorth, D. Clark, A. Carlsson, H. Wikström, B. Andersson and D. Sanchez: Abstr. VIIIth Int. Symposium on Medicinal Chemistry, Uppsala, Sweden, August 27–31, 1984, p. 91.
188 D. B. Rusterholz, J. P. Long, J. R. Flynn, J. G. Cannon, T. Lee, J. P. Pease, J. A. Clemens, D. T. Wong and F. P. Bymaster: Eur. J. Pharmacol. *55*, 73 (1979).
189 M. G. P. Feenstra, H. Rollema, D. Dijkstra, C. J. Grol, A. S. Horn and B. H. C. Westerink: Naunyn-Schmiedebergs Arch. Pharmacol. exp. Pathol. *313*, 213 (1980).
190 R. W. Fuller, J. C. Baker and B. B. Molloy: J. Pharm. Sci. *66*, 271 (1977).
191 M. Ilhan, J. P. Long and J. G. Cannon: Arch. int. Pharmacodyn. Thér. *271*, 213 (1984).

192 J. G. Cannon, J. A. Perez, J. P. Pease, J. P. Long, J. R. Flynn, D. B. Rusterholz and S. E. Dryer: J. Med. Chem. *23,* 745 (1980).
193 D. E. Nichols, K. P. Jadhav and R. A. Buzdor: Acta Pharm. Suecica, suppl. *1983,* 2, 65.
194 J. G. Cannon, Z. Perez, J. P. Long and M. Ilhan: J. Med. Chem. *26,* 813 (1983).
195 P. A. Crooks, R. Szyndler and B. Cox: Pharm. Acta Helv. *55,* 134 (1980).
196 E. E. Costakis and G. A. Tsatsas: Chem. Chron *4,* 59 (1975) [C. A. *86,* 5351 a (1977)].
197 R. M. DeMarinis, D. H. Shah, R. F. Hall, J. P. Hieble and R. G. Pendleton: J. Med. Chem. *25,* 136 (1982).
198 S. P. Arneric, J. P. Long, D. B. Goodale, J. Mott, J. M. Laboski and G. F. Gebhardt: J. Pharmacol. exp. Ther. *224,* 161 (1983).
199 Z. J. Vejdelek, A. Dlabac and M. Protiva: Coll. Czech. Chem. Commun. *39,* 2819 (1974).
200 J. G. Cannon, J. P. Pease, J. P. Long and J. Flynn: J. Med. Chem. *27,* 922 (1984).
201 J. G. Cannon, J. P. Pease, R. L. Hamer, M. Ilhan, R. K. Bhatnagar and J. P. Long: J. Med. Chem. *27,* 186 (1984).
202 U. Hacksell, L.-E. Arvidsson, U. Svensson, J. L. G. Nilsson, H. Wikström, P. Lindberg, D. Sanchez, S. Hjorth, A. Carlsson and L. Paalzow: J. Med. Chem. *24,* 429 (1981).
203 A. S. Horn, B. Kaptein, T. B. A. Mulder, J. B. DeVries and H. Wynberg: J. Med. Chem. *27,* 1340 (1984).
204 J. A. Grosso, D. E. Nichols, J. D. Kohli and D. Glock: J. Med. Chem. *25,* 703 (1982).
205 A. S. Horn, D. Dijkstra, M. G. P. Feenstra, C. J. Grol, H. Rollema and B. H. C. Westerink: Eur. J. Med. Chem. *15,* 387 (1980).
206 B. H. C. Westerink, D. Dijkstra, M. J. P. Feenstra, C. J. Grol, A. S. Horn, H. Rollema and E. Wirix: Eur. J. Pharmacol. *61,* 7 (1980).
207 H. C. Cheng, J. P. Long, L. S. VanOrden, J. G. Cannon and J. P. O'Donnell: Res. Commun. Chem. Path. Pharmac. *15,* 89 (1976).
208 J. G. Cannon, J. A. Perez, R. K. Bhatnagar, J. P. Long and F. M. Sharabi: J. Med. Chem. *25,* 1442 (1982).
209 J. G. Cannon, R. G. Dushin, J. P. Long, M. Ilhan, N. D. Jones and J. K. Swartzendruber: J. Med. Chem., *28,* 515 (1985).
210 W. R. Jackson, C. H. McMullen, R. Spratt and P. Bladon: J. Organometall. Chem. *4,* 392 (1965).
211 W. E. Rosen, L. Dorfman and M. Linfield: J. Org. Chem. *29,* 1723 (1964).
212 R. D. Sindelar, J. Mott, C. F. Barfknecht, S. P. Arneric, J. R. Flynn, J. P. Long and R. K. Bhatnagar: J. Med. Chem. *25,* 858 (1982).
213 S. P. Arneric and J. P. Long: J. Pharm. Pharmacol. *36,* 318 (1984).
214 J. T. Clark, E. R. Smith, M. L. Stefanick, S. P. Arneric, J. P. Long and J. M. Davidson: Physiol. Behav. *29,* 1 (1982).
215 M. L. Stefanick, E. R. Smith, J. T. Clark and J. M. Davidson: Physiol. Behav. *29,* 973 (1982).
216 J. T. Clark, M. L. Stefanick, E. R. Smith and J. M. Davidson: Pharmacol. Biochem. Behav. *19,* 781 (1983).
217 M. Gaino, S. Yamamura, J. Saito and M. Okashi: Japan. Pat. 7, 805, 146, 1978 [C. A. *88,* 169822 (1978)].
218 J. D. McDermed and R. J. Miller: Annu. Rep. Med. Chem. *14,* 12 (1979).
219 R. H. Fuller, J. A. Clemens, E. C. Kornfeld, H. R. Snoddy, E. B. Smalstig and N. J. Bach: Life Sci. *24,* 375 (1979).
220 M. O. Thorner, E. F. Flückiger and D. B. Calne: Bromocriptine: A Clinical and Pharmacological Review, p. 56–123. Raven Press, New York 1980.
221 J. A. Clemens, E. B. Smalstig and C. J. Schaar: Acta Endocr. *79,* 230 (1975).

222 R. Horowski and H. Wachtel: Eur. J. Pharmacol. *36,* 373 (1976).
223 D. B. Calne, P. N. Leigh, P. F. Teychenne, A. N. Bamji and J. A. Greenacre: Lancet *2,* 1355 (1974).
224 A. N. Lieberman, M. Leibowitz, A. Neophtides, M. Kupersmith, S. Mehl, D. Kleinberg, M. Serby and A. M. Goldstein: Lancet *2,* 1129 (1979).
225 L. Lemberger and R. E. Crabtree: Science *205,* 1151 (1979).
226 G. E. Martin, M. Williams and D. R. Haubrick: J. Pharmacol. exp. Ther. *223,* 298 (1982).
227 K. Fuxe, B. B. Fredholm, S. O. Ogren, L. F. Agnati, T. Hökfelt and A. Gustafsson: Fed. Proc. Am. Soc. exp. Biol. *37,* 2181 (1978).
228 M. Goldstein, J. Y. Lew, S. Nakamura, A. F. Battista, A. Leiberman and K. Fuxe: Fed. Proc. Fed. Am. Soc. exp. Biol. *37,* 2202 (1964).
229 I. Creese, D. R. Burt and S. H. Snyder: Life Sci. *17,* 1715 (1975).
230 M. Pieri, R. Schaffner, L. Pieri, M. Da Prada and W. Haefely: Life Sci. *22,* 1615 (1978).
231 J. R. Walters, M. D. Baring and J. M. Lakoski, in : Dopaminergic Ergot Derivatives and Motor Function, p. 207–221. Eds. K. Fuxe and D. B. Calne, Plenum Press, New York 1979.
232 D. M. Loew, E. B. Van Deusen and W. Meier-Ruge, in: Ergot Alkaloids and Related Compounds, p. 421. Eds. B. Berd and H. O. Schild. Springer-Verlag, Berlin 1978.
233 D. T. Wong and F. P. Bymaster: Abstracts of Papers, Joint Central-Great Lakes Regional Meeting of the American Chemical Society, Indianapolis, Ind., May, 1978. American Chemical Society, Washington, D. C. Abstr. MEDI 25.
234 A. Rubin, L. Lemberger, P. Dhahir, P. Warrick, R. E. Crabtree, B. D. Obermeyer, R. L. Wolen and H. Rowe: Clin. Pharmacol. Ther. *23,* 272 (1978).
235 J. Parli, B. Schmidt and C. J. Shaar: Biochem. Pharmacol. *27,* 1405 (1978).
236 W. Kehr, H. Wachtel and H. H. Schneider: Acta Pharm. Suecica, suppl. *1983,* 2, 98.
237 P. Stütz, P. Fehr, P. A. Stadtler, J.-M. Vigouret and A.-L. Jaton: Acta Pharm. Suecica, suppl. *1983,* 2, 111.
238 P. L. Stütz, P. Stadtler, J.-M. Vigouret and A. Jaton: Eur. J. Med. Chem. *17,* 537 (1982).
239 J. Smidrkal and M. Semonsky: Coll. Czech. Chem. Commun. *47,* 622 (1982).
240 A. Enz, W. Frick, A. Closse and R. Nordmann: 13th CINP Congress (Jerusalem, June 20–25, 1982). Abstr. 199. Cited by T. de Paulis, Annu. Rep. Med. Chem. *18,* 28 (1983).
241 H. Wachtel, W. Kehr and G. Sauer: Life Sci. *33,* 2583 (1983).
242 J. Benes and M. Semonsky: Coll. Czech. Chem. Commun. *47,* 1235 (1982).
243 J. Rutsohmann and P. A. Stadtler, in: Ergot Alkaloids and Related Compounds, p. 29. Eds. B. Berd and H. O. Schild, Springer-Verlag, Berlin 1978.
244 P. A. Stadtler and A. Hofmann: Helv. Chim. Acta *45,* 2005 (1961).
245 A. Stoll and A. Hofmann: Helv. Chim. Acta *26,* 944 (1943).
246 N. Camerman, L. Y. Y. Chan and A. Camerman: Molec. Pharmacol. *16,* 729 (1979).
247 D. R. Burt, I. Creese and S. H. Snyder: Molec. Pharmacol. *12,* 800 (1976).
248 A. Hofmann, in: Drugs Affecting the Central Nervous System, p. 169. Ed. A. Burger. Marcel Dekker, New York 1968.
249 E. Flückiger, J.-M. Vigouret and H.-R. Wagner, in: Progress in Prolactin Physiology and Pathology, p. 383. Eds. C. Robyn and H. Harter, Elsevier/North Holland, Amsterdam 1978.
250 P. L. Stütz, P. A. Stadtler, J.-M. Vigouret and A. Jaton: J. Med. Chem. *21,* 754 (1978).
251 D. E. Nichols: J. Theor. Biol. *59,* 167 (1976).

252 P. S. Anderson, J. J. Baldwin, D. E. McClure, G. F. Lundell, J. H. Jones, W. C. Randall, G. E. Martin, M. Williams, J. M. Hirshfield, B. V. Clineschmidt, P. K. Lumma and D. C. Remy: J. Med. Chem. *26,* 363 (1983).
253 L. Nedelec, A. Pierdet, P. Fauveau, C. Euvrard, L. Proulx-Ferland, C. Dumont, F. Labrie and J. R. Boissier: J. Med. Chem. *26,* 522 (1983).
254 G. E. Martin, M. Williams, B. V. Clineschmidt, G. G. Yarbrough, J. H. Jones and D. R. Haubrich: Life Sci. *30,* 1847 (1982).
255 P. S. Anderson, J. J. Baldwin, D. E. McClure, G. F. Lundell and J. H. Jones: J. Org. Chem. *47,* 2184 (1982).
256 M. Williams, J. H. Jones and K. J. Watling: Drug Dev. Res. *3,* 573 (1983).
257 J. R. Boissier, C. Euvrard, C. Oberlander, J. Laurent, C. Dumont and F. Labrie: Eur. J. Pharmacol. *87,* 183 (1983).
258 G. E. Martin, M. Williams, D. J. Pettibone, G. G. Yarbrough, B. V. Clineschmidt and J. H. Jones: J. Pharmacol. exp. Ther. *230,* 569 (1984).
259 A. S. Horn, B. Hazelhoff, D. Dijkstra, J. B. de Vries, T. B. A. Mulder, P. Timmermans and H. Wynberg: J. Pharm. Pharmacol. *36,* 639 (1984).
260 J. H. Jones, P. S. Anderson, J. J. Baldwin, B. V. Clineschmidt, D. E. McClure, G. F. Lundell, W. C. Randall, G. E. Martin, M. Williams, J. M. Hirschfield, G. Smith and P. K. Lumma: J. Med. Chem. *27,* 1607 (1984).
261 A. S. Horn, D. Dijkstra, B. Hazelhoff, J. B. de Vries, T. B. A. Mulder and H. Wynberg: Abstr. VIIIth Int. Symposium on Medicinal Chemistry, Uppsala, Sweden, August 27–31, 1984, p. 78.
262 R. Perrone, F. Berardi and V. Tortella: Farmaco, Ed. Sci. *39,* 255 (1984).
263 J. R. Boissier, L. Nedelec and C. Oberlander: Acta Pharm. Suecica, suppl. *1983,* 2, 120.
264 N. J. Bach, E. C. Kornfeld, J. A. Clemens and E. B. Smalstig: J. Med. Chem. *23,* 812 (1980).
265 E. C. Kornfeld, N. J. Bach, R. D. Titus, C. L. Nichols and J. A. Clemens: Acta Pharm. Suecica, suppl. *1983,* 2, 83.
266 N. J. Bach, E. C. Kornfeld, N. D. Jones, M. O. Chaney, D. E. Dorman, J. W. Paschal, J. A. Clemens and E. B. Smalstig: J. Med. Chem. *23,* 481 (1980).
267 R. A. Hahn and B. R. MacDonald: J. Pharmacol. exp. Ther. *230,* 558 (1984).
268 K. Fuxe, M. Goldstein, L. F. Agnati, C. Köhler, J. Y. Lew and K. Okada: Acta Physiol. Scand. *117,* 303 (1983).
269 R. D. Titus, E. C. Kornfeld, N. D. Jones, J. A. Clemens, E. B. Smalstig, R. W. Fuller, R. A. Hahn, M. D. Hynes, N. R. Mason, D. T. Wong and M. M. Freeman: J. Med. Chem. *26,* 1112 (1983).
270 J. M. Armstrong, N. Duval and S. Z. Langer: Eur. J. Pharmacol. *87,* 165 (1983).
271 J. M. Schaus, E. C. Kornfeld, R. D. Titus, C. L. Nichols, J. A. Clemens, D. T. Wong and E. B. Smalstig: Abstr. VIIIth Int. Symposium on Medicinal Chemistry, Uppsala, Sweden, August 27–31, 1984, p. 90.
272 A. A. Asselin and L. G. Humber: Eur. Pat. Appl. EP 55,043, June 30, 1982 [C. A. *97,* 162820x (1982)].
273 D. Berney and K. Schuh: Helv. Chim. Acta *65,* 1304 (1982).
274 J. G. Cannon, J. P. Long and B. J. Demopoulos: Adv. Biosci. *37,* 189 (1982).
275 J. G. Cannon, B. D. Demopoulos, J. P. Long, J. R. Flynn and F. M. Sharabi: J. Med. Chem. *24,* 238 (1981).
276 M. Ilhan, J. P. Long, R. K. Bhatnagar, J. R. Flynn, J. G. Cannon and T. Lee: J. Pharmacol. exp. Ther. *231,* 56 (1984).
277 J. G. Cannon, T. Lee, M. Ilhan, J. Koons and J. P. Long: J. Med. Chem. *27,* 386 (1984).
278 L. Nedelec, J. Guillaume, C. Oberlander, C. Euvrard, F. Labrie, A. Allais and J. R. Boissier: Abstr. VIIth Int. Symposium on Medicinal Chemistry, Torremolinos, Spain, September 2–5, 1980, p. P168.
279 N. C. Cohen, P. Colin and G. Lemoine: Tetrahedron *37,* 1711 (1981).

280 W. F. Huffman, R. F. Hall, J. A. Grant, J. W. Wilson, J. P. Hieble and R. A. Hahn: J. Med. Chem. *26,* 933 (1983).

281 H.-H. Hausberg, H. Boettcher, A. Fuchs, R. Gottsschlich, V. Koppe, K.-O. Minck, E. Poetsch, O. Saiko and C. Seyfried: Acta Pharm. Suecica, suppl. *1983,* 2, 213.

282 W. Dimpfel, J. Harting, H. Decker and B. von Schilling: Acta Pharm. Suecica, suppl. *1983,* 2, 198.

283 C. A. Seyfried, K. Fuxe, H.-P. Wolf and L. F. Agnati: Acta Pharm. Suecica, suppl. *1983,* 2, 243.

284 L. A. Chiodo and B. S. Bunney: Neuropharmacology *22,* 1087 (1983).

285 C. A. Seyfried and K. Fuxe: Arzneimittel-Forsch. *32,* 892 (1982).

286 V. Neuser, H. Jacobi and H. Schwarz: Arzneimittel-Forsch. *32,* 892 (1982).

287 J. G. Cannon, G. J. Hatheway, J. P. Long and F. M. Sharabi: J. Med. Chem. *19,* 987 (1976).

288 J. G. Cannon, C. Suarez-Gutierrez, T. Lee, J. P. Long, B. Costall, D. H. Fortune and R. J. Naylor: J. Med. Chem. *22,* 341 (1979).

289 F. M. Sharabi, J. P. Long, J. G. Cannon and G. J. Hatheway: J. Pharmacol. exp. Ther. *199,* 630 (1976).

290 J. G. Cannon, T. Lee, H. D. Goldman, J. P. Long, J. R. Flynn, T. Verimer, B. Costall and R. J. Naylor: J. Med. Chem. *23,* 1 (1980).

291 J. G. Cannon, R. L. Hamer, M. Ilhan, R. K. Bhatnagar and J. P. Long: J. Med. Chem. *27,* 190 (1984).

292 H. Wikström, D. Sanchez, P. Lindberg, L.-E. Arvidsson, U. Hacksell, A. Johansson, J. L. G. Nilsson, S. Hjorth and A. Carlsson: J. Med. Chem. *25,* 925 (1982).

293 H. Wikström: Ph. D. Thesis, Uppsala University, Sweden, 1983, (a) p. 26; (b) p. 35–38.

294 H. Wikström, B. Andersson, K. Svensson, S. Hjorth and A. Carlsson: Poster presentation, VIIIth Int. Symposium on Medicinal Chemistry, Uppsala, Sweden, August 27–31, 1984.

295 J. G. Cannon, C. Koble-Suarez, J. P. Long, M. Ilhan and R. K. Bhatnagar: J. Pharm. Sci., *74,* 672 (1985).

296 M. Ilhan, J. P. Long and J. G. Cannon: J. Pharmacol. exp. Ther. *231,* 361 (1984).

297 J. G. Cannon, T. Lee, F.-L. Hsu, J. P. Long and J. R. Flynn: J. Med. Chem. *23,* 502 (1980).

298 J. G. Cannon, J. A. Beres, T. Lee and J. P. Long: Med. Chem. Adv. Pergamon Press, Oxford 1981, p. 369–381.

299 B. Costall, R. J. Naylor, J. G. Cannon and T. Lee: Eur. J. Pharmacol. *41,* 307 (1977).

300 B. Costall, R. J. Naylor, J. G. Cannon and T. Lee: J. Pharm. Pharmacol. *29,* 337 (1977).

301 D. Kocjan and D. Hadzi: J. Pharm. Pharmacol. *35,* 780 (1983).

302 R. Nordmann and T. J. Petcher: Abstr. VIIIth Int. Symposium on Medicinal Chemistry, Uppsala, Sweden, August 27–31, 1984, p. 86.

303 R. Nordmann, U. Briner, A. Closse, W. Frick, T. J. Petcher, H.-R. Wagner and A. Widmer: Poster presentation, VIIIth Int. Symposium on Medicinal Chemistry, Uppsala, Sweden, August 27–31, 1984.

304 U. Hacksell, L.-E. Arvidsson, U. Svensson, J. L. G. Nilsson, D. Sanchez, H. Wikström, P. Lindberg, S. Hjorth and A. Carlsson: J. Med. Chem. *24,* 1475 (1981).

305 W. Arnold, J. J. Daly, R. Imhof and E. Kyburz: Tetrahedron Lett. *24,* 343 (1983).

306 S. Hjorth, A. Carlsson, D. Clark, K. Svensson, H. Wikström, D. Sanchez, P. Lindberg, U. Hacksell, L.-E. Arvidsson, A. Johansson and J. L. G. Nilsson: Psychopharmacology *81,* 89 (1983).

307 P. Plantje, A. H. Mulder and J. C. Stoof: Pharm. Weekbl. Sci. Ed. *5,* 265 (1983).
308 J. Arnt, K. P. Boegesoe, A. V. Christensen, J. Hyttel, J.-J. Larsen and O. Svendsen: Psychopharmacology *81,* 199 (1984).
309 M. Schorderet, S. Hjorth and U. Hacksell: J. Neurotransm. *59,* 1 (1984).
310 H. Wikström, D. Sanchez, P. Lindberg, U. Hacksell, L.-E. Arvidsson, A. M. Johansson, S.-O. Thorberg, J. L. G. Nilsson, K. Svensson, S. Hjorth, D. Clark and A. Carlsson: J. Med. Chem. *27,* 1030 (1984).
311 J. Arnt, K. P. Boegesoe, J. Hyttel, J. J. Larsen and O. Svendsen: Eur. J. Pharmacol. *102,* 91 (1984).
312 H. Rollema and D. Mastebroek: Abstr. VIIIth Int. Symposium on Medicinal Chemistry, Uppsala, Sweden, August 27–31, 1984, p. 88.
313 J. Guillaume, C. Oberlander, G. Lemoine, J. Laurent, C. Dumont and L. Nedelec: Abstr. VIIIth Int. Symposium on Medicinal Chemistry, Uppsala, Sweden, August 27–31, 1984, p. 75.
314 J. Weinstock, J. W. Wilson, D. L. Ladd, C. K. Brush, F. R. Pfeiffer, G. Y. Kuo, K. G. Holden, H. C. F. Yim, R. A. Hahn, J. R. Wardell, Jr., A. J. Tobia, P. E. Setler, H. M. Sarau and P. T. Ridley: J. Med. Chem. *23,* 973 (1980).
315 P. E. Setler, H. M. Sarau, C. L. Zirkle and H. L. Saunders: Eur. J. Pharmacol. *50,* 419 (1978).
316 R. G. Pendleton, L. Samler, C. Kaiser and P. T. Ridley: Eur. J. Pharmacol. *51,* 19 (1978).
317 J. W. Kebabian and D. B. Calne: Nature, Lond. *277,* 93 (1979).
318 C. Kaiser, P. Dandridge, J. Weinstock, D. M. Ackerman, H. M. Sarau, P. E. Setler, R. L. Webb, J. W. Horodniak and E. D. Matz: Acta Pharm. Suecica, suppl. *1983,* 2, 132.
319 C. Kaiser, P. A. Dandridge, E. Garvey, R. A. Hahn, H. M. Sarau, P. E. Setler, L. S. Bass and J. Clardy: J. Med. Chem. *25,* 697 (1982).
320 A. G. Molloy and J. L. Waddington: Psychopharmacology *82,* 409 (1984).
321 N. E. Anden, H. Nilsson, E. Ros and U. Thornström: Acta Pharmacol. Tox. *52,* 51 (1983).
322 F. R. Pfeiffer, J. W. Wilson, J. Weinstock, G. Y. Kuo, P. A. Chambers, K. G. Holden, R. A. Hahn, J. R. Wardell, Jr., A. J. Tobia, P. E. Setler and H. M. Sarau: J. Med. Chem. *25,* 352 (1982).
323 A. L. Blumberg, J. P. Hieble, Jr., J. McCafferty, R. A. Hahn and J. Smith, Jr.: Fed. Proc. Fed. Am. Soc. exp. Biol. *41,* 1345 (1982).
324 J. F. Plantje, F. J. Daus, H. A. Hansen and J. C. Stoof: Naunyn-Schmiedebergs Arch. Pharmacol. exp. Pathol. *327,* 180 (1984).
325 J. F. Plantje, H. A. Hansen, F. J. Daus and J. C. Stoof: Eur. J. Pharmacol. *105,* 73 (1984).
326 P. H. Volkman, J. D. Kohli, L. I. Goldberg, J. G. Cannon and T. Lee: Proc. natl. Acad. Sci. USA *74,* 3602 (1977).
327 B. Costall and R. J. Naylor: Psychopharmacologia *41,* 57 (1975).
328 B. Costall and R. J. Naylor: Adv. Neurol. *9,* 285 (1975).
329 B. Costall, D. M. Kelly and R. J. Naylor: Psychopharmacologia *41,* 153 (1975).
330 B. Costall, R. J. Naylor and C. J. Pycock: J. Pharm. Pharmacol. *27,* 943 (1974).
331 D. McKillop and H. F. Bradford: Biochem. Pharmacol. *30,* 2753 (1981).
332 W. Pöldinger and K. Taeubner (Eds.): Int. Pharmacopsychiatr. *17,* suppl. 1 (1982).
333 I. Hoffmann: Int. Pharmacopsychiatr. *17,* suppl. 1, 4 (1982).
334 G. N. Woodruff and C. Sumners: Adv. Biosci. *20,* 57 (1979).
335 J. A. Poat, G. N. Woodruff and K. J. Watling: J. Pharm. Pharmacol. *30,* 495 (1978).

336 B. Costall and R. J. Naylor: J. Pharm. Pharmacol. *30,* 514 (1978).
337 J. D. Kohli and L. I. Goldberg: J. Pharm. Pharmacol. *32,* 225 (1980).
338 P. A. Dandridge, C. Kaiser, M. Brenner, D. Gaitanopoulos, L. D. Davis, R. L. Webb, J. J. Foley and H. M. Sarau: J. Med. Chem. *27,* 28 (1984).
339 J. Presthus, J. Ankerhus, A. Buren, J. Bottcher, R. Holmsen, B. Mikkelsen, B. O. Mikkelsen, R. Nyberg-Hansen, E. Riman, B. Severin and J. A. Aarli: Acta Neurol. Scand. *58,* 77 (1978).
340 J. Tuomisto and R. Voutilainin: Abstr. Proceedings of the Joint Meeting of the Scandinavian and British Pharmacological Societies, Stockholm, July 5–6, 1982, p. P27.
341 E. Zára-Kaczian, G. Deák and L. Györgyi: Abstr. VIIIth Int. Symposium on Medicinal Chemistry, Uppsala, Sweden, August 27–31, 1984, p. 186.
342 K. Brown, R. A. Brown, R. C. Brown, J. Dixon, S. E. O'Connor, G. W. Smith and A. C. Tinker: Abstr. VIIIth Int. Symposium on Medicinal Chemistry, Uppsala, Sweden, August 27–31, 1984, p. 69.
343 K. Brown, R. C. Brown, J. Dixon, A. C. Tinker, R. A. Brown, S. E. O'Connor and G. W. Smith: Poster presentation, VIIIth Int. Symposium on Medicinal Chemistry, Uppsala, Sweden, August 27–31, 1984.
344 H. L. Komiskey, J. F. Bossart, D. D. Miller and P. N. Patil: Proc. natl. Acad. Sci. USA *75,* 2641 (1978).
345 E. L. Eliel, N. L. Allinger, S. J. Angyal and G. A. Morrison: Conformational Analysis, p. 199–200. Interscience Press, New York 1965.
346 P. W. Erhardt, R. J. Gorczynski and W. G. Anderson: J. Med. Chem. *22,* 907 (1979).
347 R. J. Gorczynski, W. G. Anderson, P. W. Erhardt and D. M. Stout: J. Pharmacol. exp. Ther. *210,* 252 (1979).
348 R. J. Borgman, P. W. Erhardt, R. J. Gorczynski and W. G. Anderson: J. Pharm. Pharmacol. *30,* 193 (1978).
349 L. I. Goldberg, P. F. Sonneville and J. L. McNay: J. Pharmacol. exp. Ther. *163,* 188 (1968).
350 L.-E. Arvidsson, A. M. Johansson, J. L. G. Nilsson, P. Lindberg, B. Andersson, H. Wikström, D. Sanchez, K. Svensson, S. Hjorth and A. Carlsson: Abstr. VIIIth Int. Symposium on Medicinal Chemistry, Uppsala, Sweden, August 27–31, 1984, p. 174.
351 H. A. J. Struyker-Boudier, L. Tepemma, A. R. Cools and J. M. Van Rossum: J. Pharm. Pharmacol. *27,* 882 (1975).
352 A. R. Cools, H. A. J. Struyker-Boudier and J. M. Van Rossum: Eur. J. Pharmacol. *37,* 283 (1976).
353 A. C. Van Oene, Sminia, A. H. Mulder and A. S. Horn: J. Pharm. Pharmacol. *35,* 786 (1983).
354 J. C. Van Oene, H. A. Houwing and A. S. Horn: Pharm. Weekbl. Sci. Ed. *4,* 205 (1982).
355 T. de Paulis: Annu. Rep. Med. Chem. *18,* 21 (1983).
356 K. Voith: Drug Dev. Res. *4,* 391 (1984).
357 J. Bagli, T. Bogri and K. Voith: Abstr. MEDI-16, 184th Am. Chem. Soc. Natl. Meeting, Kansas City, Mo., August 1982.
358 F. R. Ahmed and J. Bagli: Can. J. Chem. *60,* 2687 (1982).
359 J. Bagli, T. Bogri and K. Voith: J. Med. Chem. *27,* 875 (1984).
360 H. Corrodi, K. Fuxe und U. Ungersted: J. Pharm. Pharmacol. *23,* 989 (1971).
361 H. Corrodi, L. Farnebo, K. Fuxe, B. Humberger and U. Ungersted: Eur. J. Pharmacol. *20,* 195 (1972).
362 A. G. Jori, E. Cecchetti, E. Dolfini, E. Monti and S. Garattini: Eur. J. Pharmacol. *27,* 245 (1974).
363 J. R. Waters, B. S. Bunney and R. H. Roth: Adv. Neurol. *9,* 273 (1975).

364 B. Costall and R. J. Naylor: Naunyn-Schmiedebergs Arch. Pharmacol. exp. Pathol. *278,* 117 (1973).
365 M. Goldstein, B. Anagnoste and C. Shirran: J. Pharm. Pharmacol. *25,* 348 (1973).
366 C. T. Dourish: Prog. Neuro-Psychopharmacol. and Biol. Psychiat. *7,* 3 (1983).
367 B. Hue, M. Pelhate and J. Chanelet: J. Pharmacol., Paris *12,* 455 (1981).
368 P. Jenner, A. R. Taylor and D. S. Campbell: J. Pharm. Pharmacol. *25,* 749 (1973).
369 I. Creese: Eur. J. Pharmacol. *28,* 55 (1974).
370 M. Schorderet: Experientia *31,* 1325 (1975).
371 J. Moragues, J. Prieto, R. G. W. Spickett, A. Vega, W. Salazar and D. J. Roberts: Farmaco, Ed. Sci. *35,* 951 (1980).
372 D. L. Temple, J. P. Yevich and J. S. New: J. Clin.Psychiat. *43,* 4 (1982).
373 H. C. Stanton, D. P. Taylor and L. A. Riblet: Neurobiol. Nucl. Accumbens (Proceedings of Symposium), p. 316–321. Eds. R. B. Chromster and J. F. DeFrance, 1981. [C. A. *98,* 10911 k (1983)].
374 S. P. Yevich, J. S. New and M. S. Eison: Annu. Rep. Med. Chem. *19,* 15 (1984).
375 B. A. McMillen and L. A. Mattiace: J. Neural Transm. *57,* 255 (1983).
376 J. P. Yevich, D. L. Temple, Jr., J. S. New, D. P. Taylor and L. A. Riblet: J. Med. Chem. *26,* 194 (1983).
377 P. Burn, P. A. Crooks, F. Heatley, B. Costall, R. J. Naylor and V. Nohria: J. Med. Chem. *25,* 363 (1982).
378 D. I. Schuster, H. E. Katerinopoulos, W. L. Holden, A. P. S. Narula, R. B. Libes and R. B. Murphy: J. Med. Chem. *25,* 850 (1982).
379 S.-J. Law, J. M. Morgan, L. W. Masten, R. F. Borne, G. W. Arana, N. S. Kula and R. J. Baldessarini: J. Med. Chem. *25,* 213 (1982).
380 M. Marko: Eur. J. Pharmacol. *101,* 263 (1984).
381 F. J. Morales-Olivas, V. Palop, E. Rubio and J. Esplugues: Arch int. Pharmacodyn. Thér. *272,* 71 (1984).
382 J. Maillard, P. Dela, M. Langlois, B. Portevin, J. Legeai and C. Manuel: Eur. J. Med. Chem. *19,* 451 (1984).
383 M. Grabowska-Andén and N.-E. Andén: J. Pharm. Pharmacol. *36,* 748 (1984).
384 M. P. Seiler and R. Markstein: Molec. Pharmacol. *26,* 452 (1984).
385 H. A. J. Struyker-Boudier and A. R. Cools: J. Pharm. Pharmacol. *36,* 859 (1984).

Tetrahydroisoquinolines and β-carbolines: putative natural substances in plants and mammals*

By H. Rommelspacher and R. Susilo
Department of Neuropsychopharmacology, Free University, Ulmenallee 30, D-1000 Berlin 19, Federal Republic of Germany

* Dedicated to Professor Dr. Helmut Coper at the occasion of his 60th birthday.

1 Introduction

The occurrence of tetrahydroisoquinolines (TIQ's) and β-carbolines (BC's) in mammals under physiological conditions is discussed controversially. A reason for some scepticism might be that the determination of TIQ's and BC's requires more sensitive methods than the measurement of known neurotransmitters like noradrenaline (NA), dopamine (DA), and 5-hydroxytryptamine (5-HT) which serve as precursors. The concentrations of the transmitters in the central nervous system (CNS) of mammals exceed those of most TIQ's and BC's by the factor 1 000 at least. Furthermore, artefactual admixtures due to contaminations of organic solvents either with the trace amines themselves or with their precursors have to be prevented during the work up procedure. Finally some foodstuff contains TIQ's and BC's which gives rise to question endogenous formation by mammals.

On the other hand, the existence of TIQ's and BC's in plants has been described by many reports. The research about these alkaloids has a long tradition among ethnopharmacologists and psychopharmacologists. The observation of pharmacological effects of plant extracts is impressive and convinces sceptic persons easier than sophisticated biochemical analyses.

The search for TIQ's and BC's in the CNS and other organs of mammals has several interesting aspects. Neurochemists believe that only about 10% of all neurotransmitters and neuromodulators are known so far. The observation of various psychopharmacological effects of TIQ's and BC's suggest an important role of the trace amines as neuromodulators in the sense of fine tuning of the action of neurotransmitters. The TIQ's and BC's may bind to their own high affinity sites on neuronal membranes associated with or located close to the receptors of neurotransmitters. They might act by inducing affinitiy changes of neurotransmitter recognition sites. An increase of the fluidity of the plasma membrane is discussed as an other mode of action [1]. A further reason for a more general interest in TIQ's and BC's might be the unusual mode of their biosynthesis. Many authors assume a non-enzymatic formation. However, this point is far from being clarified.

Research on TIQ's and BC's is stimulated also by their possible role in pathological conditions especially parkinsonism, alcoholism and phenylketonuria. Furthermore, the observation that some BC's can induce a status of anxiety has attracted the interest of many psychopharmacologists.

The present review intends to compile the evidence for the natural occurrence of TIQ's and BC's by comparing the conditions of the chemical formation with the biosynthesis in plants and mammals. This approach should be helpful to assess critically the controversial issue about the natural existence of these trace amines. Detailed reviewing papers concerning other aspects of the TIQ's and BC's have appeared, e. g. the route and significance of endogenous synthesis of alkaloids in animals [2], the neurobiology of tetrahydro-β-carbolines [3], biogenic amine-aldehyde condensation products: tetrahydroisoquinolines and tryptolines (β-carbolines) [4], β-carbolines, psychoactive compounds in the mammalian body [5], tetrahydroisoquinolines: a review [6], false transmitter substances as a mechanism of drug action: a reappraisal [7], the β-carbolines (harmanes) – a new class of endogenous compounds [8], and several reviews about the role of TIQ's and BC's on alcoholism [9–14].

2 Principles of the condensation reaction

2.1 Pictet-Spengler reaction

The Pictet-Spengler reaction of β-arylethylamines (I A) with carbonyl compounds is generally considered to proceed via a Schiff's base (I B) formation and cyclization to tetrahydroisoquinolines (I C; scheme 1;

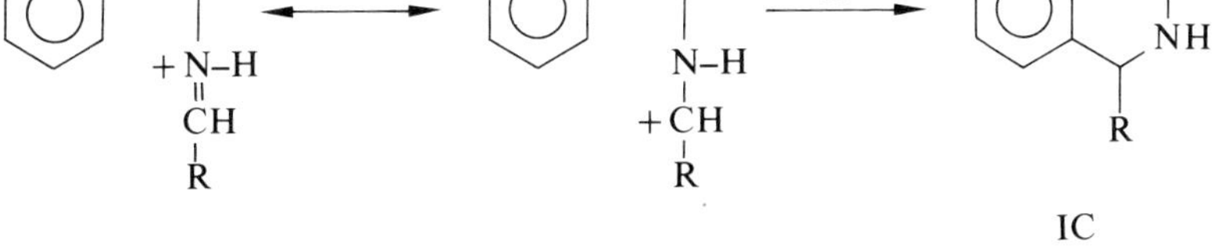

Scheme 1

[15]). In a similar reaction mechanism β-indolylethylamines (II A) condense with carbonyl compounds (II B) via a Schiff's base intermediate and then cyclize in a separate reaction to form tetrahydro-β-carbolines (scheme 2). The Schiff's bases (V) are formed via III and IV. Some of the bases have been isolated as products of chemical syntheses [16, 17].

II A + II B →

III $\xrightarrow{+H^+}$ IV $\xrightarrow{-H_2O}$

V

B A

VI VII A

↓ $-H^+$

VII B $\xrightarrow{-H^+}$ VIII

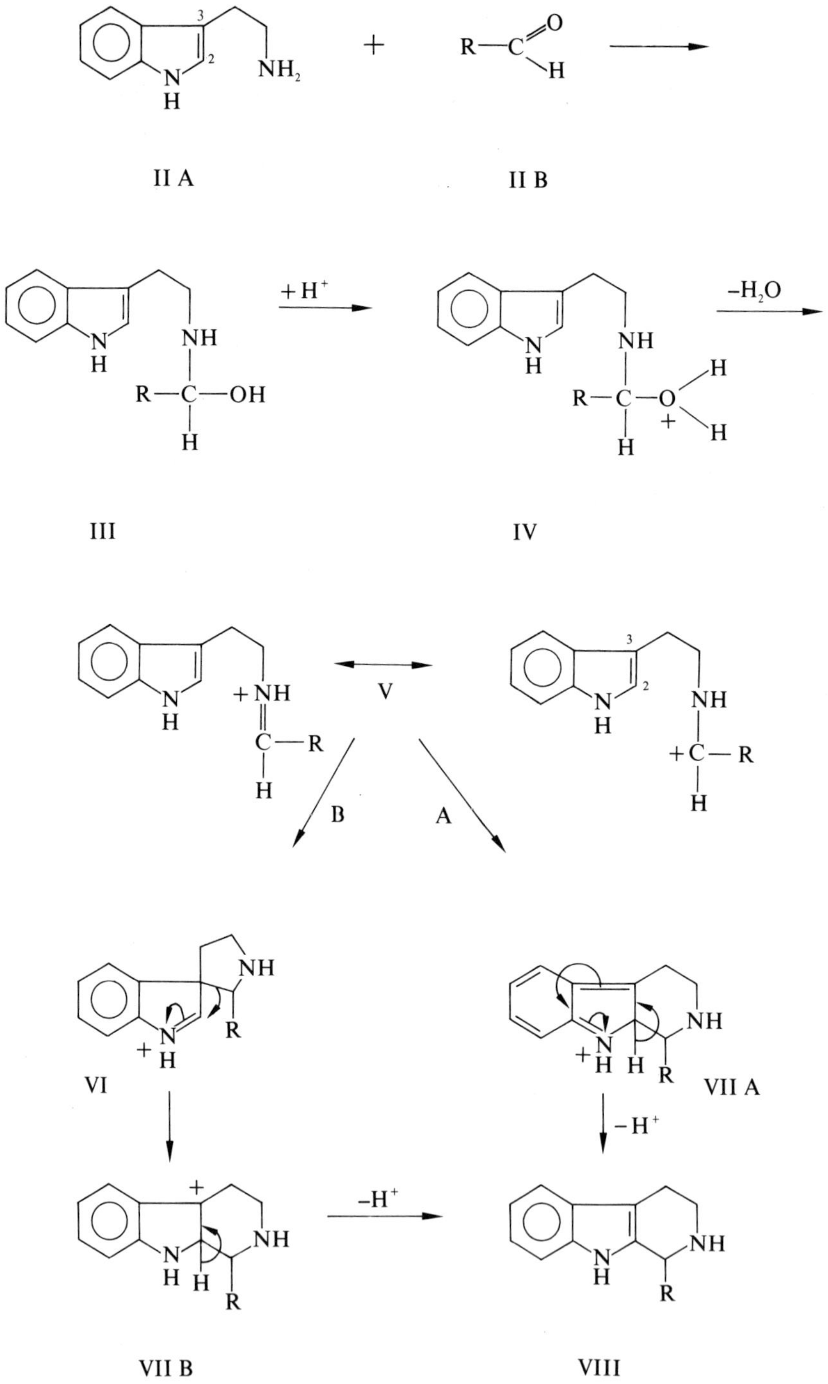

Scheme 2

For the further cyclization reaction, two mechanistic pathways have been proposed in principle: the direct cyclization (A) and the cyclization proceeding via a spiro-imine intermediate (VI, B; [18]). The more simple pathway (A), a geometrically favoured cyclization, seems to be more probable for the Pictet-Spengler cyclization, to produce BC's. The reaction is triggered by electrophilic attack of the Schiff's base (V) at the position two of the indole nucleus resulting in the cationic intermediate (VII A). However, several findings suggest that the electrophilic initial attack seems to occur at the 3-position [19–23]. The resultant spiroindolenine (VI) undergoes a Wagner-Meerwein type of rearrangement by forming the tetrahydro-β-carboline (VIII; [18]). The spiroindolenine hypothesis was suggested first by Woodward [24]. He proposed such a compound as intermediate substance of the biosynthesis of Strychnos alkaloids (scheme 3).

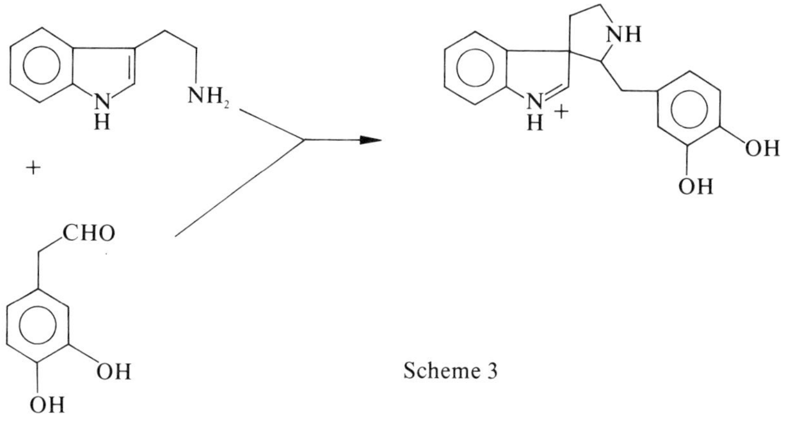

Scheme 3

The hypothesis of the intermediate spiroindolenine has been supported by the isolation of the spiroindoline (IX) as the yield of a Pictet-

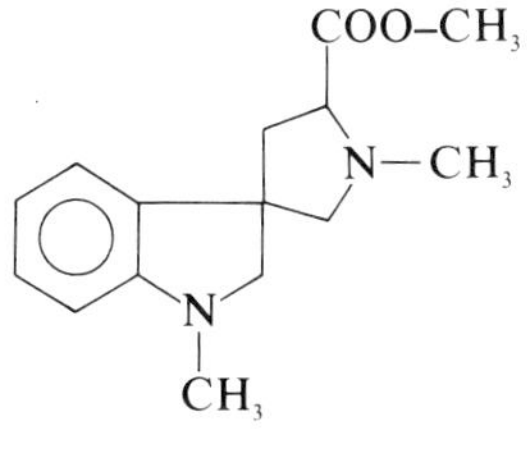

IX

Spengler cyclization in the presence of Raney nickel [25]. On the other hand it was argued that the isolation of spiroindoline (IX) from a Pictet-Spengler cyclization is not an unequivocal support for a spirointermediate in the tetrahydro-β-carboline formation, since the spiroindolenine precursor of (IX) could participate in ring-chain equilibration without leading to tetrahydro-β-carboline [16].

2.2 Bischler-Napieralski reaction

Heating of N-acyl-β-phenethylamine (X) with phosphorus pentoxide or anhydrous zinc chloride resulted in an intramolecular cyclization of the molecule yielding, 3,4-dihydroisoquinolines (XI; scheme 4; [26]).

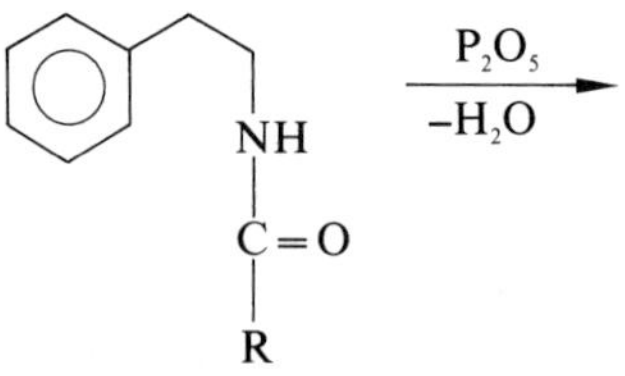

X XI

Scheme 4

An analogous reaction, the ring closure of the N-acyl-β-indolylethylamines (XII) to 3,4-dihydro-β-carbolines (XIII) ([27, 28]; scheme 5)

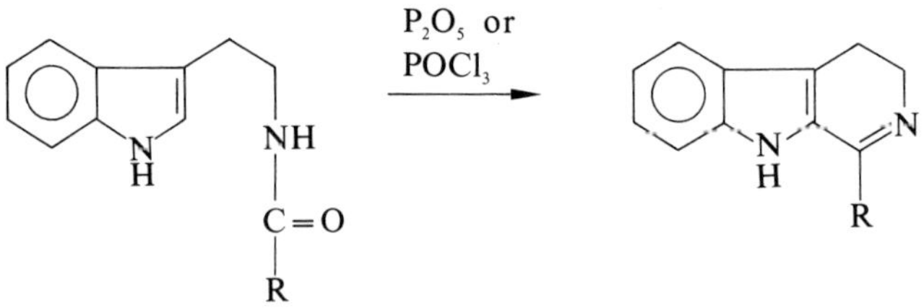

XII XIII

Scheme 5

proceeds in general more easy than the cyclization of N-acyl-β-phenethylamines. The cyclodehydration of (XIV) with phosphorus oxychlo-

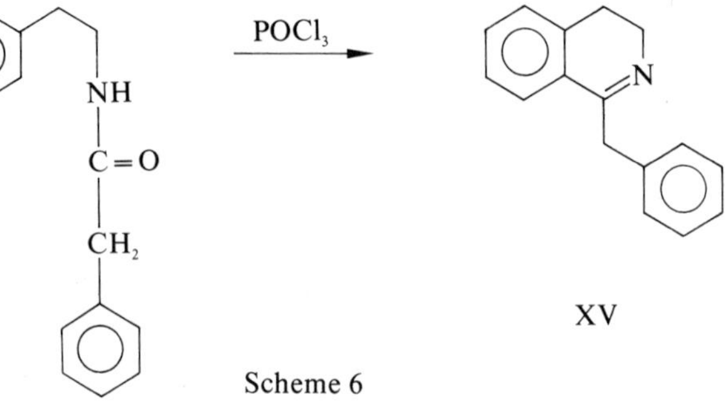

XV

XIV

Scheme 6

ride afforded 9 % yield of 1-benzyl-3,4-dihydroisoquinoline (scheme 6; XV), whereas that to the corresponding 1-benzyl-3,4-dihydro-β-carboline yielded 90 % under comparable conditions [29].

2.3 Influence of substituents on the cyclization reaction

The electrophilic ring closure of β-phenethylamines are facilitated by electron-donating substituents (e. g. alkoxyl or hydroxyl groups) in the appropriate position. In general, alkoxyl groups direct the cyclization to the *para*-position. The cyclization of *m*-methoxy-β-phenethylamide (XVI) via the intermediate (XVII) yields only 6-methoxy-3,4-dihydroisoquinoline (XVIII), but not 8-methoxy TIQ which would be expected if the carbon in *ortho*-position had been activated ([29]; scheme 7).

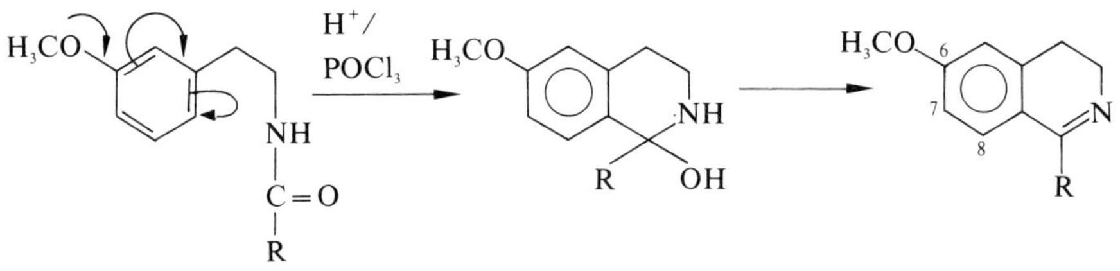

XVI XVII XVIII

Scheme 7

The same result was obtained with the Pictet-Spengler condensation of 3-methoxyphenylethylamine and formaldehyde namely 6-methoxytetrahydroisoquinoline [30]. Späth and Kruta [31] found that replacement of the alkoxyl group by a hydroxyl group abolished the orientation rule, the ring closure proceeded to both *ortho*- and *para*-positions with equal facility. Apparently the presence of free hydroxyl groups activates the benzene carbon atom at *ortho*-position as well. These findings have been confirmed by several other authors [32, 33]; for example condensation of XIX with acetaldehyde yielded the possible isomers in each instance (scheme 8; [32]).

In contrast to this finding Kovacs and Fodor [34] showed that *meta*-hydroxyl group in dopamine (XX; R′ = R″ = OH) directed the ring closure only to the *para*-position under mild reaction conditions. Methylation of *meta*--hydroxyl group of dopamine to the methyl ether derivative decreased the facility of the cyclization (scheme 9). Unfortunately, the identity of the product was not further established by oxydation or by derivatisation.

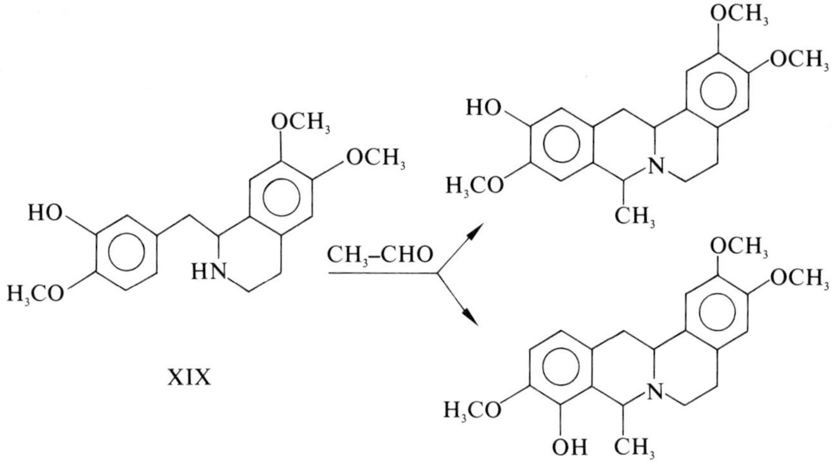

XIX

Scheme 8

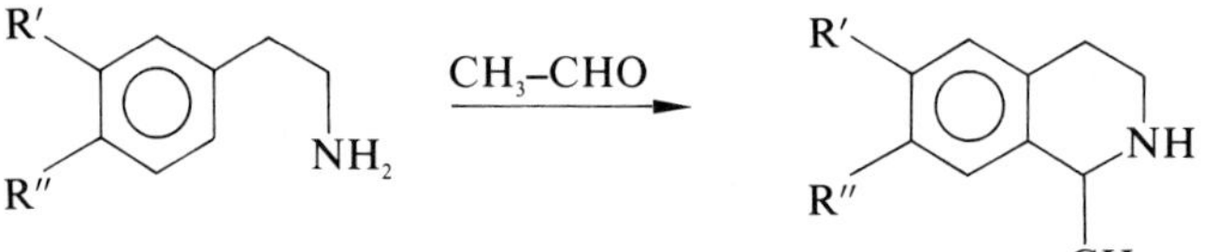

XX

R′	OH	OH	OCH_3	OCH_3
R″	OH	OCH_3	OH	OCH_3
Yield	79%	89%	O	O

Scheme 9

The course of the Bischler-Napieralski cyclization of N-acyl-β-indolylethylamines to 3,4-dihydro-β-carbolines proceeds similar as the cyclization of β-phenethylamine. Substituents on the indole nucleus which are able to influence the nucleophilicity of C-2 should facilitate the reaction. The electron shift in the indole nucleus (XXI) shows that substituents at C-6 should be able to exert a mesomeric influence on C-2 analogous to that of *para*-substituents on the benzene ring. These considerations have been substantiated by experimental data from Späth and Lederer [27] who showed that 6-methoxy derivative of N-acetyltryptamine cyclized considerably easier than 5-methoxy or 7-methoxy derivatives (scheme 10).

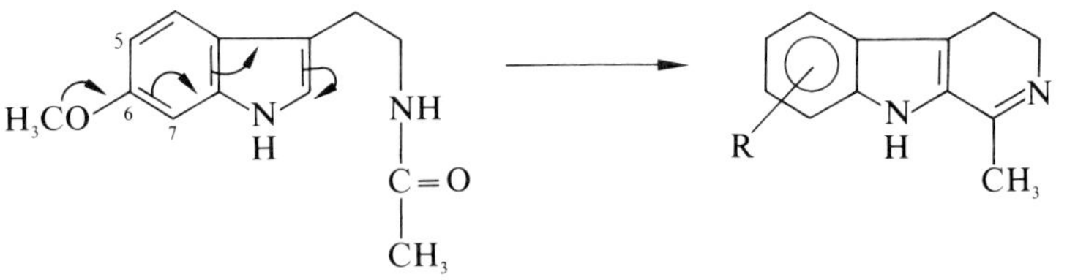

XXI

Substituent	5–OCH_3	6–OCH_3	7–OCH_3
Yield	58%	78%	32%

Scheme 10

The influence of the alkoxyl-substituents on the facilitation of ring closure (– –→) and the direction of the cyclization (——→) is shown in (XXII) and (XXIII). Hester (35) observed in a Pictet-Spengler reac-

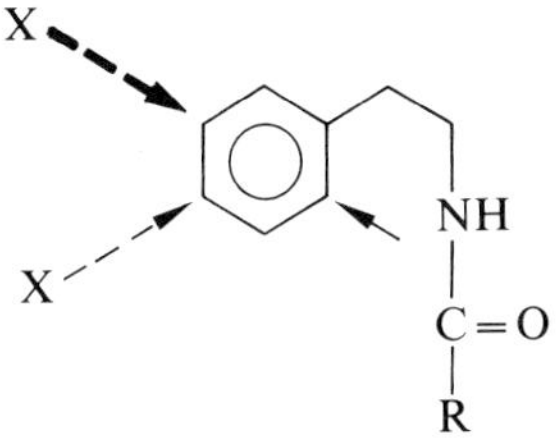

XXII

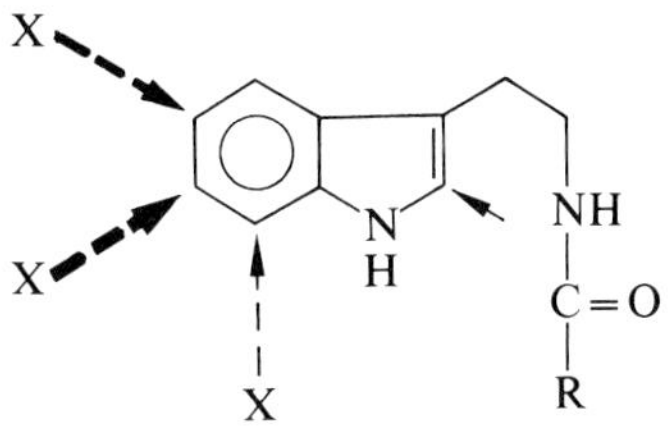

XXIII

X = Alkoxyl group

——→ position of ring closure

– –→ influence of substituents on the reactivity of the cyclisation

tion that 6-methoxytryptamine condensed with acetone yielding 92% of the expected condensed product whereas 5-methoxytryptamine and non substituted tryptamine did not react with acetone (scheme 11).

Substituent	H	5–OCH_3	6–OCH_3
Yield	O	O	92%

Scheme 11

The finding suggests that the increase of electron density at C-2 was due to the mesomeric contribution of the C-6 substituent and not a result of a purely inductive effect.

3 Tetrahydroisoquinolines

3.1 Chemical formation under so-called physiological conditions

3.11 Dopamine, dopa and epinine as precursors

3.111 Condensation reaction with simple aldehydes

In 1910, Winterstein and Trier [36] formulated the general pathway of the biosynthesis of isoquinolines in plants: dioxyphenethylamine condenses with aldehydes yielding isoquinolines (XXV). The authors used dimethoxyphenethylamine (XXIV) as precursor (scheme 12). As

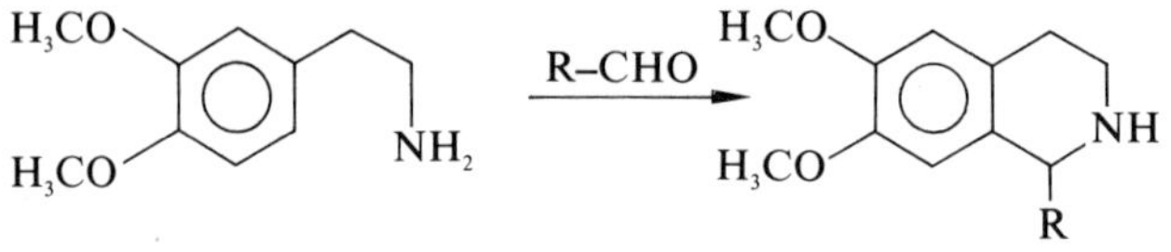

XXIV XXV

Scheme 12

postulated later by Schöpf and Bayerle [37] the dimethoxy as well as the methoxy-substituted phenylethylamines react with aldehydes only in the presence of a catalytic agent like 20% hydrochloric acid or of heat. On the other hand hydroxylated phenylethylamines react under so-called physiological conditions namely approximately neutral pH-values. The *p*-substituted (with respect to the attacked ring-carbon) hydroxy-group seems to be of special importance (for details see chapter 2). Schöpf and Bayerle [37] incubated dopamine (0.04 *M*; XLI) and acetaldehyde (0.08 *M*) at pH 3–5 and 25° C for 3 days. They measured a decrease of the acetaldehyde concentration. The yield of the synthesized salsolinol (XXVI) was approximately 50% at pH 3 and 100% at pH 5. The authors assumed that salsolinol (XXVI) is methylated in plants to salsoline (XXVII) and carnegine (XXVIII). Both compounds have been detected in plants as racemate suggesting a chemical con-

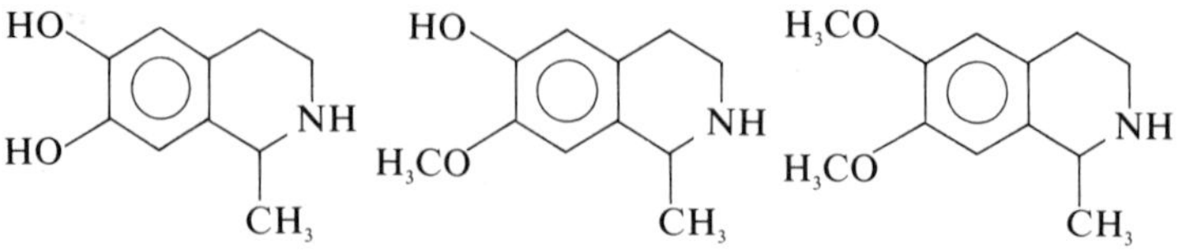

XXVI XXVII XXVIII

densation reaction without a catalyzing enzyme according to these investigators. The incubation of epinine (XXIX) (0.04 *M*) with acetaldehyde (0.08 *M*) resulted in the formation of 1,2-dimethyl-6,7-dihydroxy TIQ (XXX; pH 4, +25° C, 3 days incubation period, yield 7 mmol; scheme 13).

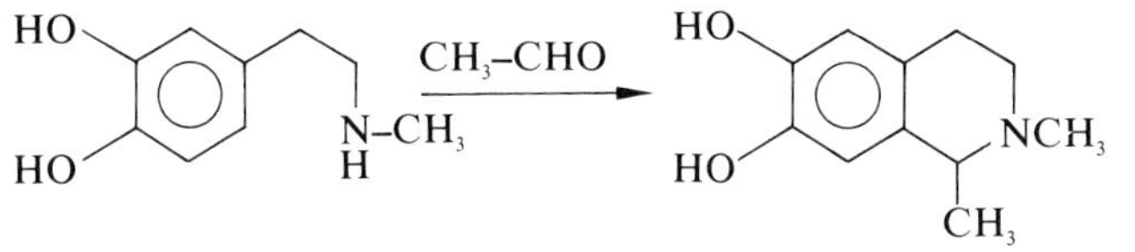

XXIX XXX

Scheme 13

Other aldehydes occurring under physiological conditions are the phenylacetaldehyde and its 3,4-dihydroxy derivative (XLII). Hahn and Stiehl [38] reported that phenylacetaldehyde reacted with DA under "physiological conditions" so slowly that other degradation products prevail. They observed similar reactions with oxyphenylacetaldehyde as substrate even more distinct. Other authors investigated the reaction of DA (0.01 *M*) with homopiperonale (XXXI; 0.01 *M*, pH 3–7, +25° C) and found a yield of 77–85% at pH 4–7 (XXXII; [39]; scheme 14). They utilized homopiperonale only not to interfere with

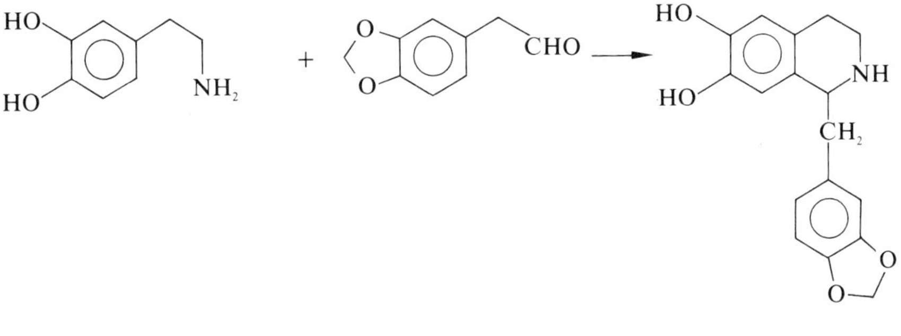

XLI XXXI XXXII

Scheme 14

the work of other investigators. The authors concluded that phenylacetaldehyde should react with DA as well. The rate of synthesis depended on pH. A yield of 30% was measured after 5 days (pH 3), 50 minutes (pH 6) and 13 minutes (pH 7). A detailed analysis revealed that at pH 6, 75% of the aldehyde had disappeared after 1 hour, but only 35% had been converted into the expected TIQ. The final yield (84%) was found only after 18 hours. Utilizing the respective α-ketoacid the same authors reported that the reaction rate was pH-dependent namely faster at higher pH. However, the reaction velocity was about 1/300 of that utilizing aldehyde (pH 7, 2.5 days, 30% disappearance of the α-keto acid; [39]).

In a more recent study [40], DA (2×10^{-3} *M)* was incubated in phosphate buffer (0.025 *M*, pH 7.4) for 30 minutes at +37° C with acetaldehyde. 43% of DA was converted into salsolinol (XXVI; table 1). The respec-

Table 1
Rate constants of the chemical synthesis of TIQ's and BC's (second order rate constants l mol^{-1} sec^{-1}).

Aldehyde	Amine	Rate*	Halftime
FA	Adrenaline	40	1.8 sec
	NA	5.83	12 sec
	DA	4.8, 5.3	14.4 sec
	5-HT	0.093	12 min
	5-Methoxytrypt-amine	0.06	
	N-Methyltryptamine	0.018	
	Tryptamine	0.016	75 min
AcAl	L-dopa	0.255, 0.16, 6.1	
	α-Methyl-dopa	0.141	
	DA	0.375, 0.38, 15.3	3.8 min
	α-Methyldopamine	0.36	
	3,4-Dihydroxyphenyl-propanolamine	0.32	
	Deoxyadrenaline	0.3	
	Adrenaline	0.10, 0.098	10–20 min
	NA	1.9, 0.10, 0.075	
	Isoproterenol	1.3×10^{-3}	
	5-HT	1.27×10^{3}	15.2 hrs
	5-Methoxytryptamine	6×10^{-4}	
	Tryptamine	8.6×10^{-5}	9.2 days
	N-Methyltryptamine	6.8×10^{-5}	12 days

FA = formaldehyde; AcAl = acetaldehyde. * Results are compiled from several authors [48–50]

tive reaction of L-dopa was slower. The authors concluded: "Considering the known concentrations of free catecholamines in extracellular body fluids and of the known physiological activity of structurally related TIQ's indicates, it is unlikely that sufficiently high concentrations of these alkaloids will form spontaneously from acetaldehyde in these fluids after ethanol ingestion to account for the major effects of acute intoxication" [48].

The reaction of formaldehyde, acetaldehyde, and anisaldehyde with dopa was investigated under different conditions by other authors. Only with formaldehyde, the desired 3-carboxy-6,7-dihydroxy-1,2,3,4-TIQ (XXXIII) was obtained (pH 6, +30° C, 72 hours yield 58%, 0.17 ml of 37% FA-solution; scheme 15; [41]). Others reported the chemical

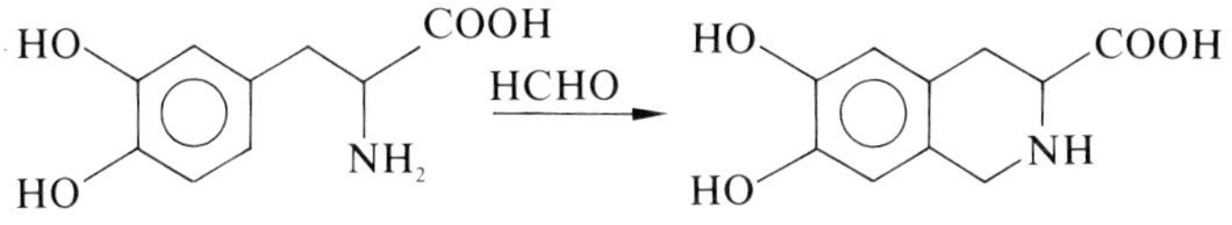

XXXIII

Scheme 15

synthesis of TIQ's from dopa as well as α-methyl-dopa and formaldehyde, acetaldehyde as well as phenylacetaldehyde [42].

3.112 Condensation reaction with simple α-keto acids

The reaction of α-keto acids proceeds easy provided the compound can be transformed into an enolic form and the amine contains a hydroxysubstituent in the *m*-position [38]. DA (189 mg) was incubated for 5 days with phenylpyruvate (XXXIV; 200 mg) in 4 ml water. The reaction proceeded faster the higher the pH was chosen (4–7.4). The yield (XXXV) was 85% at pH 7.4 (scheme 16). The relative yield and

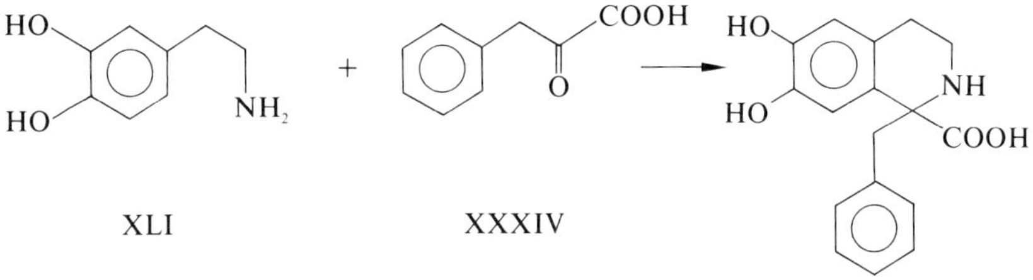

XLI XXXIV

Scheme 16 XXXV

the velocity of the reaction was reduced with *p*-hydroxy-phenyl-pyruvate. The reaction was pH-dependent as well, however the appearance of decomposition products made the assessment of the rate of reaction impossible. The reaction of 4-methyl ether derivative of dopamin and 3,4,5-trihydroxy-β-phenethylamine, a precursor of mescaline, with 11 α-keto acids was investigated to assess whether other organic acids react under physiological conditions as well [43]. Trimethylphenylpyruvate as well as phenylglyoxylic acid did not condense with the amines at all. Further experiments demonstrated that α-keto-*n*-valeric acid (XXXVI) in aqeous solution at pH 5–6 afforded 1-carboxy-6,7-dihydroxy-1-propyl-1,2,3,4-TIQ (XXXVII; 96 hours incubation time; [41]; scheme 17). Attempts to condense DA with oxalacetic acid under different condition (e. g. water, dioxane) proved unsuccessful [41].

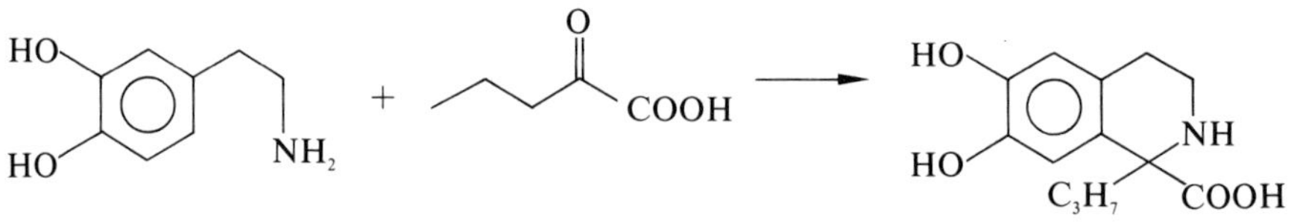

XLI XXXVI XXXVII

Scheme 17

The condensation of the amino acid dopa with pyruvic acid, phenylpyruvic acid or α-keto-*n*-valeric acid did not seem to occur [41].

3.113 Condensation reaction with pyridoxal-5-phosphate

*D*OPA reacts with pyridoxalphosphate to the expected TIQ (+22° C, 0.5 *M* sodiumphosphate buffer, pH 6.5). The turnover ended after less than 20 minutes [44]. The authors found that only primary amines with a phenolic hydroxysubstituent in *meta*-position like dopa and NA reacted with pyridoxalphosphate but not other amines like tyramine (no hydroxy-substituent in *m*-position) and adrenaline (secundary amine).

The chemical reaction between pyridoxal-5-phosphate and various substrates was investigated in some detail by Schott and Clark [45]. They noted that simple aliphatic aminoacids and amines do not react with pyridoxal-5-phosphate (codecarboxylase) at +38° C and pH 6.8 to any appreciable extent. Phenylalanine and phenylethylamine derivatives react only when the amino group ist free and there ist a *meta*-phenolic hydroxy group (table 2). Histidine, histamine and tryptophan reacted at a lower rate than *m*-hydroxyphenyl compounds. Utilizing a model substance the authors found a TIQ condensation product.

Table 2
Second order rate constants (1 mol^{-1} min^{-1}) of the reaction of pyridoxalphosphate and amines.

Noradrenaline	4.8
Dopa	3.2
3,4-Dihydroxyphenylserine	2–3
m-Tyrosine	2.2
Histamine	ca. 2
2,5-Dihydroxyphenylalanine	0.12
Adrenaline	0.13
Tryptophan	0.12
Phenylalanine	0
Tyrosin	0

These findings underline the importance of the phenolic *m*-hydroxy group to facilitate the ring closure, as elaborated above for other aldehydes.

3.12 Noradrenaline, adrenaline and its derivatives as precursors

Incubation of adrenaline and NA (5–10 m*M*) at room temperature under nitrogen in 1 *M* acetate buffer pH 6.0 and in 0.1 *M* phosphate buffer pH 7.0 caused formation of TIQ's (XXXVIII A and B) within 1–2

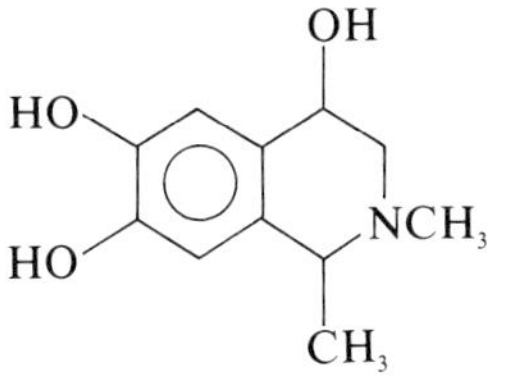

XXXVIII A

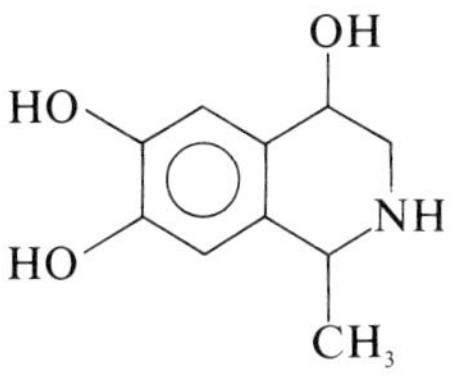

XXXVIII B

minutes if the mixture contained formaldehyde (2 *M*). The yield was 70–80%. The product was not identified with certainty. Acetaldehyde (2 *M*) instead of formaldehyde slowed the reaction (40–50 minutes) with a lower yield (70% with NA, 50% with adrenaline) [46]. Incubation of 300 mg (–)-adrenaline, in 13 ml 0.1 *N* HCl, 2.5 ml acetaldehyde, 8 ml H_2O at +25° C under nitrogen for 24 hours yielded the formation of 1,2-dimethyl-4,6,7-trihydroxy-1,2,3,4-TIQ (XXXVIII B; [47]).
The relative rates of spontaneous non-enzymatic reaction of several catecholamines and indoleamines with different carboxyl compounds was investigated systematically by several authors [48–50]. The results are compiled in table 1.

The formation of TIQ's was investigated utilizing other precursors as well. The conditions of the chemical reaction of phenylephrine, norphenylephrine and N-ethylnorphenylephrine were elaborated with glyoxylic acid as aldehyde precursor. The three amines condensed readily [51].

3.2 Occurence of TIQ's in plants

The largest group of alkaloids in the flora are those with a TIQ nucleus [52]. The TIQ's are precursors of a good array of alkaloids, e. g. protoberberine, papaverine, morphine, aporphine, thebaine, benzophenanthrene, and phthalideisoquinoline classes. Salsoline (XXVII) was the first TIQ, isolated from Salsola Richteri, collected in the southern desert of Turkmenistan [53]. The synthesis of the natural (+)- and the (–)-enantiomer has been published [54]. Salsolinol (XXVI) has been measured in the pulp (40 μ/g w.w.) and the peel (260 μ/g w.w.) of postclimacteric (10 days) banana but not in the climacteric. The authors suggest that salsolinol is oxidized to a quinone which couples with the melanin and is then responsible for the black appearance of overripe bananas. Tetrahydropapaveroline (THP, norlaudanosoline) has not been detected in banana nor O-methylation, side chain oxidation and 4-hydroxysalsolinol [55]. TIQ's with a 1-methyl substituent occur also in *Cactus* species [56]. An example for plants growing in the amazonian forest and containing TIQ's are several species of *Guatteria* (Annonaceae). Extensive reviews covering the isoquinoline alkaloids research of the seventies has been published [58, 59]. More recent investigations of plants containing TIQ's have used *Eschscholtzia tenuifolia* [60], *Papaver orientale* [61], *P. somniferum* [62], *Guatteria ouregou* [63], and *Magnolia salicifolia* [57]. It is interesting that plants containing TIQ's seem to prefere extreme climata like the desert and tropics. A similar however not exclusive distribution is reported about plants containing the *β*-carbolines.

3.3 Biosynthesis in plants

Little is known about the biosynthesis of TIQ's in plants [64]. However, there ist good evidence that TIQ's serve as precursors for opiates. As shown in scheme 18 *l*-tyrosine (XXXIX) is converted to L-dopa (XL) and dopamine (XLI, DA). DA reacts with 3,4-dihydroxyphenylacet-

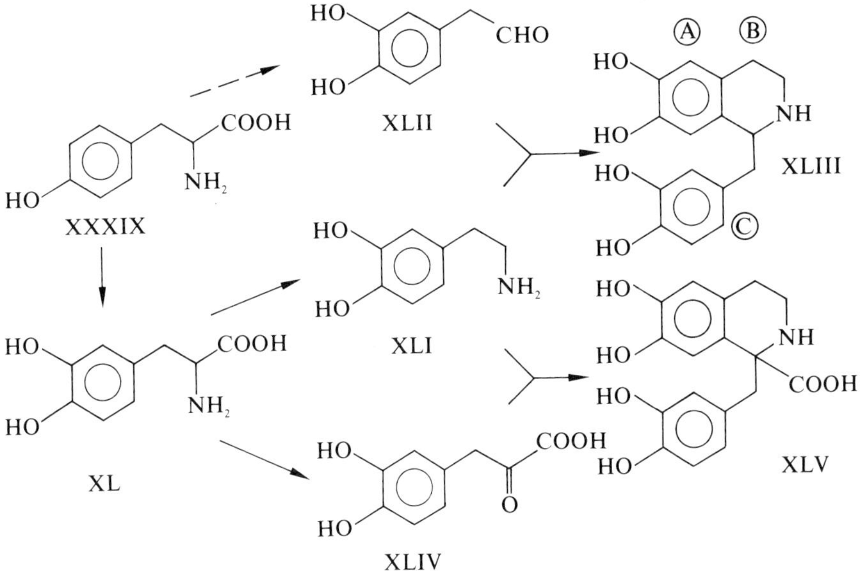

Scheme 18

aldehyde (XLII) as postulated by Winterstein and Trier in 1910 [36] yielding (–)-norlaudanosoline (= THP; XLIII). Whether the aldehyde derives in plants from DA is discussed controversially [65]. Some authors suggest an unknown pathway with tyrosine as precursor [60]. The pathway in plants seems to be different from rats in which monoamine oxidase converts DA to the aldehyde readily. This view ist based on feeding experiments with radiolabelled tyrosine as well as dopa. Tyrosine provides portions of all three rings A, B and C of THP whereas dopa, contributes to ring A and B only [66–71]. A precursor-role of α-keto acids have been denied by these authors. However, despite all presented findings the possibility should borne in mind that 3,4-dihydroxyphenylpyruvate (XLIV) might be formed by a transaminase from tyrosine or dopa in plants also, then serving as an alternative substrate of the aldehyde. To continue the report about the biosynthesis of opiates, the first TIQ intermediate is norlaudanosoline (XLIII; 72, 74, 77). This view is valid only if plants are not able to utilize α-keto acids like phenylpyruvic acid or pyruvic acid for the formation of TIQ's. These findings are remarkable because several investigators demonstrated the formation of carboxylated TIQ's in preparations from plants. A cell-free suspension from *Lupinus polyphyllus* was incubated with cadaverine (XLVI) and pyruvate. Tetracyclic lupine alkaloids were formed suggesting a carboxylated TIQ as intermediate [75]. An other study de-

$H_2N\text{–}(CH_2)_5\text{–}NH_2$

XLVI

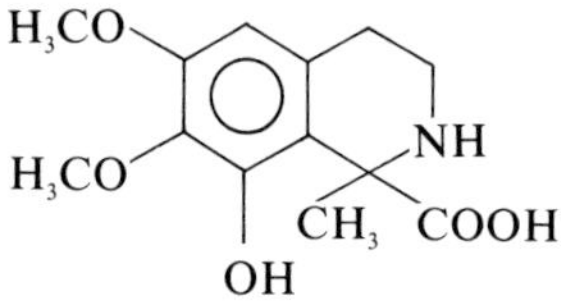

XLVII

monstrated the precursor role of 3,4-dihydroxyphenylpyruvate and the intermediacy of norlaudanosoline-1-carboxylic acid in TIQ-biosynthesis [76]. Other studies demonstrating the existence of the condensation product of pyruvic acid (XLVII) and the decarboxylated products in *Cactus* supporting the view of a precursor role of α-keto acid in *Cactus* [77]. In plants producing morphine alkaloids THP-1-carboxylic acid (XLV) has been detected [69, 76, 78].
Other authors again dispute the role of α-keto acids in the biosynthesis of TIQ's [60]. They demonstrated a (S)-norlaudanosoline synthase in a large number of Papaveraceae, Berberidaceae, and Ranunculaceae cell cultures and they found no support for enzymatic formation of norlaudanosoline-1-carboxylic acid (XLV) under cell-free conditions [79]. There was an appreciable chemical non-enzymatic condensation reaction as assessed by the appearance of tritiumlabelled water deriving from ring-2,6[^{3}H] DA. The authors found four isoenzymes. It is most remarkable that none of them used 3,4-dihydroxyphenylpyruvate (XLIV) as substrate but only the aldehyde (XLII). No evidence was obtained that (R)-norlaudanosoline could be formed by the crude enzyme mixture. There ist, however, evidence for the natural occurrence of R-coclaurine (XLVIII), a common precursor for proaporphines and aporphines [80].

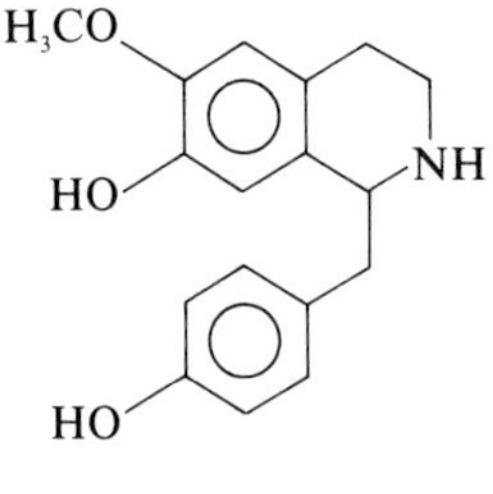

XLVIII

Whether in plants only aldehydes are utilized in the biosynthesis of TIQ's and morphine alkaloids respectively remains to be established.

Brossi points to the observation that the Papaver cell cultures utilized to investigate the topic produced neither thebaine nor morphine two products of the pathway. He concluded that the possibility still exists that the latter group of alkaloids may originate by the pyruvic acid path following the sequence 1-carboxy-TIQ – 3,4-dihydroisoquinoline – (R)-TIQ [81].
Besides the two controversial issues namely the biosynthesis of dopaldehydes (XLII) and the precursor role of α-keto acids a third point should be mentioned. Several authors demonstrated the formation of salsolinol and 3-carboxysalsolinol in plants. There ist no doubt that acetate serves as carbon donor whereas pyruvate is not utilized [126]. The authors presented evidence that the methyl-substituent in position 1 does not derive from acetate. It has been rationalized that the callus of *Stizolobium hassjoo* (belonging to a subfamily of Leguminosae) contains no enzyme capable of utilizing acetate directly for formation of two-carbon units of 3-carboxysalsolinol. Battersby and co-workers [127] found that the administration of 1-[^{14}C]acetate to the peyote cactus, *Lophophora williamsii* yielded radiolabelled pellotine, but they suggested that acetic acid was not a direct precursor for the two-carbon units. It should be realized that acetate provides one carbon for the B-ring of TIQ's. Interestingly enough, evidence is presented that acetyl CoA does not serve as precursor in plants [126].
Davis and co-workers showed that norlaudanosoline and radiolabelled SAM gave rise to three procucts utilizing tissue from animals: 2,3,10,11-tetrahydroxyberberine, and 2- or 3-monomethyl tetrahydroxyberberine [82–85]. As will be described later, Kametani and co-workers demonstrated the formation of aporphine alkaloids by incubation of rat liver with labelled reticuline. These investigations have enriched the discussion of the pathogenesis of dependence. Davis' as well as Kametani's group suggested the formation of morphine-like alkaloids in men. In view of these findings the biosynthesis of protoberberins and morphine-alkaloids will be briefly reported.
(—)-Norlaudanosoline (XLIII) serves as precursor of (+)-reticuline (XLIX) as well as (—)-reticuline (LI; scheme 19). The latter compound ist converted to salutaridine and then to the opioid thebaine in Papaver orientale [61].

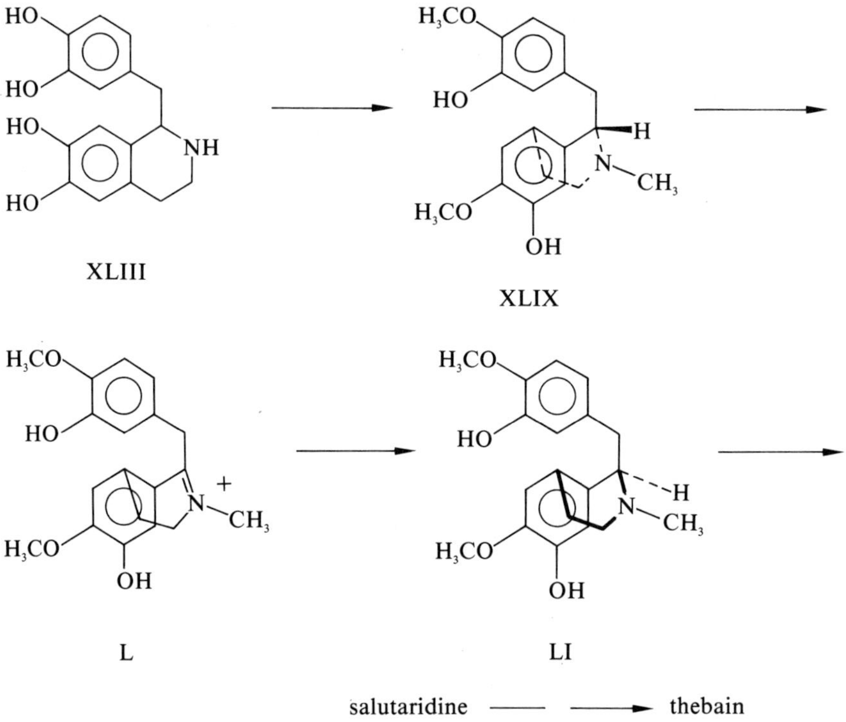

Scheme 19

3.4 Biosynthesis in mammals

3.41 In-vitro studies

3.411 Tetrahydropapaveroline (THP, norlaudanosoline)

The earliest investigation of the pharmacology of TIQ's was carried out by Laidlaw in 1910 [86]. More than 50 years later, Holtz' group incubated the dopa derived aldehyde, DA and monoamineoxidase (MAO). They demonstrated the formation of THP by chemical and pharmacological analysis [87, 88]. The findings were confirmed using rat liver and rat brain stem homogenates. The amount of THP formed depended on NAD-cofactor. In the presence of NAD, the production of THP decreased about 10-fold in rat liver [82]. This was explained by the observation that incorporation of NAD and NADH into incubation mixtures of liver homogenates essentially abolished THP production and markedly enhanced the formation of acids or neutral metabolites of DA [83].

The metabolism of THP in vitro by rat brain and liver was investigated in some detail [84]. THP is known to be an excellent substrate for catechol O-methyl-transferase (COMT) in vitro (V_{max} 5 times and 3 times higher than that for DA and NA, respectively, K_m 0.03 m*M* approximately one tenth of that of NA and DA) [85]. Incubation of tissue homogenate with THP (XLIII) and [^{14}C] SAM in phosphate buffer (pH 8.0) for 1 hour at 37° C yielded 2,3,9,10-tetrahydroxyberberine (LII), 2,3,10,11-tetrahydroxyberberine (LIII), and 2- or 3-monomethylated derivative of tetrahydroxyberberine (coreximine; LIV; scheme 20).

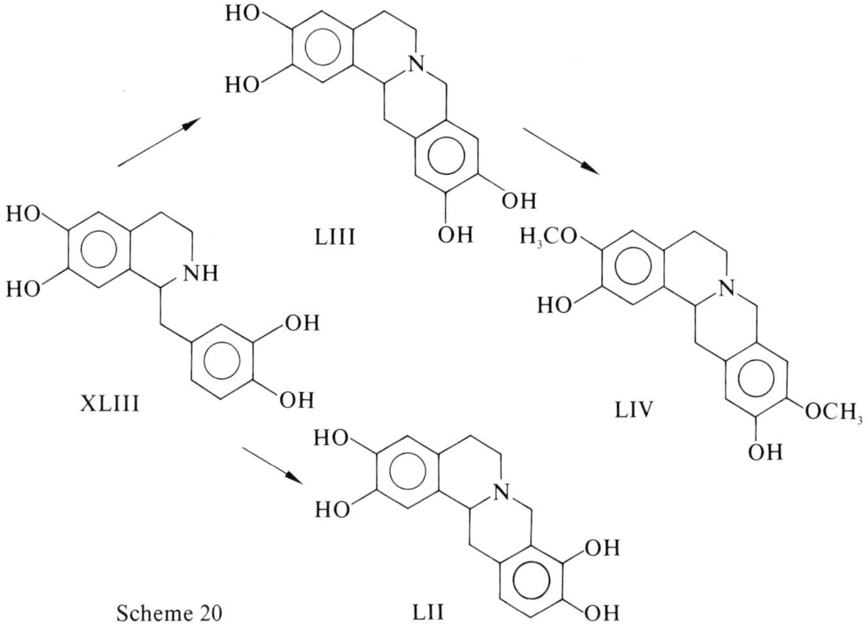

Scheme 20

The compounds were identified by GC/MS measurement. The 2,3,10,11-tetrahydroxyberberine was the major THPB (tetrahydroprotoberberine) metabolite found. Thus, mammalian systems are able to insert one carbon unit forming the "berberine bridge". This reaction is catalyzed by enzymes since boiled tissue was without effect. Some doubt has been cast on the methods used for the identification of THP formed [89]. Other investigations have confirmed the in vitro formation of THP using rat brain homogenate and mass fragmentation for the identification [90].

Other metabolism studies used radiolabelled (±)-reticuline (LV) as precursor. As mentioned above, reticuline serves as precursor in the biogenesis of opium alkaloids in plants such as berberine, morphine, benzophenanthridine and phthalideisoquinoline [91–93]. Rat liver homogenate (pH 7.4, phosphate buffer, 37° C, 2 hours) was incubated with (±)-reticuline and without cofactor yielding small amounts of coreximine (LIV), scoulerine (LVII), and N-norreticuline (LVIII;

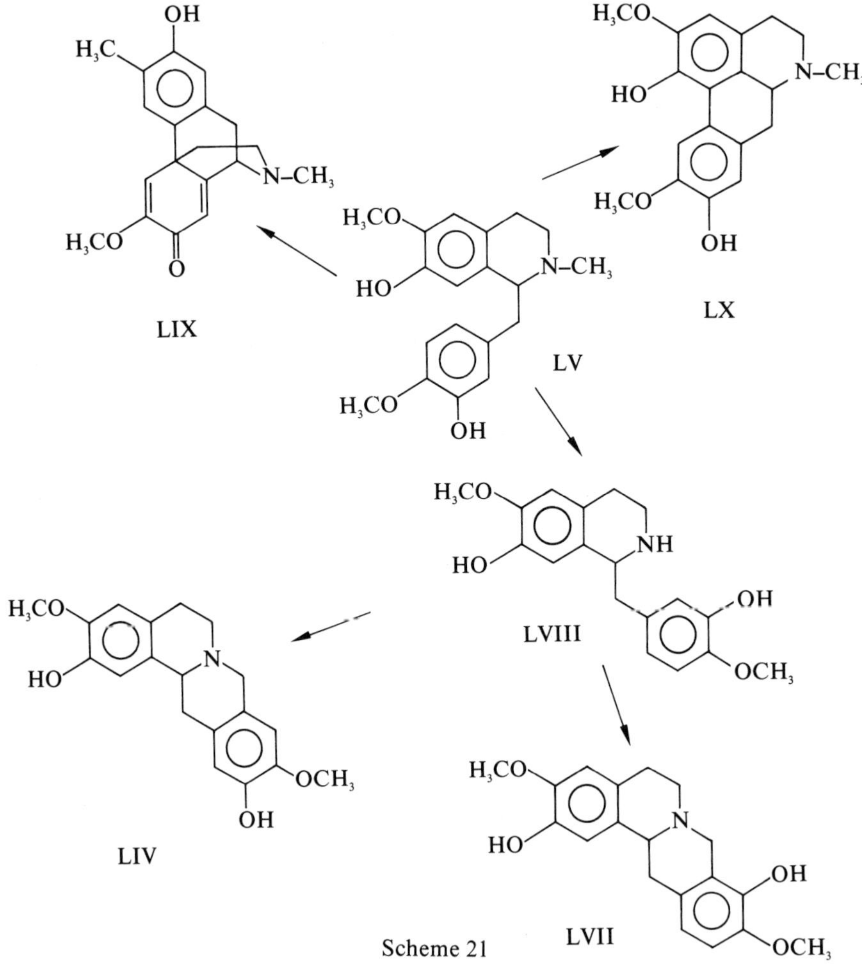

Scheme 21

scheme 21). Addition of NADPH and $MgCl_2$ to the incubation medium increased the amount of product. The amount of formed alkaloids was too small to measure the optical rotation [94]. In a consecutive study the authors investigated the formation of other alkaloids deriving from (±)-reticuline under the same conditions with rat liver homogenate. Besides the protoberberines coreximine (LIV) and scoulerine (LVII) they demonstrated the formation of the morphinandienone al-

kaloid pallidine (LIX = isosalutaridine; [138–140]) and the aporphine alkaloid isoboldine (LX; [95, for the chemistry see: 141–143]).
These in vitro studies are most remarkable since they demonstrate for the first time phenolic oxidative ring coupling of reticuline to pallidine (LIX). A second important finding shows that reticuline had to be demethylated to give N-norreticuline (see scheme 21) before the ring closure proceeds to the protoberberine alkaloids scoulerine and coreximine.
It ist most remarkable that in some plants *(Berberis beaniana,* Schneid.) this reaction occurs differently [136]. The authors isolated the berberine bridge enzyme [137] to homogenity and demonstrated that the enzyme is specific for substrates with (S)-configuration. The methyl-substituent was integrated into the berberine bridge whereas the (S)-norreticuline-N-oxide was not utilized by the enzyme [136].
Incubations under physiological conditions (0.2 *M* phosphate buffer, pH 6.0, +37° C, 30 min) of 3,4-dihydroxyphenylpyruvate (XLIV; 14 m*M*) and DA (XLI; 18 m*M*) afforded 2–3% THP-1-carboxylic acid (norlaudanosolinecarboxylic acid; XLV) in the presence of cell free homogenates of rat brain tissue. Comparable yields were realized in the absence of brain tissue or with boiled homogenates proving that the condensation was not enzyme dependent [96].

3.412 Salsolinol

Cohen and Collins perfused cow adrenal with buffered solutions of formaldehyde or acetaldehyde (1 μg/ml) at +37° C [97]. They demonstrated TIQ's deriving from NA (XXXVIII B) and adrenaline (XXXVIII A) in the efflux and suggested that the alkaloids, had been either actively secreted or leaked from the nerve terminals. The concentration of [^{14}C]acetaldehyde was in the range observed in blood during ingestion of ethanol in man. In a second study the authors perfused isolated cow adrenal glands for one hour with 23 m*M* acetaldehyde (1 mg/ml). This procedure resulted in the synthesis of the condensation products of acetaldehyde with endogenous NA and adrenaline. The concentration increased after stimulation, and decreased subsequently to depletion of calcium [98].
Yamanaka and co-workers incubated DA (3.3 m*M*), acetaldehyde (0.5–4 m*M*), NAD (4 m*M*), and homogenate from rat brain or rat liver at +37° C, pH 7.4, for 30minutes. Without homogenate 14 to 67% of

DA was converted into salsolinol dependent on the concentration of acetaldehyde. With homogenate 10 to 72% salsolinol was detected. Addition of ethanol to rat brain homogenate instead of acetaldehyde prevented the formation of salsolinol possibly due to fast oxidation of the formed acetaldehyde into acetic acid. In ethanol metabolism NAD-linked alcohol dehydrogenase is the rate limiting enzyme. Once acetaldehyde is formed the substance is rapidly metabolized [99].
Incubation of pyridoxal-5-phosphate in 0.07 *M* sodium phosphate buffer, pH 6.5 with dopa, DA or NA together with brain homogenate from cow or guinea pig resulted in a TIQ derivative. Evidence was presented that the amines react with pyridoxalphosphate within 12 minutes. The authors assumed a non-enzymatic reaction, since the rate of formation agreed approximately with the rate of the chemical reaction [87].

3.413 Effects of drugs on the formation of TIQ's

The metabolism of DA and NA proceeds in the CNS different following oxidative deamination by monoamine oxidase. The aldehyde deriving from DA ist oxidized to an acid whereas the aldehyde deriving from NA is reduced to the glycol derivative. Sedatives like phenobarbital and pentobarbital submit the reductive pathway of NA. Incubation of brain stem homogenate with [^{14}C] NA yielded the formation of hydroxy THP (12.29% of NA metabolized). The amount of metabolized [^{14}C] NA increased if the incubation mixture contained phenobarbital to 22.61% or pentobarbital to 24.32%. The author assumed that the barbiturates like many sedatives competitively inhibited aldehyde oxidoreductase, which caused an increase of the steady-state lel of the NA-derived aldehyde to the account of 4-hydroxysalsolinol, a a TIQ which had been detected as well [100].
The metabolism of DA was affected by drugs also. Substances were investigated which inhibited aldehyde dehydrogenase like ethanol, acetaldehyde, paraldehyde and chloralhydrate. Incubation of brain homogenate with ethanol (100 m*M*) increased the formation of THP from 58% of DA concentration to 60.3%, with acetaldehyde (1 m*M*) to 65% and with chloralhydrate (1 m*M*) to 72.35%. Furthermore, the authors found also an increase of the formation of salsolinol if acetaldehyde was added to the incubation mixture. This observation was explained by the higher rate of formation with DA as substrate (6.13 μmol salsoli-

nol) as compared with NA (3.75 μmol hydroxysalsolinol under the same conditions).

3.42 In-vivo studies
3.421 Tetrahydropapaveroline
3.421.1 Occurrence

In pharmacokinetic studies, a half-life of THP-disappearance of 17.3 minutes in rat brain after intraventricular injection was calculated. Administration of pyrogallol (COMT inhibitor) extended the half-life of THP to 69.3 minutes, underscoring the fact that O-methylation is a route of metabolism for this compound [110, 159]. The greatest amount of the applied THP was found in the striatum, the thalamus, and the posterior section of the brain which included both the mes- and metencephalon [102, 103]. THP hat not been detected in rat brain without pretreatment (sensitivity of the method: 2 ng/g). After pretreatment with pyrogallol TIQ-like derivatives were detectable in brain and adrenal tissue [104]. After administration of large amounts of L-dopa in combination with ethanol, about 8 ng THP/g tissue were measured by GC-MS-method [90]. Similar results were obtained following intoxication of rats with ethanol.
Determination of THP in human tissue or urine of untreated subjects has not been reported. Parkinsonian patients, treated with L-dopa (3–4 g) excreted THP into the urine [105]. 1-Carboxy-THP has been measured in the urine of parkinsonian patients on L-dopa treatment ([96]; see scheme 18, conversion of XLI + XLIV → XLV).

3.421.2 Metabolism of THP and reticuline

The hypothesis concerning the formation of morphine-like alkaloids in mammals presumes further metabolism to alkaloids as depicted in scheme 22 (adapted from [100]). 189 mg/kg THP were injected i. p. into rats. Urine was collected for 10 hours [52]. 1.23 mg of alkaloids was recovered from the urine of three rats. GC identification revealed 2,3,10,11-tetrahydroxyberberine (64.2%; LIII). Minor constituents included 2,3,9,10-tetrahydroxyberberine (5.7%; LII) and two methylated derivatives tentatively identified as coreximine (6.5%; LIV) and 2- or 3-monomethylated derivative of 2,3,10,11-tetrahydroxyberberine

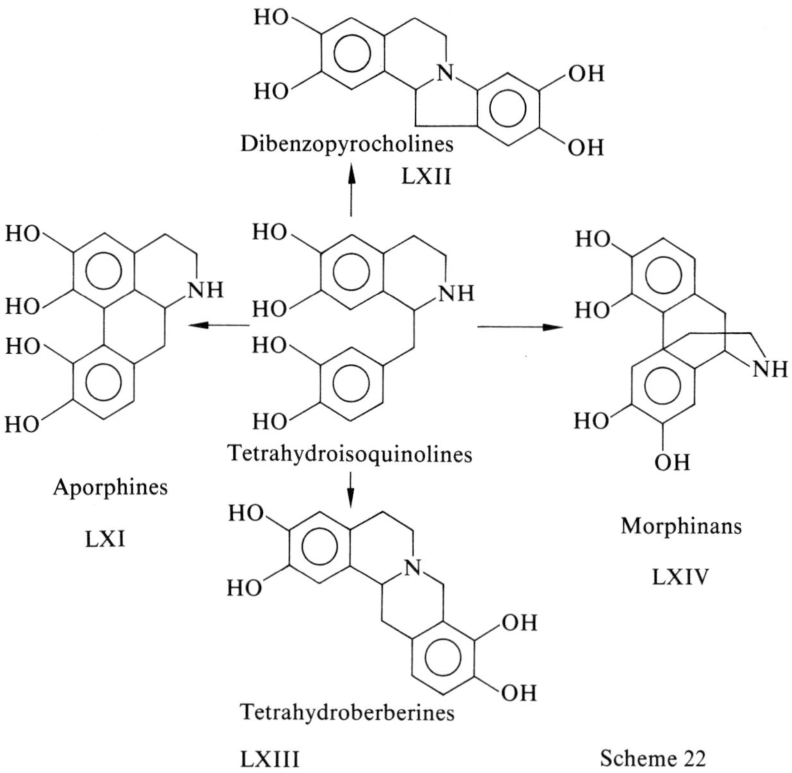

Scheme 22

(23.6%). None of these alkaloids was detected in the urine of control rats [84].

Others investigated the biotransformation of (+)-reticuline in rats [94]. In contrast to their in-vitro findings the authors observed no formation of morphinandienone-type alkaloids in vivo on the basis of the inspection by TLC or GC of crude products. It seems that there is no enzyme in rats which is responsible for the phenolic oxidative coupling of reticuline.

In conclusion, despite the in-vitro biotransformation of THP (norlaudanosoline) and reticuline to aporphine and morphine type alkaloids, such compounds have not been detected in vivo. There is only evidence for formation of berberine type alkaloids in vivo after a load with THP. There are no reports concerning the pharmacology of these berberine alkaloids. Derivatives closely related to them are known to produce sedation and to potentiate barbiturate hypnosis [107]. Additionally, a large number of substituted tetrahydroprotoberberine alkaloids have been synthesized and patented as tranquilizers [108].

3.422 Salsolinol

Endogenous salsolinol has been reported in the brain of neonatal rats [109] and in medial basal hypothalami of adult male Wistar rats [110]. Trace levels of salsolinol were measured in hypothalamus and striatum of rats [111]. The half-life of intracerebroventricular injected salsolinol was determined to be 12.5 minutes. Administration of 250 mg/kg i. p. of pyrogallol prior to the injection of salsolinol increased the half-life to 23.1 minutes [103].
The regional distribution pattern following intracerebroventricular injection of salsolinol and some of its derivatives showed higher levels in the hypothalamus than in striatum or hippocampus [112]. The biosynthesis of salsolinol in vivo has been investigated under various conditions [113]. Injections of 250 mg/kg of L-dopa i. p. caused no change of the endogenous concentration of salsolinol in the striatum of rats 45 minutes post injectionem. 3-Carboxysalsolinol (LXV A) could not be detected. On the other hand evidence was found for 1-carboxysalsolinol (LXV B) both in striatum from control and L-dopa treated animals.

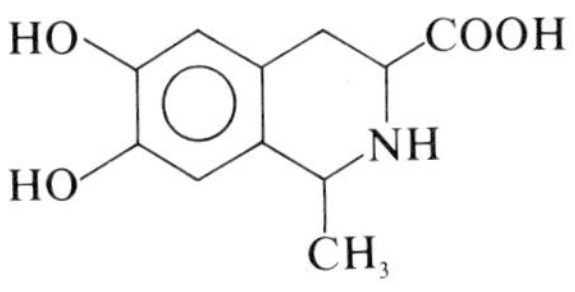

LXV A

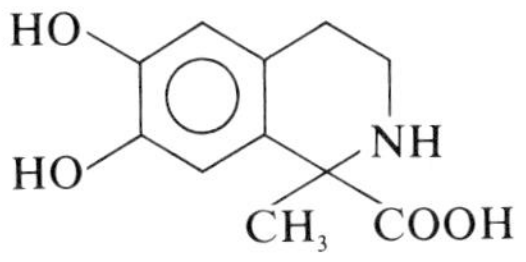

LXV B

In a second series of experiments rats were deprived of food 3 days prior to i. p. injection of anaesthetic doses of ethanol, saline and equicaloric glucose, respectively. In the striatum of rats that received glucose, the salsolinol concentration was elevated from 16.6 ± 4.6 pmol/g to 67.8 ± 27.5 pmol/g (p 0.05). These findings suggest that under physiological circumstances dopamine does not seem to be the limiting factor for salsolinol formation. Secondly, L-dopa does not seem to condense with acetaldehyde to form 3-carboxysalsolinol. However, dopamine condenses with pyruvic acid to 1-carboxysalsolinol. Thirdly, the finding of increased salsolinol concentration in animals treated with glucose indicated that the carbohydrate metabolism is of importance for salsolinol formation.

Salsolinol and THP have been detected in the urine of parkinsonian patients on L-dopa treatment (3–4 g; [105]). Without pretreatment, salsolinol has been found in human caudate nucleus and putamen [114] as well as 1-carboxysalsolinol in human caudate nucleus [113]. Salsolinol has been detected in the urine of controls (197 nmol/d), of untreated and L-dopa-treated parkinsonian patients (39; 1409 nmol/d) and alcoholics (130 nmol/d; increased after ingestion of 250 ml wine for 7 days to 573 nmol/d). The methylated derivative of salsolinol was detected in the urine of some alcoholics. The authors detected both compounds, 5-methyl and 6-methyl ether derivatives, in the CSF of all three groups [115].

Norlaudanosolinecarboxylic acids were detected in the urine of parkinsonian patients on L-dopa and in the brain as well as in urine of rats [96]. Methylnorlaudanosolinecarboxylic acid (LXVI) was detectable

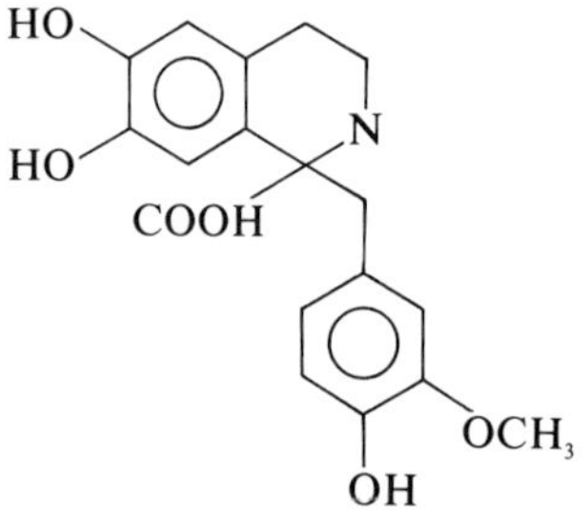

LXVI

in the urine of controls. However, the level was too close to minimal levels of detection to be conclusive. Evidence is presented that the condensation reaction was not enzyme dependent [96].

3′, 4′-Deoxynorlaudanosolinecarboxylic acid (LXVII) a TIQ derived

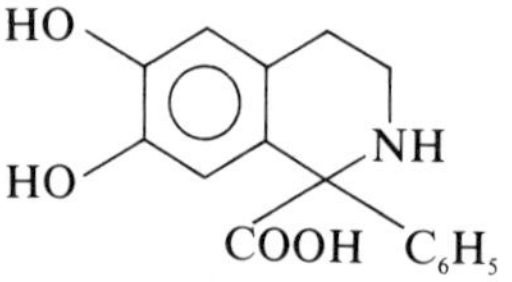

LXVII

from DA and phenylpyruvic acid, has been detected by MS in urine of phenylketonuric children and in urine and brain of rats with experimentally induced hyperphenylalaninemia. Levels were more than 10-fold higher than controls [116].

4 β-Carbolines

4.1 Non-enzymatic formation

4.11 The condensation reaction under so-called physiological conditions

The conditions of the chemical formation of BC's have been described in section 2 of this review. As in case of the TIQ's, several investigators examined the possibility of the formation of BC's under "physiological conditions". Hahn and Ludewig [28] incubated tryptamine with acetaldehyde as well as phenylacetaldehyde at pH 5.6 and 7 and +25° C. They found almost 100% conversion into the expected BC's after 24 hours (scheme 23). The same authors reported the formation

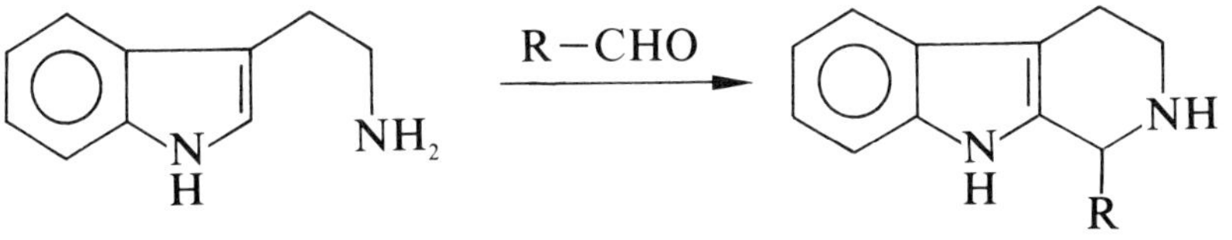

Scheme 23

of the amid by heating a solution of tryptamine and phenylacetic acid. The amid could be converted into the dihydro-β-carboline and the tetrahydro-β-carboline [28]. The phenylethylamine did not react under comparable conditions [120, 121]. The yield of BC's declined if phenylacetaldehydes are substituted with hydroxy or methoxy groups. No condensation reaction was found for *o*- and *p*-hydroxybenzaldehyde, piperonal, vanillin, glyoxal, methylglyoxal and *d*-(+)-glucose [122].

Späth and Lederer [117] demonstrated the conversion of N-acetylindolethylamines to the BC's under so-called physiological conditions supporting the possibility of a Bischler-Napieralski type of cyclodehydration ([26], scheme 5). These findings stimulated Mc Isaac to incubate radiolabelled 5-MeO-tryptamine in phosphate buffer (pH 7.0) with acetaldehyde at 37° C [118]. He demonstrated the condensation reaction yielding 6-MeO-THH (LXXIV, C) and observed, that 50% of the product had been formed in 3.5 hours. Other authors determined rate constants and half times for the condensation reactions of indoleamines and aldehydes [40, 48–50]. Second-order rate constants at pH 7.4 and +37.5° C (so-called physiological conditions) were assessed by high performance liquid chromatography with electrochemical detection to monitor utilization of the amine reactant. Values are compiled in table 1. It is noteworthy that the rate constants for indoles are ten to several hundred times smaller than those for catecholamines with respect to formaldehyde, and hundred to thousand times smaller with respect to acetaldehyde.

As elaborated for TIQ's, a third mode of formation is conceivable utilizing α-keto acids. Hahn and co-workers were the first to demonstrate condensation reactions between tryptamine and α-keto acids under so-called physiological conditions ("zellmögliche Bedingungen"; [122]). They observed that the reactions proceeded faster with α-keto acids than with the respective aldehydes. The following substrates were examined: pyruvic acid, *p*- and *m*-hydroxyphenylpyruvic acid, 4-hydroxy, 8-methoxy-phenylpyruvic acid, 8,4-dimethoxyphenylpyruvic acid, 3,4-methylendihydroxyphenylpyruvic acid and 3,4,5-trimethoxyphenylpyruvic acid. The investigation of the pH-dependence of the condensation reaction revealed two optima, one at a low pH and the other near neutral values, whereby the exact optimum depended on the α-keto acid utilized. Furthermore, the reaction rate decreased with increasing number of methoxy-substituents and increased with light [123]. The same authors demonstrated the condensation reaction of α-keto-dicarboxylic acid like α-keto-glutaric acid with tryptamine under physiological conditions [124]. As pointed out by Schöpf and Salzer, Hahn used higher concentrations than were likely to obtain in living cells and that the reaction mixtures were not homogenous but that he used a suspension. Self-condensation of the substituted phenylacetaldehydes was a natural result. Repetition of the experiments at proper dilutions showed that the aldehydes reacted hundreds of times faster than pyruvic acids [39]. On the other hand, the substrate concentrations used by these authors were still much higher than those in the brain. Therefore, some of the experiments were repeated comparing substrate concentrations which can be expected to occur in rat brain with higher ones. Incubations were performed in 0.1 *M* buffer at pH 5.0 and 7.4 and 37° C for two hours. 6 nmoles [^{3}H] tryptamine were incubated with acetaldehyde (0.1–20 m*M*) or with pyruvic acid (0.5–50 m*M*). No indication of a condensation reaction could be obtained. Only at 100 m*M* pyruvate, whereby the pH had shifted to 3.8, 7% tryptamine was converted into the BC [195, 196]. Thus, under in vivo conditions no chemical formation of BC's occurs with pyruvate as precursor. Furthermore, these findings strongly support the notion of no artifactual formation of BC's from acetaldehyde and indoleamines during

the work up of the measurement of endogenous concentrations. Such a possibility has been discussed by some authors [119].
Proteins like ovalbumin and lysozyme which contain tryptophan, develop green fluorescence (peak max. 530 nm) in strongly acidic solvents (trifluoroacetic acid) even in the absence of oxygen. Whether the fluorescence is emitted by a BC is unclear. Under the same conditions N-acetyl-*DL*-tryptophan was converted into 1-methyl-3-carboxy-3,4-dihydro-β-carboline (LXXII A). Thus, N-acetyl-tryptophan is converted to a BC under non-physiological conditions (strongly acidic, heating for several hours, yield 10%). If analogous reactions would proceed in vivo, this would be of interest with respect to the high affinity of BC's substituted in position 3 for specific benzodiazepine binding sites. However, it should be kept in mind that the above compound contains a methyl group in position 1 which causes a substantial decrease of the affinitiy.

4.12 The decarboxylation of BC's and TIQ's

As depicted in scheme 24, the step after the cyclization reaction of

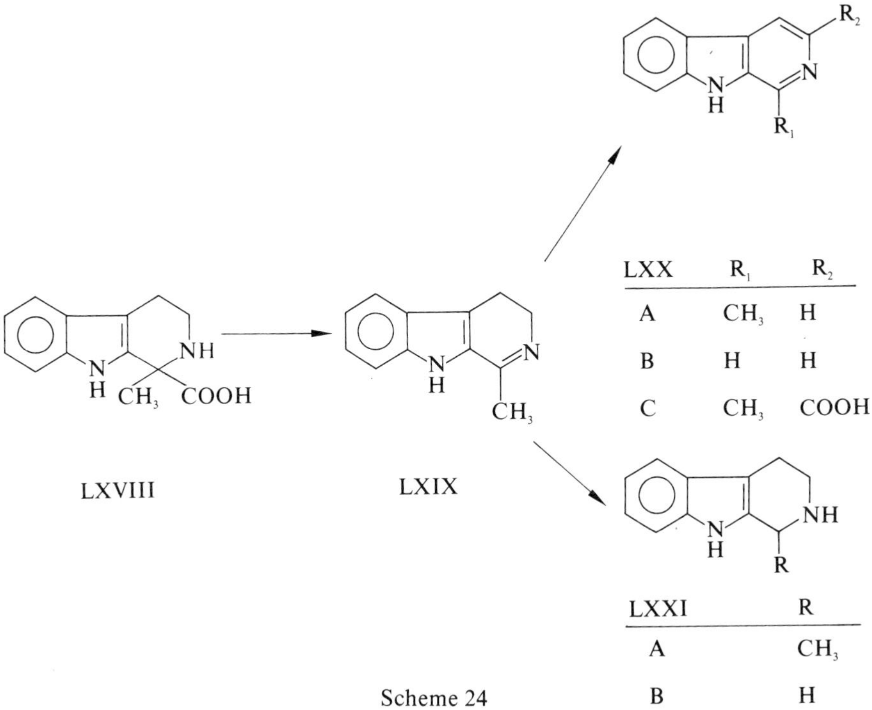

LXX	R_1	R_2
A	CH_3	H
B	H	H
C	CH_3	COOH

LXXI	R
A	CH_3
B	H

Scheme 24

α-keto acids and indoleamines comprises decarboxylation. Hahn et al. [122] observed that BC's, carboxylated in position 1, could be decarboxylated by heating in methanol and addition of hydrochloric acid within a few minutes. The yield depended on the substituent (phenyl = *m*- and *p*-hydroxyphenyl = 3,4-methylene-dioxyphenyl (100%) > 4-OH-3-methoxyphenyl = 3,4-dimethoxyphenyl (50 %) > 3,4,5-trimethoxyphenyl). No decarboxylation at all was detected with pyruvic acid as substrate under these conditions. Spenser demonstrated the oxidative decarboxylation of 1-CTHH (LXVIII) to 3,4-dihydro-β-carboline (harmalan; LXIX) with 85% yield in 24 hours [125]. We have observed that a similar reaction takes place readily in slightly basic as well as acidic solutions [128]. Under all conditions examined harmalan (LXIX) was obtained but not 1-methyl-THBC (THH; eleagnine LXXI A).

The same properties showed TIQ-1-carboxylic acid. When such a compound (XLVII) was incubated with fresh cactus slices, the 3,4-dihydroisoquinoline was detected, suggesting oxidative decarboxylation as found with BC's [129]. The reaction requires the loss of two electrons which are removed through the ring [130]. Therefore, the decarboxylation depended upon the relative electron density in the aromatic ring and correlated directly with the number of free phenol groups present in the ring. The possible ability of laccase to act as two-electron oxidant without hydroxylating action suggested its suitability in such an investigation [133]. Incubation (35° C) in phosphate buffer (0.1 *M*, pH 6.0) with a crude laccase preparation and 6-hydroxy-acids of various TIQ's resulted in the formation of the expected dihydro-TIQ in a few minutes with a yield of 80%, whereas 5% reaction occurred when the acids were left in phosphate buffer for 24 hours without laccase present. Incubation of the acids with hydrogenperoxide and horseradish peroxidase gave results similar to those obtained with laccase [132]. It should be noted that substrates without hydroxyl substituents did react neither with nor without enzymes. The hypothesis that 1-CTHH (LXVIII) serves as prime intermediate metabolite of BC's was supported by the finding or Herbert et al. who showed that the compound acted as a precursor for harman (LXX) and tetrahydroharman (LXXI) in plants [134]. The decarboxylation reaction in rats will be described in a later section.

4.13 The condensation reaction with pyridoxal-5-phosphate

The condensation reaction of 5-hydroxytryptamine and 5-hydroxytryptophan with pyridoxalphosphate (PLP) has been investigated under pseudo-physiological conditions more recently (phosphate buffer, pH 5–8, 37° C, 120 min; [135]. At 10^{-4} *M* only 4% (at 10^{-2} *M* 80%) of the indoles were converted into BC's (LXXIII A and LXXII B). With sub-

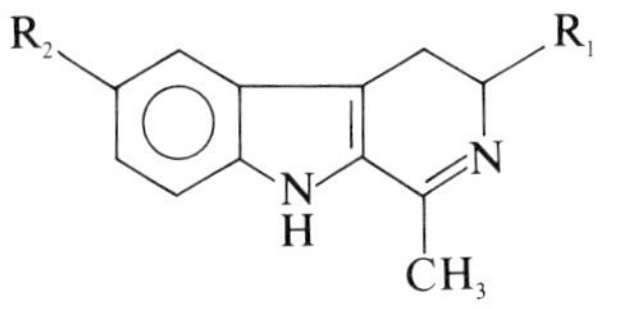

LXXII	R_1	R_2
A	COOH	H
B	H	OCH_3

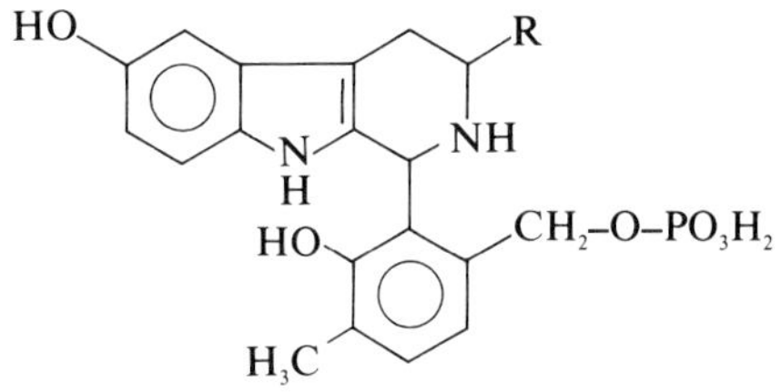

LXXIII	R
A	H
B	COOH

strate excess, velocity of the cyclization was increased. The optimal pH for PLP-5-HT was 6.8, that for PLP-5-HTP 5.4–5.6. The authors suggested that cyclization with PLP may be a significant factor in regulation of amine levels in tissues since many amines react with PLP like catecholamines [44, 45], histidine and histamine [45].

4.2 β-Carbolines in plants

4.21 Occurrence in plants and food-stuff

The β-carbolines are widespread in plants. In an excellent review Allen and Holmstedt compiled reports up to 1979 about "simple BC's" which were defined as compounds comprising the tricyclic pyrido (3,4-b) indole ring system with alkyl C_1 substituents [145]. This definition excludes compounds containing addional fused-ring systems like yohimbine, ajmaline and reserpine. The authors described 64 different BC's in 112 plants. Meanwhile only a few reports have appeared adding 5 other simple BC's and 4 plants to this list.
BC's are found in alcoholic beverages. Only in beer 6-OH-THBC (6-OH-THN; LXXIV A) was present in detectable and quantitable

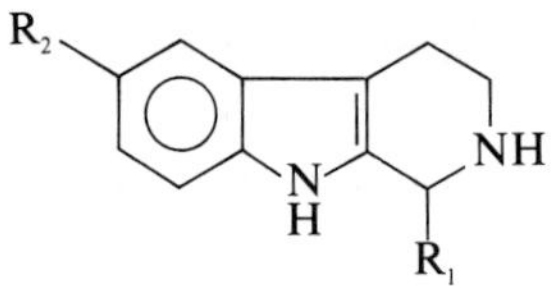

LXXIV	R_1	R_2
A	H	OH
B	CH_3	OH
C	CH_3	OCH_3

amounts (16–235 nmol/l). The concentration did not correlate with the ethanol content (1.8–4.5% w/w) and was lower than that of 6-OH-1methyl-THBC (6-OH-THH; LXXIV B). The authors assume that the generally higher levels of 6-OH, 1-methyl-THBC reflect a greater concentration of acetaldehyde vs. formaldehyde [146]. 6-OH-1-methyl-THBC was detected in beer (427–18 nmol/l). The ethanol concentration (4.5–1.8% (w/w)) did trend to correlate with the amount of the BC. In wine samples the concentration was much lower (1.1 nmol/l) up to the limit of detection (1 pmol/g; [147, 148]). Other BC's were detected in beer and wine as well. The concentration of 3-COOH-THBC (LXXV A) in beer ranged from 2–6 μg/ml, in wine from 0.5–1.0 μg/ml.

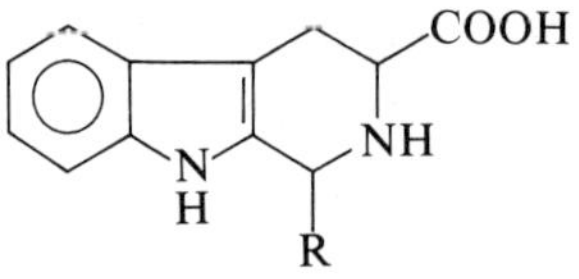

LXXV	R
A	H
B	CH_3

The concentrations of 3-COOH-1-methyl-THBC (3-COOH THH; LXXV B) ranged from 0.5–2.5 μg/ml in beer and 3.5–7.8 μg/ml in wine [149].

Other sources of 6-OH-1-methyl-THBC in dietary products are banana and plum. In most other fruits and vegetables only traces or no BC's at all were detected. In blue cheese a high concentration was measured in contrast to other milk products like yogourt, camembert, brie, or hard cheese [147].

BC's can be formed by pyrolysis from tryptophan [150]. This explains the occurrence (mostly harman and norharman) in tobacco smoke [150–154], marihuana smoke [155], charred insects [156] and well-cooked foodstuffs [157, 158].

4.22 Biosynthesis in plants

The three biosynthetic pathways of BC's (considering the non-indole moiety) have been described previously [128]. Indolealkylamines condense either with aldehydes, α-keto acids or acetyl-CoA forming BC's.Several authors investigated the precursor of the indolyl moiety in plants. In feeding experiments with either unlabelled or isotope-labelled tryptophan, the formation of BC's were demonstrated with *Eleagnus angustifolia* [106, 160], *Passiflora edulis* [162], *Carex brevicollis* [163], *Peganum harmala* [164–168], and *Phaseolus vulgaris* [169, 170]. Harman (LXX A) and norharman (LXX B) were detected in the medium and cells of culture suspension of *Phaseolus vulgaris* [169, 170]. The efficiency of the biotransformation of tryptophan to the two alkaloids judging by the yield per quantity of tryptophan (100 µg) utilized by *Phaseolus vulgaris* suspension culture was very low (1 µg harman, 4 µg norharman). The scanning of thin layer chromatograms of medium extract indicated the presence of other indole compounds with unknown identity. Unfortunately, the percursor role of tryptamine was not examined nor the non-indolic precursor identified. Only a few BC's are known in plants with carboxyl, carboxylmethylester, and methanol-substituents in position 3 (3-COOH-harman ([171, 172]; LXX C), 3-COOH-tetrahydroharman ([173]; LXXV B), 3-COO-methylester of harman [174], 3-COO-methylester of tetrahydroharmine ([175]; LXXVI and pyridindolol ([176]; LXXVII).

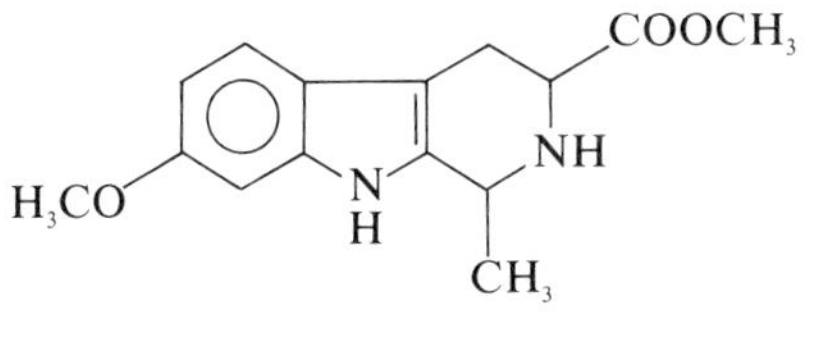

LXXVI

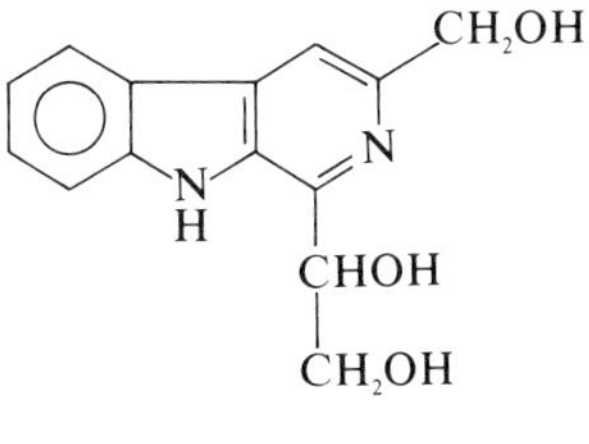

LXXVII

An alternative biosynthetic pathway of BC's with N-acetyltryptamine as intermediate has been proposed [117]. Tryptophan is decarboxylat-

ed to tryptamine which after N-acetylation and subsequent cyclo dehydration yields the 3,4-dihydro-β-carboline harmalan. This intermediate compound can then be oxidized to harman or reversibly reduced to THH. The presented evidence for decarboxylation of tryptophan prior to the cyclization reaction might explain that only a few carboxylated BC's were detected in plants. However, the subsequently demonstrated inoperation of this pathway in *Eleagnus angustifolia* [160, 177] and the apparent scarcity of the naturally occurring 3,4-dihydro forms [145] suggest that cyclodehydration is not the only pathway for the biosynthesis of these alkaloids in plants.
The origin of the "non-indolyl" biosynthetic condensation adducts are not clear as well. Incorporation of [^{14}C]pyruvate and [^{14}C]acetate into simple BC's has been demonstrated in *Elaeagnus angustifolia* [160, 161] and *Peganum harmala* [165].

4.3 β-Carbolines in mammals

4.31 In-vitro studies

Incubation of 5-HT and acetaldehyde with rat brain homogenate produced compounds tentatively identified on TLC and color reactions as 6-OH-1-methyl-THBC (6-OH-THH; LXXIV B), the methoxylated compound, its N-oxide, and 6-methoxyharmalan ([178]; LXXII B). When the duration of incubation was extended to 6 hours or more, 6-OH-THH disappeared whereas some unidentified spots were intensified. In the absence of S-adenosylmethionine the N-oxide of 6-OH-THH was detected but not the methoxylated BC's. The BC's were formed enzymatically during the short time of incubation (2 hours) since the rate of spontaneous chemical reaction was slow (+37° C, pH 7, detectable amounts after 3 hours, maximum after approximately 24 hours, yield 78%; [179]).
Incubation of N,N-dimethyltryptamine (LXXVIII A) with rat brain homogenate resulted in the formation of 2-methyl-1,2,3,4-tetrahydro-β-carboline (2-Me THN; LXXVIII C) and THBC (THN; LXXIII D; scheme 25). The concentrations of the BC's reached a maximum after 30 minutes and decreased during longer incubation periods [180]. Addition of dimedone to the incubation mixture did decrease although not eliminate the formation of 2-Me THN (LXXVIII C) and THN (LXXVIII D), indicating that a portion of the BC's was formed in vitro by mechanisms others than those requiring the dissociation of HCHO.

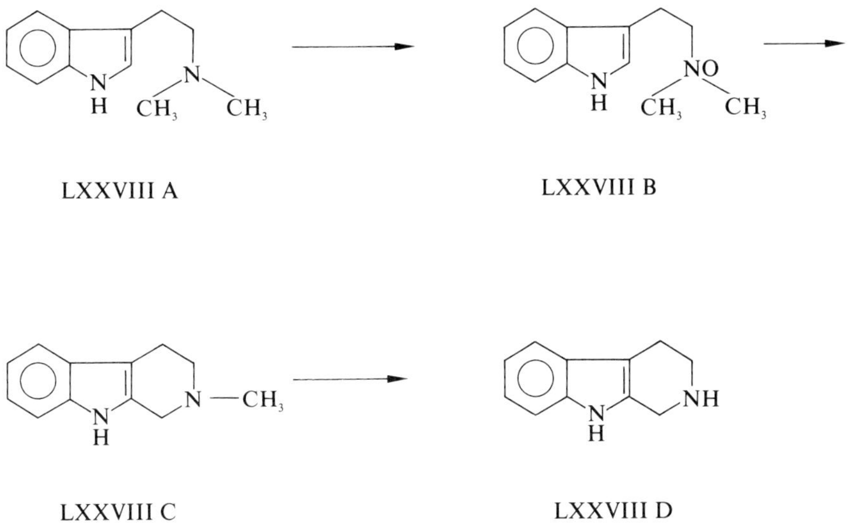

Scheme 25

It is noteworthy that the formation of 2-Me THN from DMT-NO (LXXVIII B) increased under anaerobic conditions [180].
Several groups demonstrated later that 5-methyl-tetrahydrofolate was a more suitable carbon-unit donor than 5-adenosylmethionine for the cyclization reaction whereas S-adenosylmethionine catalyzed preferentially N-methylation of indolealkylamines. These results questioned the findings of Saheb and Dajani [179] who used S-adenosylmethionine as cosubstrate. 5-MTHF was utilized as a cofactor of an enzymatic preparation of rat brain to methylate the nitrogen of phenylethylamines and indolamines as well [181, 182]. Careful investigation revealed that the reaction with indoleamines actually involved a carbon unit transfer from 5-MTHF to the tryptamine followed by a cyclization yielding THBC's. Enzyme activity was found in various tissues [183–186]. In human platelets from male volunteers a formation of THBC of 105 nmol/mg protein/hr, from females of 165 nmol/mg protein/hr was measured [183]. In the striatum of rats a rate of formation of 210 pmol THBC/mg protein was detected; in other brain regions the rate was much lower [186]. In-vitro evidence for enzymatic formation was confirmed by several authors [187–190]. Methylenetetrahydrofolate reductase (EC 1.1.1.68) was suggested to catalyze the reaction [191–193]. Involvement of free formaldehyde was demonstrated by inhibition of the reaction by semicarbazide [194]. Other conceivable cofactors beside 5-MTHF are discussed in the isoquinoline-section.

There are no reports about enzymatic formation of BC's in vivo with pyruvate as precursor. Theoretical considerations suggest 1-carboxy tetrahydroharman (1-CTHH; LXVIII) as the first product of the cyclization reaction (scheme 24). 1-CTHH (LXVIII) could serve as precursor BC to other BC's like harmalan (LXIX), harman (LXX A) and tetrahydroharman (LXXI A). To investigate this possibility, radiolabelled 1-CTHH (LXVIII) was incubated with liver or brain homogenate in phosphate buffer pH 7.4 and 37° C for 90 minutes. Harmalan (LXIX) was identified as the reaction product by TLC [195, 196].

4.32 In-vivo studies

The conception that BC's occur in vivo in mammals originates from the tentative identification of 6-methoxy-1-methyl-THBC (6-Me0-THH; LXXIV C) in bovine pineal gland by Farrell and Mc Isaak [198]. Mc Isaac demonstrated the formation of 6-methoxy-1-methyl-THBC from 5-methoxytryptamine and acetaldehyde [198]. Rats were pretreated with iproniazide and disulfiram to block metabolism of methoxytryptamine and acetaldehyde, respectively. Then, animals were treated with either ethanol or acetaldehyde. The urine was collected over a period of 24 hours. 0.5% of the injected radioactivity (5-β-[^{14}C]methoxytryptamine) was identified as 6-Me0-THH (LXXIV C) by TLC. These findings were the first suggesting in-vivo formation of BC's with acetaldehyde as precursor. These experiments, however, were not followed up by Mc Isaac and were neglected until the 1970's.
The physiologically occurring THBC increased in rat brain 90 minutes after a load with 150 mg/kg tryptamine and in tendency after a load with 150 mg/kg L-tryptophan [197]. These findings suggest a precursor function of tryptamine for the biogenesis of BC's.
The precursors of the non-indolic part of the 1-methyl-BC's might be either acetaldehyde or pyruvate. Furthermore, cyclization of N-acetyl-tryptamine derivatives to BC's might occur. Ethanol, the precursor of acetaldehyde induced an increase of harman (1-methyl BC) in the brain of rats. Subchronic treatment with ethanol caused a continous increase of the excretion of harman into the urine which was reversed following withdrawal [199].
Pyruvic acid, a key substance of the energy metabolism, was injected intraventricularly 2 minutes after tritium labelled tryptamine. 1-carboxy-tetrahydroharman (1,2,3,4-tetrahydro-1-methyl-β-carboline-1-

carboxylic acid, 1-CTHH; LXVIII) was formed in a time- and dose-dependent manner. Pretreatment with high doses (60 and 75 mg/kg) of pargyline prevented the formation of the BC suggesting an enzymatic reaction [128, 195]. The in-vivo occurrence of an analogous substance, was presented by Collins et al. who detected 1-CTHH hydroxylated in position 6 in CSF of monkeys as well as in the striatum and hypothalamus of postmortem human brain [200, 201]. After intraventricular administration of [^{3}H] 1-CTHH (LXVIII) into rats, we observed 1 hour later that equal concentrations of THH (LXXI A) and harmalan (LXIX) appeared in the brain whereas harman (LXX A) was not detectable. All three oxidation states were found in the liver with harmalan as the major metabolite. In the urine, traces of harmalan and harman were detected but not tetrahydrohaman [196]. These findings suggest that after decarboxylation reductive reactions predominate in the brain whereas in the liver both oxidation as well as reduction reactions occur.

Another pathway of the formation of BC's was suggested by Kveder and Mc Isaac [202]. They injected radiolabelled melatonin into rats and speculated that one of the metabolites in urine might be 6-methoxyharmalan (LXXII B). The authors reported that "chromatography and isotope dilution studies failed to produce unequivocal evidence for its formation". The formation of the dihydro-β-carboline could not be replicated by others or by the authors themselves later and it was thought to be some other β-carboline with a related structure [5]. We performed similar experiments using radiolabelled N-acetyltryptamine [196]. Several metabolites were observed in the urine, tentatively identified as conjugates of N-acetyltryptamine. No BC's were detected. Thus, despite feeding experiments in plants support the formation of BC's via N-acetylation of tryptamine with further cyclodehydration (Bischler-Napieralski reaction) no evidence exists for this mechanism in mammals.

To summarize the present status of the knowledge of the biosynthesis of BC's there ist good evidence for a cyclization reaction in vivo although a specific enzyme has not been described yet. The precursors are both aldehyde as well as pyruvate whereas N-acetylated indoleamines seem to be less probable.

References

1 P. Peura, P. Mackenzie, U. Koivusaari and M. Lang: Molec. Pharmac. *22,* 721 (1982).
2 C. Melchior and M. A. Collins: CRC Crit. Rev. Toxicol. *9,* 313 (1982).
3 N. S. Buchkholz: Life Sci. *27,* 893 (1980).
4 R. Deitrich and V. Erwin: Ann. Rev. Pharmac. Toxic. *20,* 55 (1980).
5 M. M. Airaksinen and I. Kari: Med. biol. *59,* 121 (1981), and *59,* 190 (1981).
6 M. E. Zaranz de Ysern and L. A. Ordoñez: Prog. Neuro-Psychopharmacol. *5,* 343 (1981).
7 E. Muscholl: Trends in autonomic pharmacology, vol. 2. Urban und Schwarzenberg, Baltimore, München 1981.
8 H. Rommelspacher: Pharmacopsychiatry *14,* 117 (1981).
9 R. D. Myers: Alc. Clin. exp. Res. *2,* 145 (1978).
10 M. A.Collins, in: Biological effects of alcohol, p. 87. Ed. H. Begleiter. Plenum Press, New York 1980.
11 R. Rahwan: Toxic. appl. Pharmac. *34,* 3 (1975).
12 C. Melchior: Alc. Clin. exp. Res. *3,* 364 (1975).
13 M. S. Hamilton and M. Hirst: Sub. Alc. Actions Misuse *1,* 121 (1980).
14 B. Sjöquist: The Sec. Malmö Sympos. on alcohol, in press.
15 W. M. Whaley and T. R. Govindachari: Organic reactions, vol. VI, p. 151. John Wiley, New York 1951.
16 R. Grigg, H. Q. N. Gunaratne and E. McNaghten: J. Chem. Soc. Perkin Trans. *I,* 185 (1983).
17 D. Soerens, J. Sandrin, F. Ungemach, P. Mokry, S. G. Wu, E. Yamanaka, L. Hutchins, M. Di Pierro and J. M. Cook: J. Org. Chem. *44,* 535 (1979).
18 F. Ungemach and J. M. Cook: Heterocycles *9,* 1089 (1978).
19 A. H. Jackson and A. E. Smith: Tetrahedron *24,* 403 (1968).
20 A. H. Jackson and P.Smith: Tetrahedron *24,* 2227 (1968).
21 A. H. Jackson, B. Naidoo and P. Smith: Tetrahedron *24,* 6119 (1968).
22 K. M. Biswas and A. H. Jackson: Tetrahedron *25,* 227 (1969).
23 A. Cipiciani, S. Clementi, P. Linda, G. Marino and G. Savelli: J. Chem. Soc. Perkin *II,* 1284 (1977).
24 R. B. Woodward: Nature *162,* 155 (1948).
25 J. R. Williams and L. R. Unger: J. Chem. Soc. Chem. Commun., 1605 (1970).
26 A. Bischler and B. Napieralski: Chem. Ber. *26,* 1903 (1893).
27 E. Späth and E. Lederer: Chem. Ber. *63,* 120 (1930).
28 G. Hahn and H. Ludewig: Chem. Ber. *67,* 2031 (1934).
29 W. M. Whaley and T. R. Govindachari: Organic Reactions, vol. VI, p. 74, John Wiley and Sons 1951.
30 H. Shirai, T. Yashiro and T. Kuwayama: J. Pharm. Soc. Jap. *93,* 1371 (1973).
31 E. Späth and E. Kruta: Monatsh. Chem. *50,* 341 (1928).
32 T. Kametani, A. Ujie, M. Ihara, K. Fukumoto and S. T. Lu: J. Chem. Soc. Perkin Trans. *I,* 1218 (1976).
33 C. Schöpf: Angew. Chem. *50,* 797 (1937).
34 O. Kovacs and G. Fodor: Chem. Ber. *84,* 795 (1951).
35 J. B. Hester, Jr.: J. Org. Chem. *29,* 2864 (1964).
36 E. Winterstein and G. Trier: Die Alkaloide, p. 307. Borntraeger, Berlin 1910.
37 C. Schöpf and H. Bayerle: Ann. Chem. *534,* 297 (1938).
38 G. Hahn and K. Stiel: Chem. Ber. *69,* 2627 (1936).
39 C. Schöpf and W. Salzer: Ann. Chem. *544,* 1 (1940).
40 J. H. Robbins: Clin. Res. *16,* 350 (1968).
41 R. J. Shah, D. D. Vaghani and J. R. Merchant: J. Org. Chem. *26,* 3533 (1961).

42 S. Rachlin, K. Worning and J. Enemark: Tetrahedron Lett. *39,* 4163 (1968).
43 G. Hahn and F. Rumpf: Chem. Ber. *71,* 2141 (1938).
44 P. Holtz and E. Westermann: Naunyn-Schmiedeberg's Arch. exp. Path. Pharmakol. *231,* 311 (1957).
45 H. F. Schott and W. G. Clark: J. Biol. Chem. *196,* 449 (1952).
46 G. Cohen and M. A. Collins: Science *167,* 1749 (1970).
47 W. Oswald, J. Polonia and M. A. Polonia: Naunyn-Schmiedeberg's Arch. Pharmac. *289,* 275 (1975).
48 J. H.Robbins: Clin. Res. *16,* 554 (1968).
49 V. E. Davis and J. L. Cashaw: Prog. Clin. Biol. Res. *90,* 99 (1982).
50 T. M. Kenyhercz and P. T. Kissinger: J. Pharm. Sci. *67,* 112 (1978).
51 J. P. Fourneau, C. Gaingnault, R. Jaquier, O. Stoven and M. Davy: Chim. Ther. *6,* 67 (1969).
52 G. A. Cordell: Introduction to alkaloids, p. 317. John Wiley, New York 1981.
53 A. Orechow and N. Proskurmina: Ber. dt. Ges. Chem. *66,* 841 (1933).
54 S. Teitel, J. O'Brian, W. Pood and A. Brossi: J. Med. Chem. *17,* 134 (1974).
55 R. M. Riggin, M. J. McCarthy and P. T. Kissinger: J. Agric. Food Chem. *24,* 189 (1976).
56 R. Mata and J. L. McLaughlin: Phytochemistry, p. 673 (1980).
57 H. Watanabe, M. Ikeda, K. Watanabe and T. Kikuchi: Planta med. *42,* 213 (1981).
58 M. Schamma and J. L. Moniot: Isoquinoline alkaloids Research 1972–1977. Plenum Press, New York 1978.
59 B. Holmstedt: Prog. Clin. Biol. Res. *90,* 3 (1982).
60 H. M. Schumacher, M. Rüffer, N. Nagakura and M. H. Zenk: Planta med. *48,* 212 (1983).
61 J. D. Phillipson, A. Scutt, A. Baytop, N. Özhatay and G. Sariyar: Planta med. *43,* 261 (1981).
62 E. Brockmann-Hanssen, C.-H. Chen, H.-C. Chiang, A. Leung and K. McMurtrey: J. Chem. Soc. Perkin *I,* 1531 (1975).
63 M. Leboeuf, D. Cortes, R. Hocquemiller and A. Cave: Planta med. *48,* 234 (1983).
64 J. D. Spenser: Comprehensive Biochemistry, vol. 20, p. 231. Elsevier, New York (1968).
65 H. L. Holland, P. W. Jeffs, T. M Capps and D. B. McLean: Can. J. Chem. *57,* 1588 (1979).
66 D. H. R. Barton, R. H. Hesse and G. W. Kirby: J. Chem. Soc., p. 6379 (1965).
67 A. R. Battersby, R. C. F. Jones, R. Kazlauskas, C. Poupat, C. W. Thornber, S. Ruchirawat and J. Staunton: J. Chem. Soc. Chem. Commun., p. 773 (1974).
68 A. R. Battersby, J. L. McHugh, J. Staunton and M. Todd: J. Chem. Soc. Chem. Commun., p. 985 (1971).
69 A. R. Battersby, R. C. F. Jones and R. Kazlauskas: Tetrahedron Lett. 1873 (1975).
70 S. Tewari, D. S. Bhakuni and R. S. Kapil: J. Chem. Soc. Chem. Commun., p. 554 (1971).
71 D. S. Bhakuni, A. N. Singh, S. Tewari and R. S. Kapil: J. Chem. Soc. Perkin *I,* 1622 (1977).
72 R. Robinson: J. Chem. Soc., p. 876 (1917).
73 A. R. Battersby and R. Binks: Proc. Chem. Soc., p. 360 (1960).
74 A. R. Battersby, R. Binks, R. J. Francis, D. J. McCaldin and H. Ramuz: J. Chem. Soc., p. 3600 (1964).
75 M. Wink and R. Hartmann: FEBS Lett. *101,* 343 (1979).
76 A. J. Scott, S. L. Lee and T. Hirata: Heterocycles *77,* 159 (1978).

77 G. J. Kapadia, G. S. Rao. E. Leete, M. B. E. Fayez, Y. N. Vaishav and H. M. Fales: J. Am. Chem. Soc., p. 6943 (1970).
78 M. L. Wilson and C. J. Coscia: J. Am. Chem. Soc., p. 431 (1975).
79 M. Rueffer, H. El-Shagi, N. Nakamura and M .H. Zenk: FEBS Lett. *129,* 5 (1981).
80 S. R. Johns, J. A. Lamberton and A. A. Sioumis: Aust. J. Chem. *20,* 1729 (1967).
81 A. Brossi: Prog. Clin. Biol. Res. *90,* 125 (1982).
82 V. E. Davis, M. J. Walsh and Y. Yamanaka: J. Pharmac. exp. Ther. *174,* 401 (1970).
83 M. J. Walsh, V. E. Davis and Y. Yamanaka: J. Pharmac. exp. Ther. *174,* 388 (1970).
84 J. L. Cashaw, K. D. McMurtrey, H. Brown and V. E. Davis: J. Chromatogr. *99,* 567 (1974).
85 A. C. Collins, J. L. Cashaw and V. E. Davis: Biochem. Pharmac. *22,* 2337 (1973).
86 P. P. Laidlaw: J. Physiol., Lond. *40,* 480 (1910).
87 P. Holtz, K. Stock and E. Westermann: Naunyn-Schmiedeberg's Arch. exp. Pathol. Pharmak. *246,* 133 (1963).
88 P. Holtz, K. Stock and E. Westermann: Nature *203,* 656 (1964).
89 P. V. Haluska and P. C. Hoffman: Science *169,* 1104 (1970).
90 A. J. Turner, K. M. Baker, S. Algeri, A. frigerio and S. Garattini: Life Sci. *14,* 2247 (1974).
91 A. R. Battersby, D. M. Foulkers, M. Hirst, G. V. Parry and J. Staunton: J. Chem. Soc., p. 210 (1968).
92 A. R. Battersby, M. Hirst, D. J. McCaldin, R. Southgate and J. Staunton: J. Chem. Soc., p. 2163 (1968).
93 D. H. R. Barton, G. W. Kirby, W. Steglich, G. M. Thomas, A. R. Battersby, T. A. Dobson and H. Ramuz: J. Chem. Soc., p. 2423 (1965).
94 T. Kametani, M. Takemura, M. Ihara, K. Takahashi and K. Fukumoto: J. Am. Chem. Soc. *98,* 1956 (1976).
95 T. Kametani, Y. Ohta, M. Takemura, M. Ihara and K. Fukumoto: Bioorgan. Chem. *6,* 249 (1977).
96 C. J. Coscia, W. Burke, G. Jamroz, J. M. Lasala, J. McFarlaine, J. Mitchell, M. M. O'Toole and M. L. Wilson: Nature *269,* 617 (1977).
97 G. Cohen and M. A. Collins: Science *167,* 1749 (1970).
98 R. S. Greenberg and G. Cohen: J. Pharmac. exp. Ther. *184,* 119 (1973).
99 Y. Yamanaka, M. J. Walsh and V. E. Davis: Nature *227,* 1143 (1970).
100 M. J. Walsh: Ann. N. Y. Acad. Sci. *215,* 98 (1973).
101 A. C. Collins, J. L. Cashaw and V. E. Davis: Biochem. Pharmak. *22,* 2337 (1973).
102 C. L. Melchior and R. A. Deitrich, in: Biological Effects of alcohol, p. 121. H. Begleiter, Plenum Press, New York 1980.
103 C. L. Melchior, A. Mueller and R. A. Deitrich: Biochem. Pharmac. *29,* 657 (1980).
104 M. Bigdeli and M. A. Collins: Trans. Am. Soc. Neurochem. *4,* 102 (1973).
105 M. Sandler, S. B. Carter, K. R. Hunter and G. M. Stern: Nature, Lond. *241,* 439 (1973).
106 I. J. Farlane and M. Slaytor: Phytochemistry *11,* 229 (1972).
107 S. M. Shanbhag, H. J. Kulkarni and D. B. Gaitonde: Jap. J. Pharmacol. *20,* 482 (1970).
108 Roussel-UCLAF patents C. A. *69* (1968) 59475 m; (1969) 70785 z, 70787 b, 91735 y, 72 (1970) 32103; 73 (1970) 15063 d.
109 C. A. Nesterick and R. G. Rahwan: J. Chromatogr., biomed. Appl. *164,* 205 (1979).
110 G. A. Smythe, M. W. Duncan and J. E. Bradshaw: IRCS Med. Sci. (Biochem.) *9,* 472 (1981).

111 B. Sjöquist and E. Magnuson: J. Chromatogr., biomed. Appl. *183,* 17 (1980).
112 M. A. Collins and T. C. Origitano: J. Neurochem. *41,* 1569 (1983).
113 B. Sjöquist, in: Aldehyde adducts in alcoholism. Ed. M. A. Collins. A. Liss, New York, 115 (1985).
114 B. Sjöquist, E. Perdahl and B. Winblad: Drug Alcohol Dep. *12,* 15 (1983).
115 G. Dordain, P. Dostert, M. Strolin Benedetti and V. Rovei, in: Monoamine oxidase and disease, p. 417. Academic Press, London 1984.
116 J. M. Lasala and C. J. Coscia: Science *203,* 283 (1979).
117 E. Späth and E. Lederer: Chem. Ber. *63,* 2102 (1930).
118 W. M. McIsaak: Biochem. biophys. Acta *52,* 607 (1961).
119 T. R. Bosin, B. Holmstedt, A. Lundman and O. Beck: Analyt. Biochem. *128,* 287 (1983).
120 G. Tasni: Chem. Zentralbl. *II,* 668 (1928).
121 S. Akabori: Chem. Ber. *63,* 2245 (1930).
122 G. Hahn, L. Bärwald, O. Schales and H. Werner: Ann. Chem. *520,* 107 (1935).
123 G. Hahn and H. Werner: Ann. Chem. *520,* 123 (1935).
124 G. Hahn and A. Hausel: Chem. Ber. *71,* 2163 (1938).
125 I. D. Spenser: Can. J. Chem. *37,* 1851 (1959).
126 H. Obata-Sasamoto, A. Komamine and K. Saito: Z. Naturforsch. *36,* 921 (1981).
127 A. R. Battersby, R. Binks and R. Huxtable: Tetrahedron Lett., p. 563 (1967).
128 R. Susilo and H. Rommelspacher, in: Aldehyde adducts in alcoholism. Ed. M. A. Collins, A. Liss, New York, 137 (1985).
129 G. J. Kapadia, G. S. Rao, E. Leete, M. B. E. Fayez, Y. N. Vaishav and H. M. Fales: J. Am. Chem. Soc. *92,* 6943 (1976).
130 J. M. Bobbitt and T. Y. Cheng: J. Org. Chem. *41,* 443 (1976).
131 J. M. Bobbitt, C. L. Kulkarni and P. Wiriyachitra: Heterocycles *4,* 1645 (1976).
132 I. G. C. Coutts, M. R. Hamblin and E. J. Tinley: J. Chem. Soc. Perkin *I,* 2744 (1979).
133 R. Malkin and B. G. Malmström: Adv. Enzymol. *33,* 177 (1970).
134 R. B. Herbert and J. Mann: J. Chem. Soc. Perkin *I,* 1523 (1982).
135 E. Koszuk, D. Kierska and C. Maslinski: Agents and Actions *8,* 185 (1978).
136 P. Steffens, N. Nagakura and M. H. Zenk: Tetrahedron Lett. *25,* 951 (1984).
137 E. Rink and H. Böhm: FEBS Lett. *49,* 396 (1975).
138 C. Chambers and K. L. Stuart: Chem. Comm., p. 328 (1968).
139 T. Kametani, K. Fukumoto, A. Kozuka and M. Koizumi: Chem. Soc. (C), p. 2034 (1969).
140 Stekol'nikov: Priroda, Moscow *5,* 107 (1970).
141 E. Späth and K. Tharrer: Chem. Ber. *66,* 904 (1933).
142 E. Schlittler: Chem. Ber. *66,* 988 (1933).
143 R. Tschesche, P. Welzel and G. Legler: Tetrahedron Lett., p. 445 (1965).
144 R. A. Uphaus, L. I. Grossweiner, J. J. Katz and K. D. Kopple: Science *129,* 641 (1959).
145 J. R. F. Allen and B. R. Holmstedt: Phytochemistry *19,* 1513 (1980).
146 T. R. Bosin, A. Lundman and O. Beck: J. Agric. Food Chem. *31,* 444 (1983).
147 O. Beck, T. R. Bosin and A. Lundman: J. Agric. Food Chem. *31,* 288 (1983).
148 O. Beck and B. Holmstedt: Food Cosmet. Toxicol. *19,* 173 (1981).
149 T. R. Bosin and C. A. Jarvis: J. Chromatogr., biomed. Appl., in press.
150 E. H. Poindexter and R. D. Carpenter: Phytochemistry *1,* 215 (1962).
151 C. Izard, J. Lacharpagne and P. Testa: C. R. Acad. Sci., Ser. D *262,* 1859 (1966).
152 J. Cuzin: Abh. dt. Akad. Wiss. Berlin, Kl. Chem., Geol., Biol. *3,* 171 (1966).
153 H.-J. Klimisch and A. Beiss: J. Chromatogr. *128,* 117 (1977).

154 J. N. Schumacher, C. R. Green, F. W. Bestand and M. P. Newell: J. Agric. Food Chem. *25,* 310 (1977).
155 J. J. Kettenes-van den Bosch and C. A. Salemink: J. Chromatogr. *131,* 422 (1977).
156 A. Ogata, K. Tagaki, A. Mizestani and S. Iijima: Yakugaku Zasski *66,* 44 (1946).
157 A. Proliac and M. Blank: Helv. Chim. Acta *59,* 2503 (1976).
158 K. Tsuhi, H. Zenda and T. Kosuge: Yakugaku Zasski *93,* 33 (1973).
159 J. L. Cashaw, S. Ruchirawat, Y. Nimit and V. E. Davis: Biochem. Pharmacol. *32,* 3163 (1983).
160 D. G. Donovan, L. Buckley and P. Geary: Proc. R. Ir. Acad., Sect. B *76,* 187 (1976).
161 D. G. O'Donovan and M. F. Kenneally: J. Chem. Soc. (C), p. 1109 (1967).
162 M. Slaytor and I. J. McFarlane: Phytochemistry *7,* 605 (1968).
163 L. Kompis, E. Grossmann, I. V. Terent'eva and G. V. Lazur'evskii: C. A. *74,* 136527 (1971).
164 D. Gröger and H. Simon: Abh. dt. Akad. Wiss. Berlin, Kl. Chem., Geol., Biol. *4,* 343 (1963).
165 K. Stolle and D. Gröger: Arch. Pharm. *301,* 561 (1968).
166 D. R. Liljegren: Phytochemistry *7,* 1299 (1968).
167 L. Nettleship and M. Slaytor: Phytochemistry *13,* 735 (1974).
168 I. Kompis, M. Hesse and H. Schmid: Lloydia *34,* 269 (1971).
169 I. A. Veliky: Phytochemistry *11,* 1405 (1972).
170 I. A. Veliky and K. M. Barber: Lloydia *38,* 125 (1975).
171 L. D. Antonaccio and H. Budzikiewicz: Monatsh. Chem. *93,* 962 (1962).
172 L. Sanchez, E. Wolfgango and K. S. Brown: Ann. Acad. Bras. Cienc. *43,* 603 (1971).
173 S. Takase and H. Murakami: Agric. Biol. Chem. Tokyo *30,* 869 (1966).
174 S. McLean and D. G. Murray: Can. J. Chem. *50,* 1478 (1972).
175 F. Faini, M. Castillo and R. Torres: Psychochemistry *17,* 338 (1978).
176 Umezawa, T. Aoyagi, M. Hamada, T. Takenaka and S. Kumagai: C. A. *83,* 41539 (1975).
177 I. J. McFarlane and M. Slaytor: Phytochemistry *11,* 229 (1972).
178 R. M. Dajani and S. E. Saheb: Ann. N. Y. Acad. Sci. *215,* 120 (1973).
179 S. E. Saheb and R. M. Dajani: Comp. Gen. Pharmacol. *4,* 225 (1973).
180 S. A. Barker, J. A. Monti and S. T. Christian: Biochem. Pharmac. *29,* 1049 (1980).
181 P. Laduron: Nature *238,* 212 (1972).
182 L. L. Hsu and A. J. Mandell: Life Sci. *13,* 847 (1973).
183 J. D. Barchas, G. R. Elliott, J. DoAmaral, E. Erdelyi, S. O'Connor, H. Bowden, H. K. H. Brodie, P. A. Berger, J. R. Renson and R. J. Wyatt: Arch. Gen. Psychiat. *31,* 862 (1974).
184 R. J. Wyatt, E. Erdelyi, J. R. DoAmaral, G. R. Elliott, J. Renson and J. D. Barchas: Science *187,* 853 (1975).
185 L. R. Mandel, A. Rosegay, R. W. Walker, W. J. A. Van den Heuvel and J. Rockach: Science *186,* 741 (1974).
186 L. L. Hsu and A. J. Mandell: J. Neurochem. *24,* 631 (1975).
187 E. Meller, H. Rosengarten, A. J. Friedhoff, R. D. Stebbins and R. Silber: Science *184,* 171 (1975).
188 J. Leysen and P. Laduron: FEBS Lett. *47,* 299 (1974).
189 H. Rosengarten, E. Meller and A. J. Friedhoff: Biochem. Pharmac. *24,* 1759 (1975).
190 W. Lauwers, J. Leysen, H. Verhoeven, P. Laduron and M. Claeys: Biomed. Mass Spectrom. *2,* 15 (1975).
191 R. T. Taylor and M. L. Hanna: Life Sci. *17,* 111 (1975).

192 L. A. Ordonez and F. Caraballo: Psychopharmacol. Commun. *1,* 253 (1975).
193 A. G. M. Pearson and A. J. Turner: Nature *258,* 173 (1975).
194 H. Rommelspacher, H. Coper and S. Strauss: Life Sci. *18,* 81 (1976).
195 R. Susilo and H. Rommelspacher: Biochem. Pharmacol., submitted.
196 R. Susilo and H. Rommelspacher, in: Second trace amines smyposium. Eds. P. R. Bieck, A. A. Boulton, L. Maitre and P. Riederer. The Humana Press, Clifton, USA, in press.
197 H. Honecker and H. Rommelspacher: Naunyn-Schmiedeberg's Arch. Pharmac. *305,* 135 (1978).
198 G. Farrell and W. M. McIsaak: Arch. Biochem. Biohys. *94,* 543 (1961).
199 H. Rommelspacher, H. Damm, S. Strauss and G. Schmidt: Naunyn-Schmiedeberg's Arch. Pharmac. *327,* 107 (1984).
200 M. A. Collins, K. Dahl, W. Nijm and L. F. Major: Soc. Neurosci. USA *8,* 277 (1982).
201 N. Ung-Chhun, B. Y. Cheng, D. A. Pronger, P. Serrano, B. Chavez, R. F. Perez, J. Morales and M. A. Collins, in: Acetaldehyde adducts in alcoholism. Ed. M. A. Collins, A. Liss, New York, 125 (1985).
202 S. Kveder and W. M. Mc Isaak: J. Biol. Chem. *263,* 3214 (1961).

Index Vol. 29

The references of the Subject Index are given in the language of the respective contribution.
Die Stichworte des Sachregisters sind in der jeweiligen Sprache der einzelnen Beiträge aufgeführt.
Les termes repris dan la Table des matières sont donnés selón la langue dans laquelle l'ouvrage est écrit.

Index of Titles
Verzeichnis der Titel
Index des titres
Vol. 1–29 (1959–1985)

Author and Paper Index
Autoren- und Artikelindex
Index des auteurs et des articles
Vol. 1–29 (1959–1984)

Petrussis agglutinins and complement fixing antibodies in whooping cough *19*, 178 (1975)	Dr. K. C. Agarwal Dr. M. Ray Dr. N. L. Chitkara Department of Microbiology, Postgraduate Institute of Medical Education and Research, Chandigarh, India
Pharmacology of clinically useful beta-adrenergic blocking drugs *15*, 103 (1971)	Prof. Dr. R. P. Ahlquist Professor of Pharmacology, School of Medicine, Medical College of Georgia, Augusta, Georgia, USA Dr. A. M. Karow, Jr. Assistant Professor of Pharmacology, School of Medicine, Medical College of Georgia, Augusta, Georgia, USA Dr. M. W. Riley Assistant Professor of Pharmacology, School of Medicine, Medical College of Georgia, Augusta, Georgia, USA
Adrenergic beta blocking agents *20*, 27 (1976)	Prof. Dr. R. P. Ahlquist Professor of Pharmacology, Medical College of Georgia, Augusta, Georgia, USA
Trial of a new anthelmintic (bitoscanate) in ankylostomiasis in children *19*, 2 (1975)	Dr. S. H. Ahmed Dr. S. Vaishnava Department of Paediatrics, Safdarjung Hospital, New Delhi, India
Development of antibacterial agents of the nalidixic acid type *21*, 9 (1977)	Dr. R. Albrecht Department of Drug Research, Schering AG, Berlin
Biological activity in the quinazolone series *14*, 218 (1970)	Dr. A. H. Amin Director of Research, Alembic Chemical Works Co. Ltd., Alembic Road, Baroda 3, India Dr. D. R. Mehta Dr. S. S. Samarth Research Division, Alembic Chemical Works Co. Ltd., Alembic Road, Baroda 3, India

Enhancement and inhibition of microsomal drug metabolism *17,* 11 (1973)	Prof. Dr. M. W. Anders Department of Pharmacology, University of Minnesota, Minneapolis, Minnesota, USA
Reactivity of rat and man to egg-white *13,* 340 (1969)	Dr. S. I. Ankier Allen & Hanburys Ltd., Research Division, Ware, Hertfordshire, England
Narcotic antagonists *8,* 261 (1965)	Dr. S. Archer Assistant Director of Chemical Research, Sterling-Winthrop Research Institute, Rensselaer, New York, USA Dr. L. S. Harris Section Head in Pharmacology, Sterling-Winthrop Research Institute, Rensselaer, New York, USA
Recent developments in the chemotherapy of schistosomiasis *16,* 11 (1972)	Dr. S. Archer Associate Director of Research, Sterling-Winthrop Research Institute, Rensselaer, New York, USA Dr. A. Yarinsky Sterling-Winthrop Research Institute, Rensselaer, New York, USA
Recent progress in the chemotherapy of schistosomiasis *18,* 15 (1974)	Prof. Dr. S. Archer Professor of Medicinal Chemistry, School of Science, Department of Chemistry, Rensselaer Polytechnic Institute, Troy, N. Y. 12181, USA
Recent progress in research on narcotic antagonists *20,* 45 (1976)	Prof. Dr. S. Archer Professor of Medicinal Chemistry, School of Science, Department of Chemistry, Rensselaer Polytechnic Institute, Troy, New York, USA Dr. W. F. Michne Sterling-Winthrop Research Institute, Rensselaer, New York, USA
Molecular geometry and mechanism of action of chemical carcinogens *4,* 407 (1962)	Prof. Dr. J. C. Arcos Department of Medicine and Biochemistry, Tulane University, U. S. Public Health Service, New Orleans, Louisiana, USA
Molecular pharmacology, a basis for drug design *10,* 429 (1966) Reduction of drug action by drug combination *14,* 11 (1970)	Prof. Dr. E. J. Ariëns Institute of Pharmacology, University of Nijmegen, Nijmegen, The Netherlands
Stereoselectivity and affinity in molecular pharmacology *20,* 101 (1976)	Prof. Dr. E. J. Ariëns Dr. J. F. Rodrigues de Miranda Pharmacological Institute, University of Nijmegen, Nijmegen, The Netherlands Prof. Dr. P. A. Lehmann F. Departamento de Farmacologia y Toxicologia, Centro de Investigación y Estudios Avanzados, Instituto Politécnico Nacional, México D. F., México

Drugs affecting the renin-angiotensin system *26,* 207 (1982)	Dr. R. W. Ashworth Pharmaceuticals Division, Ciba-Geigy Corporation, Summit, New Jersey, USA
Tetanus neonatorum *19,* 189 (1975) Tetanus in children *19,* 209 (1975)	Dr. V. B. Athavale Dr. P. N. Pai Dr. A. Fernandez Dr. P. N. Patnekar Dr. Y. S. Acharya Department of Pediatrics, L. T. M. G. Hospital, Sion, Bombay 22, India
Toxicity of propellants *18,* 365 (1974)	Prof. Dr. D. M. Aviado Professor of Pharmacology, Department of Pharmacology, School of Medicine, University of Pennsylvania, Philadelphia, USA
Neuere Aspekte der chemischen Anthelminticaforschung *1,* 243 (1959)	Dr. J. Bally Wissenschaftlicher Mitarbeiter der Sandoz AG, Basel, Schweiz
Problems in preparation, testing and use of diphtheria, pertussis and tetanus vaccines *19,* 229 (1975)	Dr. D. D. Banker Chief Bacteriologist, Glaxo Laboratories (India) Ltd., Bombay 25, India
Recent advances in electrophysiology of antiarrhythmic drugs *17,* 33 (1973)	Prof. Dr. A. L. Bassett and Dr. A. L. Wit College of Physicians and Surgeons of Columbia University, Department of Pharmacology, New York, N. Y., USA
Stereochemical factors in biological activity *1,* 455 (1959)	Prof. Dr. A. H. Beckett Head of School of Pharmacy, Chelsea College of Science and Technology, Chelsea, London, England
Industrial research in the quest for new medicines *20,* 143 (1976) The experimental biologist and the medical scientist in the pharmaceutical industry *24,* 83 (1980)	Dr. B. Berde Head of Pharmaceutical Research and Development, Sandoz Ltd., Basle, Switzerland
Newer diuretics *2,* 9 (1960)	Dr. K. H. Beyer, Jr. Vice-President, Merck Sharp and Dohme Research Laboratories, West Point, Pennsylvania, USA Dr. J. E. Bear Director of Pharmacological Chemistry, Merck Institute für Therapeutic Research, West Point, Pennsylvania, USA

Recent developments in 8-amino-quinoline antimalarials *28,* 197 (1984)	Dr. A. P. Bhaduri, Scientist B. K. Bhat, M. Seth, Central Drug Research Institute, Lucknow, 226001 India
Studies on diphtheria in Bombay *19,* 241 (1975)	M. Bhaindarkar Y. S. Nimbkar Haffkine Institute, Parel, Bombay 12, India
Bitoscanate in children with hookworm disease *19,* 6 (1975)	Dr. B. Bhandari Dr. L. N. Shrimali Department of Child Health, R. N. T. Medical College, Udaipur, India
Recent studies on genetic recombination in *Vibrio cholerae* *19,* 460 (1975)	Dr. K. Bhaskaran Central Drug Research Institute, Lucknow, India
Interbiotype conversion of cholera vibrios by action of mutagens *19,* 466 (1975)	Dr. P. Bhattacharya Dr. S. Ray WHO International Vibrio Reference Centre, Cholera Research Centre, Calcutta 25, India
Experience with bitoscanate in hookworm disease and trichuriasis in Mexico *19,* 23 (1975)	Prof. Dr. F. Biagi Departamento de Parasitología, Facultad de Medicina, Universidad Nacional Autónoma de Mexico, Mexico
Analysis of symptoms and signs related with intestinal parasitosis in 5,215 cases *19,* 10 (1975)	Prof. Dr. F. Biagi Dr. R. López Dr. J. Viso Departamento de Parasitología, Facultad de Medicina, Universidad Nacional Autónoma de Mexico, Mexico
Untersuchungen zur Biochemie und Pharmakologie der Thymoleptika *11,* 121 (1968)	Dr. M. H. Bickel Privatdozent, Medizinisch-Chemisches Institut der Universität Bern, Schweiz
The role of adipose tissue in the distribution and storage of drugs *28,* 273 (1984)	Prof. Dr. M. H. Bickel Universität Bern, Pharmakologisches Institut, 3008 Bern, Schweiz
The β-adrenergic blocking agents, pharmacology, and structure-activity relationships *10,* 46 (1966)	Dr. J. H. Biel Vice-President, Research and Development, Aldrich Chemical Company Inc., Milwaukee, Wisconsin, USA Dr. B. K. B. Lum Department of Pharmacology, Marquette University School of Medicine, Milwaukee, Wisconsin, USA
Prostaglandins *17,* 410 (1973)	Dr. J. S. Bindra and Dr. R. Bindra Medical Research Laboratories, Pfizer Inc., Groton, Connecticut, USA

The red blood cell membrane as a model for targets of drug action *17*, 59 (1973)	Prof. Dr. L. Bolis Università degli Studi di Roma, Istituto di Fisiologia Generale, Roma, Italia
Epidemiology and public health. Importance of intestinal nematode infections in Latin America *19*, 28 (1975)	Prof. Dr. D. Botero R. School of Medicine, University of Antioquia, Medellin, Colombia
Clinical importance of cardiovascular drug interactions *25*, 133 (1981)	Dr. D. C. Brater Division of Clinical Pharmacology, Departments of Pharmacology and Internal Medicine, The University of Texas, Health Science Center at Dallas, 5323 Harry Hines Boulevard, Dallas, Texas, USA
Update of cardiovascular drug interactions *29*, 9 (1985)	D. Craig Brater, M. D. Michael R. Vasko, Ph. D. Departments of Pharmacology and Internal Medicine, The University of Texas Health Science Center at Dallas and Veterans Administration Medical Center, 4500 Lancaster Road, Dallas, TX 75216
Some practical problems of the epidemiology of leprosy in the indian context *18*, 25 (1974)	Dr. S. G. Browne Director, Leprosy Study Centre, 57a Wimpole Street, London, England
Die Ionenaustauscher und ihre Anwendung in der Pharmazie und Medizin *1*, 11 (1959) Wert und Bewertung der Arzneimittel *10*, 90 (1966)	Prof. Dr. J. Büchi Direktor des Pharmazeutischen Institutes der ETH, Zürich, Schweiz
Cyclopropane compounds of biological interest *15*, 227 (1971) The state of medicinal science *20*, 9 (1976)	Prof. Dr. A. Burger Professor Emeritus, University of Virginia, Charlottesville, Virginia, USA
Human and veterinary anthelmintics (1965–1971) *17*, 108 (1973)	Dr. R. B. Burrows Mount Holly, New Jersey, USA
The antibody basis of local immunity to experimental cholera infection in the rabbit ileal loop *19*, 471 (1975)	Dr. W. Burrows Dr. J. Kaur University of Chicago, P.O.B. 455, Cobden, Illinois, USA
Les dérivés organiques du fluor d'intérêt pharmacologique *3*, 9 (1961)	Prof. Dr. N. P. Buu-Hoï Directeur de Laboratoire à l'Institut de chimie des substances naturelles du Centre National de la Recherche Scientifique, Gif-sur-Yvette, France

Teaching tropical medicine *18,* 35 (1974)	Prof. Dr. K. M. Cahill Tropical Disease Center, 100 East 77th Street, New York City 10021, N.Y., USA
Anabolic steroids *2,* 71 (1960)	Prof. Dr. B. Camerino Director of the Chemical Research Laboratory of Farmitalia, Milan, Italy Prof. Dr. G. Sala Department of Clinical Chemistry and Director of the Department of Pharmaceutical Therapy, Farmitalia, Milan, Italy
Immunosuppression agents, procedures, speculations and prognosis *16,* 67 (1972)	Dr. G. W. Camiener Research Laboratories, The Upjohn Company, Kalamazoo, Michigan, USA Dr. W. J. Wechter Research Head, Hypersensitivity Diseases Research, The Upjohn Company, Kalamazoo, Michigan, USA
Dopamine agonists: Structure-activity relationships *29,* 303 (1985)	Joseph G. Cannon The University of Iowa, Iowa City, Iowa 52242
Analgesics and their antagonists: recent developments *22,* 149 (1978)	Dr. A. F. Casy Norfolk and Norwich Hospital and University of East Anglia, Norwich, Norfolk, England
Chemical nature and pharmacological actions of quaternary ammonium salts *2,* 135 (1960)	Prof. Dr. C. J. Cavallito Professor, Medicinal Chemistry, School of Pharmacy, University of North Carolina, Chapel Hill, North Carolina, USA Dr. A. P. Gray Director of the Chemical Research Section, Neisler Laboratories Inc., Decatur, Illinois, USA
Contributions of medicinal chemistry to medicine – from 1935 *12,* 11 (1968) Quaternary ammonium salts – advances in chemistry and pharmacology since 1960 *24,* 267 (1980)	Prof. Dr. C. J. Cavallito Professor, Medicinal Chemistry, School of Pharmacy, University of North Carolina, Chapel Hill, North Carolina, USA
Changing influences on goals and incentives in drug research and development *20,* 159 (1976)	Prof. Dr. C. J. Cavallito Ayerst Laboratories, Inc., New York, N. Y., USA
Über Vorkommen und Bedeutung der Indolstruktur in der Medizin und Biologie *2,* 227 (1960)	Dr. A. Cerletti Direktor der medizinisch-biologischen Forschungsabteilung der Sandoz AG, Basel, Schweiz

Cholesterol and its relation to atherosclerosis *1*, 127 (1959)	Prof. Dr. K. K. Chen Department of Pharmacology, University School of Medicine, Indianapolis, Indiana, USA Dr. Tsung-Min Lin Senior Pharmacologist, Division of Pharmacologic Research, Lilly Research Laboratories, Indianapolis, Indiana, USA
Effect of hookworm disease on the structure and function of small bowel *19*, 44 (1975)	Prof. Dr. H. K. Chuttani Prof. Dr. R. C. Misra Maulana Azad Medical College & Associated Irwin and G. B. Pant Hospitals, New Delhi, India
The psychotomimetic agents *15*, 68 (1971)	Dr. S. Cohen Director, Division of Narcotic Addiction and Drug Abuse, National Institute of Mental Health, Chevy Chase, Maryland, USA
Implementation of disease control in Asia and Africa *18*, 43 (1974)	Prof. Dr. M. J. Colbourne Department of Preventive & Social Medicine, University of Hong Kong, Sassoon Road, Hong Kong
Structure-activity relationships in certain anthelmintics *3*, 75 (1961)	Prof. Dr. J. C. Craig Department of Pharmaceutical Chemistry, University of California, San Francisco, California, USA Dr. M. E. Tate Post Doctoral Fellow, University of New South Wales, Department of Organic Chemistry, Kensington, N. S. W., Australia
Contribution of Haffkine to the concept and practice of controlled field trials of vaccines *19*, 481 (1975)	Dr. B. Cvjetanovic Chief Medical Officer, Bacterial Diseases, Division of Communicable Diseases, WHO, Geneva, Switzerland
Antifungal agents *22*, 93 (1978)	Prof. Dr. P. F. D'Arcy Dr. E. M. Scott Department of Pharmacy, The Queen's University of Belfast, Northern Ireland
Some neuropathologic and cellular aspects of leprosy *18*, 53 (1974)	Prof. Dr. D. K. Dastur Dr. Y. Ramamohan Dr. A. S. Dabholkar Neuropathology Unit, Grant Medical College and J. J. Group of Hospitals, Bombay 8, India
Autonomic dysfunction as a problem in the treatment of tetanus *19*, 245 (1975)	Prof. Dr. F. D. Dastur Dr. G. J. Bhat Dr. K. G. Nair Department of Medicine, Seth G. S. Medical College and K. E. M. Hospital, Bombay 12, India

Studies on *V. parahaemolyticus* infection in Calcutta as compared to cholera infection *19,* 490 (1975)	Dr. B. C. Deb Senior Research Officer, Cholera Research Centre, Calcutta, India
Biochemical effects of drugs acting on the central nervous system *8,* 53 (1965)	Dr. L. Decsi Specialist in Clinical Chemistry, University Medical School, Pécs, Hungary
Some reflections on the chemotherapy of tropical diseases: Past, present and future *26,* 343 (1982)	Dr. E. W. J. de Maar
Drug research – whence and whither *10,* 11 (1966)	Dr. R. G. Denkewalter Vice-President for Exploratory Research, Merck Sharp & Dohme Research Laboratories, Rahway, New Jersey, USA Dr. M. Tishler President, Merck Sharp & Dohme Research Laboratories, Rahway, New Jersey, USA
Hypolipidemic agents *13,* 217 (1969)	Dr. G. De Stevens Vice-President and Director of Research, CIBA Pharmaceutical Company, Summit, New Jersey, USA Dr. W. L. Bencze Research Department, CIBA Pharmaceutical Company, Summit, New Jersey, USA Dr. R. Hess CIBA Limited, Basle, Switzerland
The interface between drug research, marketing, management, and social, political and regulatory forces *20,* 181 (1976)	Dr. G. de Stevens Executive Vice President & Director of Research, Pharmaceuticals Division, CIBA-GEIGY Corporation, Summit, New Jersey, USA
Antihypertensive Agents *20,* 197 (1976)	Dr. G. De Stevens Dr. M. Wilhelm Pharmaceuticals Division, CIBA-GEIGY Corporation, Summit, New Jersey, USA
Medicinal Research: Retrospectives and Perspectives *29,* 97 (1985)	George DeStevens Department of Chemistry, Drew University, Madison, N.J., USA
Transport and accumulation in biological cell systems interacting with drugs *20,* 261 (1976)	Dr. W. Dorst Dr. A. F. Bottse Department of Pharmacology, Vrije Universiteit, Amsterdam, The Netherlands Dr. G. M. Willems Biomedical Centre, Medical Faculty, Maastricht, The Netherlands

Immunization of a village, a new approach to herd immunity *19*, 252 (1975)	Prof. Dr. N. S. Deodhar Head of Department of Preventive and Social Medicine, B. J. Medical College, Poona, India
Surgical amoebiasis *18*, 77 (1974)	Dr. A. E. deSa Bombay Hospital, Bombay, India
Epidemiology of pertussis *19*, 257 (1975)	Dr. J. A. D'Sa Glaxo Laboratories (India) Limited, Worli, Bombay 25, India
Profiles of tuberculosis in rural areas of Maharashtra *18*, 91 (1974)	Prof. Dr. M. D. Deshmukh Honorary Director Dr. K. G. Kulkarni Deputy Director Dr. S. S. Virdi Senior Research Officer Dr. B. B. Yodh Memorial Tuberculosis Reference Laboratory and Research Centre, Bombay, India
The Pharmacology of the immune system: Clinical and experimental perspectives *28*, 83 (1984)	Prof. Dr. Jürgen Drews, Director Sandoz Ltd., Pharmaceutical Research and Development, CH-4002 Basel, Switzerland
An Overview of studies on estrogens, oral contraceptives and breast cancer *25*, 159 (1981)	Prof. Dr. V. A. Drill Department of Pharmacology, College of Medicine, University of Illinois at the Medical Center, Chicago, Ill. 60680, USA
Aminonucleosid-nephrose *7*, 341 (1964)	Dr. U. C. Dubach Privatdozent, Oberarzt an der Medizinischen Universitäts-Poliklinik Basel, Schweiz
Impact of researches in India on the control and treatment of cholera *19*, 503 (1975)	Dr. N. K. Dutta Director, Vaccine Institute, Baroda, India
The perinatal development of drugmetabolizing enzymes: What factors trigger their onset? *25*, 189 (1981)	Prof. Dr. G. J. Dutton Dr. J. E. A. Leakey Department of Biochemistry, The University Dundee, Dundee, DD1 4HN, Scotland
Laboratory methods in the screening of anthelmintics *19*, 48 (1975)	Dr. D. Düwel Helminthology Department, Farbwerke Hoechst AG, Frankfurt/Main 80, Federal Republic of Germany
Progress in immunization *19*, 274 (1975)	Prof. Dr. G. Edsall Department of Microbiology, London School of Hygiene and Tropical Medicine, London W.C.1, England

Host factors in the response to immunization *19,* 263 (1975)	Prof. Dr. G. Edsall Department of Microbiology, London School of Hygiene and Tropical Medicine, London, W.C.1, England M.A. Belsey World Health Organization, Geneva, Switzerland Dr.R. LeBlanc Tulane University School of Public Health and Tropical Medicine, New Orleans, La., USA L. Levine State Laboratory Institute, Boston, Mass., USA
Drug-macromolecular interactions: implications for pharmacological activity *14,* 59 (1970)	Dr. S. Ehrenpreis Associate Professor and Head Department of Pharmacology, New York Medical College, Fifth Avenue at 106th Street, New York, N.Y. 10029, USA
Betrachtungen zur Entwicklung von Heilmitteln *10,* 33 (1966)	Prof. Dr. G. Ehrhart Farbwerke Hoechst AG, Frankfurt a. M.-Höchst, BR Deutschland
Progress in malaria chemotherapy. Part 1. Repository antimalarial drugs *13,* 170 (1969) New perspectives on the chemotherapy of malaria, filariasis and leprosy *18,* 99 (1974)	Dr. E. F. Elslager Section Director, Chemistry Department, Parke, Davis & Company, Ann Arbor, Michigan, USA
Recent research in the field of 5-hydroxytryptamine and related indolealkylamines *3,* 151 (1961)	Prof. Dr. V. Erspamer Institute of Pharmacology, University of Parma, Parma, Italy
Bacteriology at the periphery of the cholera pandemic *19,* 513 (1975)	Dr. A. L. Furniss Public Health Laboratory, Maidstone, England
Iron and diphtheria toxin production *19,* 283 (1975)	Dr. S. V. Gadre Dr. S. S. Rao Haffkine Institute, Bombay 12, India
Effect of drugs on cholera toxin induced fluid in adult rabbit ileal loop *19,* 519 (1975)	Dr. B. B. Gaitondé Dr. P. H. Marker Dr. N. R. Rao Haffkine Institute, Bombay 12, India
Drug action and assay by microbial kinetics *15,* 519 (1971) The pharmacokinetic bases of biological response quantification in toxicology, pharmacology and pharmacodynamics *21,* 105 (1977)	Prof. Dr. E. R. Garrett Graduate Research Professor The J. Hillis Miller Health Center, College of Pharmacy, University of Florida, Gainesville, Florida, USA

The chemotherapy of enterobacterial infections *12,* 370 (1968)	Prof. Dr. L. P. Garrod Department of Bacteriology, Royal Postgraduate Medical School, Hammersmith Hospital, London, England
Metabolism of drugs and other foreign compounds by enzymatic mechanisms *6,* 11 (1963)	Dr. J. R. Gillette Head, Section on Enzymes Drug Interaction, Laboratory of Chemical Pharmacology, National Heart Institute, Bethesda 14, Maryland, USA
The art and science of contemporary drug development *16,* 194 (1972)	Dr. A. J. Gordon Associate Director, Department of Scientific Affairs, Pfizer Pharmaceuticals, 235 East 42nd Street, New York, USA Dr. S. G. Gilgore President, Pfizer Pharmaceuticals, 235 East 42nd Street, New York, USA
Basic mechanisms of drug action *7,* 11 (1964) Isolation and characterization of membrane drug receptors *20,* 323 (1976)	Prof. Dr. D. R. H. Gourley Department of Pharmacology, Eastern Virginia Medical School, Norfolk, Virginia, USA
Zusammenhänge zwischen Konstitution und Wirksamkeit bei Lokalanästhetica *4,* 353 (1962)	Dr. H. Grasshof Forschungschemiker in Firma M. Woelm, Eschwege, Deutschland
Das Placeboproblem *1,* 279 (1959)	Prof. Dr. H. Haas Leiter der Pharmakologischen Abteilung Knoll AG, Ludwigshafen, und Dozent an der Universität Heidelberg Dr. H. Fink und Dr. G. Härtefelder Forschungslaboratorien der Knoll AG, Ludwigshafen, Deutschland
Clinical field trial of bitoscanate in *Necator americanus* infection, South Thailand *19,* 64 (1975)	Dr. T. Harinasuta Dr. D. Bunnag Faculty of Tropical Medicine, Mahidol University, Bangkok, Thailand
Pharmacological control of reproduction in women *12,* 47 (1968) Contraception – retrospect and prospect *21,* 293 (1977)	Prof. Dr. M.J.K. Harper The University of Texas, Health Science Center at San Antonio, San Antonio, Texas, USA
Drug latentiation *4,* 221 (1962)	Prof. Dr. N. J. Harper Head of the Department of Pharmacy, University of Aston, Birmingham 4, England
Chemotherapy of filariasis *9,* 191 (1966) Filariasis in India *18,* 173 (1974)	Dr. F. Hawking Clinical Research Centre, Watford Road, Harrow, Middlesex, England

Recent studies in the field of indole compounds 6, 75 (1963)	Dr. R. V. Heinzelman Section Head, Organic Chemistry, The Upjohn Company, Kalamazoo, Michigan, USA Dr. J. Szmuszkovicz Research Chemist, The Upjohn Company, Kalamazoo, Michigan, USA
Neuere Entwicklungen auf dem Gebiete therapeutisch verwendbarer organischer Schwefelverbindungen 4, 9 (1962)	Dr. H. Herbst Forschungschemiker in den Farbwerken Hoechst, Frankfurt a.M., Deutschland
The management of acute diarrhea in children: an overview 19, 527 (1975)	Dr. N. Hirschhorn Consultant Physician and Staff Associate, Management Sciences for Health, One Broadway, Cambridge, Mass., USA
The tetracyclines 17, 210 (1973)	Dr. J. J. Hlavka and Dr. J. H. Booth Lederle Laboratories, Pearl River, N. Y., USA
Relationship of induced antibody titres to resistance to experimental human infection 19, 542 (1975)	Dr. R. B. Hornick Dr. R. A. Cash Dr. J. P. Libonati The University of Maryland School of Medicine, Division of Infectious Diseases, Baltimore, Maryland, USA
Recent applications of mass spectrometry in pharmaceutical research 18, 399 (1974)	Mag. Sc. Chem. G. Horváth Research Chemist, Research Institute for Pharmaceutical Chemistry, Budapest, Hungary
Recent developments in disease-modifying antirheumatic drugs 24, 101 (1980)	Dr. I. M. Hunneyball Research Department, Boots Co. Ltd., Pennyfoot Street, Nottingham, England
The pharmacology of homologous series 7, 305 (1964)	Dr. H. R. Ing Reader in Chemical Pharmacology, Oxford University, and Head of the Chemical Unit of the University Department of Pharmacology, Oxford, England
Progress in the experimental chemotherapy of helminth infections. Part 1. Trematode and cestode diseases 17, 241 (1973)	Dr. P. J. Islip The Wellcome Research Laboratories, Beckenham, Kent, England
Pharmacology of the brain: the hippocampus, learning and seizures 16, 211 (1972)	Prof. Dr. I. Izquierdo Dr. A. G. Nasello Departamento de Farmacología, Facultad de Ciencias Químicas, Universidad Nacional de Córdoba, Estafeta 32, Córdoba, Argentina

Cholinergic mechanism – monoamines relation in certain brain structures *16,* 334 (1972)	Prof. Dr. J. A. Izquierdo Department of Experimental Pharmacology, Facultad de Farmacia y Bioquímica, Buenos Aires, Argentina
The development of antifertility substances *7,* 133 (1964)	Prof. Dr. H. Jackson Head of Department of Experimental Chemotherapy, Christie Hospital and Holt Radium Institute, Paterson Laboratories, Manchester 20, England
Agents acting on central dopamine receptors *21,* 409 (1977)	Dr. P. C. Jain Dr. N. Kumar Medicinal Chemistry Division, Central Drug Research Institute, Lucknow, India
Recent advances in the treatment of parasitic infections in man *18,* 191 (1974) The levamisole story *20,* 347 (1976)	Dr. P. A. J. Janssen Director, Janssen Pharmaceutica, Research Laboratories, Beerse, Belgium
Recent developments in cancer chemotherapy *25,* 275 (1981)	Dr. K. Jewers Tropical Product Institute, 56/62, Gray's Inn Road, London, WC1X8LU, England
Search for pharmaceutically interesting quinazoline derivatives: Efforts and results (1969–1980) *26,* 259 (1982)	Dr. S. Johne Institute of Plant Biochemistry, The Academy of Sciences of the German Democratic Republic, DDR-4010 Halle (Saale), PSF 250
A review of Advances in prescribing for teratogenic hazards *29,* 121 (1985)	E. Marshall Johnson, Ph. D. Daniel Baugh Institute, Jefferson College, Thomas Jefferson University, 1020 Locust Street, Philadelphia, PA 19107
A comparative study of bitoscanate, bephenium hydroxynaphthoate and tetrachlorethylene in hookworm infection *19,* 70 (1975)	Dr. S. Johnson Department of Medicine III, Christian Medical College Hospital, Vellore, Tamilnadu, India
Tetanus in Punjab with particular reference to the role of muscle relaxants in its management *19,* 288 (1975)	Prof. Dr. S. S. Jolly Dr. J. Singh Dr. S. M. Singh Department of Medicine, Medical College, Patiala, India
Virulence-enhancing effect of ferric ammonium citrate on *Vibrio cholerae* *19,* 546

Toxoplasmosis 18, 205 (1974)	Prof. Dr. B. H. Kean The New York Hospital – Cornell Medical Center, 525 East 68th Street, New York, N. Y., USA
Tabellarische Zusammenstellung über die Substruktur der Proteine 16, 364 (1972)	Dr. R. Kleine Physiologisch-Chemisches Institut der Martin-Luther-Universität, 402 Halle (Saale), DDR
Experimental evaluation of antituberculous compounds, with special reference to the effect of combined treatment 18, 211 (1974)	Dr. F. Kradolfer Head of Infectious Diseases Research, Biological Research Laboratories, Pharmaceutical Division, Ciba-Geigy Ltd., Basle, Switzerland
The oxidative metabolism of drugs and other foreign compounds 17, 488 (1973)	Dr. F. Kratz Medizinische Kliniken und Polikliniken, Justus-Liebig-Universität, Giessen, BR Deutschland
Die Amidinstruktur in der Arzneistofforschung 11, 356 (1968)	Prof. Dr. A. Kreutzberger Wissenschaftlicher Abteilungsvorsteher am Institut für pharmazeutische Chemie der Westfälischen Wilhelms-Universität Münster, Münster (Westfalen), Deutschland
Present data on the pathogenesis of tetanus 19, 301 (1975) Tetanus: general and pathophysiological aspects; achievements, failures, perspectives of elaboration of the problem 19, 314 (1975)	Prof. Dr. G. N. Kryzhanovsky Institute of General Pathology and Pathological Physiology, AMS USSR, Moscow, USSR
Lipophilicity and drug activity 23, 97 (1979)	Dr. H. Kubinyi Chemical Research and Development of BASF Pharma Division, Knoll AG, Ludwigshafen/Rhein, Federal Republic of Germany
Klinisch-pharmakologische Kriterien in der Bewertung eines neuen Antibiotikums. Grundlagen und methodische Gesichtspunkte 22, 327 (1978)	Prof. Dr. H. P. Kuemmerle München/Eppstein, BR Deutschland
Über neue Arzneimittel 1, 531 (1959), 2, 251 (1960), 3, 369 (1961), 6, 347 (1963), 10, 360 (1966)	Dr. W. Kunz Forschungschemiker in Firma Dr. Schwarz GmbH, Monheim (Rheinland), BR Deutschland
Die Anwendung von Psychopharmaka in der psychosomatischen Medizin 10, 530 (1966)	Dr. F. Labhardt Privatdozent, stellvertretender Direktor der psychiatrischen Universitätsklinik, Basel, Schweiz

Therapeutic measurement in tetanus *19,* 323 (1975)	Prof. Dr. D. R. Laurence Department of Pharmacology, University College, London, and Medical Unit, University College Hospital Medical School, London, England
Physico chemical methods in pharmaceutical chemistry, 1. Spectrofluorometry *6,* 151 (1963)	Dr. H. G. Leemann Head of the Analytical Department in the Pharmaceutical Division of Sandoz Ltd, Basle, Switzerland Dr. K. Stich Specialist for Questions in Ultraviolet and Fluorescence Spectrophotometry, Analytical Department, Sandoz Ltd., Basle, Switzerland Dr. Margrit Thomas Research Chemist in the Analytical Department Research Laboratory, Sandoz Ltd., Basle, Switzerland
Biochemical acyl hydroxylations *16,* 229 (1972)	Dr. W. Lenk Pharmakologisches Institut der Universität München, Nussbaumstrasse 26, München, BR Deutschland
Cholinesterase restoring therapy in tetanus *19,* 329 (1975)	Prof. Dr. G. Leonardi Department of Medicine, St. Thomas Hospital, Portogruaro, Venice, Italy Dr. K. G. Nair Prof. Dr. F. D. Dastur Department of Medicine, Seth G. S. Medical College and K. E. M. Hospital, Bombay 12, India
Biliary excretion of drugs and other xenobiotics *25,* 361 (1981)	Prof. Dr. W. G. Levine Department of Molecular Pharmacology, Albert Einstein College of Medicine, Yeshiva University, 1300 Morris Park Avenue, Bronx, New York 10461, USA
Structures, properties and disposition of drugs *29,* 67 (1985)	Eric J. Lien Biomedicinal Chemistry, School of Pharmacy, University of Southern California, Los Angeles, Calif. 90033, USA
Interactions between androgenic-anabolic steroids and glucocorticoids *14,* 139 (1970)	Dr. O. Linèt Sinai Hospital of Detroit, Department of Medicine, 6767 West Outer Drive, Detroit, Michigan 48235
Drug inhibition of mast cell secretion *29,* 277 (1985)	R. Ludowyke D. Lagunoff Department of Pathology, St. Louis University, School of Medicine, 1402 S. Grand Blvd. St. Louis, Mo 63104

Reactivity of bentonite flocculation, indirect haemagglutination and casoni tests in hydatid disease *19,* 75 (1975)	Dr. R. C. Mahajan Dr. N. L. Chitkara Division of Parasitology, Department of Microbiology, Postgraduate Institute of Medical Education and Research, Chandigarh, India
Epidemiology of diphtheria *19,* 336 (1975)	Dr. L. G. Marquis Glaxo Laboratories (India) Limited, Worli, Bombay 25, India
Biological activity of the terpenoids and their derivatives *6,* 279 (1963)	Dr. M. Martin-Smith Reader in Pharmaceutical Chemistry, University of Strathclyde, Department of Pharmaceutical Chemistry, Glasgow, C. 1, Scotland Dr. T. Khatoon Lecturer in Chemistry at the Eden Girls College, Dacca, East Pakistan
Biological activity of the terpenoids and their derivatives – recent advances *13,* 11 (1969)	Dr. M. Martin-Smith Reader in Pharmaceutical Chemistry, University of Strathclyde, Glasgow, C. 1, Scotland Dr. W. E. Sneader Lecturer in Pharmaceutical Chemistry, University of Strathclyde, Glasgow, C. 1, Scotland
Antihypertensive agents 1962–1968 *13,* 101 (1969) Fundamental structures in drug research – Part I *20,* 385 (1976) Fundamental structures in drug research – Part II *22,* 27 (1978) Antihypertensive agents 1969–1980 *25,* 9 (1981)	Prof. Dr. A. Marxer Dr. O. Schier Chemical Research Department, Pharmaceuticals Division, Ciba-Geigy Ltd., Basle, Switzerland
Relationships between the chemical structure and pharmacological activity in a series of synthetic quinuclidine derivatives *13,* 293 (1969)	Prof. Dr. M. D. Mashkovsky All-Union Chemical Pharmaceutical Research Institute, Moscow, USSR Dr. L. N. Yakhontov All-Union Chemical Pharmaceutical Research Institute, Moscow, USSR
Further developments in research on the chemistry and pharmacology of synthetic quinuclidine derivatives *27,* 9 (1983)	Prof. M. D. Mashkovsky Prof. L. N. Yakhontov Dr. M. E. Kaminka Dr. E. E. Mikhlina S. Ordzhonikidze All-Union, Chemical Pharmaceutical Research Institute, Moscow, USSR

On the understanding of drug potency *15*, 123 (1971) The chemotherapy of intestinal nematodes *16*, 157 (1972)	Dr. J. W. McFarland Pfizer Medical Research Laboratories, Groton, Connecticut, USA
Zur Beeinflussung der Strahlenempfindlichkeit von Säugetieren durch chemische Substanzen *9*, 11 (1966)	Dr. H.-J. Melching Privatdozent, Oberassistent am Radiologischen Institut der Universität Freiburg i.Br., Freiburg i.Br., Deutschland Dr. C. Streffer Wissenschaftlicher Mitarbeiter am Radiologischen Institut der Universität Freiburg i.Br., Freiburg i.Br., Deutschland
Analgesian and addiction *5*, 155 (1963)	Dr. L. B. Mellett Assistant Professor of Pharmacology, University of Michigan Medical School, Ann Arbor, Michigan, USA Prof. Dr. L. A. Woods Department of Pharmacology, College of Medicine, State University of Iowa, Iowa City, USA
Comparative drug metabolism *13*, 136 (1969)	Dr. L. B. Mellett Head, Pharmacology & Toxicology, Kettering-Meyer Laboratories, Southern Research Institute, Birmingham, Alabama, USA
Pathogenesis of amebic disease *18*, 225 (1974) Protozoan and helminth parasites – a review of current treatment *20*, 433 (1976)	Prof. Dr. M. J. Miller Tulane University, Department of Tropical Medicine, New Orleans, Louisiana, USA
Synopsis der Rheumatherapie *12*, 165 (1968)	Dr. W. Moll Spezialarzt FMH Innere Medizin – Rheumatologie, Basel, Schweiz
On the chemotherapy of cancer *8*, 431 (1965) The relationship of the metabolism of anticancer agents to their activity *17*, 320 (1973) The current status of cancer chemotherapy *20*, 465 (1976)	Dr. J. A. Montgomery Kettering-Meyer Laboratory, Southern Research Institute, Birmingham, Alabama, USA
Der Einfluss der Formgebung auf die Wirkung eines Arzneimittels *10*, 204 (1966) Galenische Formgebung und Arzneimittelwirkung. Neue Erkenntnisse und Feststellungen *14*, 269 (1970)	Prof. Dr. K. Münzel Leiter der galenischen Forschungsabteilung der F. Hoffmann-La Roche & Co. AG, Basel, Schweiz

A field trial with bitoscanate in India *19,* 81 (1975)	Dr. G. S. Mutalik Dr. R. B. Gulati Dr. A. K. Iqbal Department of Medicine, B. J. Medical College and Sassoon General Hospital, Poona, India
Comparative study of bitoscanate, bephenium hydroxynaphthoate and tetrachlorethylene in hookworm disease *19,* 86 (1975)	Dr. G. S. Mutalik Dr. R. B. Gulati Department of Medicine, B. J. Medical College and Sassoon General Hospital, Poona, India
Ganglienblocker *2,* 297 (1960)	Dr. K. Nádor o. Professor und Institutsdirektor, Chemisches Institut der Tierärztlichen Universität, Budapest, Ungarn
Nitroimidazoles as chemotherapeutic agents *27,* 163 (1983)	Dr. M. D. Nair Dr. K. Nagarajan Ciba-Geigy Research Centre, Goreagon East, Bombay 400063
Recent advances in cholera pathophysiology and therapeutics *19,* 563 (1975)	Prof. Dr. D. R. Nalin Johns Hopkins School of Medicine and School of Public Health. Guest Scientist, Cholera Research Hospital, Dacca, Bangladesh
Preparing the ground for research: importance of data *18,* 239 (1974)	Dr. A. N. D. Nanavati Assistant Director and Hcad, Department of Virology, Haffkine Institute, Bombay, India
Mechanism of drugs action on ion and water transport in renal tubular cells *26,* 87 (1982)	Prof. Dr. Yu. V. Natochin I. M. Sechenov Institute of Evolutionary Physiology and Biochemistry, Leningrad, USSR
Recent advances in drugs against hypertension *29,* 215 (1985)	Neelima B. K. Bhat A. P. Bhaduri Central Drug Research Institute, Lucknow – 226001, India
Antibody response to two cholera vaccines in volunteers *19,* 554 (1975)	Y. S. Nimbkar R. S. Karbhari S. Cherian N. G. Chanderkar R. P. Bhamaria P. S. Ranadive Dr. B. B. Gaitondé Haffkine Institute, Parel, Bombay 12, India
Die Chemotherapie der Wurmkrankheiten *1,* 159 (1959)	Prof. Dr. H.-A. Oelkers Leiter der pharmakologischen und parasitologischen Abteilung der Firma C. F. Asche & Co., Hamburg-Altona, Deutschland

Drug research and human sleep *22*, 355 (1978)	Prof. Dr. I. Oswald University Department of Psychiatry, Royal Edinburgh Hospital, Edingburgh, Scotland
An extensive community outbreak of acute diarrhoeal diseases in children *19*, 570 (1975)	Dr. S. C. Pal Dr. C. Koteswar Rao Cholera Research Centre, Calcutta, India
Drug and its action according to Ayurveda *26*, 55 (1982)	Dr. Shri Madhabendra Nath Pal
3,4-Dihydroxyphenylalanine and related compounds *9*, 223 (1966)	Dr. A. R. Patel Post-Doctoral Research Assistant. Department of Chemistry, University of Virginia, Charlottesville, Virginia, USA Prof. Dr. A. Burger Department of Chemistry, University of Virginia, Charlottesville, Virginia, USA
Mescaline and related compounds *11*, 11 (1968)	Dr. A. R. Patel Post-Doctoral Research Assistant, Department of Chemistry, University of Virginia, Charlottesville, Virginia, USA
Experience with bitoscanate in adults *19*, 90 (1975)	Dr. A. H. Patricia Dr. U. Prabakar Rao Dr. R. Subramaniam Dr. N. Madanagopalan Madras Medical College, Madras, India
Monoaminoxydase-Hemmer *2*, 417 (1960)	Prof. Dr. A. Pletscher Direktor der medizinischen Forschungsabteilung F. Hoffmann-La Roche & Co. AG, Basel, und Professor für Innere Medizin an der Universität Basel Dr. K. F. Gey Medizinische Forschungsabteilung F. Hoffmann-La Roche & Co. AG, Basel Schweiz Dr. P. Zeller Chefchemiker in Firma F. Hoffmann-La Roche & Co. AG, Basel, Schweiz
What makes a good pertussis vaccine? *19*, 341 (1975) Vaccine composition in relation to antigenic variation of the microbe: is pertussis unique? *19*, 347 (1975) Some unsolved problems with vaccines *23*, 9 (1979)	Dr. N. W. Preston Department of Bacteriology and Virology, University of Manchester, Manchester, England

Antibiotics in the chemotherapy of malaria *26,* 167 (1982)	Dr. S. K. Puri Dr. G. P. Dutta Division of Microbiology, Central Drug Research Institute, Lucknow 226001, India
Clinical study of diphtheria, pertussis and tetanus *19,* 356 (1975)	Dr. V. B. Raju Dr. V. R. Parvathi Institute of Child Health and Hospital for Children, Egmore, Madras 8, India
Epidemiology of cholera in Hyderabad *19,* 578 (1975)	Dr. K. Rajyalakshmi Dr. P. V. Ramana Rao Institute of Preventive Medicine, Hyderabad, Andhra Pradesh, India
Problems of malaria eradication in India *18,* 245 (1974)	Dr. V. N. Rao Joint Director of Health Services (Health), Maharashtra, Bombay, India
The photochemistry of drugs and related substances *11,* 48 (1968)	Dr. S. T. Reid Lecturer in Chemical Pharmacology, Experimental Pharmacology Division, Institute of Physiology, The University, Glasgow, W.2, Scotland
Orale Antikoagulantien *11,* 226 (1968)	Dr. E. Renk Dr. W. G. Stoll Wissenschaftliche Laboratorien der J. R. Geigy AG, Basel, Schweiz
Tetrahydroisoquinolines and ß-carbolines: putative natural substances in plants and animals *29,* 415 (1985)	H. Rommelspacher R. Susilo Department of Neuropsychopharmacology, Free University, Ulmenallee 30, D-1000 Berlin 19, F R G
Functional significance of the various components of the influenza virus *18,* 253 (1974)	Prof. Dr. R. Rott Institut für Virologie, Justus-Liebig-Universität, Giessen, Deutschland
Role of beta-adrenergic blocking drug propranolol in severe tetanus *19,* 361 (1975)	Prof. Dr. G. S. Sainani Head, Upgraded Department of Medicine, B. J. Medical College and Sassoon General Hospitals, Poona, India Dr. K. L. Jain Prof. Dr. V. R. D. Deshpande Dr. A. B. Balsara Dr. S. A. Iyer Medical College and Hospital, Nagpur, India
Studies on *Vibrio parahaemolyticus* in Bombay *19,* 586 (1975)	Dr. F. L. Saldanha Dr. A. K. Patil Dr. M. V. Sant Haffkine Institute, Parel, Bombay 12, India

Pharmacology and toxicology of axoplasmic transport *28,* 53 (1984)	Dr. Fred Samson, Ph. D., Director Ralph L. Smith Research Center, The University of Kansas Medical Center, Department of Physiology Dr. J. Alejandro Donoso Ralph L. Smith Research Center, The University of Kansas Medical Center, Department of Neurology, Kansas City, Kansas 66103, USA
Clinical experience with bitoscanate *19,* 96 (1975)	Dr. M. R. Samuel Head of the Department of Clinical Development, Medical Division, Hoechst Pharmaceuticals Limited, Bombay, India
Tetanus: Situational clinical trials and therapeutics *19,* 367 (1975)	Dr. R. K. M. Sanders Dr. M. L. Peacock Dr. B. Martyn Dr. B. D. Shende The Duncan Hospital, Raxaul, Bihar, India
Epidemiological studies on cholera in non-endemic regions with special reference to the problem of carrier state during epidemic and non-epidemic period *19,* 594 (1975)	Dr. M. V. Sant W. N. Gatlewar S. K. Bhindey Haffkine Institute, Parel, Bombay 12, India
Epidemiological and biochemical studies in filariasis in four villages near Bombay *18,* 269 (1974)	Dr. M. V. Sant, W. N. Gatlewar and T.U.K. Menon Department of Zoonosis and of Research Divison of Microbiology, Haffkine Institute, Bombay, India
Hookworm anaemia and intestinal malabsorption associated with hookworm infestation *19,* 108 (1975)	Prof. Dr. A. K. Saraya Prof. Dr. B. N. Tandon Department of Pathology and Department of Gastroenterology, All India Institute of Medical Sciences, New Delhi, India
The effects of structural alteration on the anti-inflammatory properties of hydrocortisone *5,* 11 (1963)	Dr. L. H. Sarett Director of Synthetic Organic Chemistry, Merck Sharp & Dohme Research Laboratories, Rahway, New Jersey, USA Dr. A. A. Patchett Director of the Department of Synthetic Organic Chemistry, Merck Sharp & Dohme Research Laboratories, Rahway, New Jersey, USA Dr. S. Steelman Director of Endocrinology, Merck Institute for Therapeutic Research, Rahway, New Jersey, USA
The impact of natural product research on drug discovery *23,* 51 (1979)	Dr. L. H. Sarett Senior Vice-President for Science and Technology, Merck & Co., Inc., Rahway New Jersey, USA

Anti-filariasis campaign: its history and future prospects *18*, 259 (1974)	Prof. Dr. M. Sasa Professor of Parasitology, Director of the Institute of Medical Science, University of Tokyo, Tokyo, Japan
Platelets and atherosclerosis *29*, 49 (1985)	Robert N. Saunders, Sandoz Research Institute, East Hanover, N. J., USA
Immuno-diagnosis of helminthic infections *19*, 119 (1975)	Prof. Dr. T. Sawada Dr. K. Sato Dr. K. Takei Department of Parasitology, School of Medicine, Gunma University, Maebashi, Japan
Immuno-diagnosis in filarial infection *19*, 128 (1975)	Prof. Dr. T. Sawada Dr. K. Sato Dr. K. Takei Department of Parasitology, School of Medicine, Gunma University, Maebashi, Japan Dr. M. M. Goil Department of Zoology, Bareilly College, Bareilly (U. P.), India
Quantitative structure-activity relationships *23*, 199 (1979)	Dr. A. K. Saxena Dr. S. Ram Medicinal Chemistry Division, Central Drug Research Institute, Lucknow, India
Phenothiazine und Azaphenothiazine als Arzneimittel *5*, 269 (1963)	Dr. E. Schenker Forschungschemiker in der Sandoz AG, Basel, Schweiz Dr. H. Herbst Forschungstechniker in den Farbwerken Hoechst, Frankfurt a. M., Deutschland
Antihypertensive agents *4*, 295 (1962)	Dr. E. Schlittler Director of Research of CIBA Pharmaceutical Company, Summit, New Jersey, USA Dr. J. Druey Director of the Department of Synthetic Drug Research of CIBA Ltd., Basle, Switzerland Dr. A. Marxer Research Chemist of CIBA Ltd., Basle, and Lecturer at the University of Berne, Switzerland
Die Anwendung radioaktiver Isotope in der pharmazeutischen Forschung *7*, 59 (1964)	Prof. Dr. K. E. Schulte Direktor des Instituts für Pharmazie und Lebensmittelchemie der Westfälischen Wilhelms-Universität Münster, Münster (Westfalen), Deutschland Dr. Ingeborg Mleinek Leiterin des Isotopen-Laboratoriums, Institut für Pharmazie und Lebensmittelchemie der Westfälischen Wilhelms-Universität Münster, Münster (Westfalen), Deutschland

Natürliche und synthetische Acetylen-Verbindungen als Arzneistoffe *14,* 387 (1970)	Prof. Dr. K. E. Schulte Direktor des Instituts für pharmazeutische Chemie der Westfälischen Wilhelms-Universität Münster, Münster (Westfalen), Deutschland Dr. G. Rücker Dozent für pharmazeutische Chemie an der Westfälischen Wilhelms-Universität Münster, Münster (Westfalen), Deutschland
Central control of arterial pressure by drugs *26,* 353 (1982)	Dr. A. Scriabine Dr. D. G. Taylor Miles Institute for Preclinical Pharmacology, P.O. Box 1956, New Haven, Connecticut 06509, USA Dr. E. Hong Instituto Miles de Terepeutica Experimental, A. P. 22026, Mexico 22, D. F.
The structure and biogenesis of certain antibiotics *2,* 591 (1960)	Dr. W. A. Sexton Research Director of the Pharmaceuticals Division of Imperial Chemical Industries Ltd., Wilmslow, Cheshire, England
Role of periodic deworming of preschool community in national nutrition programmes *19,* 136 (1975)	Prof. Dr. P. M. Shah Institute of Child Health Dr. A. R. Junnarkar Reader in Preventive and social Medicine Dr. R. D. Khare Research Assistant, Institute of Child Health, J. J. Group of Government Hospitals and Grant Medical College, Bombay, India
Chemotherapy of cestode infections *24,* 217 (1980)	Dr. Satyavan Sharma Dr. S. K. Dubey Dr. R. N. Iyer Medicinal Chemistry Division, Central Drug Research Institute, Lucknow 226001, India
Chemotherapy of hookworm infections *26,* 9 (1982)	Dr. Satyavan Sharma Dr. Elizabeth S. Charles Medicinal Chemistry Division, Central Drug Research Institute, Lucknow 226001, India
The benzimidazole anthelmiticschemistry and biological activity *27,* 85 (1983)	Dr. Satyavan Sharma Dr. Syed Abuzar Central Drug Research Institute, Lucknow 226001, India
Ayurvedic medicine – past and present *15,* 11 (1971)	Dr. Shiv Sharma 'Baharestan', Bomanji Petit Road, Cumballa Hill, Bombay, India

Mechanisms of anthelmintic action *19,* 147 (1975)	Prof. Dr. U. K. Sheth Seth G. S. Medical College and K. E. M. Hospital, Parel, Bombay 12, India
Some often neglected factors in the control and prevention of communicable diseases *18,* 277 (1974)	Dr. C. E. G. Smith Dean, London School of Hygiene and Tropical Medicine, Keppel Street, London, England
Tetanus and its prevention *19,* 391 (1975)	Dr. J. W. G. Smith Epidemiological Research Laboratory, Central Public Health Laboratory, London England
Growth of *Clostridium tetani in vivo* *19,* 384 (1975)	Dr. J. W. G. Smith Epidemiological Research Laboratory, Central Public Health Laboratory, London England Dr. A. G. MacIver Department of Morbid Anatomy, Faculty of Medicine, Southampton University, Southampton, England
The biliary excretion and enterohepatic circulation of drugs and other organic compounds *9,* 299 (1966)	Dr. R. L. Smith Senior Lecturer in Biochemistry at St. Mary's Hospital Medical School (University of London), Paddington, London, W.2, England
Noninvasive pharmacodynamic and bioelectric methods for elucidating the bioavailability mechanisms of ophthalmic drug preparations *25,* 421 (1981)	Dr. V. F. Smolen President and Chief Executive Officer Pharmacontrol Corp. 661 Palisades Ave., P.O. Box 931, Englewood Cliffs, New Jersey, 07632
On the relation between chemical structure and function in certain tumor promoters and anti-tumor agents *23,* 63 (1979) Relationships between structure and function of convulsant drugs *24,* 57 (1980)	Prof. Dr. J. R. Smythies Department of Psychiatry, University of Alabama in Birmingham Medical Center, Birmingham, Alabama, USA
Gram-negative bacterial endotoxin and the pathogenesis of fever *19,* 402 (1975)	Dr. E. S. Snell Glaxo Laboratories Limited, Greenford, Middlesex, England
Strukturelle Betrachtungen der Psychopharmaka: Versuch einer Korrelation von chemischer Konstitution und klinischer Wirkung *9,* 129 (1966)	Dr. K. Stach Stellvertretender Leiter der Chemischen Forschung der C. F. Boehringer & Söhne GmbH, Mannheim-Waldhof, Deutschland Dr. W. Pöldinger Oberarzt für klinische Psychopharmakologie an der Psychiatrischen Universitätsklinik Basel, Basel, Schweiz

Chemotherapy of intestinal helminthiasis *19,* 158 (1975)	Dr. O. D. Standen The Welcome Research Laboratories, Beckenham, Kent, England
The leishmaniases *18,* 289 (1974)	Dr. E. A. Steck Department of the Army, Walter Reed Army Institute of Research, Division of Medicinal Chemistry, Washington, D.C., USA
The benzodiazepine story *22,* 229 (1978)	Dr. L. H. Sternbach Research Department, Hoffmann-La Roche Inc., Nutley, New Jersey, USA
Progress in sulfonamide research *12,* 389 (1968) Problems of medical practice and of medical-pharmaceutical research *20,* 491 (1976)	Dr. Th. Struller Research Department, F. Hoffmann-La Roche & Co. Ltd., Basle, Switzerland
Antiviral agents *22,* 267 (1978)	Dr. D. L. Swallow Pharmaceuticals Division, Imperial Chemical Industries Limited, Alderley Park, Macclesfield, Cheshire, England
Antiviral agents 1978–1983 *28,* 127 (1984)	Dr. D. L. Swallow, M. A., B. Sc., D. Phil., F.R.S.C. Imperial Chemical Industries PLC, Pharmaceutical Division, Alderley Park, Macclesfield, Cheshire SK 10 4 TG, England
Ketoconazole, a new step in the management of fungal disease *27,* 63 (1983)	Dr. J. Symoens Dr. G. Cauwenbergh Janssen Pharmaceutica, B–2340 Beerse, Belgium
Antiarrhythmic compounds *12,* 292 (1968)	Prof. Dr. L. Szekeres Head of the Department of Pharmacology, School of Medicine, University of Szeged, Szeged, Hungary Dr. J. G. Papp Senior Lecturer, University Department of Pharmacology, Oxford, England
Practically applicable results of twenty years of research in endocrinology *12,* 137 (1968)	Prof. Dr. M. Tausk State University of Utrecht, Faculty of Medicine, Utrecht, Netherlands
Age profile of diphtheria in Bombay *19,* 412 (1975)	Prof. Dr. N. S. Tibrewala Dr. R. D. Potdar Dr. S. B. Talathi Dr. M. A. Ramnathkar Dr. A. D. Katdare Topiwala National Medical College, BYL Nair Hospital and Kasturba Hospital for Infectious Diseases, Bombay 11, India

Antibakterielle Chemotherapie der Tuberkulose *7,* 193 (1964)	Dr. F. Trendelenburg Leitender Arzt der Robert-Koch-Abteilung der Medizinischen Universitätskliniken, Homburg, Saar, Deutschland
Diphtheria *19,* 423 (1975)	Prof. Dr. P. M. Udani Dr. M. M. Kumbhat Dr. U. S. Bhat Dr. M. S. Nadkarni Dr. S. K. Bhave Dr. S. G. Ezuthachan Dr. B. Kamath The Institute of Child Health, J. J. Group of Hospitals, and Grant Medical College, Bombay 8, India
Biologische Oxydation und Reduktion am Stickstoff aromatischer Amino- und Nitroderivate und ihre Folgen für den Organismus *8,* 195 (1965) Stoffwechsel von Arzneimitteln als Ursache von Wirkungen, Nebenwirkungen und Toxizität *15,* 147 (1971)	Prof. Dr. H. Uehleke Pharmakologisches Institut der Universität Tübingen, 74 Tübingen, Deutschland
Mode of death in tetanus *19,* 439 (1975)	Prof. Dr. H. Vaishnava Dr. C. Bhawal Dr. Y. P. Munjal Department of Medicine, Maulana Azad Medical College and Associated Irwing and G. B. Pant Hospitals, New Delhi, India
Comparative evaluation of amoebicidal drugs *18,* 353 (1974) Comparative efficacy of newer anthelmintics *19,* 166 (1975)	Prof. Dr. B. J. Vakil Dr. N. J. Dalal Department of Gastroenterology, Grant Medical College and J. J. Group of Hospitals, Bombay, India
Cephalic tetanus *19,* 443 (1975)	Prof. Dr. B. J. Vakil Prof. Dr. B. S. Singhal Dr. S. S. Pandya Dr. P. F. Irani J. J. Group of Hospitals and Grant Medical College, Bombay, India
Methods of monitoring adverse reactions to drugs *21,* 231 (1977) Aspects of social pharmacology *22,* 9 (1978)	Prof. Dr. J. Venulet Division of Clinical Pharmacology, Department of Medicine, Hôspital Cantonal and University of Geneva, Geneva, Switzerland. Formerly: Senior Project Officer, WHO Research Centre for International Monitoring of Adverse Reactions to Drugs, Geneva, Switzerland

The current status of cholera toxoid research in the United States *19,* 602 (1975)	Dr. W. F. Verwey Dr. J. C. Guckian Dr. J. Craig Dr. N. Pierce Dr. J. Peterson Dr. H. Williams, Jr. The University of Texas Medical Branch, Galveston, State University of New York Medical Center (Downstate), and Johns Hopkins University School of Medicine, USA
Cell-kinetic and pharmacokinetic aspects in the use and further development of cancerostatic drugs *20,* 521 (1976)	Prof. Dr. M. von Ardenne Forschungsinstitut Manfred von Ardenne, Dresden, GDR
The problem of diphtheria as seen in Bombay *19,* 452 (1975)	Prof. Dr. M. M. Wagle Dr. R. R. Sanzgiri Dr. Y. K. Amdekar Institute of Child Health, J. J. Group of Hospitals and Grant Medical College, Bombay 8, India
Cell-wall antigens of *V. cholerae* and their implication in cholera immunity *19,* 612 (1975)	Dr. Y. Watanabe Dr. R. Ganguly Bacterial Diseases, Division of Communicable Diseases, World Health Organization, Geneva 27, Switzerland
Where is immunology taking us? *20,* 573 (1976)	Dr. W. J. Wechter Dr. Barbara E. Loughman Hypersensitivity Diseases Research, The Upjohn Company, Kalamazoo, Michigan, USA
Immunology in drug research *28,* 233 (1984)	Dr. W. J. Wechter, Ph. D., Research Manager Dr. Barbara E. Loughman, Ph. D., Research Head The Upjohn Company, Kalamazoo, Michigan 49001, USA
Metabolic activation of chemical carcinogens *26,* 143 (1982)	Dr. E. K. Weisburger Division of Cancer Cause and Prevention, National Cancer Institute, Bethesda, Maryland 20205, USA
A pharmacological approach to allergy *3,* 409 (1961)	Dr. G. B. West Reader in the School of Pharmacy, Department of Pharmacology, University of London, London, England

A new approach to the medical interpretation of shock *14,* 196 (1970)	Dr. G. B. West Scientific Secretary, The British Industrial Biological Research Association, Woodmansterne Road, Carshalton, Surrey, England Dr. M. S. Starr Department of Pharmacology, St. Mary's Hospital Medical School, University of London, London, England
Adverse reactions of sugar polymers in animals and man *23,* 27 (1979)	Dr. G. B. West Department of Paramedical Sciences, North-East London Polytechnic, London, England
Biogenic amines and drug research *28,* 9 (1984)	Dr. G. B. West Department of Paramedical Sciences, North-East London Polytechnic, England
Some biochemical and pharmacological properties of anti-inflammatory drugs *8,* 321 (1965)	Dr. M. W. Whitehouse Lecturer in Biochemistry at the University of Oxford, Oxford, England
Wirksamkeit und Nebenwirkungen von Metronidazol in der Therapie der Trichomonasis *9,* 361 (1966)	Dr. K. Wiesner Tierarzt, wissenschaftlicher Mitarbeiter der Pharmawissenschaftlichen Literaturabteilung, Farbenfabriken Bayer AG, Leverkusen, Deutschland Dr. H. Fink Leiter der Pharmawissenschaftlichen Literaturabteilung, Farbenfabriken Bayer AG, Leverkusen, Deutschland
Carcinogenicity testing of drugs *29,* 155 (1985)	G. M. Williams, J. H. Weisburger Naylor Dana Institute for Desease Prevention, American Health Foundation, Valhalla, N. Y. 10595
Drug treatment of asthma *28,* 111 (1984)	Prof. Dr. Archie F. Wilson, M. D., Ph. D. University of California, Irvine Medical Center, Orange, CA 92683, USA
Nonsteroid antiinflammatory agents *10,* 139 (1966)	Dr. C. A. Winter Senior Investigator Pharmacology, Merck Institute for Therapeutic Research, West Point, Pennsylvania, USA
A review of the continuum of drug-induced states of excitation and depression *26,* 225 (1982)	Prof. Dr. W. D. Winters Departments of Pharmacology and Internal Medicine, School of Medicine, University of California, Davis, California 95616, USA
Basic research in the US pharmaceutical industry *15,* 204 (1971)	Dr. O. Wintersteiner The Squibb Institute for Medical Research, New Brunswick, New Jersey, USA

The chemotherapy of amoebiasis *8,* 11 (1965)	Dr. G. Woolfe Head of the Chemotherapy Group of the Research Department at Boots Pure Drug Company Ltd., Nottingham, England
Antimetabolites and their revolution in pharmacology *2,* 613 (1960)	Dr. D. W. Woolley The Rockefeller Institute, New York, USA
Noise analysis and channels at the postsynaptic membrane of skeletal muscle *24,* 9 (1980)	Dr. D. Wray Lecturer, Pharmacology Department, Royal Free Hospital School of Medicine, Pond Street, London NW3 2QG, England
Krebswirksame Antibiotika aus Actinomyceten *3,* 451 (1961)	Dr. Kh. Zepf Forschungschemiker im biochemischen und mikrobiologischen Laboratorium der Farbwerke Hoechst, Frankfurt a.M., Deutschland Dr. Christa Zepf Referentin für das Chemische Zentralblatt, Kelkheim (Taunus), Deutschland
Fifteen years of structural modifications in the field of antifungal monocyclic 1-substituted 1H-azoles *27,* 253 (1983)	Dr. L. Zirngibl Siegfried AG, Zofingen, Switzerland
Lysostaphin: model for a specific anzymatic approach to infectious disease *16,* 309 (1972)	Dr. W. A. Zygmunt Department of Biochemistry, Mead Johnson Research Center, Evansville, Indiana, USA Dr. P. A. Tavormina Director of Biochemistry, Mead Johnson Research Center, Evansville, Indiana, USA